Sustainable Biofuels Development in India

Anuj K. Chandel • Rajeev K. Sukumaran
Editors

Sustainable Biofuels Development in India

Springer

Editors
Anuj K. Chandel
Centro de Tecnologia Canavieira
(Sugarcane Technology Centre, CTC)
Piracicaba, Fazenda Santo Antonio
Santo Antonio, Sao Paulo
Brazil

Rajeev K. Sukumaran
Centre for Biofuels Microbial Processes
& Technology Division
CSIR, National Institute for Interdisciplinary
Science and Technology
Thiruvananthapuram, Kerala
India

ISBN 978-3-319-50217-5 ISBN 978-3-319-50219-9 (eBook)
DOI 10.1007/978-3-319-50219-9

Library of Congress Control Number: 2017932890

Printed on acid-free paper

This Springer imprint is published by Springer Nature
The registered company is Springer International Publishing AG
The registered company address is: Gewerbestrasse 11, 6330 Cham, Switzerland

Quotes for Biofuels Book

The use of vegetable oils for engine fuels may seem insignificant today. But such oils may in the course of time become as important as petroleum and the coal tar products of present time

Rudolf Diesel in 1912

The fuel of the future is going to come from fruit like that sumac out by the road, or from apples, weeds, sawdust—almost anything. There is fuel in every bit of vegetable matter that can be fermented. There's enough alcohol in one year's yield of an acre of potatoes to drive the machinery necessary to cultivate the fields for a hundred years.

Henry Ford in 1925

Foreword

Biomass (food and feed crops and other grassy plants, wood, and algae) is available in abundance on Earth in a variety of forms that can be converted into fuels, food/feed, chemicals, and biomaterials. These conversions can be brought out by employing physical, chemical, thermal, biological, and biotechnological processes. Bioenergy is one of the most significant forms among alternate energy forms, which offers potential benefits for sustainable development. The most potential way to valorize such biomass is the principles of biorefinery, where each component of plant biomass is utilized, generating zero- or near-zero waste. The concept of biorefinery has emerged fast, which has huge flexibility. This has led to the development of relevant strategies to valorize the biomass for environmental, social, and economic development, with sustainability as the keyword. Biorefinery is a metaphor for the petroleum refinery where biomass is considered as a possible substitute to petroleum and capable of generating several products similar to petroleum refinery. As we know, sustainable supply of energy is required for the overall human development, which ultimately decides the holistic growth of any society or for that matter any country. Renewable energy or bioenergy is definitely an apt technological solution because it can be produced directly from the natural resources and is 100% renewable.

I understand that this will be the first book covering the key features on Indian Bioenergy program at one point, providing useful source of information for the students and researchers involved in energy and environmental programs primarily but also a pivotal source of information for the researchers, academicians, economists, policy analysts, and policy makers. I also understand that this book specifically is a first uniquely designed scientific and technical literature on bioenergy production in the context of India, which covers technological updates on biomass processing, system biology, microbial fermentation, catalysis, regeneration, and monitoring of renewable energy and recovery process and also presents proximate techno-economic analysis, climate change, geopolitical analysis of bioenergy, and green transportation fuels at industrial scale, thus making it a unique source of wealth of knowledge.

India has a biofuel policy and has a large number of government-supported initiatives and programs for the research and technological development of biomass-based economy development. The country advocates blending of 10% ethanol in petrol and also looks forward to the scenario of blending of biodiesel with diesel. Policy frameworks are in place with ample opportunities to develop and set up not only start-ups but also SMEs and large industries for the development and production of fuels and chemicals from agro-industrial residues, primarily which are available as surplus.

I have great pleasure in presenting this book to the readers, who I am confident would find it a very valuable and unique source of information. With their vast experience and global expertise, the editors of the book are among the champions of biomass researchers not only in India but internationally, who I would say are among the best to do this job. My compliments to them and also to the authors of the various chapters for their contributions.

Prof. Ashok Pandey

Center of Innovative and Applied Bioprocessing (A national institute under Department of Biotechnology, Govt of India)
Mohali, Punjab
India

Preface

Energy has a central role in meeting the economic development goals set by Indian government in all the declared schemes, whether it is "Make in India,", "Start-up India," or "Swachh Bharat." India is one of the fastest growing economies in the world today with over 8% annual economic growth rate. To maintain or surpass this growth rate, energy would be in prime demand, and the consumption is projected to double by 2020. India is already very behind as measured by per capita primary consumption, which is linked with the human development index (HDI) set by United Nations. In India, total energy (primary and secondary) demand is up to 400 gigawatt (GW). Coal-based thermal power generation supply the primary energy generation and more than 70% of the imported crude petroleum supply 80% of the transportation liquid energy. Both of these energy sources are nonrenewable, adding greenhouse gas emissions and the rising costs impacting the nation's foreign exchange reserves. Replacement at any level of both conventional energy sources (coal and petroleum) with renewable alternatives, be it from solar, wind, or biomass, will be a paradigm shift in India's energy security, along with reduction in carbon emissions, eventually impacting the economy of India.

Renewable energy sources are diverse and broad in range. It can be categorized mainly into solar, wind, hydrothermal, and biomass derived. For the sustainable energy drive, India's Government is aiming at an ambitious plan for installing 175 GW of solar energy by 2030. Similarly, wind energy is integral part of this massive energy drive, with judicious utilization of the current manufacturing capacity of 9500 MW. Amongst all the renewable resources, biomass energy is one of the most promising answers to cater to the increasing demand for transportation fuels. India is a large agricultural nation and is blessed by nature with appropriate fertile land, rain, sunlight, and water. India holds a strong position in the production of agricultural commodities in the world. These factors position the country as having enormous potential for renewable energy generation. In the last three decades, the Government of India has taken various initiatives in order to implement a bioenergy program in the country. Several research programs are currently ongoing nationwide to develop solar energy, biodiesel, cellulosic ethanol, and other renewable energy options into reality, with the financial help from Ministry of New and Renewable Energy; Ministry of Petroleum and Natural Gas; Ministry of Science and Technology; Ministry of Agriculture; Ministry of Environment, Forest and Climate Change; and private investments as well. Despite these efforts under way today, India is still very far behind in demonstrating the judicious use of its natural resources for renewable energy production and dangerously dependent upon gasoline, which is majorly imported from the oil producing countries. Gasoline sources are limited and nonrenewable and limit economic growth. The use of renewable energy sources may improve the socioeconomic and environmental status, along with savings in foreign exchange reserves, while making the country more energy stable.

This book aims to disseminate the most current advances in bioenergy programs of India, starting from the feedstock analysis and availability, chemical composition of various raw materials, technical aspects and challenges, techno-economic analysis, and finally government policies. The book is a unique collection of four Parts encompassing 22 book chapters written by the specialists of the related field, who offer critical insights into several topics, review current research, and discuss future progress in the respective fields. Broadly, this book intends to provide critical insight and background research analysis on raw materials in India, feedstock processing, product synthesis, recovery and applications for energy uses, their commercialization, and overview of India's Government policies on biofuels in addition to sustainable design of biofuels.

Part I has six chapters which deal with conventional fuels and generalized scenario of new and advanced biofuels.

Chapter 1 by Tuli and Gupta provides the latest available information on the production and consumption of petroleum and compressed natural gas in India and oil and gas import and export data. Various Indian Government initiatives for increasing oil and gas production in the country, along with the development of alternative sources of energy, are addressed. Chapter 2, by Singh et al., provides analyses of the future of bioethanol market in India based on biomass availability, policy barriers, and perspectives along with the outline of the lignocellulosic ethanol technology research institutes in India. In Chap. 3, Pal et al. describe the renewable energy potential of the country, various forms of renewable energy, existing technologies, policies, and programs and finally bring out prospects and challenges of renewable energy in the country. Chapter 4 by Sharma presents the current status and potential of geothermal energy in India. Kalita and colleagues in Chap. 5 discuss the factors which strongly influence the efficiency of biogas digesters and fuel cells on using as vehicle fuels. Chapter 6 by Kataki et al. reviews the current potential of biomass and biowastes in the country and also the technologies available for biomass-based electricity generation with or without heat recovery. This chapter also presents a case study on biomass-based energy generation in India including the prospective of bioenergy technology.

Part II has six chapters which explore various raw materials' availability, feedstock, compositional analysis, by-products, and application.

Alam et al. in Chap. 7 discuss various applications of glycerol, a major by-product of biodiesel production. Punia et al. in Chap. 8 describe the availability of crop residues and forestry waste for gasification, combustion, and power generation in India. About 500 million tons of crop residues are generated out of total agriculture produce, which are either used as animal feed, livestock bedding, packaging material, cooking in households, paper making, etc., or burnt in the farms and cause environmental pollution along with deterioration of the farm fertility. The judicious use of these agro-residues along with forestry residues could provide a sustainable source for biofuels or electricity generation. Similarly, Pathak in Chap. 9 summarizes the recent advances in mechanization and economic analysis concerning with biomass feedstock transportation, agricultural processing, and logistics from farm to biorefinery. Chapter 10 by Sukumaran et al. comprehensively reviews the availability, composition, conversion efficiencies, and potential ethanol yields from the first- and second-generation feedstock in the country. Chapter 11, by Pohit and Biswas, highlights the current status, prospects, and shortcomings in India's biodiesel program and suggests that principal changes are required considering the multi-feed feedstock approach, the attractive incentive mechanism both at feedstock stage and biodiesel production stage, and R&D for increasing the yield from feedstock. Chapter 12, by Thomas et al., discusses various feedstock options for biofuel production and biotechnological routes for conversion of biomass into biomethane, biohydrogen, and bio-butanol.

Part III contains five chapters which gives details about technical aspects, commercial view, and capture and storage of biofuels.

Borse and Sheth in Chap. 13 summarize the recent trends and developments in current and emerging ethanol technologies in India. This chapter is particularly focused on Praj Industries' technical diligence toward cellulosic ethanol development. Kumar et al. in Chap. 14 comprehensively discuss the global technological advancements and its prospective role in the development of economically feasible cellulosic ethanol. Particular emphasis is placed on the recent techniques including protein engineering, imaging, fermentation technology, chromatography, enzyme technology, chemical engineering, genome sequencing, bioinformatics, and synthetic biology. Chapter 15 by Prasad presents the cutting-edge catalytic based approach (heterogeneous metal-based catalysts, carbon catalysts, and biocatalysts) for the conversion of vegetable oils, waste cooking oils, and animal fats to biodiesel. Chapter 16, by Roy et al., describes in detail the different fermentation methods and various parameters for biohydrogen production including the mathematical modeling using integrated approaches. Basic technologies for solar energy development, current research programs, and the role of Indian Government in solar energy development have been apprised by Baig et al. in Chap. 17.

Part IV has five chapters providing outline information on proximate techno-economic analysis, biofuel policies, environmental scenario, and sustainable design in Indian context.

Prasad et al. in Chap. 18 provide an overview on substrates used for biodiesel production and technologies available for biodiesel production including life cycle analysis. Chandel et al. in Chap. 19 entail the biofuel policies in major biofuels supporting countries, key sustainable indicators, and socioeconomic aspects of biofuels from an Indian perspective. In Chap. 20, Kataki et al. present the details of the indicators and roles played by India's government for promotion of biofuels, particularly biodiesel and bioethanol production from nonedible oil seeds and other second-generation feedstocks. Various research programs supported by India's Government and challenges have also been highlighted in this chapter. Chapter 21 by Sheetal et al. presents the implications on the production and use of bioethanol and biodiesel in India in the past decade based on relevant studies. Impacts of biofuels on land, water use, ecosystems, and carbon accumulation have also been apprised. In the last Chap. 22, Satlewal and colleagues present the biofuel progress in India with respect to ethanol and biodiesel from involvement of the private sector. Greenhouse gas emissions in India and their comparisons with the world leading countries have also been discussed in detail.

We greatly appreciate the scholarly contribution of authors who added highly informative chapters in this book. We would also like to thank Isabel Ullmann, Hanna Hensler, Sumathy Thanigaivelu, Sulata Kumari Nayak, and the production staff of Springer Verlag, Germany, for their timely suggestions and support to publish this book. We would like to thank Sachin Kumar from SSNIRE, Kapoorthala-Punjab; Om V. Singh from University of Pittsburgh-Bradford; Ashok Pandey from Centre of Innovative and Applied bioprocessing, Mohali-Punjab; and Michael L. Cook from Centro de Tecnologia Canaviera, Piracicaba-Brazil, for the completion of this book. We are indebted to our families and express our sincere thanks to our families for their unconditional support and cooperation while editing this book. Last but not least, we would like to appreciate the reader's suggestions because we think that readers' benefits are the best reward for editors, contributing authors, and publishers.

Piracicaba, Sao Paulo, Brazil — Anuj K. Chandel
Thiruvananthapuram, Kerala, India — Rajeev K. Sukumaran

Contents

About the Editors

Dr. Anuj K. Chandel is working as a lead scientist for cellulosic ethanol production at Centro de Tecnologia Canavieira (Sugarcane Technology Centre, CTC), Piracicaba, Brazil. He is responsible for scientific leadership for deployment of cellulosic ethanol process at demonstration plant and scale-up activities. Overall, he has 16 years' research experience working in industries and universities on biomass conversion into ethanol, industrial enzyme production, and membrane-based separations. He has published 55 articles in peer-reviewed journals and 20 book chapters. He has also coedited 5 books on D-xylitol, lignocellulose degradation, extremophiles, and Brazilian bioenergy development. His primary research interest is to develop the sustainable process for bioconversion of lignocellulosics into renewable energy and biochemicals. Proactive research lead with an entrepreneurial mind-set, he brings hands-on strategic experience in research projects and start-up of piloting to full commercial scale operations.

Centro de Tecnologia Canavieira (Sugarcane Technology Centre, CTC), Piracicaba, Fazenda Santo Antônio—Santo Antônio, São Paulo Brazil

Dr. Rajeev Kumar Sukumaran is heading the Biofuels and Biorefineries Section at the CSIR-National Institute for Interdisciplinary Science and Technology's Microbial Processes and Technology Division, in Trivandrum, India. He is also the activity lead for the Centre for Biofuels (CBF) at CSIR-NIIST. He has 20 years of research experience spanning diverse fields including fermentation technology, embryonic stem cell biology, and transcriptomics, though currently he is working mainly on biomass conversion technologies. He played the leading role in setting up the CBF's lignocellulosic ethanol pilot plant and the solid-state enzyme production pilot plant at CSIR-NIIST. His current research interests include developing enzymes for biomass conversion, glucose-tolerant β-glucosidases, heterologous protein expression in fungi, and fungal genomics. He has served as consultant for some leading Biotech companies in India. He has also published widely with 75 publications in International Journals and has several conference papers, book chapters, and reports.

Centre for Biofuels, CSIR-National Institute for Interdisciplinary Science and Technology, Thiruvananthapuram Kerala, India

Part I
Conventional Fuels, Generalized Scenario of New and Advanced Biofuels

Chapter 1
Production and Consumption of Petroleum and Compressed Natural Gas in India, Oil and Gas Import, Development of Alternative Energy Sources, Price Escalation Pattern, and Its Effect on Indian Economy

Deepak K. Tuli and Ravi Gupta

Abstract Crude oil and natural gas prices play a very significant role in the economy of any country. India's growth story hovers around the import of crude oil as India imports more than 75% of its crude requirements and a significant amount of natural gas. India spends a very significant part of its foreign exchange for import of crude oil and natural gas, and this leads to impact on balance of payments. The oil and gas after coal is the major source of energy, and its availability is linked to the growth in all sectors of the economy. Traditionally, the Government of India has been shielding the variation and escalations of crude prices by providing huge financial subsidies to make available petroleum products to masses at a lower rate. However, during the last few years significant changes have taken place in India's oil and gas industry, resulting in the gradual withdrawal of subsidies. Over the period of last two decades, India has witnessed a quantum jump in the refining capacity. New refining capacity was established in both public and private sectors. Currently, India has excess of refining capacity, and after meeting domestic needs, large volumes of transportation fuels are exported to Europe and North America. Major fuel quality upgradation in transport fuel has been undertaken in almost all refineries in India. Under a planned program, fuel quality improvement has taken place for gasoline and diesel. This chapter contains the latest available information on the production and consumption of petroleum and compressed natural gas in India and oil and gas import/export data. This data has been taken from the official site of Ministry of Petroleum and Natural Gas. Brief on initiatives by the Government of India for enhancing oil and gas production in the country, taking oil assets abroad, and the development of alternative sources of energy is also given.

D.K. Tuli (✉) • R. Gupta
DBT-IOC Center for Advanced Bio-Energy Research, Indian Oil, R&D Center, Sector-13, Faridabad 121007, India
e-mail: tulidk@indianoil.in; tulidk@gmail.com

A.K. Chandel, R.K. Sukumaran (eds.), *Sustainable Biofuels Development in India*,
DOI 10.1007/978-3-319-50219-9_1

Keywords Oil • Natural gas • Economy • Price escalation • Subsidy • Alternative energy • Exploration • GDP

1 Introduction

India imports more than 203 million tons of crude oil (2014–2015) and other petroleum products and exports refined products, and crude for its own consumption is imported to the extent of >75%. The indigenous production of crude oil has almost stagnated and seen very small increase, ~34 MMTA in 2007–2008 to ~38 MMTA in 2012–2013. Natural gas did show increase in production but slowed down after a couple of years. Forecasts by industry experts do not predict any appreciable increase in domestic oil and gas production, and dependency on import is likely to increase in the next decade as the country is showing robust annual growth for consumption of petroleum products.

Crude oil remains as a commodity with most unpredictable pricing, and most of the long-range price predictions have failed miserably. Only a few years ago, the crude prices touched $147 per barrel and then defying predictions saw it crashing to as low as $30 per barrel. The high crude prices had a significant impact on the Indian economy, and the Government of India had to subsidize the price petroleum fuels so as to minimize the impact on users. However, this huge budgetary support led to deficit financing and indirect inflation.

Crude oil and gas not only serves as a source of energy but also as a major raw material to various petrochemical industries. Due to tremendous increase in oil and gas prices, the dependent petrochemical industries of importing countries came under severe pressure as the gas producing countries supplied cheaper gas to their own petrochemical industry.

Major growth of oil and gas consumption has been mainly from China and India as the economies of these countries showed decent GDP growth. The demand increment in Europe, Japan, and the USA has been almost nil during the last few years.

The Indian economy is at a critical stage of development. During 2013–2014, the growth rate of Gross Domestic Product (GDP) at current prices is estimated to have increased by 4.9%, with growth in subsequent years at 6–9%. The slowdown in growth of the industrial sector had an impact on the demand for petroleum products that increased by only 0.7% over the previous year 2012–2013.

2 Production of Crude Oil and Natural Gas

India had about 780 million metric tons of oil equivalent (MMTOE) of proven oil reserves as of April 2010 or 5.8 billion barrels as per EIA estimate for 2009, which is the second largest amount in the Asia-Pacific region behind China. Indian gas reserves are in the range of 1075–1400 billion cubic meters (BCM). Most of India's crude oil reserves are located in the western coast (Mumbai High) and in the northeastern parts of the country, although considerable undeveloped reserves are also located in the offshore Bay of Bengal and in the state of Rajasthan. During the last decade, deep water drilling has been successfully undertaken. The combination of rising oil consumption and fairly unwavering production levels leaves India highly dependent on imports to meet the consumption needs. India's oil sector is dominated by state-owned enterprises, although the government has taken steps in recent years to deregulate the hydrocarbons' industry and support greater foreign involvement. India's state-owned Oil and Natural Gas Corporation is the largest oil company. ONGC is the leading player in India's upstream sector, accounting for roughly 75% of the country's oil output during 2006, as per Indian government estimates. The other national oil exploration company is OIL India Ltd which operates mainly in the north east of the country. As a net importer of all oil, the Indian Government has introduced policies aimed at growing domestic oil production and oil exploration activities. As part of the effort, the Ministry of Petroleum and Natural Gas crafted the New Exploration License Policy (NELP) in 2000, which permits foreign companies to hold 100% equity possession in oil and natural gas projects. However, to date, only a handful of oil fields are controlled by foreign firms. Slow decision making, environmental clearance, and lack of long-term policy were main reasons for MNC companies staying away. However, in subsequent years the Government of India made clear, transparent, and internationally acceptable long-term policy for exploration of oil and gas. India's downstream sector is also dominated by state-owned entities, though private companies have enlarged their market share in recent years.

As per the Ministry of petroleum, Government of India, India has 1437 billion cubic meters (50.7 $\times$ 1012 cu ft) of confirmed natural gas reserves as of April 2010. A huge mass of India's natural gas production comes from the western offshore regions, particularly the Mumbai High complex. The onshore fields in Assam, Andhra Pradesh, and Gujarat states are also major producers of natural gas. As per EIA data, India produced 996 billion cubic feet (2.82×10^{10} m^3) of natural gas in 2004. India imports small amounts of natural gas. In 2004, India consumed about 1089 $\times$ 109 cu ft (3.08×10^{10} m^3) of natural gas, the first year in which the country showed net natural gas imports. During 2004, India imported 93 $\times$ 109 cu ft (2.6 $\times$ 109 m^3) of liquefied natural gas (LNG) from Qatar.

As in the oil sector, India's state-owned companies account for the bulk of natural gas production. ONGC and Oil India Ltd. (OIL) are the leading companies with respect to production volume, while some foreign companies take part in upstream developments in joint ventures and production sharing contracts. Reliance Industries, a privately owned Indian company, discovered a large natural gas find in 2002 in the Krishna Godavari basin. The Gas Authority of India Ltd. (GAIL) holds an effective control on natural gas transmission and allocation activities, though private companies like Adnani have also joined in a big way. In December 2006, the Minister of Petroleum and Natural Gas issued a new policy that allows foreign investors, private domestic companies, and Government oil companies to hold up to

100% equity stakes in pipeline projects. While GAIL's domination in natural gas transmission and allocation is not ensured by statute, it will continue to be the leading player in the sector because of its existing natural gas infrastructure.

The crude oil production for the year 2013–2014 was 37.788 million metric tons (MMT) as against 37.862 MMT for the previous year viz. 2012–2013, showing a marginal decrease of about 0.20%. The majority of crude oil production is from aging fields except new fields, viz., Rajasthan and KG deep-water blocks.

Natural Gas Production for the year 2013–2014 was 35.407 billion cubic meters (BCM) as against 40.679 BCM in the previous year viz. 2012–2013, showing decline of about 13%. The trend in the production of crude oil and natural gas since 2007–2008 to 2013–2014 has been shown in Table 1.1 and Fig. 1.1.

3 Refining Capacity and Production of Petroleum Products

Projected high domestic demand for petroleum products coupled with large export market has seen big investments in the refining sector. India, with 18 refineries, currently has a surplus refining capacity which has placed India among net petroleum product exporter countries. Increasingly stringent fuel specifications have put pressure on the old and noncompliant refineries to upgrade their refinery configurations to produce compliant fuels. Most of the Indian refineries have undergone series of upgrades, are now producing fuel better than Euro IV, and are thus able to export this to Europe. The Government is seriously considering promoting India as a competitive refining destination to service export market for petroleum products. In order to increase their profitability, more and more Indian refineries are integrating with the petrochemical and chemicals businesses (Fig. 1.2 and Table 1.2).

Table 1.1 Crude oil and natural gas production

Year	Crude oil production (MMT)	% growth in crude oil production	Natural gas production (BCM)	% growth in natural gas production
2007–2008	34.118	0.38	32.417	2.11
2008–2009	33.508	−1.79	32.845	1.32
2009–2010	33.690	0.54	47.496	44.61
2010–2011	37.684	11.85	52.219	9.94
2011–2012	38.090	1.08	47.559	−8.92
2012–2013	37.862	−0.60	40.679	−14.47
2013–2014	37.788	−0.19	35.407	−12.96

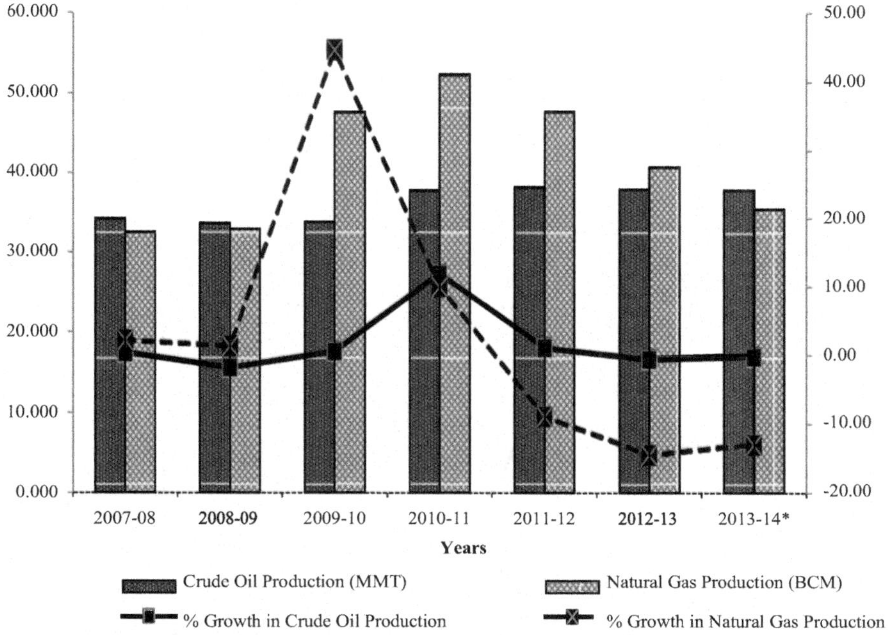

Fig. 1.1 Crude oil and natural gas production

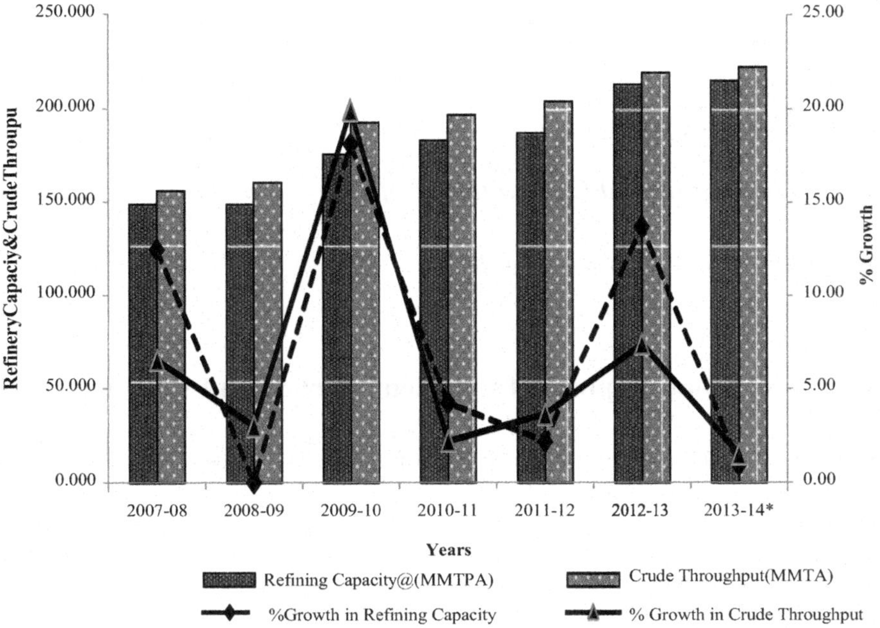

Fig. 1.2 Refinery capacity and refinery crude throughput

Table 1.2 Refinery capacity and refinery crude throughput (in terms of crude oil processed)

Year	Refining capacity @(MMTPA)	% growth in refining capacity	Crude throughput (MMTA)	% growth in crude throughput
2007–2008	148.968	12.46	156.103	6.52
2008–2009	148.968	0.00	160.772	2.99
2009–2010	175.956	18.12	192.768	19.90
2010–2011	183.386	4.22	196.989	2.19
2011–2012	187.386	2.18	204.121	3.62
2012–2013	213.066	13.70	219.212	7.39
2013–2014	215.066	0.94	222.497	1.50

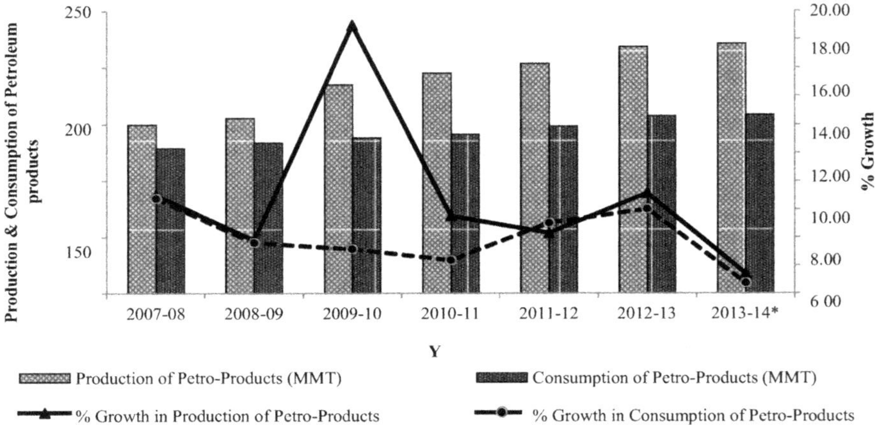

Fig 1.3 Production and consumption of petroleum products

3.1 Production and Consumption of Petroleum Products

See Figs. 1.3 and 1.4; Tables 1.3 and 1.4

4 Imports and Exports of Petroleum Crude and Products

India imports 80% of its crude oil and 18% of its natural gas needs. The country was the third largest energy consumer in the world in 2013, followed by China and the USA. It ranked as the fifth largest energy producer and fourth largest energy importer in the same year. At current year prices, India's energy imports total some USD 145 billion a year and are expected to reach USD 300 billion by 2030. The crude import will continue to be the biggest drain on the foreign exchange

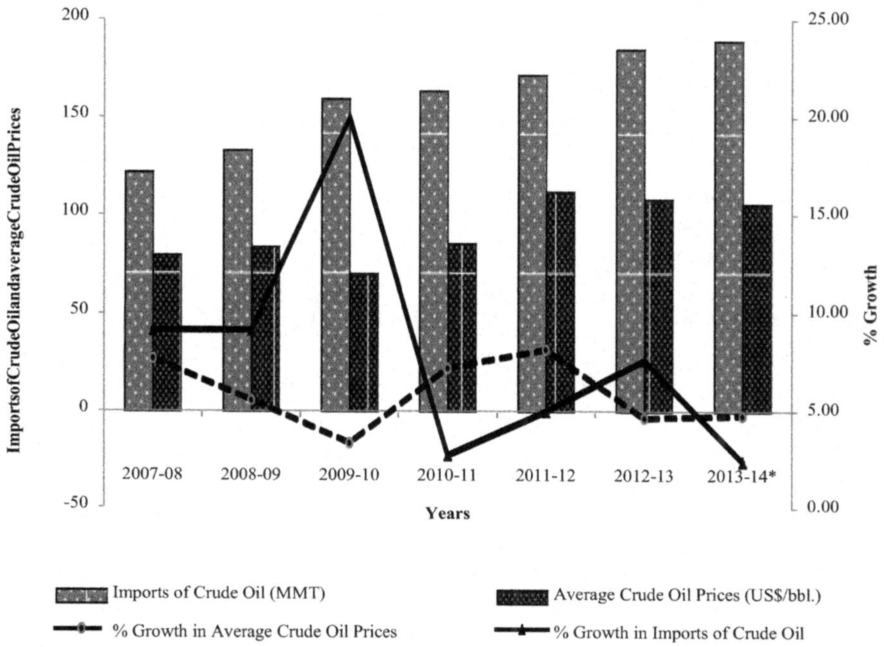

Fig. 1.4 Quantity of crude oil imports and average international crude oil prices

Table 1.3 Production and consumption of petroleum products

Year	Production of petro-products (MMT)	% growth in production of petro-products	Consumption of petro-products (MMT)	% growth in consumption of petro-products
2007–2008	149.472	7.02	128.946	6.79
2008–2009	155.148	3.80	133.599	3.61
2009–2010	184.608	18.99	137.808	3.15
2010–2011	194.821	5.53	141.040	2.35
2011–2012	203.202	4.30	148.132	5.03
2012–2013	217.736	7.15	157.057	6.02
2013–2014	220.756	1.39	158.197	0.73

holdings of the country. Natural gas demand is seen to grow at an average of 6.8% a year from FY 2013 to FY 2030, and according to some estimates, this demand could still rise due to environmental reasons.

Analysis of data presented by Petroleum Planning and Analysis Cell (PPAC) attached to the Ministry of Petroleum and Natural Gas, Govt. of India (ppac.org.in), makes a very interesting reading. Looking at the historic data presented, India imported crude oil approximately 40 MMT (1998–1999), 99 MMT (2005–2006), 189 MMT (2013–2014), and 203 MMT in 2015–2016. India also imported finished

Table 1.4 Imports of crude oil and average crude oil prices

Year	Imports of crude oil (MMT)	% growth in imports of crude oil	Average crude oil prices (US$/bbl.)	% growth in average crude oil prices
2007–2008	121.672	9.12	79.250	26.88
2008–2009	132.775	9.13	83.570	5.45
2009–2010	159.259	19.95	69.760	−16.53
2010–2011	163.595	2.72	85.090	21.98
2011–2012	171.729	4.97	111.890	31.50
2012–2013	184.795	7.61	107.970	−3.50
2013–2014[a]	189.238	2.40	105.520	−2.27

[a]Provisional

products, mainly diesel, lubes, and fuel oils, and total imports of products were 23 MMT (198–199), 13 MMT (2005–2006), 16 MMT (2013–2014), and 21 MMT in 2015–2016. Clearly, the import of finished products has reduced very substantially from 1998 to 2016 period. This was possible by vertical expansion of the refining capacity in India during this period.

During this period, India not only became self-sufficient in refined petroleum products but also had very large export surplus. This is clearly evident from the data. India exported refined petroleum products of 0.7 MMT in 1998–1999, 23 MMT in 2005–2006, and 68 MMT in 2013–2014 which is an extremely large jump.

India is the fifth largest Liquefied Natural Gas (LNG) importer after Japan, South Korea, the United Kingdom, and Spain and accounts for 5.5% of the total global trade. The LNG imports had increased by 43.38% year on year in May 2016 to 2.08 billion cubic meters (BCM). Domestic LNG demand is expected to grow at a CAGR of 16.89% to 306.54 million metric standard cubic meter per day (MMSCMD) by 2021 from 64 MMSCMD in 2015 (Table 1.5).

5 Equity Oil and Gas from Abroad

In a major policy shift, the Government of India decided to hold oil and gas equity both in-production fields and also fields which are under development. As a first step, the domestic major oil and gas producer, Oil and Natural Gas Corporation (ONGC), was tasked to take oil and gas equity abroad through its wholly owned subsidiary called ONGC Videsh Ltd. Later Govt. also encouraged Oil India limited and downstream oil majors like Indian Oil Corporation, BPCL, and HPCL to acquire oil assets abroad. Indian companies in the last few years have taken oil equity in existing producing fields as well as in underdeveloped fields in Sudan, Venezuela, Syria, Vietnam, Russia, Colombia, etc. During 2014–2015, these efforts brought home 8.4 MMT of oil equivalent. ONGC Videsh has stake in 33 oil and gas

Table 1.5 Imports and exports of petroleum products

Year	Imports of petroleum products (MMT)	Exports of petroleum products (MMT)	Imports of LNG (MMT)
2007–2008	22.462	40.779	8.320
2008–2009	18.586	38.944	8.060
2009–2010	14.665	51.155	9.148
2010–2011	17.379	59.077	9.931
2011–2012	15.849	60.837	13.214
2012–2013	15.774	63.408	13.136
2013–2014	16.718	67.864	13.032

projects in 16 Countries, viz. Vietnam (2 projects), Russia (2 projects), Sudan (2 projects), South Sudan (2 projects), Iran (1 project), Iraq (1 project), Libya (1 project), Myanmar (2 projects), Syria (2 projects), Cuba (1 project), Brazil (2 projects), Nigeria (1), Colombia (8 projects), Venezuela (2 projects), Kazakhstan (1 project), and Azerbaijan (2 projects).

As per Perspective Plan 2030, OVL's oil and gas production should increase from the existing level of 8.26 MMT to 20 MMT by 2017–2018 and 60 MMT by 2029–2030.

5.1 *Foreign Direct Investment Inflows*

In initial years after independence, oil sector witnessed a heavy dose of nationalization, and major oil MNCs like Shell, Burma Oil, and Caltex left the country. During this phase, National oil companies like ONGC, IOC, BPC, and HPC were created, and for three decades, there was no private sector refinery in the country. Starting 1990, along with other sectors oil and gas sector was also liberalized which later resulted in the establishment of two mammoth refineries in private sector by Reliance Industries Ltd. Later other private players like Essar and Mittal also established refineries. Earlier Foreign Direct Investment (FDI) was allowed in the sector to the extent of 49%. The policy of FDI was relaxed, and since Aug 2013 in petroleum and natural gas sector, 100% FDI is allowed in exploration and production, refining by the private companies, and marketing of petroleum products, among other areas. FDI in the petroleum sector in 2012–2013 was $198 million, while in 2014–2015 it went to $1078 million. It is expected that FDI in this sector will be ~$10 billion by 2018–2019.

Year-wise inflow of FDI in petroleum sector is given in Fig. 1.5 and Tables 1.6 and 1.7.

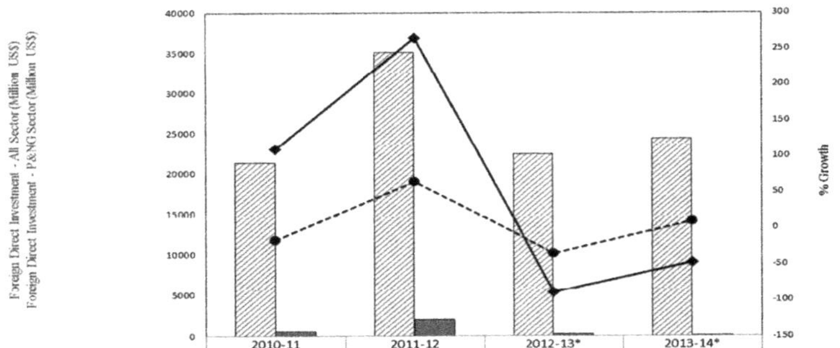

	2010-11	2011-12	2012-13*	2013-14*
Foreign Direct Investment - All Sector (Million US$)	21383	35121	22424	24299
Foreign Direct Investment - P&NG Sector (Million US$)	556	2030	215	112
% Annual Growth - All Sector (Million US$)	-17.23	64.25	-36.15	8.37
% Annual Growth - P&NG Sector (Million US$)	109.55	264.82	-89.42	-47.75

Fig. 1.5 Year-wise FDI inflows under petroleum and gas sector

6 Oil Price Escalation Pattern and Its Effect on Indian Economy

When oil prices are high and given the very large dependence on imported oil, the Government of India is left with some difficult choices to make. The choices are (a) passing on the price increase to the consumer, (b) rationalizing taxes and other levies on petroleum products, (c) making the National Oil Companies (NOCs) bear the burden, and (d) providing subsidized products by financing through the budget. Although the Government has resorted to a combination of all above four options in the past, each of these options has its own drawbacks. Passing price increase causes hardship to lower income group. Lower taxes will impact Government revenue collection, and letting companies bear the burden fully has a very negative impact on their commercial viability. In the long run, the only viable policy to deal with high international oil prices is to rationalize the tax burden on oil products over time, pass on some increase to consumers, and protect specially those who are below the poverty line.

Oil prices are dependent on several factors the major being political situation in and around major oil producing countries, OPEC policies to restrict oil production, weather severity in Europe and the USA, growth of economy of major countries, etc. In the past due to its conflict, the USA had declared oil embargo on Iran, and coupled with OPEC resolve to cut production and heavy demand from China and India, global level tight supplies resulted in shooting up of prices. In June 2014, the price of Brent crude was at USD 115 per barrel (June 23, 2014) and went up further to touch $147 per barrel. This high price of oil resulted in activation of several low producing fields. Meanwhile, the US shale oil exploration reached new scales. Currently, US oil output levels are at their highest in almost 30 years. The oil-drilling boom in the USA has increased crude production by over 70% since 2008.

Table 1.6 Indian economy-selected indicators

Indicator	Unit/base	2007–2008	2008–2009	2009–2010	2010–2011	2011–2012	2012–2013	2013–2014
1. Geographical area	M.Sq.Km.	3.29	3.29	3.29	3.29	3.29	3.29	3.29
2. Population	Million	1138	1154	1170	1186	1202	1213	1229
3. *Gross domestic product*								
At factor cost								
(i) At current prices	Rs.'000Crore	4582	5304	6109	7249	8392	9389	10,540
(ii) At constant prices	Rs.'000Crore	3897	4159	4516	4919	5248	5482	5749
4. *Net national product*								
At factor cost								
(i) At current prices	Rs. '000Crore	4077	4705	5411	6407	7435	8256	9238
(ii) At constant prices	Rs. '000Crore	3452	3664	3966	4294	4573	4729	4927
5. *Per capita net*								
National product								
(i) At current prices	Rupees	35,825	40,775	46,249	54,021	61,855	67,839	74,920
(ii) At constant prices	Rupees	30,332	31,754	33,901	36,202	38,048	38,856	39,961
6. *Foreign exchange reserves*								
(i) Gold	Rs.Crore	40,124	48,793	81,188	10,2572	13,8250	139,737	129,616
(ii) SDR	Rs.Crore	74	6	22,596	20,401	22,866	23,538	26,826
(iii) Foreign currency asse	Rs.Crore	119,6023	1,230,066	1,149,650	1,224,883	1,330,511	1,412,631	1,660,914
7. *Foreign trade*								
(i) Import	Rs.Crore	1,012,312	1,374,436	1,363,704	1,683,467	2,345,463	2,669,162	2,714,182
(ii) Export	Rs.Crore	655,864	840,755	845,534	1,142,922	1,465,959	1,634,319	1,894,182
(iii) Balance of trade	Rs.Crore	−356,448	−533,681	−518,170	−540,545	−879,504	−1,034,843	−820,000
8. *Index of production*								
(i) Agricultural	2007–2008 = 100	107	107	103	121	125	124	129
(ii) Industrial	2004–2005 = 100	142	145	153	166	170	172	172

(continued)

Table 1.6 (continued)

Indicator	Unit/base	2007–2008	2008–2009	2009–2010	2010–2011	2011–2012	2012–2013	2013–2014
9. Wholesale price index	2004–2005 = 100	117	125	131	143	156	168	178
10. *Consumer price index*								
(i) Industrial workers	2001 = 100	133	145	163	180	195	215	236
(ii) Agricultural laborers	1986–1987 = 100	409	450	513	564	611	672	750
(iii) Rural laborers	1986–1987 = 100	409	451	513	564	611	673	751
11. *Energy generation(gross)*								
Coal	MMT	457	493	532	533	540	556	566
Lignite	MMT	34	32	34	38	42	46	44
Natural Gas	BCM	32	33	47	52	48	41	35
Crude oil	MMT	34	34	34	38	38	38	38
Petroleum products	MMT	149	155	185	195	203	218	221
12. *Electricity generated (gross)*								
(i) Total utilities	B.KWH	723	741	800	884	973	964	1023
Hydel	B.KWH	120	110	104	114	131	114	135
Thermal	B.KWH	585	616	677	704	759	760	792
Nuclear	B.KWH	17	15	19	26	32	33	34
R.E.S.	B.KWH	–	–	–	39	51	57	61
(ii) Non-utilities	B.KWH	91	100	106	121	128	144	151
Total electricity	*B.KWH*	*813*	*841*	*906*	*1005*	*1102*	*1108*	*1174*
13. POL traffic by rail	MMT	36	35	39	39	40	41	42

Note: Figures reconciled for previous years, provisional for 2013–2014
Source: economic survey, RBI, ministry of statistics/industry/finance

Table 1.7 Trends in Indian petroleum industry at a glance

Item	Unit	2007–2008	2008–2009	2009–2010	2010–2011	2011–2012	2012–2013	2013–2014[a]
1. *Production*								
(i) Crude oil	MMT	34.12	33.51	33.69	37.68	38.09	37.86	37.79
(ii) Natural gas	BCM	32.42	32.84	47.50	52.22	47.56	40.68	35.41
(iii) Petroleum products	MMT	149.47	155.15	184.61	194.82	203.20	217.74	220.76
2. *Consumption*								
(i) Crude oil (in terms of refinery crude processed)	MMT	156.10	160.77	192.77	196.99	204.12	219.21	222.50
(ii) Natural gas	BCM	31.48	31.75	46.52	51.25	46.48	39.78	34.64
(iii) Petroleum products	MMT	128.95	133.60	137.81	141.04	148.13	157.06	158.20
3. *Reserves (balance recoverable)*[b]								
(i) Crude oil	MMT	770.12	773.34	774.66	757.40	759.59	758.44	762.74
(ii) Natural gas	BCM	1089.97	1115.26	1148.54	1278.06	1330.24	1354.76	1427.15
4. *Imports and exports*								
(i) Gross imports								
(a) Crude oil	MMT	121.67	132.78	159.26	163.60	171.73	184.80	189.24
LNG	MMT	8.32	8.06	9.15	9.93	13.21	13.14	13.03
Pol. products	MMT	22.46	18.59	14.67	17.38	15.85	15.77	16.72
Total(a)	MMT	152.45	159.42	183.07	190.91	200.79	213.70	218.99
(b) Crude oil	Rs. Billion	2726.99	3483.04	3752.77	4552.76	6722.20	7846.52	8648.75
LNG	Rs. Billion	71.97	95.48	106.95	143.62	317.18	417.31	533.07
Pol. products	Rs. Billion	541.00	611.56	336.87	559.98	680.91	683.63	746.05
Total(b)	Rs. Billion	3339.96	4190.08	4196.59	5256.37	7720.29	8947.46	9927.87

(continued)

Table 1.7 (continued)

Item	Unit	2007–2008	2008–2009	2009–2010	2010–2011	2011–2012	2012–2013	2013–2014[a]
(ii) Exports								
(a) Qty: Pol. products	MMT	40.78	38.94	51.15	59.08	60.84	63.41	67.86
(b) Value: Pol. products	Rs. Billion	1107.89	1224.75	1446.87	1968.61	2846.43	3200.90	3682.79
(iii) Net imports								
(a) Qty: crude oil	MMT	121.67	132.78	159.26	163.60	171.73	184.80	189.24
Pol. products	MMT	−18.32	−20.36	−36.49	−41.70	−44.99	−47.63	−51.15
Total(a)	MMT	111.67	120.48	131.92	131.83	139.96	150.30	151.12
(b) Value: crude oil	Rs. Billion	2726.99	3483.04	3752.77	4552.76	6722.20	7846.52	8648.75
Pol. products	Rs. Billion	−566.89	−613.19	−1110.00	−1408.62	−2165.52	−2517.27	−2936.74
Total(b)	Rs. Billion	2160.10	2869.85	2642.77	3144.14	4556.68	5329.26	5712.01
(iv) Unit value of crude oil imports(gross)	Rs./MT	22,413	26,233	23,564	27,829	39,144	42,461	45,703
5. India's total imports	Rs. Billion	10,123	13,744	13,637	16,835	23,455	26,692	27,174
6. India's total exports	Rs. Billion	6559	8408	8455	11,429	14,660	16,343	18,997
7. Petroleum imports as % of India's total imports	%	32.99	30.49	30.77	31.22	32.92	33.52	36.53
8. Petroleum exports as % of India's total exports	%	16.89	14.57	17.11	17.22	19.42	19.59	19.39
9. *Contribution of oil sector to center/state resources*								
(i) Royalty from crude oil	Rs. Crore	6544	7155	6155	8958	15,921	18,083	20,114
(ii) Royalty from gas	Rs. Crore	1487	1624	1800	2355	3294	3880	3483

(iii) Oil development cess	Rs. Crore	7156	6886	7096	6783	8119	15,784	16,181
(iv) Excise and customduties[c]	Rs. Crore	78,373	70,557	71,767	102,828	95,229	98,603	104,163
(v) Sales tax	Rs. Crore	59,890	62,962	63,949	80,709	100,375	111,438	124,984
(vi) Dividend	Rs. Crore	9974	9860	13,811	13,329	13,042	14,064	14,994

Note: Figures reconciled for previous years
[a]provisional for 2013–2014
[b]As on 1st April of initial year
[c]Includes cess on crude oil

Later in early 2015, US demand of imported crude diminished to a great extent and the USA became net large exporter of Oil and Gas due to large shale oil discoveries. Coupled with slowdown of Japanese and European economies, resolution of Iran issue, and lower demand from China, it resulted in net surplus in market and prices slashed. Oil producers in Gulf pumped more oil for maintaining revenue, resulting in OPEC disagreement and more surplus oil in market. While the global benchmark Brent crude had hit a low of USD 47.22 per barrel, US WTI crude is trading at around USD 45 per barrel. Oil prices have already plummeted 60% from their 2014 peak. In the last one year, Brent has weakened by nearly 54.04%. Crude oil prices are now nearly at a 6-year low in the international market.

For Saudi Arabia's economy, the current level might not be disastrous in the short term due to its large reserve fund of around USD 750 billion which it will use to finance its deficits. However, in the long term, the country needs oil prices to be at around USD 80 per barrel. If the prices sustain at their current levels or see more decline, then it will be very difficult for the US shale oil producers to continue their operations. Some major Gulf oil producers will increase oil production so as to lower prices in a hope to increase their market share in the longer run by throttling the US oil boom. As per experts, shale oil exploration is not competitive below the level of USD 60 per barrel. Therefore, international oil diplomacy is yet another factor of price fluctuation of crude oil.

As far as India is concerned, falling crude is certainly a blessing for the economy as it helps macro-economic management of inflation, fiscal deficit, and current account deficit. However, the Indian exploration companies suffer due to lower oil prices, and they generally shut down their low yielding marginal oil fields. A fall of one dollar in the price of oil saves the country about Rs 40 billion. Moreover, lower crude price will surely facilitate the Reserve Bank of India in adopting growth-centric approach while reviewing monetary policy. It is estimated that a fall of USD 10 in crude could reduce the Current Account Deficit by roughly 0.5% of GDP and the fiscal deficit by around 0.1% of GDP.

But on the flip side, analysts are also highlighting the potential downside risks associated with lower oil prices. Companies slash their exploration budgets and new finds become impossible and some may find it to be very tough for them to continue production under these circumstances. Russia is one of the world's largest oil producers, and its dramatic interest rate hike to 17% in support of its troubled Rouble underscores how heavily its economy depends on energy revenues, with oil and gas accounting for 70% of export incomes. For the first time, Saudi Arabia had to present a deficit budget and is actively thinking of increasing taxes.

6.1 Increasing the Energy Basket

The above discussions lead to two major points: (1) energy demand will continue to grow in India for the next few decades and will closely match the growth of GDP and (2) the import dependency of India on crude oil will not decrease, and if at all, it will not be by any significant amount. As the Government is duty bound to provide affordable and sustainable energy to all sections of society, it has planned several initiatives for increasing the energy basket.

The Ministry of Petroleum and Natural Gas has announced a new "Marginal Fields Policy," which aims to bring into production 69 marginal oil and gas fields with 89 million tons or US $11.12 billion worth of reserves, by offering various incentives to oil and gas explorers such as exemption from payment of oil cess and customs duty on machinery and equipment.

In 2013, the Indian government approved a shale gas exploration policy that allows two state-run national oil companies (NOCs)—ONGC and OIL—to drill for shale gas within their respective conventional fuel exploration blocks.

The policy has provided the NOCs with a number of financial incentives including income tax and customs exemptions, while keeping royalties and taxes the same as for conventional extraction in a particular area. Carrying out shale gas/oil exploration by private and joint venture companies is envisaged under a proposed uniform licensing policy (ULP). However, the hydraulic fracturing technology used in shale gas extraction is feared to be a powerful pollutant, capable of devastating India's water resources, thus offsetting any social benefit from increased energy production.

In 2014, the country moved a step closer to market-driven fuel pricing by lifting diesel price controls and raising the cost of natural gas by some 30%. The new price of locally produced gas came into effect on November 1, 2014. This will increase revenue of oil exploration companies, and they will spend more on new explorations to increase oil pool.

Removal the bottlenecks in CBM exploration and production is necessary while safeguarding the environment. Coal bed methane (CBM) is formed in association with coal at shallow depths. Its extraction does not entail horizontal drilling and requires a much smaller degree of fracturing compared to shale gas. However, a considerable amount of water associated with the gas needs to be removed to allow the gas to flow. This water can contain dissolved solids and pollutants, which will need to be treated or disposed of safely. Although 33 blocks have been awarded since 2001, mainly in east India, production is currently around just 3 billion cubic meters per annum.

6.2 *Exploring Gas Hydrates*

These are methane and water molecules in seabed sediments that get frozen into ice due to low temperatures and high pressures. India's offshore reserves have been tentatively estimated at around 66,000 tcf or 1500 times more than the known conventional gas reserves. Though the government formulated a National Gas Hydrate Program in 1997 and under an Indo-US initiative a drilling ship explored four seabed areas in 2006, nothing much has happened since.

7 Conclusion

India is highly dependent on imported crude and natural gas, and the estimates indicate that this dependency is likely to increase in the near future. The Government of India has taken several steps in the past to insulate common users, especially poor, from the impact of very high crude prices. However, it is only a short-term remedy as it leads to unacceptable subsidies which are met from revenue budget. National oil companies were encouraged to acquire oil and gas assets abroad and bring home additional oil and gas as equity. This strategy is paying dividends. India has over the years become a refining hub and presently exports huge quantities of refined petroleum fuels, after meeting domestic demands. More refineries are being built in both public and private sectors, Government has opened oil and gas sector to 100% foreign direct investment (FDI), and huge investments are flowing. Alternatives like shale oil, coal bed methane, and gas hydrates are being encouraged. Secure and affordable oil and gas is essential if India has to maintain a healthy GDP growth of the past few years.

Bibliography

1. Ministry of petroleum and natural gas. http://petroleum.nic.in
2. Natural resources—3rd five-year plan. Planning commission, Government of India
3. Basic statistics on indian petroleum and natural gas. Ministry of petroleum and natural gas, Indian Government. 2011–2012
4. https://en.wikipedia.org
5. http://www.nmims.edu
6. Report of the working group on petroleum and natural gas sector for the XI Plan (2007–2012)

Chapter 2
Bioethanol Production Scenario in India: Potential and Policy Perspective

Surender Singh, Anurup Adak, M. Saritha, Sonia Sharma, Rameshwar Tiwari, Sarika Rana, Anju Arora, and Lata Nain

Abstract Energy crisis has become a fundamental issue over mankind across the globe. To meet the energy demand for transport, blending of bioethanol with gasoline has been mandated in India like several other developing and developed countries. Moreover, in spite of urbanisation and industrial development, agriculture sector has long been the backbone of Indian economy. Currently in India, sugarcane molasses is being used for the production of bioethanol but cannot fulfil the demand of ethanol for blending targets. Therefore, the surplus crop residues may be explored for second-generation biofuel to meet the demand for alternative and renewable energy sources. However, the national policy on biofuels has no clear long-term mandated targets and penalties to ensure its successful execution. To safeguard the long-term sustainability and economic viability of Indian bioethanol market, it is indispensable to diversify the feedstock basket and gain insights from the reality of pilot-scale studies. India's present bioethanol production scenario can be boosted by introduction of new low-cost technology for the bioconversion of lignocellulosics. This chapter is a comprehensive analysis of the future of bioethanol market in India based on biomass availability, policy barriers and perspectives along with the possible solutions.

Keywords Bioethanol • Crop residue • Biofuel policy • Sustainability • Bioethanol market • Second generation biofuel

1 Introduction

Energy security is of prime importance for the economic growth and sustainable development of a country, whether developing or developed. Most of the world's energy demand is primarily met from non-renewable resources like coal, petroleum and natural gas which take thousands of years to form and cannot be replaced at sustainable rates. The unequal geographical distribution of these natural resources

---

S. Singh • A. Adak • M. Saritha • S. Sharma • R. Tiwari • S. Rana • A. Arora • L. Nain (✉)
Division of Microbiology, Indian Agricultural Research Institute, New Delhi 110012, India
e-mail: latarajat@yahoo.co.in

A.K. Chandel, R.K. Sukumaran (eds.), *Sustainable Biofuels Development in India*,
DOI 10.1007/978-3-319-50219-9_2

warrants nations to import crude oil from the Middle East in order to meet their energy requirements. India is the fifth largest consumer of energy after the USA, China, Russia and Japan, accounting for 4% of the global energy consumption [1]. With an annual energy demand growing steadily at a rate of 4.8%, India is projected to become the world's third biggest energy consumer by 2030 [2]. Economic development is greatly linked to industrial energy usage, as the sector demands the highest energy in India, followed by the transportation sector, which depends on petroleum products to meet 40% of the energy requirement [3]. However, the sharp increase in human population along with urbanisation and the rising demand for energy has resulted in a huge gap between the demand and supply of crude oil in India. The domestic demand of crude oil in India is expected to reach more than 8 million barrels per day in 2035 [4], whereas the current scenario reflects the capacity of domestic production to account for only 18% of the national requirement. India is, thus, heavily dependent on crude oil imports, with petroleum crude accounting for about 34% of the total inward shipments [5]. The oil imports during April–May, 2015–2016, were valued at US $15981.59 million [5], and the ever-increasing imports have been leaving the country with a growing balance of payment deficits. The growing energy demand along with the rapid depletion of petroleum and associated resources and concerns about climate and environmental deterioration have led to resurgence in the development of sustainable and renewable energy alternatives. Though a number of alternatives like ocean water power, geothermal energy, wind energy and solar energy are being explored, a lot of emphasis has been given world over to the alternative fuels as a strong source of energy in the coming years.

Biofuels are considered the only foreseeable, feasible and sustainable energy resource. Biofuels derived from biomass may be in solid, liquid or gaseous forms, of which liquid biofuels are the most imperative, given the current scenario of world transportation sector and their ability to reduce/replace the non-renewable petroleum fuels. Biofuels have several advantages such as:

- They address the issues related to energy, food security and integrated rural development by creating job opportunities.
- Their production and development influence the policies that affect multiple sectors involved in the socio-economics of the country.
- Their consumption addresses the concerns about containment of carbon emission as it improves combustion and reduces the hydrocarbon emission.
- The easy adaptability of bioethanol to the existing engines makes this fuel one of the most important alternatives.
- These are cleaner fuels with higher octane rating than gasoline [6].

In the last decade of biofuel history, countries like the USA, Brazil and Canada and some European nations were the pioneers in large-scale biofuel production. The total bioethanol production of the world was 88.68 billion litres in 2013 (Table 2.1), of which 50.34 billion litres produced by the USA and 23.72 billion litres produced by Brazil, the two leading bioethanol producers in the world [7].

Table 2.1 Status of bioethanol production in different countries

	2007	2008	2009	2010	2011	2012	2013
Country	Billion litres						
USA	24.68	35.23	41.40	50.33	52.79	50.34	50.34
Brazil	19.00	24.50	24.90	26.20	21.09	21.11	23.72
Europe	2.16	2.78	3.94	4.57	4.42	4.46	5.19
China	1.84	1.90	2.05	2.05	2.10	2.10	2.63
Canada	0.80	0.90	1.10	1.35	1.75	1.70	1.98
Rest of world	1.19	1.47	3.46	3.73	2.64	2.85	4.81
Total	49.67	66.78	76.85	88.23	84.80	82.56	88.68

Source: Licht [8]

However, the present global bioethanol production scenario is not very promising. Brazil, the largest producer of sugarcane in the world, uses it as the primary feedstock for bioethanol production, while the USA uses 80% of the global corn produced for bioethanol production [9]. More importantly, the use of food grains is expected to increase at a rate of 10% annually for this purpose, indicating their scarcity for human consumption [10]. For example, the US corn-bioethanol programme disrupts the international corn markets which can have serious impacts on the food security of several developing countries with billions of people dependent on corn as the staple food source. The diversion of food crops for energy production is also a major factor for global food price hike. The "food vs fuel" debate has forced a search on alternative feedstocks for advanced generation biofuels, which do not compete directly with food crops.

The bioethanol market in India is expected to increase dramatically with the steadily growing transportation sector and the consequent increase in expenditure of petroleum products. Continued economic growth coupled with urbanisation and the fast lifestyle of Indian people push them to go in favour of road transport. About 142 million motor vehicles were registered in India during the financial year 2010–2011, and new vehicle registration is expected to reach 167 million by the end of the current financial year [11]. Moreover, the growing concern about environmental pollution through carbon emission has led the Government of India transport policy to target Euro III and IV vehicle norms [12], which in turn require more clean and green fuel. The market value of potable alcohol industry is approximately Rs 300 billion and has been growing at a fast rate of 7–10% per year [13]. This highly growing potable industry provides an alternative source to gasoline when devoted to the biofuel sector.

Taking lessons from the global experience, India's bioethanol programme exclusively depends upon non-edible feedstocks like sugarcane molasses. Although being the second largest sugarcane producer, India accounts for only 1% of the global biofuel production [9]. In India, most of the bioethanol production and its blending with gasoline have been largely driven by the National Biofuel Policy (http://mnre.gov.in/file-manager/UserFiles/biofuel_policy.pdf). In the year 2003, the Govt. of India launched the Ethanol Blended Petrol Programme (EBPP) to promote bioethanol and proposed a 5% blending with gasoline [14]. Initially, it was not implemented because of the shortage of sugarcane molasses. To compromise the initial cost of technology innovation and market development, the Government of India subsidised different sectors involved in the process. In 2010, the National Biofuel Policy set a biethanol blending target of 20% by 2017, and this induced search for alternative biomass and development of new and useful bioprocess techniques. The rigorous initiative by the Government of India has failed to achieve the blending targets mainly due to the cyclical nature of sugarcane production and consequent shortage of molasses [15]. To overcome this situation, alternative and effective biomass sources are to be tried. In a country like India with vast amount of agricultural land, the biomass residues can be employed as alternative source of renewable energy [16]. India produces about 686 million tonnes of crop biomass per year, of which about 34% surplus biomass can be used for bioenergy generation [17]. Alternative feedstocks like cane juice, *Miscanthus*, sweet sorghum and other available lignocellulosic biomass need to be promoted for realising the blending target. India should also improvise new technologies for better conversion of this biomass for sustainable development of the bioenergy sector of the country.

2 Overview of the Biofuel Production Process

The uniform and abundant availability of biomass resource in India has potential prospects as the future energy source. Once advanced conversion technologies for biofuels are in place, it would be possible to utilise forest and agricultural residues to produce enough ethanol to meet the country's entire gasoline needs. Considering the importance of biofuels, Indian ministries are currently involved in the promotion, development and policymaking for the optimal development and exploitation of indigenous biomass. On the basis of the biomass sources, biofuels are grouped into the following types.

(i) First-generation biofuels: These are made by converting grains and sugar to ethanol or vegetable oils to biodiesel. These include the sugarcane-ethanol, starch-based or corn ethanol.
(ii) Second-generation biofuels: They are obtained by converting lignocellulosic biomass to alcohols and green diesel. Second-generation biofuel technologies gained importance because of the limitations of first-generation biofuels as they cannot be produced beyond a threshold level without threatening food security.
(iii) Third- and fourth-generation biofuels: The third-generation biofuels include algal biodiesel, algal hydrogen and hydrogen produced from conversion of biomass, whereas the fourth-generation biofuels include fuels from high solar efficiency cultivations. These are, but, still in the early stages of development.

Of these, second-generation biofuels obtained from biomass are attractive as alternative fuels because these are produced from renewable bio-based sources.

Each lignocellulosic substrate is a complex mix of cellulose, hemicellulose and lignin, bound in a matrix. While cellulose and hemicellulose yield fermentable sugars, lignin is the most recalcitrant polymer, consisting of phenyl-propanoid units. The production of bioethanol from lignocellulosic biomass is a multistep process as depicted in Fig. 2.1. Each step is important in determining the overall efficiency of the process of conversion of biomass to bioethanol. These steps are:

1. Pretreatment of the biomass for lignin removal and increased enzyme access to the polysaccharides.
2. Saccharification of polysaccharides (cellulose and hemicellulose) in the biomass for the release of fermentable sugars.
3. Fermentation of the sugars, whereby the sugar is converted into ethanol and CO_2.
4. Distillation and rectification for concentration and purification of the ethanol produced.

Commercialisation of lignocellulosic ethanol, however, is largely limited due to the lack of cost-effective processing technologies and high cost of enzymes. As suggested by Moreno et al. [18], future biorefineries will have to be able to produce biofuels and high-value compounds at low cost with minimal downstream wastes for sustainable social, economic and environmental development. Modernising biomass technologies is, therefore, a viable option for the creation of economic packages for generating "wealth from wastes".

3 The Availability of Biomass for Bioethanol in India

Biomass is a natural source of polysaccharides that store solar energy via photosynthesis in the form of chemical energy. Plant biomass has great potential to provide renewable energy for the future because of its abundance and inexpensiveness than fossil fuel. Being an agriculturally dominant nation, the strength of India's

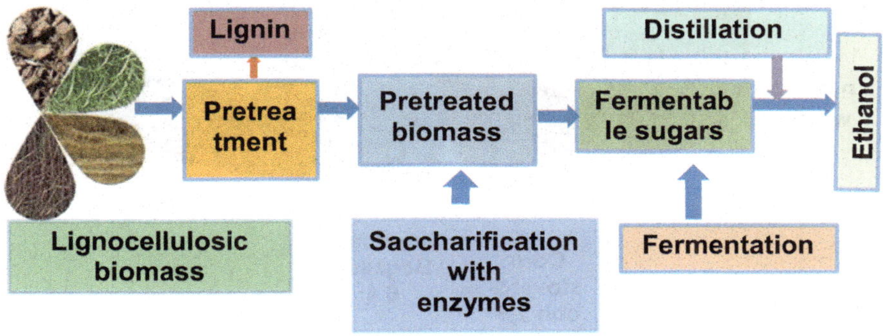

Fig. 2.1 Second-generation biofuel production process from lignocellulosic biomass

bioenergy programme is largely influenced by the agricultural sector. In India, with an extraordinary diversity of climatic conditions, from tropical in the south to temperate and alpine in the north, about 60% of the land is used for vigorous agricultural practices [17]. These generate huge amounts of primary (rice straw, wheat straw, barley straw, legume stover and sugarcane tops) and secondary (rice husk, bagasse) agricultural residues. While the primary crop residues are mainly utilised for fodder purposes, the secondary residues, which are less important as animal feed, can be processed for second-generation biofuel production. The availability of these biomass resources along with their biofuel production potential needs to be calculated in order to meet the national bioethanol demand.

Though a large amount of agricultural and forest products is generated in India, which may be potent bioethanol feedstocks, their availability for bioethanol production needs to be looked with caution. All the potential biomass is not available for bioethanol production. In some regions of India, where forest resources are not sufficient, the available agro residues are used for cooking and burning, thereby limiting their availability for biofuel production. Interestingly, a survey conducted by the NIST–TIFAC [19] observed that the availability of surplus residues would be sufficient to support the projected bioethanol demand for 2020 (Fig. 2.2). Moreover, it was suggested that the more surplus residues which are almost always burnt in the field itself should be explored [19].

Singh and Gu [20] calculated an approximate gross potential of 1050 MT biomass per year including plantation crops such as rubber and coffee. Recently, Hiloidhari et al. [17] estimated that about 686 MT gross residue is available in India of which 545 MT is contributed by cereals, oilseed, pulses and sugarcane together.

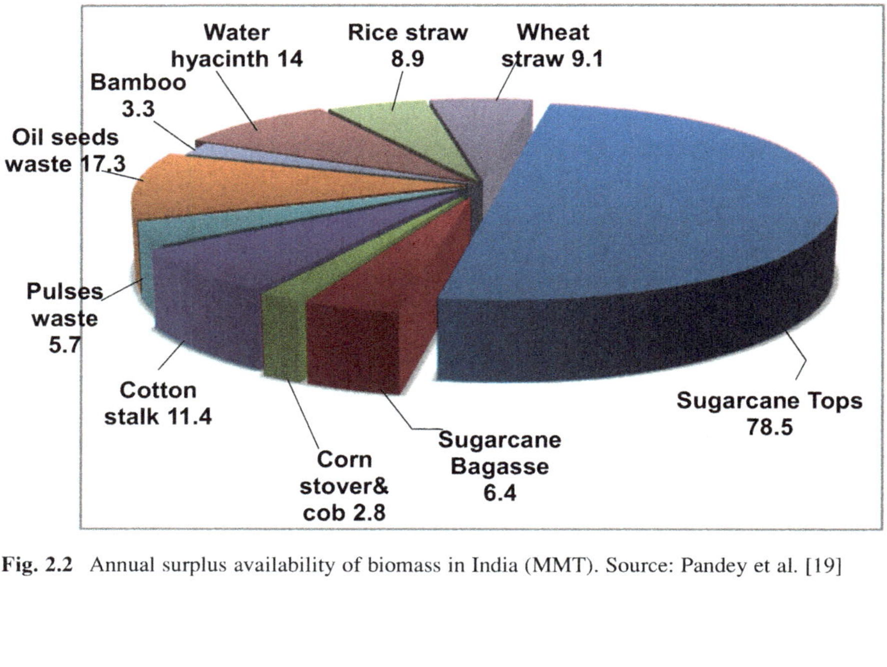

Fig. 2.2 Annual surplus availability of biomass in India (MMT). Source: Pandey et al. [19]

It is noticeable that rice straw, sugarcane tops [21], bagasse, wheat straw and corn stover [22] followed by other cereal biomass represent the most compelling feedstocks for bioenergy. Other promising crop residues like biomass from cotton and chilli cultivation (18.9 and 0.6 MMT, respectively) can also be explored for the same (Table 2.2).

The gradual spread of bioethanol all over India demands exploration of other biomass sources in the near future, and as a result, several potential sources of energy, besides agricultural residues, are being identified. The North-East and Himalayan forest biomass including bamboo, reed and pine can serve as powerful bioenergy feedstocks. It is estimated that about 1.6 MMT of pine needles are produced in the Himalayan region which can be explored for their biofuel generation potential [23]. Bamboo processing waste is a major resource to be reckoned considering the 3.3 MMT annual surplus [21, 24] availability of this feedstock (Fig. 2.2). Compared to other biomass, bamboo with high cellulose and lignin content makes it alternative bioresource for biofuel production [25]. Sweet sorghum, with an annual production of 7.3 million tonnes, is another attractive feedstock for bioethanol production given its high photosynthetic efficiency [26]. Interestingly research in sorghum, through the past few decades, have resulted in new hybrid varieties that can meet the sugar demand for production of bioethanol [27].

From reports, the national level production potential of crop biomass is available, but the data on agro-residual biomass availability in different states is presently insufficient because the state-wise crop residue production and surplus vary

Table 2.2 Region-wide distribution of available agro-feedstocks in India

Indian region	States	Available agro-feedstocks	App. cost Range (Rs/t)
North-west	Rajasthan, Gujarat	Stalks of mustard, juliflora, maize, dhaniya, cotton, soybean and tuver	1300–2500
Central and South-west	Madhya Pradesh, Maharashtra	Cotton stalk, soy husk and mustard, maize stalks, rice husk, juliflora and bamboo	1500–2800
South	Andhra Pradesh, Kerala, Karnataka, Tamilnadu	Rice husk, juliflora, groundnut and coconut shell, bengal gram stover, cane trash and maize	1200–2500
North-east	Jharkhand, West Bengal	Wood chips, rice husk and sugarcane trash	1100–2600
North	Punjab, Haryana, Himachal Pradesh, Uttaranchal, Uttar Pradesh	Rice husk and straw, mustard stalk, wheat straw and cane trashes	1550–3000
Central and South-east	Orissa, Chattisgarh	Rice husk, cotton stalk, sawdust and juliflora	1100–2600

Source: EIA "A market intelligence report" September 2012

from season to season [28, 29]. Identification of major locations with the highest concentration of feedstock can give a better understanding of sustainable availability of biomass for bioenergy purposes. Hence, there should be a proper documentation of the nation's biomass resources and their availability for fuel ethanol production along with computing industrial applications. This documentation of the surplus biomass sources can help the policymakers as well as scientists to decide the future prospects of lignocellulosic bioethanol production.

4 Outlines of the National Biofuel Policy of India

In recent years, with fast economic growth, India's energy consumption has been increasing due to its population growth and rapid urbanisation. The energy demand, especially of transportation sector, is gradually forcing the country to search for alternate fuels. To lookout for a sustainable option, the Government of India has undertaken several policy measures to boost up the production and use of biofuels during the past decade [14, 30]. The National Biofuel Mission was launched in 2003 under the support of the Planning Commission, to develop different projects and policies regarding biofuel efforts in the country. The aim of the policies is to accelerate the development and promotion of cultivation, production and use of biofuels to replace or reduce the dependency on crude oil. Initiative was taken in this regard to make biofuel blending a binding obligation on the states. Primarily, in 2003, the Ministry of Petroleum and Natural Gas (MoPNG) made 5% ethanol blending in petrol mandatory in nine states and across five union territories. In the second phase of Ethanol Blended Petrol Programme (EBPP) in 2006, the blending mandate was further extended to cover 20 states and eight union territories. However, there were shortcomings as a result of which the established targets couldn't be achieved. As a result, commencement of large-scale production unit and the implementation of modern technologies to achieve the target were sought after. In September 2007, the Cabinet Committee on Economic Affairs (CCEA) recommended 5% ethanol blending across the country with the exception of Jammu and Kashmir, the North-East and island territories. Subsequently, the Ministry of New and Renewable Energy (MNRE) formulated the "National Biofuel Policy" in September 2008 which was ultimately released in December 2009. However, it was executed partially because of the scarcity of suitable biomass for conversion to ethanol. Indian Government also endowed many types of subsidies for promoting biofuel production all over India. The blending level of bioethanol at 5% with petrol by 2012, 10% by 2017 and 20% after 2017 was suggested in the policy.

Biofuel development in India has been focused on Jatropha-based biodiesel and molasses-based bioethanol technologies. Molasses, though available across only nine states in India, is the main feedstock used for bioethanol. India is the second largest producer of sugarcane after Brazil, but only 25% of the production is put to uses other than sugar manufacture. For producing ethanol, India has about

330 distilleries with the annual production capacity of over 4.0 billion litres [12]. However, ethanol production is highly volatile in India due to periodic market gluts/deficit of sugarcane, sugar and molasses. India produced around 2.15 billion litres of ethanol in 2008, of which 280 million litres were used for blending purpose. In 2009, ethanol production went down to 1.07 billion litres and blending to 100 million litres. Blending was further down to 50 million litres in 2010 [15]. The current ethanol production allows a blending of only ~1%, and this volatile blending of ethanol with petrol has made the Indian targets "highly unrealistic" [31].

In order to achieve the policy objectives, the government is considering the creation of a National Biofuel Fund (NBF) for providing financial inducement such as subsidies and grants for new and second-generation feedstocks, advanced technologies and conversion processes and production units based on new and second-generation feedstocks. Moreover, the biofuel technologies and projects would be allowed 100% foreign equity through automatic approval routes to attract foreign direct investment (FDI), provided such biofuels are put only to domestic use.

5 Outline of the Lignocellulosic Ethanol Technology Research/Process Innovations Underway in India

Realising the significance of biofuels in securing the energy demands of the nation by reducing the dependence on non-renewable energy sources, many major institutes, universities and industries in India are devoted to research in this area. The following is a list of a few major initiatives undertaken by various organisations.

1. Development of novel multi feedstock pretreatment strategies—Indian Oil Corporation (IOC), NCL and Co., DBT-ICT-CEB, NIIST, Praj industries, University of Delhi South Campus, ICAR- IARI
2. Designing of novel enzymes for saccharification—DBT-ICT-CEB, ICGEB, University of Delhi South Campus, ICAR-IARI
3. Designing of novel biosystems for consolidated bioprocessing—DBT-ICT-CEB, ICGEB, University of Delhi South Campus
4. Development of novel fermentation strategies—India Glycols, Fermenta Biotech, DBT-ICT-CEB, University of Delhi South Campus, ICAR-IARI
5. Development of butanol and other fermentation strategies using synthetic biology—IIT-B/Praj, DBT-ICT-CEB, ICGEB, Tata Chemicals

The country's foremost objective is the development of more pilot plants and infrastructure to produce bioethanol in quantity needed to meet the proposed blending target. As given in Table 2.3, the projected demand of biofuel in India has increased gradually with time.

The Petroleum Conservation Research Association (http://www.pcra.org/), an autonomous body of the Ministry of Petroleum and Natural Gas, Government of

Table 2.3 Projected demand for petrol, diesel and biofuel requirements

Fiscal year	Petroleum demand in Mt	Ethanol blending requirement in Mt			Diesel demand in Mt	Biodiesel blending requirement in Mt		
		@5%	@10%	@20%		@5%	@10%	@20%
2011–2012	12.85	0.64	1.29	2.57	66.91	3.35	6.69	13.38
2016–2017	16.4	0.82	1.64	3.82	83.58	4.18	8.36	16.72

Source: Planning Commission [14]

India, has been promoting research and development in biofuels by sponsoring related projects to reputed research and development institutes of India. The National Informatics Centre on Biofuels, a part of PCRA, provides all the necessary information related to biofuels to all the consumers and other agencies in this field. Further, the Ethanol India initiative (http://www.ethanolindia.net/) of the Government of India is involved in research and development in establishing the usage of ethanol-doped petrol in vehicles and provides information on various biodiesel and bioethanol projects in India.

These strategies have strengthened the research and development of bio-based alternative fuels in India. The New Delhi-based research organisation, The Energy and Resources Institute (TERI) (http://www.teriin.org), works on bioenergy projects and publishes information related to biofuels and the utilisation of biomass, agriculture residues and industrial wastes as fuels. The institute plays a significant role in the country's bioenergy sector by carrying out biofuels' resource and consumption surveys, pre-feasibility studies, technology assessment, preparation of investment proposals and techno-economic feasibility studies. Similarly, an India-based consortium, Biodiesel Technologies (http://www.biodieseltechnologiesindia.com), is engaged in the study, improvement and research and development of indigenous biodiesel. It designs process machineries for supplying custom-built biodiesel reactors taking into account the requirements of the operators.

Another pioneer in the field of biofuel technology is Praj Industries Limited (http://www.praj.net/), a Pune-based company involved in fuel-ethanol production utilising oil extraction through molecular sieve dehydration plants, fermentation, wastewater utilisation and heat-transfer systems. Praj Industries Ltd. has emerged as the first company in South Asia to set up an integrated second-generation cellulosic ethanol plant in Sangli District in Maharashtra (India). This demo plant will also enable the development of an entire value chain including biomass handling and its impact on the operations. The demo plant operates on different varieties of biomass, with a capacity of 100 metric tons of dry biomass per day, which includes agricultural wastes such as corn stover, cobs and bagasse. The same plant could also enable the co-production of various biochemicals and bioproducts. Praj expects the project cost to be about $25 million.

6 Next-Generation Ethanol: Possible Upsides

From the perspective of the vast amount of agricultural residue generated, the development of next-generation ethanol could provide considerable economic and environmental benefits for India, some of which are listed below:

6.1 Revenues and Exports

The development of next-generation ethanol could have a great impact on the Indian economy. Assuming a cost of crude at $100 per barrel, India is forecast to spend approximately $19.4 billion on gasoline imports by 2020, accounting for 80% of its gasoline requirement. On the other hand, Indian producers of next-generation ethanol could theoretically generate $15–20 billion in revenue. The domestically produced next-gen ethanol could also theoretically supplant crude oil imports, thus strengthening India's security of energy supply.

6.2 Environmental Benefits

In the absence of a productive use of crop residues, farmers have traditionally burned excess residues as a means of quick disposal. The burning of wastes from crop residues emitted about 6.6 million tonnes CO_2 in 2007. Usage of residues in other useful activities like ethanol conversion could reduce these emissions (Table 2.4).

6.3 Possible Job Creation

A new industry could be created that would generate thousands of new jobs. The employment opportunities would include both the building of new biofuel plants

Table 2.4 Comparison of reduction in carbon footprint by using bioethanol produced from different feedstocks

Fuel	Feedstock types	Reduction in carbon emission
Bioethanol	Sugarcane juice	70–80%
	Molasses	70–75%
	Sweet sorghum	80–90%
	Grains (maize, cereals)	19–52%
	Lignocellulosic feedstock	70–80%

Source: U.S. Department of Energy, Alternative Fuels Data Center. (http://www.afdc.energy.gov/vehicles/flexible_fuel_emissions.html)

and their operation in the longer term. Baling, hauling and transportation of residues as well as operation of plants set up for ethanol production will create permanent jobs. This increases rural employment in areas surrounding the ethanol production facility. Since the plant itself would be set up in a region of biomass availability, it would likely be located on the outskirts of a city or in rural areas. Employment opportunities for low skilled workers, medium skilled personnel in the plant operation and administration as well as for engineers and the scientific community would be open.

7 Barriers to Development of Next Generation Ethanol in India

Sustainable development of bioenergy for the economic, social and environmental security of a nation is influenced by several factors like the availability and accessibility of new and renewable resources, the affordability of their utilisation and acceptability among the consumers. Various barriers to the development of advanced generation biofuels in India include the following.

7.1 Policy: Lack of Enforcement and Coherence

India does not have a strong track record of fulfilling its biofuel blending ambitions. This erratic success in introducing ethanol blending does not set a good long-term precedent for potential biofuel investors. Today, there is no clear signal on the long-term pricing of ethanol. Transporting the fuel within and across the state borders is a barrier due to various taxes and regulations and needs streamlining. There is no manufacturer producing flex-fuel vehicles in India that can run on a variation of gasoline/ethanol blends. For potential investors, all this adds up to an uncertain environment which is likely to impede the development of India's biofuel industry at least for the short term.

7.2 Unstable Production of Molasses

In India, alcohol production mainly depends on molasses, which in turn depends on the sugar industry. About 60% of the sugar production is from two states—Maharashtra and Uttar Pradesh, followed by Karnataka and Tamil Nadu. As a consequence, these states are the only ethanol suppliers of India and nearly about 3 billion litres of ethanol is produced annually (Fig. 2.3). Moreover, as inter-state movement of molasses is not freely permitted across India, it is not viable to procure molasses

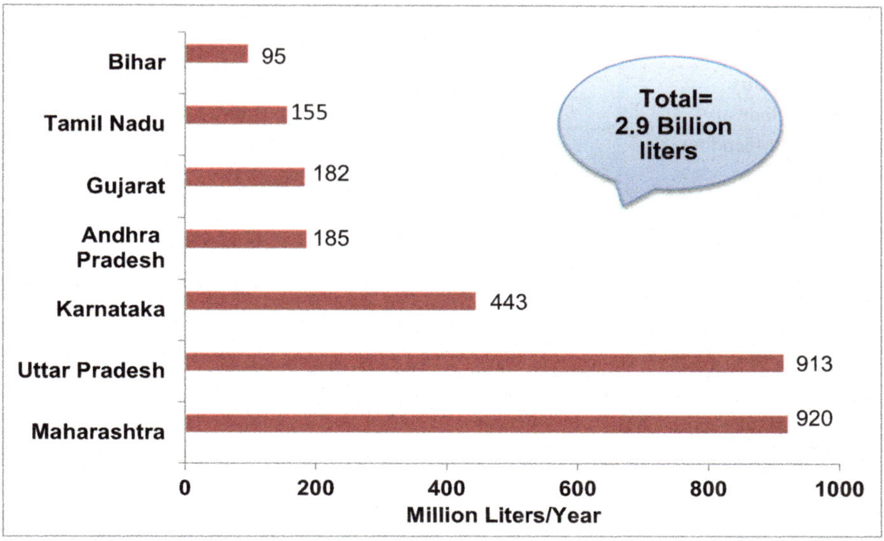

Fig. 2.3 Present situation of state-wise bioethanol production. Source: Chaudhary [32]

from different parts of the country. As a result, most of the distilleries are attached to the sugar factories that produce molasses. Considering all these factors, an alternative source for production of bioethanol has to be examined. Lignocellulosic biomass and sweet sorghum are currently the viable options for bioethanol production in India.

7.3 *Competition for Distillery End Products*

The liquor industry, chemical industry and fuel industry are the three end users which compete among themselves for the availability of molasses and distillery end products (Table 2.5). This necessitates the production of surplus molasses in order to produce enough ethanol for all the three sectors.

7.4 *Lack of Market for Residues*

The current National Policy on Biofuels states the importance of intensive research and development for the development of next-generation biofuel technologies. But currently there is no policy mechanism incentivising farmers to collect and deliver biomass residues to a next-generation ethanol plant. If the policy makers seek to foster an industry longer term, they could create a mechanism whereby

Table 2.5 Sector-wise utilisation of alcohol in India (in million litres)

Sector	2010	2011	2012
Liquor industry	900	950	1010
Chemical industry	720	750	775
Ethanol for blending	50	250	300
Total	1670	1950	2085

Source: Singh [33]

Table 2.6 Some sustainable and economical resources for second-generation fuel ethanol production

Crops	Crop duration (months)	Total production[a] (million tonnes)	Biomass tonnes/ha	Carbohydrate (%)	Calculated ethanol yield (litres/ha)
Sugarcane tops	12–16	97.8	20–25	60.50	5600
Sugarcane bagasse	12–16	101.3	25–30	67.15	7000[d]
Sweet sorghum	4	7.3	40–55	71.60	3400[b]
Corn stover	3–4	22.7	2.6–3.0	63.29	900[c]
Rice straw	3–4	130	5.0–6.0	57.00	1680[d]
Wheat straw	3–4	112	9.5–10.0	54.00	2755[d]

[a]Source: Crop production statistics information system, Department of Agriculture Cooperation and Farmers' Welfare. Government of India, http://apy.dacnet.nic.in/
[b]75 tonne/ha over two crops per year and ethanol yield @ 40 L/tonne
[c]Ethanol yield calculated @ 300 L/tonne [35]
[d]Kim and Dale [36]

cooperatives or farming communities could be involved in the process of collection, storage and delivery of residues.

7.5 Conversion Economics

In addition to policy, the development of next-generation ethanol in India will be driven by its production economics, which should be compared with those of the first-generation industry. According to an analysis, it costs \$0.56 to produce a litre of first generation of fuel when the molasses is priced at \$60–65/tonne. By comparison, next-generation ethanol would cost \$0.70 per litre although this figure should fall over the next decade with technology improvements [34]. The price of molasses has varied significantly in the past years. Prices of certain biomass residues such as rice husks have consistently been lower than that of molasses. Some sustainable biomass feedstocks for bioethanol production are given in Table 2.6 with estimated ethanol yields. This highlights the potential of new substrates for future bioethanol production. Technology improvements should also lower conversion and capital costs. But scaling up is contingent on making the residues available which would need an effective policy mechanism to incentivise the agricultural industry or bring about a significant breakthrough in the technology.

7.6 Technical Blending Hurdles

It would require scientific studies demonstrating the optimal mix of ethanol in gasoline in order to make it possible to formulate a common international consensus on the appropriate ethanol–gasoline proportions.

7.7 Involvement of Multiple Stakeholders

As biofuels involve fuel and agriculture perspectives, the list of stakeholders involved in determining India's long-term policies in this area is long and varied. It includes a myriad of ministries, departments and industrial associations, involved in potential ethanol demand, supply and pricing. But their perspectives and interests often differ significantly. The Ministries of Petroleum and Natural Gas, Agriculture and Farmers' Welfare, New and Renewable Energy and Chemicals and Fertilizers focus on the ethanol issue with different agendas. Bringing all the stakeholders to a common platform and addressing the issues productively is a challenging task in itself and it inevitably delays reaching any consensus.

7.8 Capital Shortage

Currently, there is only one next-generation ethanol manufacturing facility operative in India and the technology is very much in the development stage. This, together with the above-mentioned factors, makes it difficult for project developers to secure financing. Investor's lack of familiarity with the technology implies that the projects are vulnerable to very steep lending rates, which in turn places a considerable burden on a technology in the development stage.

8 Conclusions

With time the second generation of biofuels will play a crucial role in the energy scenario in years to come. However, India, with its abundant source of renewable energy, lacks a cost-effective, proven technology for biomass bioconversion.

Bioethanol derived from lignocellulosic biomass, with its several advantages, is a potential option to pursue. Statistics indicate that sugarcane bagasse and tops, rice straw and wheat straw may be explored for second-generation biofuel production in India. Although a lot of research is being carried out in this area, India still lacks large-scale pilot plants for the production of this fuel. It is important to bring fuel-ethanol programmes to the attention of big investors, especially the petroleum companies, since this can hasten the commercialisation of second-generation biofuel technology. Private investors should be encouraged to invest in biofuel programmes and government policies should be conducive for their participation. A look at the steps taken by the government from a long-term perspective and the participation of non-governmental organisations makes the future of second-generation bioethanol in India quite promising.

References

1. Federation of Indian Chambers of Commerce and Industry – Pricewaterhouse Coo Report (2013) Forging ties securing energy supply for a stronger economy. https://www.pwc.in/en_IN/in/assets/pdfs/publications/2013/forging-ties-Securing-energy-supply-for-a-stronger-economy.pdf
2. International Energy Agency (IEA) (2010) World Energy Outlook. http://www.worldenergyoutlook.org/media/weo2010.pdf
3. Dasgupta S, Roy J (2015) Understanding technological progress and input price as drivers of energy demand in manufacturing industries in India. Energy Policy 83:1–13
4. International Energy Agency (IEA) (2013) Annual report. https://www.iea.org/publications/freepublications/publication/2013_AnnualReport.pdf
5. Ministry of Commerce and Industry, Government of India, Press release, 16th June 2015. http://commerce.nic.in/tradestats/filedisplay.aspx?id=1
6. Grad P (2006) Biofuelling Brazil: an overview of the bioethanol success story in Brazil. Refocus 7:56–59
7. RFA, Ethanol industry outlook (2015) www.ethanolrfa.org/outlook
8. Licht FO (2013) RFA (Renewable Fuels Association) Washington, DC, USA. http://www.ethanolrfa.org/page/PDFs/RFA%202013%20Ethanol%20Industry%20Outlook.pdf?nocdn=1
9. Shinoj P, Raju SS, Joshi PK (2011) India's biofuel production programme: need for prioritizing the alternative options. Indian J Agric Sci 81:391–397
10. Chand R (2009) Challenges to ensuring food security through wheat. CAB Rev Prospect Agric Vet Sci Nutr Nat Resour 4:1–13
11. Ministry of Road Transport and Highways (MORTH), Government of India. http://morth.nic.in/
12. Aradhey A (2013) India biofuels annual. In: Global agricultural information network (GAIN) report no IN3073. USDA-Foreign Agricultural Services. http://gain.fas.usda.gov/Recent%20GAIN%20Publications/Biofuels%20Annual_New%20Delhi_India_8-13-2013.pdf
13. ICRIER Policy Report (2011) Ethanol blending policy in India: demand and supply issues. http://icrier.org/pdf/Policy_Series_No_9.pdf
14. Planning Commission (2003) Government of India, report of the committee on development of biofuel http://planningcommission.nic.in/reports/genrep/cmtt_bio.pdf
15. Raju SS, Shinoj P, Chand R, Joshi PK, Kumar P, Msangi S (2012) Biofuels in India: potential, policy and emerging paradigms. Policy paper, NCAP, New Delhi

16. Prasad S, Singh A, Joshi HC (2007) Ethanol as an alternative fuel from agricultural, industrial and urban residues. Resour Conserv Recycl 50:1–39
17. Hiloidhari M, Das D, Baruah DC (2014) Bioenergy potential from crop residue biomass in India. Renew Sust Energ Rev 32:504–512
18. Moreno AD, Ibarra D, Alvira P, Tomás-Pejó E, Ballesteros M (2014) A review of biological delignification and detoxification methods for lignocellulosic bioethanol production. Crit Rev Biotechnol 35(3):342–354
19. Pandey A, Biswas S, Sukumaran RK, Kaushik N (2009) Study on availability of Indian biomass resources for exploitation: a report based on a nationwide survey. TIFAC, New Delhi
20. Singh J, Gu S (2010) Biomass conversion to energy in India. A critique. Renew Sust Energ Rev 14:1367–1378
21. Sukumaran RK, Surender VJ, Sindhu R, Binod P, Janu KU, Sajna KV, Rajasree KP, Pandey A (2010) Lignocellulosic ethanol in India: prospects, challenges and feedstock availability. Bioresour Technol 101:4826–4833
22. Karp A, Shield I (2008) Bioenergy from plants and the sustainable yield challenge. New Phytol 179:15–32
23. Ghosh MK, Ghosh UK (2011) Utilization of pine needles as bed material in solid state fermentation for production of lactic acid by *Lactobacillus* strain. BioResources 6:1556–1575
24. Sindhu R, Kuttiraja M, Binod P, Sukumaran RK, Pandey A (2014) Bioethanol production from dilute acid pretreated Indian bamboo variety (*Dendrocalamus sp.*) by separate hydrolysis and fermentation. Ind Crop Prod 52:169–176
25. Kuttiraja M, Sindhu R, Varghese PE, Sandhya SV, Binod P, Vani S, Pandey A, Sukumaran RK (2013) Bioethanol production from bamboo (*Dendrocalamus sp.*) process waste. Biomass Bioenerg 59:142–150
26. Reddy BVS, Ramesh S, Sanjana RP, Ramaih B, Salimath PM, Rajashekhar K (2005) Sweet sorghum—a potential alternative raw material for bioethanol and bioenergy. International Sorghum and Millets Newsletter 46:79–86
27. Ratnavathi CV, Suresh K, Kumar BSV, Pallavi M, Komala VV, Seetharama N (2010) Study on genotypic variation for ethanol production from sweet sorghum juice. Biomass Bioenerg 34:947–952
28. Chauhan S (2010) Biomass resources assessment for power generation: a case study from Haryana state, India. Biomass Bioenerg 34:1300–1308
29. Das S, Jash T (2009) District-level biomass resource assessment: a case study of an Indian state West Bengal. Biomass Bioenerg 33:137–143
30. Ministry of New and Renewable Energy (MNRE) Government of India, Annual report 2009–2010. http://mnre.gov.in/file-manager/annual-report/2009-2010/EN/index.htm
31. Biswas AK, Manocha N (2014) India's biofuel policies: progress or boondoggle? NGI Knowl Exch 5:6–8
32. Chaudhary A (2012) Ethanol blending program in India—one step forward and two backwards. http://petrofed.winwinhosting.net/upload/19-20%20April%202012/Presentations/Session%201/2_Abhay%20Chaudhari.pdf
33. Singh S (2012). India grain and feed annual. In: Global Agricultural Information Network (GAIN) report no IN2026. USDA-Foreign Agricultural Services. http://gain.fas.usda.gov/Recent%20GAIN%20Publications/Grain%20and%20Feed%20Annual_New%20Delhi_India_2-23-2012.pdf
34. Bloomberg New Energy Finance Report (2011) http://www.unep.org/pdf/BNEF_global_trends_in_renewable_energy_investment_2011_report.pdf
35. Luo L, van der Voet E, Huppes G (2009) An energy analysis of ethanol from cellulosic feedstock–corn stover. Renew Sust Energ Rev 13(8):2003–2011
36. Kim S, Dale BE (2004) Global potential bioethanol production from wasted crops and crop residues. Biomass Bioenerg 26:361–375

Chapter 3
Renewable Sources in India and Their Applications

K. Pal, P. Yadav, and S. K. Tyagi

Abstract India holds seventh largest share of land and second largest share of human population on the earth. Unique geographical location and bountiful natural resources that have once made India the cradle of human civilisation have been redefined as renewable resources in the wake of fast depleting fossils and impending doom of global warming. Moreover, as the issue of sustainability rises up every now and then, India seeks to find solution in its veteran trusted resources, the sun, the wind, the oceans and flora–fauna, yet following a well-defined strategic trajectory. Due to consistent endeavours and strong policy support, currently, India is amongst top six countries of the world, in terms of highest renewable additions in the past decade, fifth in terms of wind power generation and second largest biogas consumer in the world. Further, a bundle of programmes to promote renewable energy, viz., Jawaharlal Nehru National Solar Mission (JNNSM), National Biomass Cookstoves Initiative (NBCI), National Biogas and Manure Management Programme (NBMMP), Village Energy Security Programme (VESP), Remote Village Electrification Programme (RVEP) and National Hydrogen Energy Programme (NHEP), are being implemented countrywide. Adding one more feather in its cap, recently, India launched its much awaited offshore wind policy. This chapter tries to investigate the renewable energy potential of the country, existing technologies, policies and programmes and finally brings out prospects and challenges of renewable energy for the country.

Keywords Renewable energy sources • Sustainable development • Solar mission • Hydrogen economy • Offshore wind • Decentralised energy systems

---

K. Pal • P. Yadav • S.K. Tyagi (✉)
Thermochemical Conversion Division, Sardar Swaran Singh National Institute of Renewable Energy, Kapurthala 144601, Punjab, India
e-mail: sudhirtyagi@yahoo.com

A.K. Chandel, R.K. Sukumaran (eds.), *Sustainable Biofuels Development in India*,
DOI 10.1007/978-3-319-50219-9_3

1 Introduction

Conceptualisation of renewable energy as the keystone of sustainable development has led to rapid adoption of the technologies by developed as well as developing economies of the world. Renewable energy sector has shown tremendous growth, contributing around 19 % of global energy supply in 2013. Moreover, in the last decade (2004–2013), participation of renewable energy as a primary energy source has risen by 30 %. Region wise, China has emerged as the world leader with the highest number of renewable energy additions during the last decade followed by European Union, the USA, Germany, India and Brazil in top six [1]. Such statistics are extremely significant for fast growing economies like China, India and Brazil facing the challenge of rapid yet sustainable development.

India has the ninth largest economy, fifth largest power generation portfolio and third largest carbon emission in the world [2–4]. Moreover, energy demands are rising at 2.8 % per annum, mainly due to steady economic growth of the country [4]. In addition to dwindling coal supply, major challenge in ensuring energy security is the climatic frameworks. Subsequently, the Government plans to revamp renewable energy and climate change objectives post-2022. However, at present, India is promoting renewable energy technologies to reduce dependence on expensive imported fuels, foster economic growth, generate employment and reduce carbon footprint [5]. Favoured by policy, renewable energy technologies have become increasingly cost competitive compared to fossil fuel-based generation and contribute significant 13 % of the total installed capacity of 245 GW of the country (March, 2014) [3].

2 Renewables in India

According to International Energy Agency, India will become single largest importer of oil after 2020. In fact, crude oil imports are prime cause of India's current account deficit. On the other hand, around 100,000 villages in India out of total 600,000 are yet not electrified [6]. Against this background, renewable energy technologies are best suited to seal the gap between demand and supply, either as grid interactive power or as an alternative to grid extension as off-grid power. Moreover, decentralised or distributed application for cooking and heating like solar water heaters, biogas plants and rooftop SPV systems has become overtly significant not only in villages but also in towns and cities [5]. The major renewable energy sources of India are:

- Solar Energy
- Wind Energy
- Hydropower
- Biomass Energy
- Geothermal Energy
- Ocean Energy

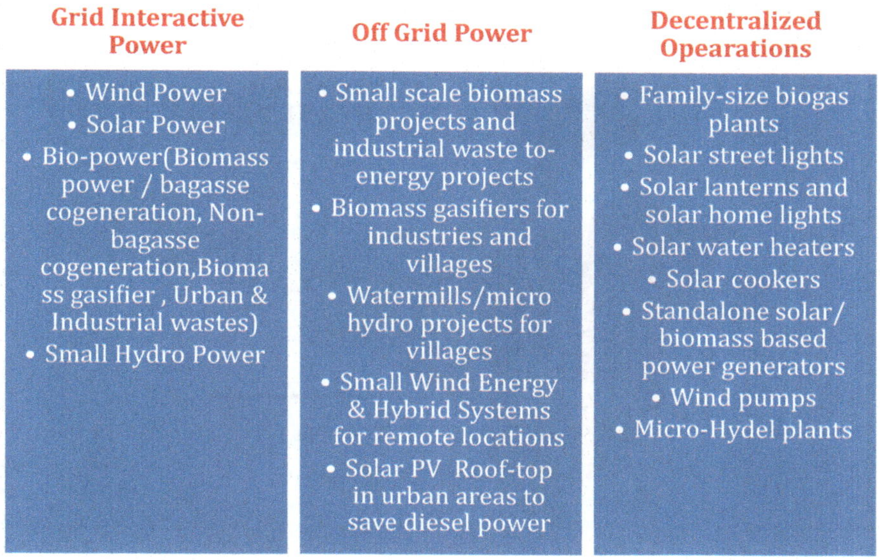

Fig. 3.1 Illustration of deployed renewable energy technologies in India

2.1 Solar Energy

Sun is the prime source of all energy on earth. The solar power technology tries to harness a part of this vast resource by capturing solar radiations. Immense potential of solar radiations could be inferred from the fact that the global average solar radiation, per square metre and per year, contains energy equivalent to one barrel of oil [4]. Geographical location endows India with ample amount of solar energy for sustaining energy demands in both grid integrated and decentralised power systems (Fig. 3.1).

2.1.1 Potential

Geographical location of India is ideal for tapping solar power; with an average of 300 sunny days in a year, India receives 1600–2200 kWh/m^2 of solar radiations annually, equivalent to an estimated potential of 6 billion GWh [5]. However, National Institute of Solar Energy (NISE) has estimated the total solar energy potential of the country as around 750 GW based on the assumption that only 3 % of the total wasteland available in a state is used for installing solar power plants. According to the estimates, states of Rajasthan, Jammu and Kashmir, Maharashtra and Madhya Pradesh possess high solar power potentials [7]. However, total installed capacity of around 3 GW is mainly concentrated in eight states,

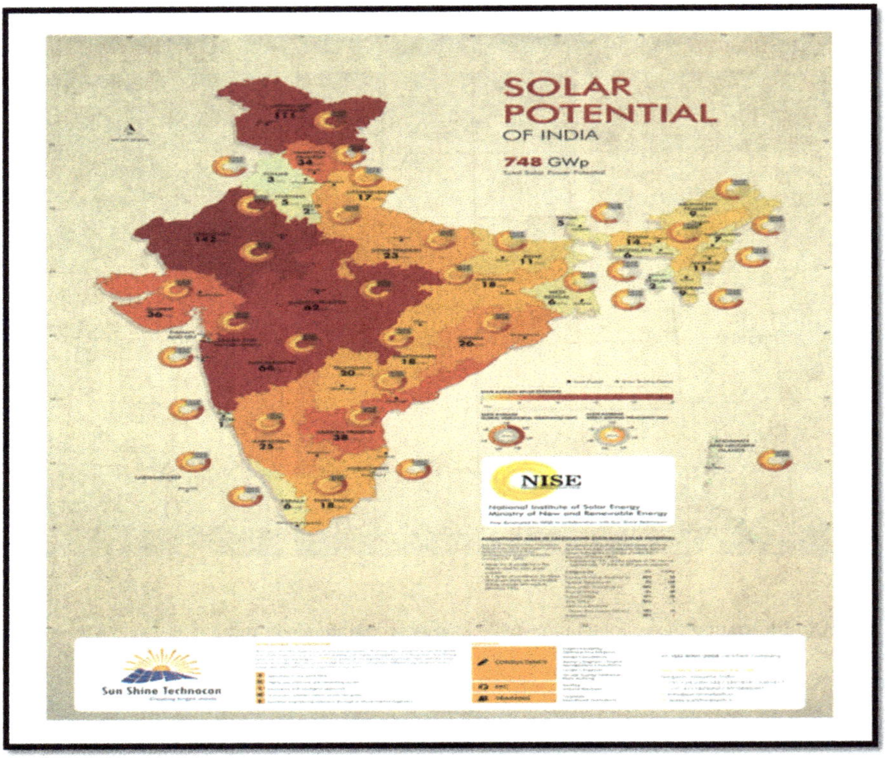

Fig. 3.2 Map showing solar energy potential of India [7]

Rajasthan, Gujarat, Madhya Pradesh, Maharashtra, Andhra Pradesh, Tamil Nadu, Karnataka and Uttar Pradesh [5] (Fig. 3.2).

Accordingly, the Government of India has set an ambitious target to attain 100 GW of installed solar capacity by 2022, including 20 GW of ultra-mega solar power projects across 12 states, with an installed capacity of 500 MW or above [5].

2.1.2 Technology

Prime solar energy technologies in India are concentrating photovoltaics and thermal collectors. The solar thermal (STE) power plants are technically similar to conventional coal-based power plants while the concentrating photovoltaics (CPV) systems concentrate sunlight onto photovoltaic surfaces for power generation.

1. *Solar Thermal Energy (STE)*

STE technology collects the solar radiation by reflection and thereafter harvests the heat energy of rays by convection devices, collector. Based on scale of project, three broad categories of collectors are recognised:

- Low Temperature Collector: Designed as flat-plate, such collectors are mostly used for drying or heating water in homes, commercial buildings and swimming pools.
- Medium Temperature Collectors: Comprising of a group of advanced flat-plate collectors or evacuated tube collectors, such devices acquire temperature above 100 °C for large commercial complexes or even residential area with higher energy demands.
- High temperature collectors: Capable of concentrating large volume of radiations, hence apt for solar power plants (Fig. 3.3)

2. *Concentrating Photovoltaic (CPV)*

CPV plant focuses solar radiation with the help of either mirrors or lenses onto a photovoltaic module that converts the radiation directly to electricity. However, the solar cells are more expensive in comparison to conventional cells used in flat-plate photovoltaic systems. Further, the CPV technology diversified into single-axis tracking, line focus CPV and two-axis tracking, point focus CPV. However, recent development has primarily been on the two-axis tracking systems [10]. Like conventional battery, desired power range can be obtained by connecting PV modules in series or parallel to each other. Photovoltaic modules are made up of either mono-crystal silicon (used in space application or in solar PV automobiles) or polycrystalline silicon, thin film amorphous silicon or cadmium telluride (used in solar photovoltaic application) [11]. In India, CPV are employed for both grid-connected or off-grid power generation.

Buildings with grid-connected PV system (Fig. 3.4) have two parallel power supplies, one from the solar PV system and the other from the power grid. Although

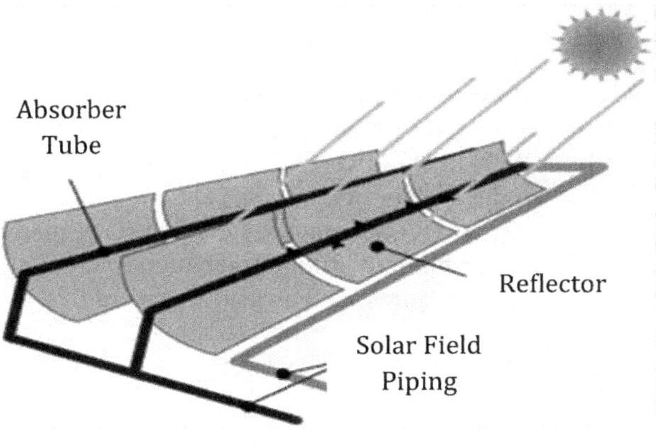

Fig. 3.3 Schematic view of parabolic trough power plant [8, 9]

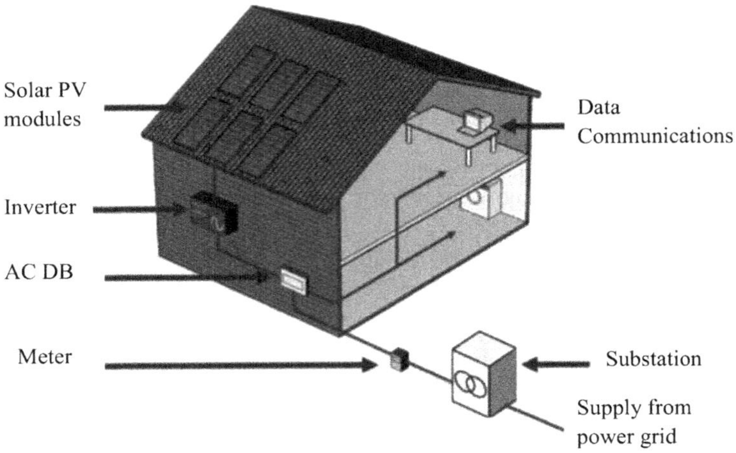

Fig. 3.4 Grid-connected solar PV system configuration [13]

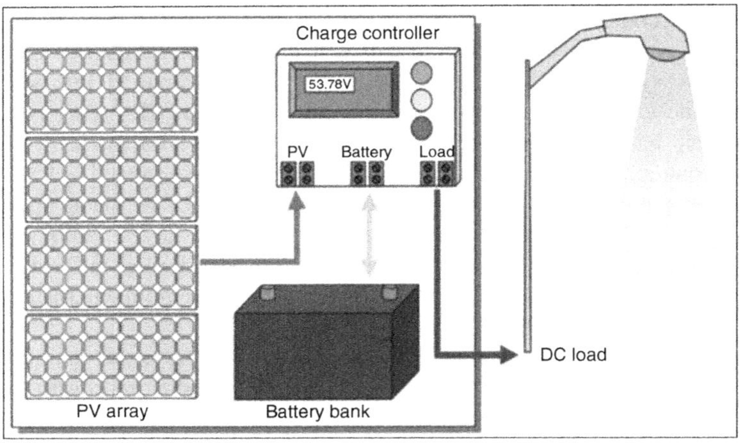

Fig. 3.5 Off-grid solar PV technology [11]

the ratio of solar PV supply to power grid varies, depending on size of the solar PV system, yet overall supply is the sum of the two, ultimately connected to the main supply ACDB. In case the solar PV supply is in surplus, excess electricity will be exported to the grid and vice versa [11].

Off-grid solar PV systems need deep cycle rechargeable batteries such as lead-acid, nickel–cadmium or lithium-ion batteries to store electricity for use at nights or on cloudy days. Figure 3.5 shows arrangement of a typical off-grid power supply.

A typical hybrid solar power system is a combination of PV modules and a complementary means of power generation like diesel, gas or wind generator

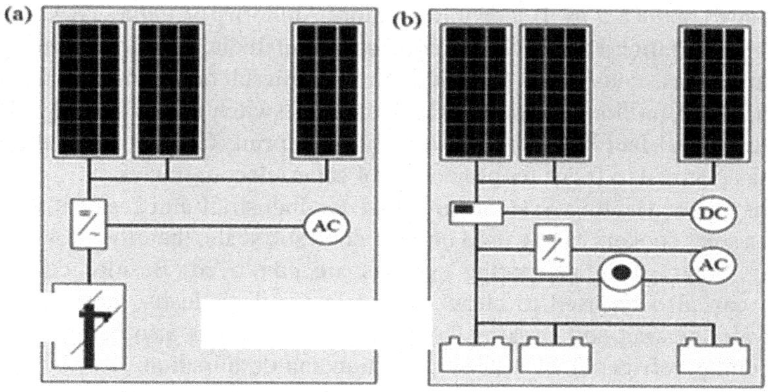

Fig. 3.6 Schematic representation of (**a**) a grid-connected PV system and (**b**) a hybrid system [11]

Table 3.1 Phase-wise targets of JNNSM [6]

	Phase I	Phase II	Phase III
Sector	2007–2012	2012–2017	2017–2022
Grid-connected power (in MW)	1100	10,000	20,000
Solar collectors (in million sq. metres)	7	15	20
Off-grid installations (in MW)	200	1000	2000

(Fig. 3.6). Such systems are similar to off-grid supply yet need more sophisticated control for ensuring smooth and efficient coordination between two generators.

2.1.3 Applications

Under National Action Plan for Climate Change (NAPCC), 2008, the Government of India launched Jawaharlal Nehru National Solar Mission (JNNSM) to establish India as a global leader in solar energy by formulating policy to ensure its large-scale deployment in a short span of time. The Mission has been planned in 3 phases (Table 3.1) [13–15].

Backed up by JNSSM, solar energy application is growing leaps and bounds in India; most common applications include rooftop solar PV, off-grid lighting systems, solar city, solar water pumps, cookers, water heaters telecom tower and solar collectors, etc.

Rooftop solar PV Programme and off-grid lighting systems are important to overcome power shortage, whereas solar PV systems are useful to pump water for domestic or agricultural needs. Moreover, the Government has started the solar city scheme in 100 cities across India during phase II of JNNSM with the objective to cut down projected demand of conventional energy in these cities by at least one-fourths at the end of five years through use of renewable energy.

A study conducted by Telecom Regulatory Authority of India (TRAI) on Green Telecommunication (March 2011) concluded that India had more than 3.10 lakh telecom towers, consuming 2 billion litres of diesel fuel annually and thereby generating 5.3 million tonnes of CO_2. However, switch to solar energy could not only save fossil fuel but also reduce carbon footprint. Consequently, the government has planned to focus on promotion of solar telecom towers.

Solar Water Heating Systems are used for industrial and residential systems, whereas solar cookers can be used only on domestic scale; therefore, for large-scale cooking, solar steam generating systems are employed. Besides cooking, this system can also be used to clean clothes in textile industry, hotels, etc. Apart from water heating and cooking, solar energy in India is also being used for air conditioning, refrigeration, water purification and desalination.

At industrial level, process heat application of solar power along with conventional boilers is a potential power source for textiles, chemicals and plastics industries. These systems can also be used in conjunction with other conventional heating systems.

2.1.4 Prospects and Challenges

Lack of proper Direct Normal Incidence (DNI) data, dusty condition and supply issues associated with heat transfer fluid (HTF) led to deployment of only around 50 MW grid-connected CSP projects against the target of 470 MW grid (JNNSM phase I). On the other hand, most of the solar PV projects were commissioned as per schedule prompted by reduction in cost of the technology. Looking at the scenario, it is evident that for developing countries like India making solar technology cost competitive is the only key to boost the deployment.

Although solar power is a potentially clean source of energy, the toxic and often carcinogenic emissions during manufacture of PV cells pose occupational and health-related risks and negligence of such issues might create problems in the near future [16].

2.2 Wind Energy

India's flagship programme for commercial application of wind energy in the 1990s opened new avenues for sustainable and clean energy option in Asia [13]. Following the swift pace, in 2014, India became the fifth largest wind energy market worldwide with an addition of 2315 MW to lift overall production to 22.5 GW [17]. Wind energy accounts for almost two-thirds of the installed renewable energy capacity of the country and up to 6.9 % of the total power production. The ambitious plan of Ministry of New and Renewable Energy (MNRE), India, to harness 60,000 MW of power from wind energy by 2022 obligates the industry to scale up production by 5000 MW each year up to 2022 [5].

2.2.1 Potential

Centre for Wind Energy Technology (C-WET), India, in February 2010 has estimated India's onshore wind potential as 49.1 GW at 50 m hub height but later revised it in 2012 to 102.8 GW at 80 m hub height, considering the fact that the recent advancements in technology can support higher hub height and yield higher conversion efficiencies [13, 18]. However, International estimates are relatively much higher; Lawrence Berkeley National Laboratory has estimated India's wind energy potential in the range of 2000–3000 GW [19]. Amongst Indian states, Gujarat owns highest wind potential while Tamil Nadu accounts for highest number of installations followed by Gujarat and Maharashtra (Fig. 3.7).

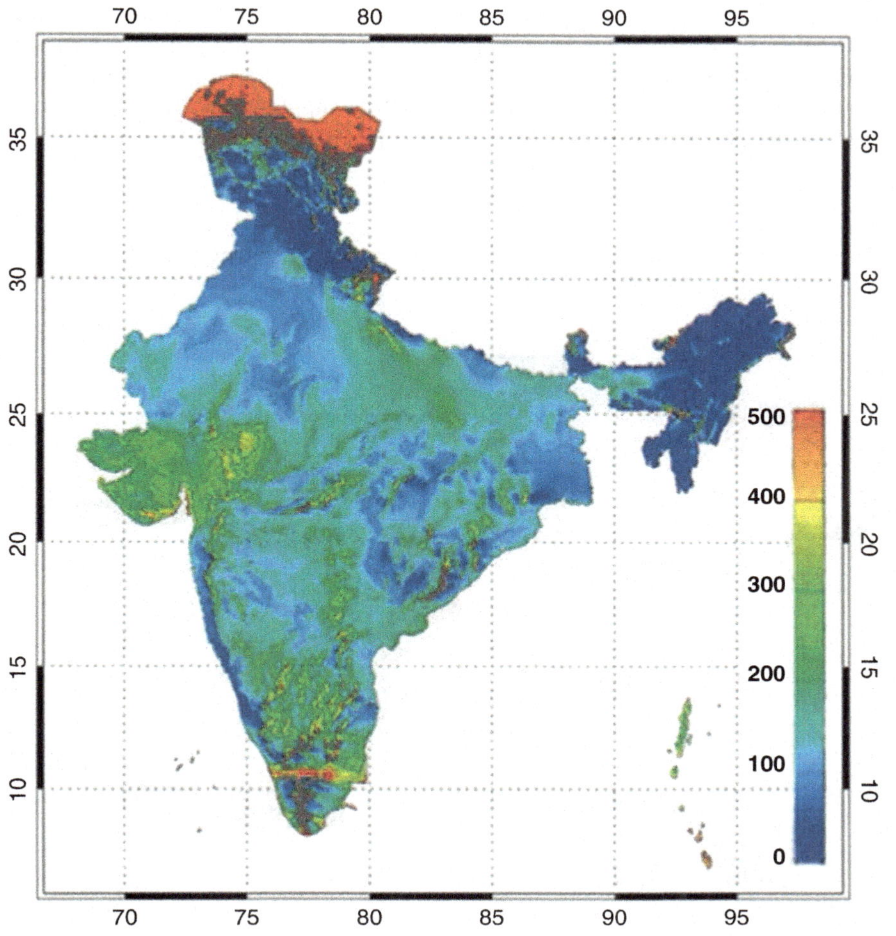

Fig. 3.7 Wind power potential of India [18]

2.2.2 Technology

Wind energy technology captures the kinetic energy of the wind blowing on land (on shore) or in the water body (off shore) and converts it to electrical energy. Various technologies exist, based on the axis of wind turbine (vertical or horizontal) and their location (onshore or offshore). Technology for harnessing onshore wind energy is in advanced stage and highly deployed in comparison to offshore ones. Power generation is determined by the nameplate capacity (in kW or MW) of the turbine, intensity of the wind, height of the turbine tower and diameter of the rotor.

Wind turbine is based on aerodynamic modelling: kinetic energy of the wind is harvested by the rotor blades and transferred to the generator via rotor shaft and gear box. To optimise the output, wind blades can be rotated and adjusted along the direction of the wind. For operating a typical large wind turbine power generator (2.3 MW), wind speed of at least 10 feet per second is required. Wind turbines are usually collected in parks and power generated is transferred to a central transformer via cables. The central transformer further supplies it to the regional power grid for distribution [20] (Fig. 3.8).

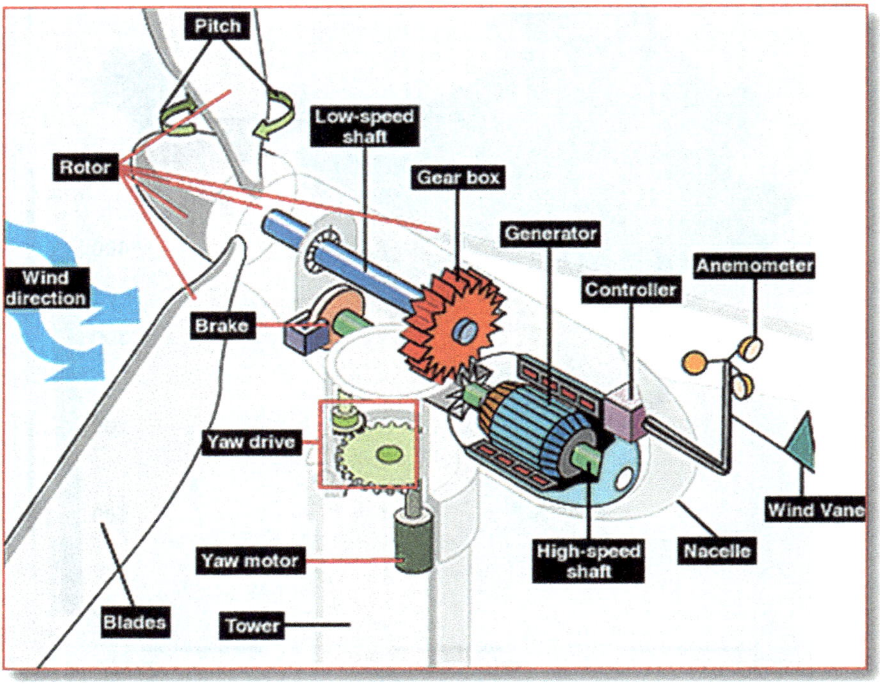

Fig. 3.8 Schematic representation of a typical wind turbine [18]

2.2.3 Applications

Large-scale projects (50–200 kW) spread over several acres of land provide electricity to the utility grid. On the other hand, small turbines (below 50 kW) are used for decentralised and off-grid purposes, to light homes, charge batteries, run water pumps, etc.

Recently, India approved National Offshore Wind Energy policy to provide the roadmap to harness huge offshore wind potential of India. Accordingly, offshore wind energy plants can come up to seaward distance of 200 nautical miles of the country from the baseline. Estimates show that Gujarat and Tamil Nadu with offshore wind energy potential of around 100,000 MW and 60,000 MW can grow into hub of offshore wind energy in India and eventually make the technology cost competitive [21].

2.2.4 Prospects and Challenges

High cost of finance, limited debt financing, inefficient supply chain and logistics are key challenges for developers as well as Original Equipment Manufacturers (OEM) in the country. In addition, inverted duty structure levying higher import duties on raw material than turbine parts discourages local manufacturing. On technical front, grid integration issues and the development of indigenous turbine for harnessing lower wind velocity are the issues.

Implementation of National Green Energy Grid will ease rapid distribution of wind power in the country. Besides, India is planning to launch National Wind Energy Mission to coordinate and facilitate a stable policy framework for strengthening the sector by regulating and incentivising onshore, offshore and small wind generation [5].

Quality wind resource, Government regulatory support, generation-based incentives and cost competitiveness are the chief factors pushing up wind power in India, yet the sector needs thorough revamping as the reports claim that the majority of Indian states fail to meet the recommended Renewable Purchase Obligation (RPO) target of 15 %, set by National Action Plan for Climate Change, 2008 [13].

2.3 Small Hydropower

Hydropower is the most deployed and generously invested amongst all renewable technologies. Turbines harness the kinetic energy of moving water to convert it to mechanical energy, which could either be used directly or converted to hydroelectricity. India is amongst the top ten largest markets for hydropower in terms of capacity worldwide [22]. Consistent power supply by hydropower projects and support policies of the Government have boosted the sector in India and currently it's the most widely deployed renewable energy [23].

Based on power generation capacity, hydro power projects can be divided into two categories: small and large hydro. In India, plants with capacity higher than 25 MW are considered as large hydro power plants, mostly controlled by public sector companies governed by Ministry of Power. Top public sector companies include National Hydroelectric Power Corporation (NHPC), Northeast Electric Power Company (NEEPCO), Satluj Jal Vidyut Nigam (SJVNL), THDC, NTPC-Hydro. On the other hand, projects up to 25 MW capacity are categorised as small hydro power (SHP) plants, controlled by MNRE, currently ventured by private sector. In this section, we will discuss only SHP, as it comes in purview of renewable energy sector in India. Compared to large hydropower plants, SHP is considered environmentally benign, as latter does not lead to habitat destruction and community displacement.

2.3.1 Potential

SHP have the potential to electrify remote and hilly terrains, where the power transmission is yet a costly affair. The estimated potential of SHP in India is about 20,000 MW, almost half of it concentrated in the Himalayan States of Himachal Pradesh, Uttarakhand, Jammu and Kashmir and Arunachal Pradesh. In addition, Maharashtra, Chhattisgarh, Karnataka and Kerala also have a considerable potential [5, 19]. In view of that, MNRE has identified and documented 6474 potential sites for SHP with cumulative capacity of 19,749.44 MW. SHP can be further classified as micro hydro (up to 100 kW), mini hydro (101–2000 kW) and small hydro (2001–25,000 kW) [5]. By March 2013, 967 SHP with installed capacity of 3632 MW have been deployed in various parts of the country. India aims to scale up SHP installed capacity to about 7000 MW by the end of 12th Plan [19].

2.3.2 Technology

A typical hydroelectric system (Fig. 3.9) comprises of water source, pipe (penstock), flow control, electric generator, turbine, electric generator and wiring for power distribution (reticulations).

Turbines can be further classified as reaction turbines (completely embedded in the fluid and derives power by pressure drop across the device) and impulse turbine (flow hits the turbine as water jet and kinetic energy of the flow is tapped) [25]. The speed of the turbine matches with the pace of power generation, usually at about 400 V AC [26]. Large capacity systems have an especially designed generator running from the same shaft as the turbine, thereby minimising power losses whereas small systems (~10 kW) use off-the-shelf generators or induction motors. In case the turbine speed is not high enough to match the generator, then gearing is used. A V-belt is a commonly used gearing mechanism in the small systems [27]. Unlike large hydropower plants, SHP do not require a large dam; only a

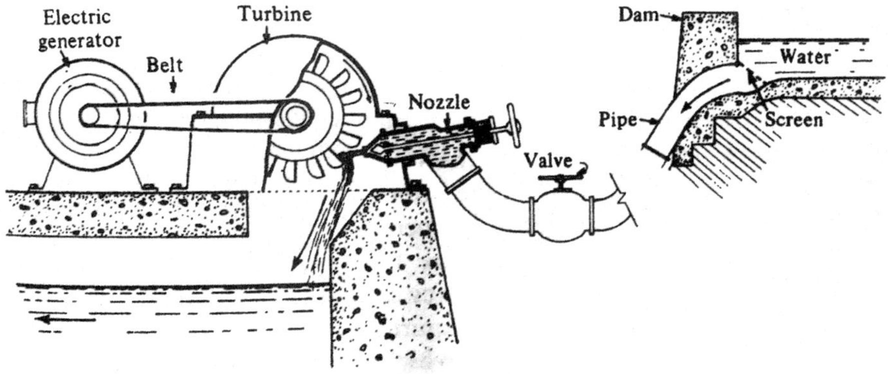

Fig. 3.9 Layout of a micro hydroelectric system using a Pelton wheel [24]

retaining wall of low height, sufficient to keep the penstock fully immersed, is needed. Conversely, SHP do not have large storage.

2.3.3 Applications

In India, hydropower technology is a mature one, capable of simultaneously meeting the requirements of large centralised urban areas as well as decentralised remote location. High power generation efficiency (up to 90 %) and rapid power generation make it suitable to supply both base load and peak demand. Multiple uses of hydropower include drinking water, irrigation, flood and drought control, navigation, over and above power supply. With the potential of SHP to meet energy demands of remote locations, MNRE is implementing the projects in Ladakh and Arunachal Pradesh to minimise dependence on fossil fuels in these regions and to meet power requirement through local renewable resources [19].

2.3.4 Prospects and Challenge

Huge untapped SHP potential in congruence to demand–supply gap in power sector and distributed electricity demand have prompted interest in this sector, both of public and private players. Besides, Government incentives and increasing cost competitiveness of renewable energy technologies further the impetus. However, SHP in India face few drawbacks due to lack of adequate transmission infrastructure, delay in project clearance, difficult terrains, small scale of projects, huge capital investment in comparison to fossil power stations and above all heavy dependence on state policies since water is a state subject [15, 19].

2.4 Bioenergy

Bioenergy is energy produced by living organisms or their metabolic products, collectively called as biomass. Since the dawn of civilisation, bioenergy has been the only renewable resource used as a substitute to fossil fuels for cooking meals, heating space and later for powering diesel or gasoline engines, too. Prognoses show that by 2050 bioenergy will satisfy around one-thirds of global primary energy demand (up to 250 EJ). However, the close nexus between bioenergy development and escalating demand for food, feed and fibre is a serious issue and requires determination of the optimal deployment level of bioenergy to ensure sustainable development [28, 29].

India is the seventh largest country of the world in terms of geographical area and more than half of it is fertile, against a global average of 11 %, making India an agro-based economy [30]. Around 32 % of total primary energy consumed in the country comes from biomass and caters to the needs of almost two-thirds of population [19]. However, high population rise has led to low land-to-man ratio resulting in decreasing size of land holdings and land degradation and has eventually complicated land resources management. Besides, alternate use of biomass for fodder and other applications results in a low surplus for fuel generation [31].

2.4.1 Potential

Estimates by MNRE, India, show that every year around 500 MT of biomass, including agro residue and forestry waste, is generated in India. However, only 25–30 % of it, capable of generating 18 GW of power, is available for power generation. Consequently, at present, a major portion (around 65 %) of biomass power is derived from bagasse-based cogeneration in sugar mill. Study further states that by raising dedicated energy plantations on two million hectares of forest and non-forest degraded land around 5 GW of power can be obtained. By March 2013, total grid-interactive installed biomass capacity in India touched 3.6 GW including 2.3 GW from bagasse based cogeneration [19].

In another nationwide survey, conducted by Indian Market Research Bureau (IMRB international) on behalf of National Institute for Interdisciplinary Science and Technology (NIIST) and Technology Information Forecasting and Assessment Council (TIFAC), Government of India, concluded that almost 80 % of agro-residues comprise of rice straw, rice husk, wheat straw, sugarcane tops and bagasse (Fig. 3.10) [31, 32]. The survey further revealed that Uttar Pradesh, Punjab, Tamil Nadu, Haryana, West Bengal and Maharashtra possess highest availability of biomass per unit area; hence, they are potent locations for developing biomass-based industries [33].

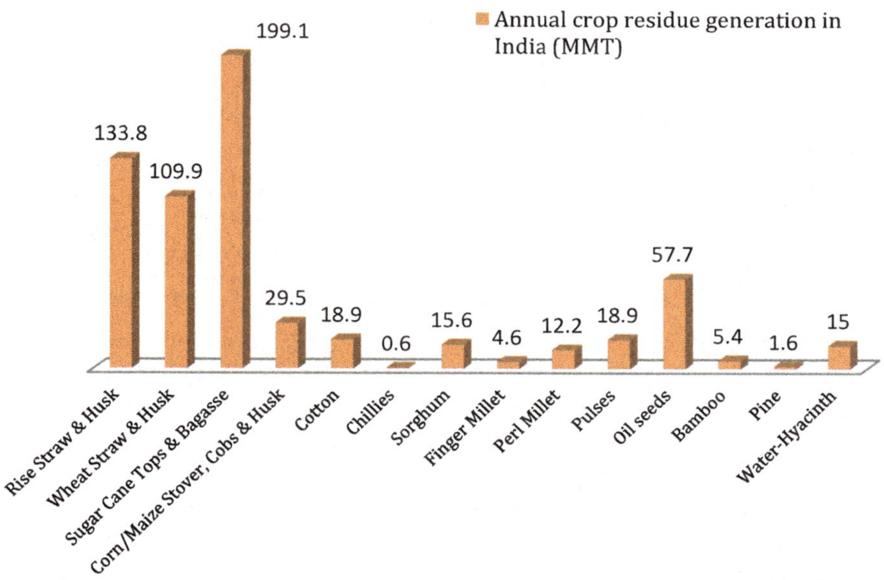

Fig. 3.10 Annual crop residue generation in India [31]

2.4.2 Technology

Bioenergy technologies are, in fact, a collection of a wide array of techniques applicable to a variety of feedstocks ranging from agro-residue to microalgae and even municipal waste. Through a number of processes, these feedstocks can be directly used to generate electricity or heat or can be converted to gaseous, liquid or solid biofuels for future applications. Bioenergy technologies can be classified based on feedstock used for biofuel production (Fig. 3.11) or conversion route employed (Fig. 3.12)

Nature and structure of biomass feedstock are the key factors for the selection of processing technologies. At the same time, a single feedstock can be used to obtain a variety of end products depending upon the conversion route employed [34]. Recently, analogous to petroleum refinery the concept of bio-refinery has emerged, whereby various biomass conversion processes and technologies can be integrated to produce a wide range of products, such as fuels, power, chemicals, etc., thereby taking advantage of the natural differences in the chemical and structural composition of biomass feedstocks (Table 3.2).

2.4.3 Applications

In India, four sets of programmes are being implemented to tap competitively priced biomass power or heat from agricultural, agro-industrial, municipal and

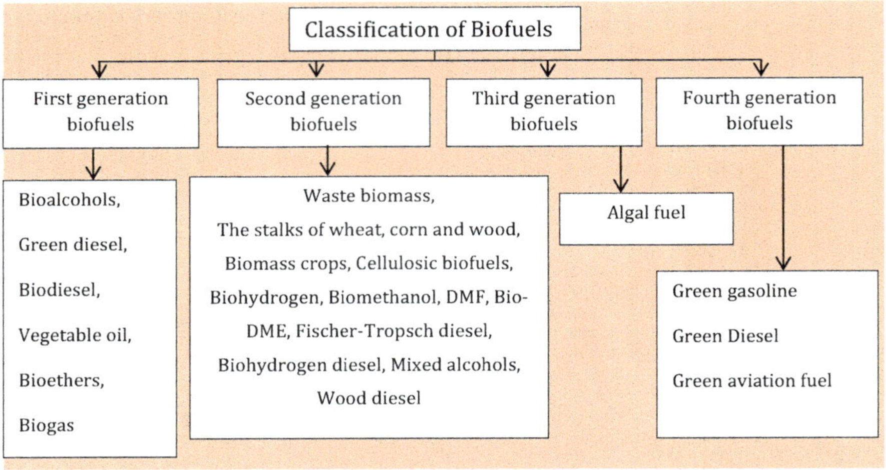

Fig. 3.11 Classification of biofuels based on feedstock

Thermochemical Conversion Routes

- Direct combustion: Immediate heat can be obtained by direct combustion of dry homogeneous input in device like cookstove equipped with excess air supply.
- Pyrolysis: Biomass is heated either in the absence of air or in a restricted air supply, ,generating a variety of products, gases, liquids, oils, solid char and ash depending up on temperature, feedstock and treatment process.
- Gasification: Heating of biomass, preferably moist, in limited air supply to obtain combustible gas as the main product, is called gasification.

Biochemical Conversion routes

- Aerobic digestion: Microbial aerobic metabolism of biomass generates heat with the emission of CO_2, but not methane. Although the process is of great significance for the biological carbon cycle, e.g. decay of forest litter, yet not used significantly for commercial bioenergy.
- Anaerobic digestion: Certain microbes can thrive on carbon compounds, in absence of air to produce both CO_2 and CH_4.The evolved mix of CO_2, CH_4 and trace gases is called biogas or sewage gas or landfill-gas and is used as fuel.
- Alcoholic fermentation: Ethanol is a volatile liquid fuel, resulting from microbial fermentation of carbohydrate rich feedstock. It may be used in place of refined petroleum.
- Bio photolysis: Photolysis is the splitting of water into hydrogen and oxygen by the action of light. Some biological organisms like bacteria can produce hydrogen by bio photolysis .

Chemical Conversion Routes

- Fuel extraction: Some plants like *Euphorbia*, produce hydrocarbons of less molecular weight that may be used as petroleum substitutes and turpentine.
- Biodiesel and esterification: Vegetable oils from plants like Soya bean can be used directly as fuel in diesel engines but high viscosity and combustion deposits of vegetable oil pose problems, when compared with standard diesel fuel mineral oil, especially at low ambient temperature. Both **difficulties can be overcome by converting the vegetable oil to the corresponding ester by transesterification. Biodiesel is considered a fuel better suited to diesel engines than conventional diesel oil.**

Fig. 3.12 Classification of biomass technologies based on conversion routes

Table 3.2 Summary of existing bioenergy technologies [34]

Bioenergy technology	Application	Supported feedstock variety	Conversion efficiency	Output flexibility	Commercial value of product	Current status
Biomass combustion	Large scale	High	Low	Low	Low	Established
Anaerobic digestion	Small	Medium	Medium	Low	Medium	Established
Fermentation	Medium	Medium	Medium	Low	High	Established
Oil extraction/ esterification	Small	Low	High	Low	High	Established
Pyrolysis	Large	High	Medium	High	Medium	Early commercial
Gasification	Large	Medium	Medium	Medium	Medium	Early commercial

industrial wastes: biomass power/bagasse cogeneration, non-bagasse cogeneration, biomass gasifier and urban and industrial wastes.

Rich in natural wealth, India heavily depends on bioenergy for meeting the basic requirement of cooking. Consequently, a large number of household still use traditional cookstoves for cooking, resulting in high emission and adverse effect on both environment and people's well-being. The issues of energy security, climate change and health can be addressed simultaneously by replacing traditional cookstoves with energy-efficient, cost-effective and easy-to-use improved biomass cookstoves. With this objective, the Government of India launched National Programme on Improved Chulhas (NPIC) in 1983 followed by National Biomass Cookstoves Initiative (NBCI) in 2009 with the primary aim to enhance the use of improved biomass cookstoves.

Another programme, National Biogas and Manure Management Programme (NBMMP), provides for setting up of Family Type Biogas Plants, specially for rural and semi-urban/households. A family type biogas plant runs by biogas produced during anaerobic digestion (AD) of organic matter as cattle dung, agro residue, kitchen waste and night soil wastes, etc. Manifold benefits of biogas-based decentralised power plants include clean gaseous fuel for cooking and lighting; digested slurry is used as bio-manure to supplement the use of chemical fertilisers and improve sanitation in villages and semi-urban areas by linking sanitary toilets with biogas plants. Matching the pace, up to March 2014, about 47.5 lakh biogas plants have already been installed in the country. Moreover, under Biogas-based Distributed/Grid Power Generation Programme (BPGP), 327 projects with a total capacity of 6 MW are being implemented in 16 states and out of which 191 projects (with a capacity of 3 MW) have been completed. Up to December 2012, 23 BPGP projects with a capacity of 656 kW have been installed successfully. India is the second largest country in the world in biogas production.

To promote biomass-based gasifier technologies, the Ministry of New and Renewable Energy (MNRE) has been setting up biomass gasification systems under various schemes like Village Energy Security Programme, Remote Village Electrification Programme and Biomass Energy and Cogeneration (non-bagasse) in Industry. As a result, up to 31 July 2013, 12 village level biomass gasification projects were installed in Bihar to provide power supply to 40 villages. Besides, 24 biomass gasification systems are at various stages of installation, which have been approved during 2011–2013. India leads the world in the total capacity of small gasifiers for electricity generation.

Keeping in view food vs. fuel enigma, the country announced National Policy on biofuels in December 2009. The policy promotes the development and utilisation of second-generation biofuels based on indigenous non-food feedstocks raised on degraded or waste lands, while giving due importance to research and development on cultivation, processing and production of biofuels, and finally envisages blending mandate of 20 % ethanol and bio-diesel by 2017.

NIIST-TIFAC survey based on the surplus availability of biomass feedstock estimated potential of 5.42 billion litre of bio-ethanol generation from the crop

Table 3.3 Ethanol production potential from major agro-residues available in surplus

Feedstock	Annual availability (MMT)	Theoretical yield (L/Dry ton)	Max production potential (billion litters)	Max production potential assuming 50 % efficiency
Rice straw	8.9	416	3.70	1.85
Wheat straw	9.1	432	3.93	1.97
Bagasse	6.4	428	2.74	1.37
Corn stover	1.1	422	0.46	0.23
Total			10.84	5.42

Table 3.4 Procurement prices for major agro-residues in India [35]

Crop	Residue	Basic material cost (Rs/Ton)	Likely price (Rs/Ton)
Rice	Straw	600–1500	700
	Husk	1500–4000	1700
Wheat	Straw	2000–2700	2500
	Bagasse	1350–1500	1500
Sugarcane	Tops	Not sold often	–
	Stover	800–1500	1000
Maize/corn	Husk	Not sold often	–
Cotton	Stalk	500–800	600
Chilli	Stalk	Not sold often	–
Millets	Stover	3000–5000	4000
	Stalk	Not sold often	–

residues (Table 3.3) and also generated data on the current selling price of the major agro-residue- related prospects (Table 3.4) [31].

2.4.4 Prospects and Challenges

India has rich bio-energy potential, especially in the form of lignocelluloses biomass which is freely available in most of the time, especially after harvesting the rice and wheat grains from the field. Most of this biomass has been burnt out in the field itself as a very small amount can be utilised in the form of cattle feed and other type of household requirements in the rural and remote areas. The burning of biomass in the field creates many environmental and health problem [31].

2.5 Hydrogen

High energy content (120.7 MJ/kg) and zero-carbon emission make hydrogen a promising fuel of future. Liquid hydrogen may be use a as fuel in space shuttles. The storage of hydrogen is still a very difficult task for most of the researchers because of its high flammability property. As it boil around 20.268 K (−252.882 °C), It requires cryogenic tank for storage. Hence the storage tank must be well insulated to prevent the boil off hazards. It can be directly used in the existing internal combustion engines or as a fuel in fuel cells for power generation. Besides industrial application, hydrogen is used for power, heat and transport. Use of hydrogen in vehicles, in comparison to the other alternatives, is more advantageous due to its on-board storage facility [31]. India too realises potential of hydrogen as clean and sustainable energy resource, and therefore, the Government of India is actively supporting RD& D projects in this field. As a result, hydrogen-based small power generating sets, vehicles (two wheelers, three wheeler and buses) and catalytic combustion systems (for industries and households) have been developed and demonstrated [5, 31, 36].

2.5.1 Technologies

Hydrogen is the basic constituent of all organic matter on the earth, including fossil fuels. Consequently, it can be obtained from both conventional and non-conventional sources. Figure 3.13 brings out the outline of commonly employed hydrogen production routes employed in India.

Amongst all the technologies, steam methane reforming (SMR) is the most cost-effective and commercially proven method. This method accounts for around 50 %

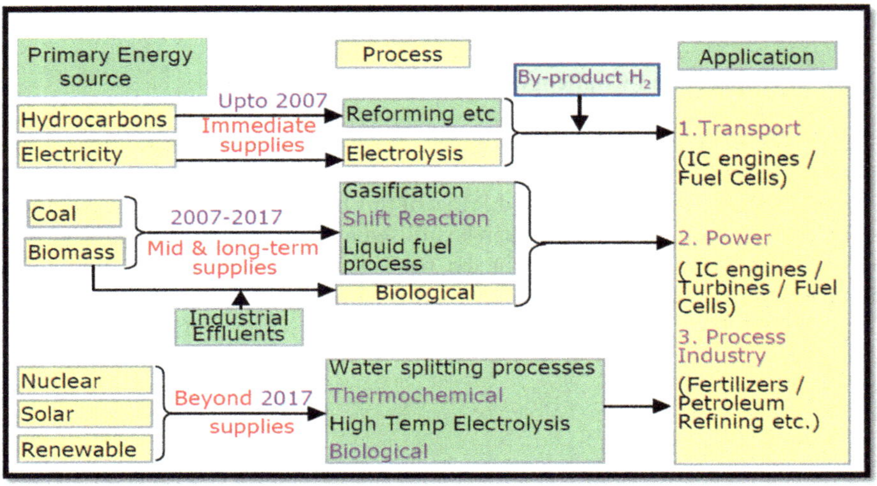

Fig. 3.13 Outline of NHEP, India, showing major hydrogen production routes and end products [37]

of hydrogen production worldwide, the biggest advantage being recovery of 99.99 % pure product. In India, fertiliser industries and petroleum refineries extensively use small- to medium-scale SMR units with a capacity of 50–24,000 kg/day for distributed production.

2.5.2 Applications

Realising the potential of Hydrogen as a clean fuel, the Government of India launched National Hydrogen Energy Programme (NHEP) based on PPP model. The programme aims not only to generate 1000 MW hydrogen-based power but also to extensively deploy the technology by putting one million hydrogen fuelled vehicles on road by 2020. Two most important initiatives of NHEP include Green Initiative for Future Transport (GIFT) and Green Initiative for Power Generation (GIP). GIFT aims RD& D of hydrogen-powered IC engine and fuel cell-based vehicles while GIP envisages RD& D of hydrogen-powered IC engine/turbine and fuel cell-based decentralised power generating systems.

2.5.3 Fuel Cell Technology

Amongst other options, hydrogen-powered fuel cells are becoming increasingly popular in India as a clean alternative energy source. The strongest motivation behind rapid growth of this technology is the direct conversion of chemical energy of hydrogen to electricity, with pure water and potentially useful heat as the only by-products (Fig. 3.14).

A fuel cell like a battery generates electricity from an electrochemical reaction, yet, unlike conventional cell, a fuel cell based on an external supply of chemical energy can run indefinitely, as long as it is supplied with fuel (source of hydrogen) and air (source of oxygen). Although all fuel cells work on the same fundamental concept of generating potential difference across the two electrodes immersed in a solid or liquid electrolyte, the nature of electrodes characterises the fuel cells and determines its applications.

For India, modular nature of fuel cells is of immense significance for decentralised operations as well as for automotive. Diverse applications including CHP mode, industrial, residential, surface transportation, electrification to remote villages, power supply to personal computers, mobiles and hospitals are possible with the help of this technology. Accordingly, the country is promoting extensive RD& D activities in this sphere; prototypes of Polymer Electrolyte Membrane Fuel Cells (PEMFCs) and Phosphoric Acid Fuel Cells (PAFCs) have been developed in kW size. Both PEMFC and PAFC are being used for power generation while in transport sectors only PEMFC is used [5].

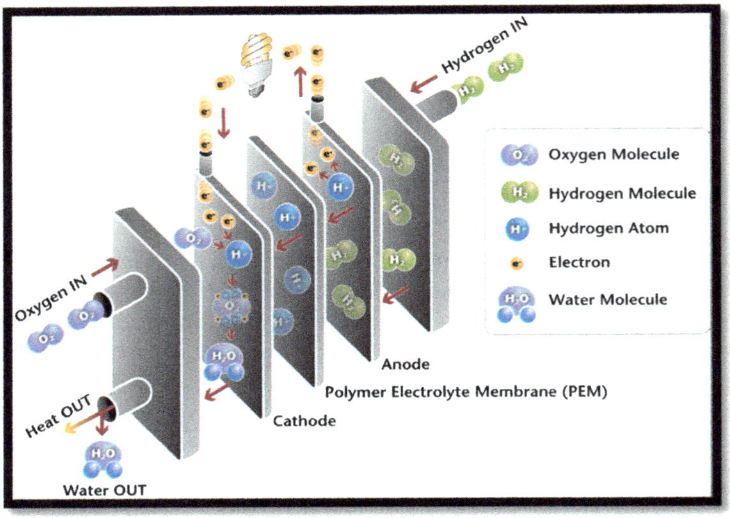

Fig. 3.14 Schematic Arrangement of a typical fuel cell [38]

High cost and low durability are the biggest challenges in fuel cell commercialisation. However, future of fuel cell seems bright in India as reliable power supply to premium power market, e.g. hotels, hospitals, information technology companies, rural areas and large stationary-power generation plants using fuel cells. Availability of hydrogen sources is also a matter of concern in case of India. Although gasoline and diesel are currently available in surplus, sulphur specifications for these fuels are much higher than tolerated by reforming catalysts or the fuel cells. The remaining fuels, hydrogen, natural gas and methanol, are all in short supply. The utilisation of these fuels in fuel cells could also be limited because of safety and infrastructure considerations [38].

2.5.4 Prospects and Challenges

Presently, India is producing hydrogen from fossil fuels (hydrocarbons), in centralised production plants for localised use only, as distribution via pressurised vessels is difficult as well as expensive. Further, on site production, technologies are still in nascent stages of development, and the lack of infrastructure to supply hydrogen for decentralised power plants and transport is another limiting factor. Alternatively, major hydrogen production techniques based on non-fossil fuels such as biomass gasification or pyrolysis, biological, photocatalytic, thermo-chemical methods using nuclear and solar energy, etc., are in the early stages of development and need intensive RD& D to compete conventional techniques.

2.6 Geothermal Energy

Geothermal Energy is a non-conventional energy source tapping the enormous heat locked in the earth crust. The volcanoes, fumaroles, geysers and steaming grounds are the surface manifestations of this inexhaustible resource. Presently, many countries are employing this clean and sustainable energy resource for fulfilling their industrial as well as domestic requirements.

Global interest in geothermal energy for power generation is relatively new, yet, in cold countries like China and Japan, ancient people used water from hot springs for cooking and bathing whereas Romans also used it for its medicinal value. In India, too, people revered the water as "God's gift" for its therapeutic effect on skin diseases and other rheumatic ailments like arthritis. At many places like Badrinath and Gangotri (Uttarakhand), Sohna (Haryana), Rajgir (Bihar), Bakreshwar (West Bengal) and Ganeshpuri (Maharashtra), temples have been built on thermal springs and the water has been channelised for public use [39, 40].

2.6.1 Potential

Around 300 thermal springs have been reported in India, falling in Himalayan (orogenic) as well as in peninsular (non-orogenic) parts of India. The systematic exploration of geothermal potential started in 1973 and by the last decade detailed study of thirty-one areas has been carried out. Out of which, sixteen areas have been drilled. The temperature of these springs ranges between 35 °C and the boiling point of water. Accordingly, based on enthalpy characteristics, the geothermal systems in India have been classified as medium (100 °C–200 °C) and low enthalpy (<100 °C). However, based on the geophysical locations, six major sites (Fig. 3.15), four in Himalayas and two in peninsular India have been recognised, having potential for electric power generation as well as for direct industrial applications. The major characteristics of these sites have been tabulated as below (Table 3.5)

2.6.2 Technology

Geothermal turbines can be classified as dry stream, binary and flash. Although, currently, these technologies are in nascent stage of development and most of geothermal energy is put to direct use, yet, India is keen to harness this untapped resource.

Amongst the commonly used technologies, the dry steam is the oldest one, where the steam is extracted directly from an underground reservoir to run the turbines. However, in flash plants, geothermal water under high temperature and pressure begins to separate into steam and water as it rises to the surface. The mixture of steam and liquid, thus obtained, is separated ("flashed") in a surface separator. The steam is delivered to the turbine and the liquid is re-injected to the reservoir.

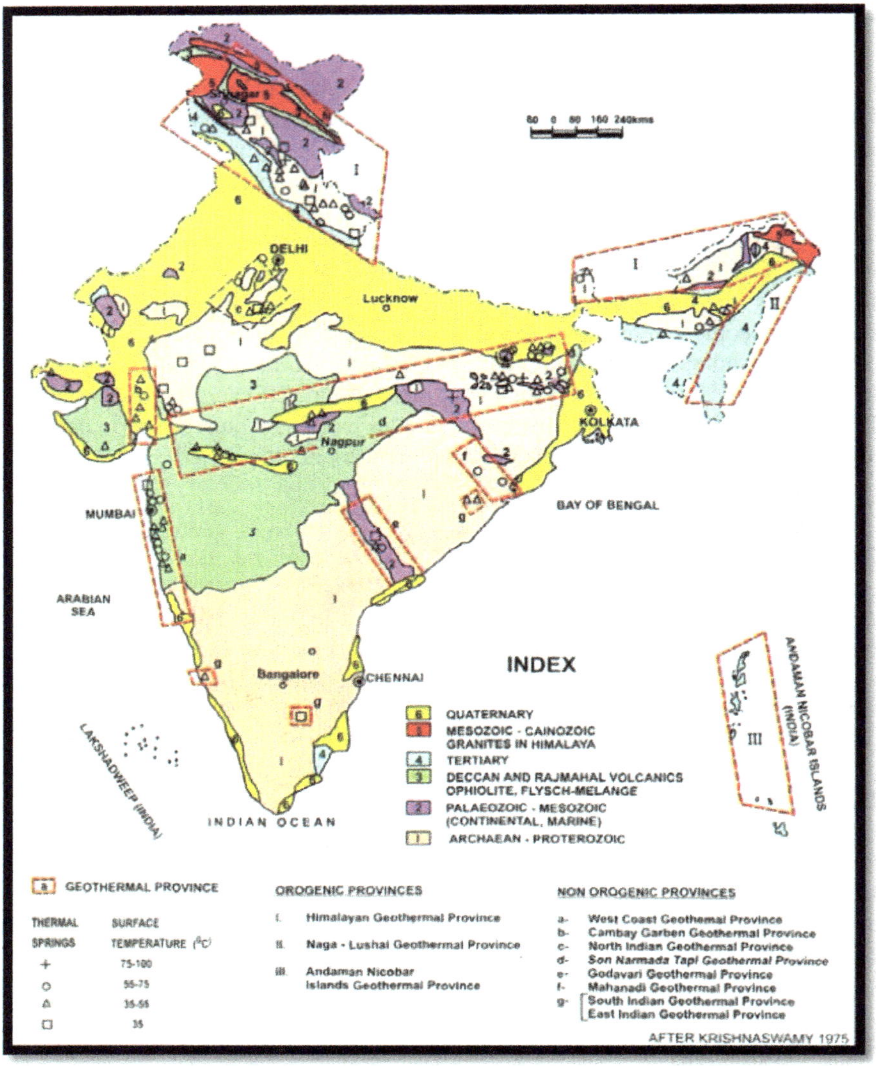

Fig. 3.15 Map of India showing major and representative geothermal provinces [41]

Table 3.5 Characteristics of major geothermal sites of India

Geothermal site	Drilling depth (in m)	Discharge (in tonnes/hr)
Puga, Ladakh, J&K	385	250
Chhumathang, Ladakh, J&K	220	50
Manikaran, Himachal Pradesh	700	100
Tapoban, Uttar Pradesh	728	150
Tattapani, Madhya Pradesh	620	120

In case of binary plants, geothermal heat is transferred to a secondary liquid or working fluid (freons) that boils at a lower temperature than water. Thereafter, vaporised working fluid turns the turbines and the geothermal water is injected back into the reservoir in a closed loop, separated from groundwater [42].

2.6.3 Applications

At present, geothermal energy in India is mainly used for non-electrical purposes: space heating, poultry farming, green house cultivation, mineral water bottling, extraction of salts and processing of sulphur and borax. A cold storage plant based on ammonia absorption, having a capacity of 7.5 tonnes, has been installed at Manikaran (Parbati Geothermal Field). Besides, experiments are also being carried out for the extraction of rare metal caesium. Efforts to utilise the medium enthalpy waters for either primary cycle power production (Puga geothermal system) or binary cycle power production (Tattapani geothermal system, Madhya Pradesh) are under way. Further, the Government of India is planning installation of a 300 kWe pilot binary cycle power plant in Madhya Pradesh [39, 43].

2.6.4 Prospects and Challenges

Although geothermal technologies in India are still in nascent stage of development, the exploration endeavours have shown that India has favourable geological—hydrogeological setup for tapping sizeable geothermal energy, capable of sustaining electrical power production on MW scale and non-electrical applications on industrial scale [41]. The greatest attraction for geothermal energy in India is that power can be provided almost continuously at full rating, independent of intermittent or purchased source of energy. Besides, these plants require moderate and inexpensive maintenance [44].

Geothermal is considered a clean energy source worldwide, yet it can sometimes produce pollutants, and even improper drilling can lead to emission of harmful gases and release of hazardous minerals. Besides, renewability of geothermal sites in the long run is also questioned [43].

2.7 Ocean Energy

Oceans hold twofold renewable energy, thermal energy due to surface heating by the sun and mechanical energy emanating from tides and waves. Globally, several technologies are under development to harness this huge resource in all its forms, i.e. tides, waves and thermal gradient. However, only 4 % of total tidal energy potential (3000 GW) has been harnessed so far. It is a significant resource, especially for Pacific Northwest region (including Alaska, British Columbia and

Washington) and contributes big share in powering many countries like the UK, France, and China. Although India has a long coastline of 6000 km with estuaries and gulfs producing strong tides, by now, it is the least harnessed renewable resource of the country [4, 5, 7].

2.7.1 Potential

Preliminary estimates state total available tidal potential of India as 40,000 MW whereas total identified potential is about 9000 MW at three major locations: Gulf of Cambay, Gulf of Kutch and Sunderbans Delta (Fig. 3.16).

India is planning to conduct tidal resource assessment studies from 2015 to 2018 to accelerate the potential analysis and to identify sites with rich potential [5].

2.7.2 Technology

Tidal Energy During a tidal cycle, the difference in the height of water during high and low tide generates potential energy, and it can be harnessed in two ways: the tidal barrage and tidal streams.

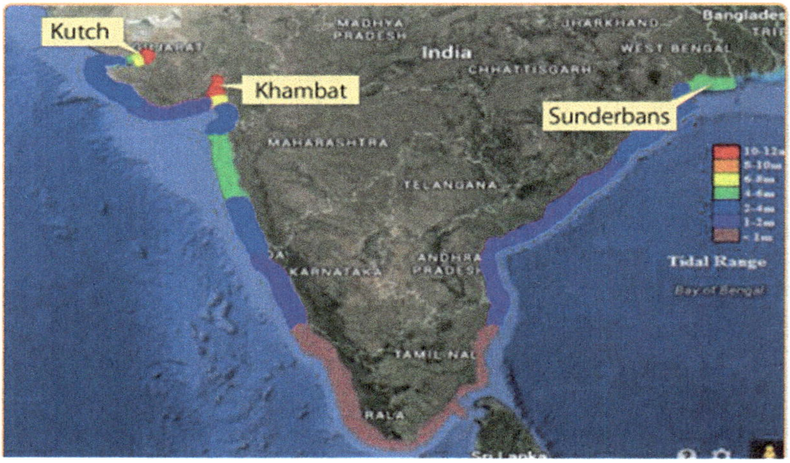

Gulf of Cambay/Khambat	Gulf of Kutch	Sunderbans Delta
• Maximum Tidal Range: 11m • Average Tidal Rang: 6.77 • Potential: 7000MW	• Maximum Tidal Range: 8m • Average Tidal Rang: 5.23 • Potential: 1200MW	• Maximum Tidal Range: 5m • Average Tidal Rang: 2.97 • Potential: 100MW

Fig. 3.16 Potential Tidal energy sites in India [12, 45, 46].

Tidal barrages are based on traditional hydroelectric technology where a dam or "barrage" is built across a tidal bay or estuary. To make technology cost-effective, the sites where a bay has a narrow opening are best suited for tapping energy, thus reducing expenditure on dams. At the time of sufficient difference in height of water (at least 16 feet) on either side of barrage, the gates are opened and the "hydrostatic head" thus obtained causes water to flow through turbines producing electricity by power generator.

Tidal Streams In coastal regions or in between islands, tidal energy emanating from tidal currents may be utilised for power generation called as tidal current or tidal stream. This technology works similar to wind mills and produces comparatively low power but is highly cost competitive for export to a utility grid or for localised applications. In contrast to wind mills, tidal power has advantage of predictable power generation due to predictable tidal flow and requires smaller scale turbines, yet small fluid velocity and harsh marine environment are problematic.

Wave Energy Wave conversion devices capture the kinetic energy of the waves and convert it to electrical energy. In many respects, this technology is similar to that of tidal or hydroelectric power. Kinetic energy thus obtained turns a turbine attached with generator to produce energy.

Ocean Thermal Energy Conversion (OTEC) This technology indirectly harnesses solar energy incident on surface of the ocean, in the form of temperature differences between the surface and depths lower than 1000 m below the surface; temperature difference up to 20 °C can yield significant energy. India has potential installed capacity of 180,000 MW for OTEC. Two types of OTEC technologies can be employed for electricity generation: closed cycle and open cycle.

Closed Cycle A working fluid, such as ammonia, is pumped through a heat exchanger and gets vaporised due to high temperature of the surface, and this vaporised steam runs a turbine generating electricity. The cold water found at the depths of the ocean condenses the vapour back to a fluid which goes back to the heat exchanger.

Open Cycle The warm water at the surface of ocean is converted to steam under pressure that in turn runs the turbine. Thereafter, the steam is condensed using cold ocean water.

Despite having high ocean energy potential, this renewable resource is most underutilised by India due to expensive technology, high cost of building, maintenance in harsh marine environment, power distribution and inconsistent supply (in case of tidal energy). The above-mentioned factor refrains India from relying and investing on this technology so far, but with entry of FDI in renewable energy sector future seems brighter for this technology [5, 46, 47] (Fig. 3.17).

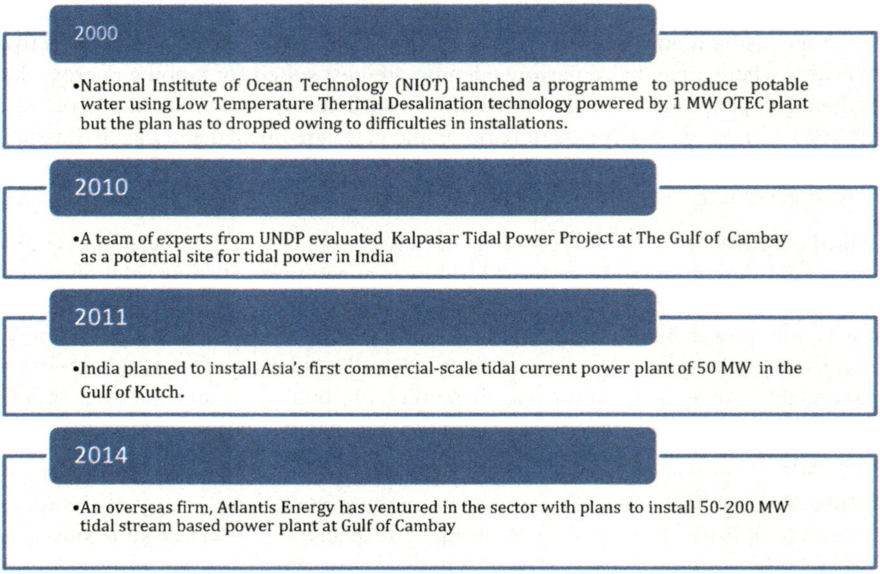

Fig. 3.17 Development of tidal power in India [12]

3 Discussion

Recognising the potential of renewables, India is amongst few nations of the world that have 9–13 policy on renewables at present [1]. The government is playing an active role in promoting the adoption of renewable energy resources by offering various incentives, such as generation-based incentives (GBIs), capital and interest subsidies, viability gap funding, concessional finance, custom concessions, income tax holiday and fiscal incentives [3] (Fig. 3.18).

As discussed earlier, renewable energy applications in India include grid-interactive power, off-grid power and other decentralised operations. By now, these policies and programmes have started paying dividends. Figure 3.19 shows cumulative grid power generated by various renewable resources up to July 2015, and it clearly indicates dominance of wind power amongst all other technologies.

Amongst off-grid applications, almost 75 % installation comes from bioenergy in the form of non-bagasse-based cogeneration and biomass installation systems. Yet another quarter is supported by SPV systems (Fig. 3.20). Above statistics clearly indicate why solar, biopower and small hydro are the focus of Make in India programme.

Systematic promotion for renewable energy in India started with the Electricity Act, 2003, and mooted by National Action Plan on Climate Change (NAPCC, June 2008). Moreover, establishment of dedicated financial institution—the Indian Renewable Energy Development Agency (IREDA)—is providing financial aid to

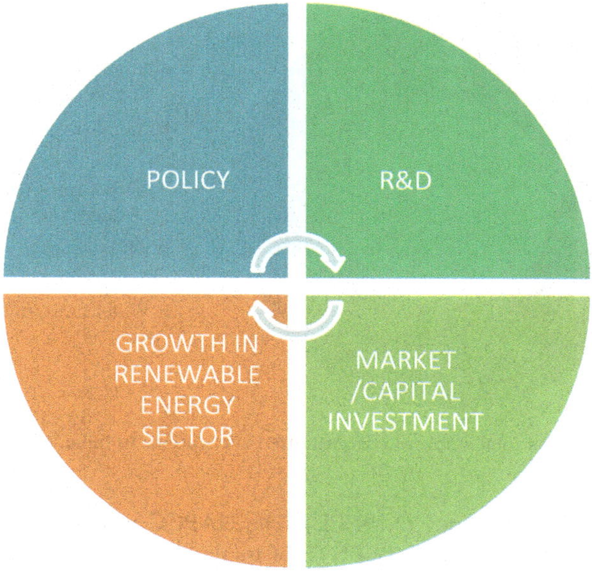

Fig. 3.18 Factors promoting renewable energy technologies in India

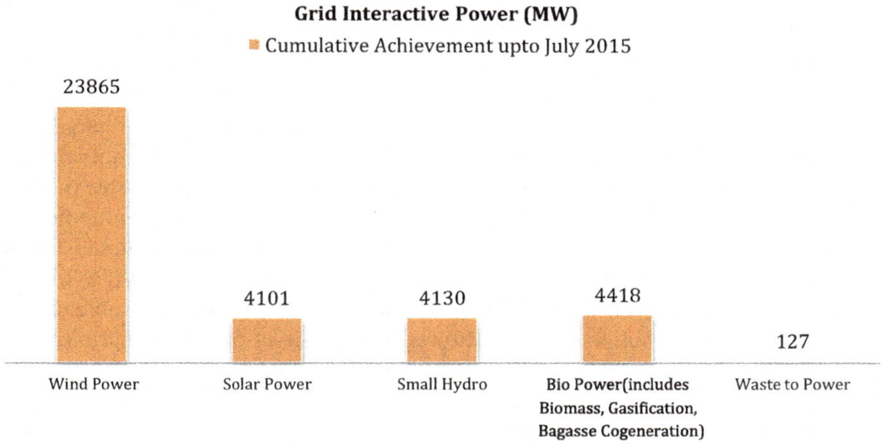

Fig. 3.19 Break-up of total grid interactive power generated in India [5]

promote renewable energy and energy efficiency/conservation projects. The electricity Act 2003 calls for optimal utilisation of resources and national policy for stand-alone systems for rural areas and promotes electricity from renewables. On the other hand, two out of eight missions, outlined in NAPCC, were dedicated to energy, viz., National Solar Mission (JNSSM) and the National Mission for

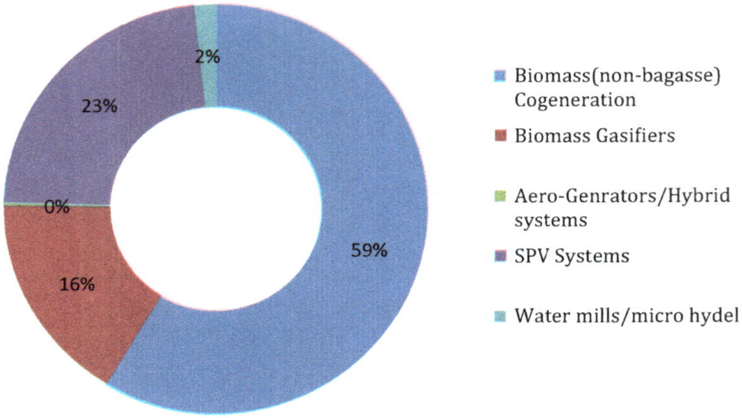

Fig. 3.20 Break-up of total employed decentralised applications in India, July 2015 [5]

Enhanced Energy Efficiency (NMEEE). The NAPCC proposed a dynamic minimum renewable purchase target of 5 % of total grid purchase, scaled up by 1 % every year for a period of 10 years. This complies that by 2020, 15 % of power in India should come from renewable energy sources [19, 48].

To promote renewable energy, the Electricity Act 2003 proposed mandatory Renewable Purchase Specification (RPS) for all states, introducing the concept of Renewable Purchase Obligations (RPOs) that requires distribution licensees, captive power consumers and open access consumers to purchase or generate a certain percentage of their total electricity requirement from appropriate renewable sources. States with low renewable energy potential can meet their targets by buying renewable energy certificates (RECs). Basically, an REC is equivalent to 1 MWh of electricity generated by a renewable energy plant. It is a tradable market-based instrument to facilitate RPO obligations, thereby catalysing development of renewable energy technologies in India. Further, policies have promoted liberal environment for foreign investment in this sector. Policy wise, up to 100 % foreign direct investment (FDI) is allowed under the automatic route for renewable energy generation and distribution projects subject to provisions of The Electricity Act, 2003. In another initiative, to check emission from fossil fuels, National Clean Energy Fund (NCEF) was established, in which clean energy tax (cess) of INR 50 (~$1) per tonne on all coal produced as well as on coal imports in India was levied.

Renewable energy in India has gradually become cost competitive in comparison to fossil fuels. Technical innovation, expanding manufacturing scale and experience curve gains have lowered the cost of wind energy equipment in the country. Consequently, the price of wind turbines has fallen by more than 25 % since 2008. Moreover, in the same span of time prices of solar modules have shown a phenomenal decline of almost 80 % [3] (Fig. 3.21).

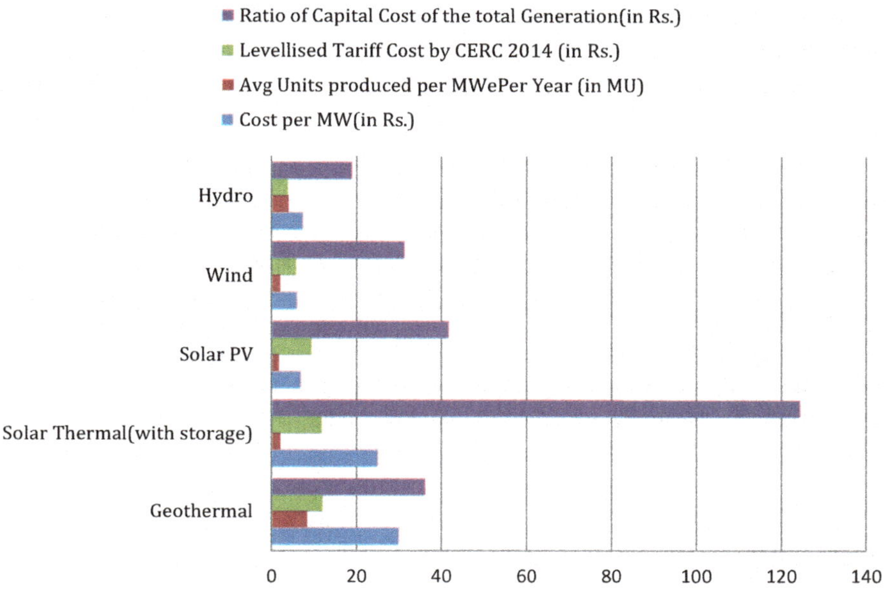

Fig. 3.21 Cost of renewable energy technology in India [5]

Maintaining the momentum, the Government of India has set a capacity addition target of 30 GW by 2017, thereby moving towards around 55 GW of the total renewable capacity. This includes 15 GW from wind power, 10 GW from solar power, 2.9 GW from biomass power and 2.1 GW from small hydro [3, 5].

Powered by Make in India Campaign, the ambitious target of Indian Government of 100 GW solar capacity by 2022 has set in indigenous manufacturing extravaganza. Congruently, in a joint venture, US-based SunEdison and its Indian counterpart, Adani Enterprises, is to develop a US$4b solar panel factory and more than 10 GW of PV capacity over the next five years. A public sector company, Coal India Ltd., is to develop 1 GW of domestic solar capacity, whereas the US Ex-Im Bank has committed US$1b to clean energy projects in India. Moreover, Gujarat is developing a 100 MW offshore wind farm, the opening venture towards India's goal of 1 GW offshore wind capacity by 2020 [49]. Accordingly, based on surveys, latest GDP forecasts and foreign direct investment (FDI) statistics, Ernst and Young LLP's renewable energy country attractiveness index, May 2015, has placed India at overall fifth place [49].

4 Conclusion

Today for India key policy drivers for renewables are to ensure energy security, accelerate economic development, address climate change and foster International collaboration. However, in the domain of low carbon technologies, niche of renewable energy is determined by the interplay of two main elements, relative cost competitiveness of various technologies and their relative impact on energy security, social equity and the environment. Dynamic markets of fossil fuels, favourable policy of the Government, technical innovation and declining cost of renewables have propelled the renewable energy sector in India; consequently, at present the country is witnessing largest renewable energy expansion programme with a number of renewable energy policies. However, such efforts have become obligatory for India citing it as the third largest carbon emitter and second most populous country of the world. Solar energy holds the highest technical potential followed by the wind, and in fact, technical potential for renewable energy as a whole is sufficient to support deployment of these technologies in the country, yet fragile market, high up-front costs, shortage of space, inadequate resource mapping, dearth of reliable data, lack of trained man-power and limited public awareness retard the pace. However, promoted by "enabling policies" renewable energy sources of India can pay higher dividends in the coming decades.

References

1. Moomaw W, Yamba F, Kamimoto M, Maurice L, Nyboer J, Urama K, Weir T (2011) Introduction. In: Edenhofer O, Pichs-Madruga R, Sokona Y, Seyboth K, Matschoss P, Kadner S, Zwickel T, Eickemeier P, Hansen G, Schlomer S, von Stechow C (eds) IPCC Special Report on Renewable Energy Sources and Climate Change Mitigation. Cambridge University Press, Cambridge
2. http://trak.in/tags/business/2015/06/26/india-economic-growth-2014-2050-top-3
3. http://www.makeinindia.com/sector/renewable-energy
4. WEC (World Energy Council), World Energy Issues Monitors (2015) Available via http://www.worldenergy.org/wp-content/uploads/2015/01/2015-World-Energy-Issues-Monitor.pdf
5. MNRE (2015) Report ministry of new and renewable energy. Government of India, New Delhi. http://mnre.gov.in
6. Murthy VAV (February 2014) India's solar energy future. Policy and institutions, a report of the CSIS Wadhwani Chair in US-India policy studies
7. State wise (2015) Estimated solar power potential in the Country, (Solar R&D Division) ministry of new and renewable energy, government of India
8. NK Bansal (2015) Non-conventional energy resources. isbn: 978-93259-7857-7
9. Solar Energy Fundamental (SEF) (2015) Available via http://www.energy-fundamentals.eu/17.htm
10. Sharma NK et al (2012) Solar energy in India: strategies, policies, perspectives and future potential. Renew Sustain Energy Rev 16:933–941
11. Tan D, Seng AK (2008) Handbook for solar photovoltaic (PV) systems. isbn: 978-981-08-4462-2, pp. 1–64
12. http://archaeologyonline.net/artifacts/gulf-of-cambay
13. Arora DS et al (October 2010) Indian renewable energy status report. Background report for DIREC
14. NAPCC. http://www.moef.nic.in/sites/default/files/Pg01-52_2.pdf
15. Novonous (2014) India renewable energy status report
16. Fthenakis VM, Moskowitz PD (2000) Photovoltaics: environmental, health and safety issues and perspectives. Prog Photovolt Res Appl 8(1):27–38
17. National Load Dispatch Centre (2014) www.posoco.in

18. http://www.indianwindpower.com
19. Mapping India's Renewable Energy Growth Potential, Status and Outlook (2013) Renewable energy India expo
20. Maurya V et al (2015) Future scope of wind energy in India. IOSR J Electr Electron Eng 10 (1 Ver. II):79–83 e-ISSN: 2278-1676,p-ISSN: 2320-3331
21. http://articles.economictimes.indiatimes.com/2015-09-10/news/66396901_1_wind-energy-offshore-windmills
22. World Energy Council (2001) Survey of energy resource 2001 (Chapter on hydropower). Available via www.worldenrgy.org/wec-geis/publicatiopns/reports/ser/hydro.asp
23. Report ES (2015) Central statistics office. National statistical organisation, ministry of statistics and programme implementation, Government of India
24. www.mospi.gov.in
25. Ramage J (2004) 2nd edn. Hydroelectricity, Chapter 5 of Boyle, G (ed) Renewable Energy: Power for a Sustainable Future, Oxford UP. (Non-Technical Survey)
26. Bhoi R et al (2014) Potential of hydro power plant in India and its impact on environment. Int J Eng Trends Technol 10(3)
27. Xu J (2002) River sedimentation and channel adjustment of the lower Yellow River as influenced by low discharges and seasonal channel dry-ups. Geomorphology 43(1–2):151–164
28. Plan S (2015–2020) IEA Bioenergy
29. Sternberg R (2008) Hydropower: dimensions of social and environmental coexistence. Renew Sustain Energy Rev 12(6):1588–1162
30. Annual Report (2014) IEA bioenergy
31. Wyman CE (1996) In: Wyman CE (ed) Handbook on bioethanol: production and utilization. Washington, DC, Taylor and Francis
32. Ghosh P, Ghose TK (2003) Bioethanol in India: recent past and emerging future. Adv Biochem Eng Biotechnol 85:1–27
33. Lin Y, Tanaka S (2006) Ethanol fermentation from biomass resources: current state and prospects. Appl Microbiol Biotechnol 69(6):627–642
34. Margeot A, Hahn-Hagerdal B, Edlund M, Slade R, Monot F (2009) New improvements for lignocellulosic ethanol. Curr Opin Biotechnol 20(3):372–380
35. Riva G. et al (2015) Handbook on renewable energy sources, EnerSupply
36. Holladay JD et al (2009) An overview of hydrogen production technologies. Catal Today 139:244–260
37. Abridged Version of National Hydrogen Energy Roadmap (2015) mnre.gov.in/file-manager/UserFiles/abridged-nherm.pdfNHEP
38. Program FCT (2010) Energy efficiency and renewable energy. USA Department of Energy
39. Thussu JL (2012) Evolution of geothermal energy in India. Akshay Urja, Lucknow
40. Roy S et al (2012) Geothermal energy: an overview. Renew Energy Akshay Urja 5(5):19–24
41. Razdan PN et al (2008) Geothermal energy resources and its potential in India. e-J Earth Sci India I(I):30–42
42. Benjamin M (2013) Geothermal power: international market overview. Geothermal energy association, Pennsylvania Avenue SE, Washington, DC. www.geo-energy.org
43. Sharma OP, Trikha P (2013) Geothermal energy and its potential in India. Akshay Urja, Lucknow
44. Electra Therm (2013) Electra therm generates power from low temperature geothermal resources. Electra Therm, Reno. Available via http://electratherm.com/docs/FCRomaniacommissioning20130409JF1.pdf
45. http://wildplacesofindia.com/sunderbans-national-park-and-tiger-reserve.html
46. www.energetica-india.net
47. www.eai.in
48. India Wind Energy Outlook (2012) GWEC
49. Renewable energy country attractiveness index, RECAI, March 2015

Chapter 4
Geothermal Energy in India: Current Status and Future Dimensions

Oum Prakash Sharma

Abstract Geothermal energy is one of the potential alternative sources of energy available in the form of the vast natural reservoir of heat energy in the earth's interior. Right from the inception of the civilisation, people from all over the world have been using geothermal energy in the form of hot springs for different purposes like bathing and washing of clothes. It is well established by now that the geothermal energy has huge potential to contribute towards meeting the increasing demand of energy. Over the last few decades, it has been successfully catering to both industrial and domestic energy requirements in many parts of the world. Besides, explaining the meaning and advantages of geothermal energy, this chapter describes the potential applications of the geothermal energy and various technologies used to capture the geothermal energy across the world. The main focus of the chapter is on highlighting the current status of geothermal energy in India and the potential sites of geothermal energy in the country. Besides, putting forward the future dimensions in the field of geothermal energy in context with India, the possible barriers and limitations have also been explained in this chapter.

Keywords Geothermal energy • Hot spring • Current status • Enhanced geothermal system • Geothermal potential sites

1 Introduction

Energy is required for all kinds of activities in our daily life. It provides comfort, enhances productivity and helps us to live our life as per our convenience. According to Fridleifsson [1] there is a strong positive correlation between per capita energy use and the issues like per capita productivity and life expectancy in the country. In fact, energy is considered to be the backbone of the economic development of any nation. So is the case with India. Presently, most of the energy demand, all over the world, is met by the conventional sources of energy mainly the

---

O.P. Sharma (✉)
National Centre for Innovations in Distance Education, Indira Gandhi National Open University, Maidan Garhi 110068, India
e-mail: opsharma@ignou.ac.in

A.K. Chandel, R.K. Sukumaran (eds.), *Sustainable Biofuels Development in India*,
DOI 10.1007/978-3-319-50219-9_4

fossil fuels such as coal, petrol, diesel, kerosene oil, natural gas, etc. However, it is estimated that we would run out of petroleum resources in another 40 years and soon after that we would run out of the natural gas also. We are not only running out of fossil fuels, but the too much consumption of these resources is adding to our environmental problems also by releasing harmful byproducts which are increasingly polluting the environment and contributing to global warming. In view of the limited stock of fossil fuels and ever-increasing gap between the demand and supply of energy, it becomes necessary to explore the new and renewable sources of energy. As India has one of the highest potential for the effective use of renewable energy, we should try to exploit it to the fullest possible. During the last 10–15 years, there has been a visible impact of renewable energy in the Indian energy scenario. However, it is clear that no single source of energy is capable of replacing the highly polluting fossil fuels in the coming years. Therefore, there is a need of exploring all local sources of energy and integrate them to meet the increasing energy demands. It is true that the greater reliance on renewable energy sources offers enormous economic, social, and environmental benefits; therefore it is necessary to explore more sources of renewable energy.

In such a situation, geothermal energy is certainly one of the potential alternative sources of energy. In fact it has been successfully catering to both industrial and domestic energy requirements all over the world for the last few decades. This renewable source of energy is available in the form of the vast natural reservoirs of heat energy in the earth's interior, and it comes out in the form of hot springs in different parts of the world. People have been using such hot springs for bathing and washing of clothes right from the inception of civilisation. A number of geothermal power plants are operational in various countries of the world and they are generating more than 10,000 MW power. In addition to it, geothermal energy is also being used directly for heating in almost 78 countries. During the last three decades, the utilisation of geothermal energy has increased rapidly. According to Lund et al. [2], there is a significant increase in direct utilisation of geothermal energy in 78 countries from the 72 reported in 2005, the 58 reported in 2000 and the 28 reported in 1995. Presently, the USA is the largest producer of this energy which generates about 3086 MW of electricity. It is estimated that the cost of electricity generation by the existing geothermal power is around $0.05–0.08/kw h.

1.1 Geothermal Energy

Basically, geothermal energy is the natural heat of the earth. Geothermal energy is generated in the earth's core, at a depth of almost 4000 miles below the earth's surface. Double-layered core of earth is made up of very hot magma made of melted rocks surrounding a solid iron centre. Inside the earth, there is a continuous decay of radioactive elements in the rocks. As a result of which, a very high temperature is produced continuously inside the earth, due to which geothermal energy is generated.

The term **geothermal** is made of two Greek words—*geo* which means "earth" and *therme* "heat". Thus, it can be said that geothermal energy is the heat obtained from the earth. It is clean, renewable and sustainable source of energy. Resources of geothermal energy include the moderate- to low-temperature hot spring system as well as the hot rocks found a few miles below the earth's surface. They are also found even down deeper to the extremely high temperatures of molten rocks. Below the earth's crust, there is a layer of hot and molten rocks called magma. Heat is continually produced in this layer, mostly because of the decay of naturally radioactive materials such as uranium and potassium. From the earth's interior heat flows outward. Normally, the crust of the earth insulates us from the which comes out of the earth's interior. It is interesting to mention here that the amount of heat within 10,000 metres of earth's surface contains 50,000 times more energy than the energy obtained from entire oil and natural gas resources in the world. It is estimated that the world average geothermal heat flow is 0.06 W/m^2. From the surface down through the crust, the normal temperature gradient is 17–30 °C/km of depth. The temperature gradient is basically the increase of temperature with increase of depth in the earth's crust. Below the crust there is the mantle which is made of highly viscous, partially molten rocks having temperature between 650 °C and 1250 °C. At the earth's core, which consists of a liquid outer core and a solid inner core, temperature varies from 4000 °C to 7000 °C.

The geothermal energy being the oldest type of natural sources of heat dates back to Roman times, when the heat energy of earth was used to heat the rooms and warm the water for bathing. Because of its inherent features, presently, it is being used as a source for producing electricity, mainly in regions of tectonic plate movement. The geothermal energy has varied uses such as generation of electricity, heating, drying appliances, melting of snow and ice on roads in winter, sterilisation, etc. The unique characteristics of geothermal energy that make it significant and reliable source of energy include its extensive global distribution, environment-friendly nature, independent on weather and climate conditions and its indigenous nature.

The geothermal energy resources can be categorised into four major types, such as hydrothermal, geopressurised brines, hot dry rocks and magma. Currently, the hydrothermal resources are being commercially used for electricity generation and for meeting thermal energy requirements. In fact, all the hydrothermal resources have two ingredients in common. They are water (hydro) and heat (thermal). According to Li and Lior [3], geothermal energy resources exist mainly in two categories: convection dominant and conduction dominant. Convection dominant resources are also called as natural hydrothermal system, and they contain sufficient natural hot fluid that can be brought directly to the surface for electric power generation, whereas conduction dominant resources do not have natural fluid and the heat is contained in hot dry rocks (HDR). Further, he explains that the systems designed to use HDR resources are called engineered or enhanced geothermal systems (EGS) by fracturing the HDR and then circulating a geofluid into it and bringing the heated geofluid to a power plant to generate electricity.

1.2 Technology for Capturing Geothermal Energy

Now, the basic question is how to capture the geothermal energy for useful purposes. Normally the geothermal energy is directly obtained from the geothermal hotspots. A hotspot is an area of reduced thickness in the mantle of the earth from which the excess internal heat of the interior of the earth comes out to the outer crust. These hotspots include the volcanic islands, the mineral deposits and geysers normally known as hot springs. In order to capture the geothermal energy, different types of plants are operating, as described below:

1.2.1 Flash Steam Plant

When the geothermal energy is available at a temperature of 150 °C and above, the fluids from the deep reservoirs can be used directly to generate electricity. In such a case when deep hole is made in the reservoir, high-pressure steam is released, which is termed as flashed stream. Such direct steam released from the geothermal reservoir is used to rotate turbines and thus generates electricity. After sometime the steam gets condensed and is converted to water again which is returned back to the reservoir.

1.2.2 Binary Power Plant

This type of plant is used when geothermal temperature is between 100 °C and 150 °C. In binary power plants, geothermal water is extracted and passed through a heat exchanger where the heat is transferred to the organic liquid low boiling point such as isobutene, iso-pentane or ammonia–water mixture. This type of liquid gets converted into high-pressure vapour, which is used to rotate turbines to generate electricity.

1.2.3 Dry Steam Power Plants

This type of plant is used when geothermal reservoirs mostly produce steam and little water. Usually geysers are the main source of dry steam. When the steam from the reservoir comes out with pressure, it is used to rotate the turbines and thus to produce electricity.

1.2.4 Hybrid Power Plants

It is found that some of the geothermal energy fields produce both boiling water and steam. In hybrid power plants, both of these heat sources are used to generate power. In fact, in this system flashed and binary systems are combined to make use hot water as well as of steam.

Bhardwaj and Tiwari [4] state that all the above-mentioned power plants are practically possible and are a part of electricity generation system across the world. Further they say that the USA is generating more geothermal energy than any other country in the world. In India first binary pilot power plant has been successfully constructed and operated by the Geological Survey of India. Similar power plants are being constructed at several other places in the country.

1.3 Potential Applications of Geothermal Energy

Geothermal energy has immense applications for us. Geothermal energy can be used very effectively both on- and off-grid development. The potential applications of geothermal energy include its direct applications as well as for electricity generation. Possibilities of using the geothermal energy for small-scale power generation and also for other thermal applications are being explored all over the world.

1.3.1 Direct Applications of Geothermal Heat

Direct application of geothermal energy involves a wide variety of end usage such as heating and cooling of space, fish farming, health spas, the use in greenhouse cultivation and crop drying, etc. When the geothermal reservoirs of hot water, found a couple of miles or more beneath the earth's surface, are used to provide heat directly, then it is called the direct use of geothermal energy. In fact, the direct use of geothermal energy is a very old method of using geothermal energy. Right from the beginning, people have been using hot springs for bathing, cooking food and other day-to-day heating purposes. In addition to it, the hot spring water is also used to heat greenhouses, dry out fish, de-ice roads and heat fish farms and spas. But, now modern systems and technology are being used for direct-use systems. In a direct use-system first, a well is drilled into a geothermal reservoir to provide a steady stream of hot water. Then the water is brought up through the well, and a mechanical system—piping, a heat exchanger and control—delivers the heat directly for its intended use. The technology, reliability, economics and environmental acceptability of direct use of geothermal energy have been demonstrated throughout the world. According to Lund and Freeston [5], the main types of direct use of geothermal heat include bathing and swimming (42%), space heating (35%), greenhouses (9%), fish farming (6%) and industry (6%). Evidently, geothermal energy is widely used for space conditioning including both heating and cooling.

Direct application of geothermal heat can use both high- and low-temperature geothermal resources. In geothermal industry, low temperature means temperature of 149 °C or less. Because of this feature, the use of geothermal energy is much more widespread in the world as compared to electricity generation. However, the direct application of geothermal energy is more site specific for market, as steam and hot water are rarely transported long distances from the geothermal site [1].

Ground source heat pumps (GSHP) system is also a direct application of geothermal energy. GSHP uses the earth's relatively constant temperature between 16 and 24 °C. It is found that the upper 10 feet of the earth maintains a nearly constant temperature. During winter this ground temperature is warmer than the air above it, whereas in the summers it is cooler than the air. In order to take advantage, the geothermal heat pumps are set up to heat and cool buildings. Geothermal heat pump systems consist of ground heat exchanger, heat pump unit and air delivery system. The heat exchanger is basically a system of pipes called a loop, which is buried in the shallow ground near the building, and they use much less energy than conventional heating systems, since they draw heat from the ground. It is a much more conventional way to tap geothermal energy to heat and cool the buildings. GSHP is effective in all kinds of climate zones and can be deployed anywhere in India on 24×7 bases. This technology is being used all over the world for more than last 50 years. As per a report published in the World Geothermal Congress 2010, the installed capacity of GSHP in the world was 52.7 GW up to the year 2013. According to the National Policy on Geothermal Energy [6], the USA, Canada, Sweden, Switzerland, Germany, Japan and China are the leading countries using this technology. According to Lund [7], geothermal ground source heat pumps have the largest energy use and installed capacity, accounting for 68.3% and 47.2% of worldwide capacity and use.

1.3.2 Application for Electricity Production

The most common way of utilising the energy from geothermal heat is to tap into naturally occurring "hydrothermal convection" systems where cooler water seeps into the earth's crust, where it is heated up and then rises to the surface. When heated, water from the hot springs is forced to come out of the surface to drive electric power generators. For setting to set up geothermal power plants, holes are drilled into the rock to capture the steam more effectively for running the electric generators. If the water comes out of the hot spring as steam, it can be used directly to generate electricity, whereas the hot water of a high enough temperature can be used as a flash system. Thus, geothermal steam is used to produce electricity in different countries of the world. However, the thermal efficiency of such geothermal power plants is low which is around 10–23%.This is because of the reason that the geothermal fluids in the boilers do not reach the high temperature of steam. Moreover, as geothermal electric power generation is site and technology specific and India being in low enthalpy region, there are not much prospects of huge generation of electric power by using geothermal energy. According to Fridleifsson [1], the top ten countries in 1999 were the USA (2228 MWe), the Philippines (1909 MWe), Italy (785 MWe), Mexico (755 MWe), Indonesia (590 MWe), Japan (547 MWe), New Zeeland (437 MWe), Iceland (170 MWe), El Salvador (161 MWe) and Costa Rica (143 MWe).

1.4 Advantages and Limitations of Geothermal Energy

Geothermal energy has a number of advantages over the conventional sources of energy. It can be used for heating homes as well as for generating electricity without producing any harmful emissions. Moreover, the geothermal energy can be used to produce electricity 24 h a day. Thus, the geothermal energy is an excellent source of clean, inexpensive and renewable energy. However, once in a while it may release some gases from deep down inside the earth which may be slightly harmful. If the geothermal energy is harnessed correctly, it leads to no harmful byproducts. Unlike traditional electric power plants, the geothermal power plants use renewable resources that need not be imported. Geothermal power plants are generally small and hardly have any on the natural landscape and the nearby environment. As no fuel is required to generate the power from the geothermal heat, the running costs for the geothermal power plants are very low. The cost of the land to build a geothermal power plant is usually less expensive as compared to constructing an oil, gas, coal or nuclear power plants. However, the initial cost may be on higher side because of exploration costs and possibility of only one in many explorations to be sustainable and useful reservoir.

Though geothermal energy has several advantages, it has some disadvantages and limitations also. For example, if the geothermal energy is not harnessed properly, it can sometime generate some pollutants also. Besides, improper drilling into the earth can also result into emission of hazardous gases and minerals. It is also feared that the geothermal power plant sites may run out of steam in the long run. In addition to it, there are certain limitations also in utilising the geothermal energy such as finding suitable location to set up power plant in a particular area. It is also a fact that other renewable energy sources like wind, solar and water are more popular and better established; therefore normally the developers prefer these sources to set up power plants as compared to geothermal energy.

1.5 Current Status of Geothermal Energy in India

India has an enormous potential to become a leading contributor in generating cost-effective and eco-friendly geothermal energy. It is estimated that around 6.5% of electricity generation in the world would be done with the help of geothermal energy and India is likely to play a bigger role in this field in the future. However, recognising the potential of the geothermal energy in India, several geothermal power plants have been established in various parts of the country. But, the power generation by using geothermal resources in India is still in initial stage. In this context, the Geological Survey of India (GSI) has prepared Geothermal Atlas of India, which gives information and data for more than 340 geothermal potential sites. This Atlas is being updated by GSI with the help of MNES. A number of studies for locating the renewable energy sources were undertaken by the Geological Survey of India over the last three decades. As a result of which, about 340 geothermal hot springs have been identified in the country. Most of them are in low surface temperature range from 37 °C to 90 °C, which is suitable for direct heat applications. These springs in India are grouped into seven geothermal provinces such as the Himalayas, Sohana, West Coast, Cambay, Son-Narmada-Tapi (SONATA), Godavari and Mahanadi. According to Ottlakan [8], the important geothermal sites being explored in India are shown in the following figure.

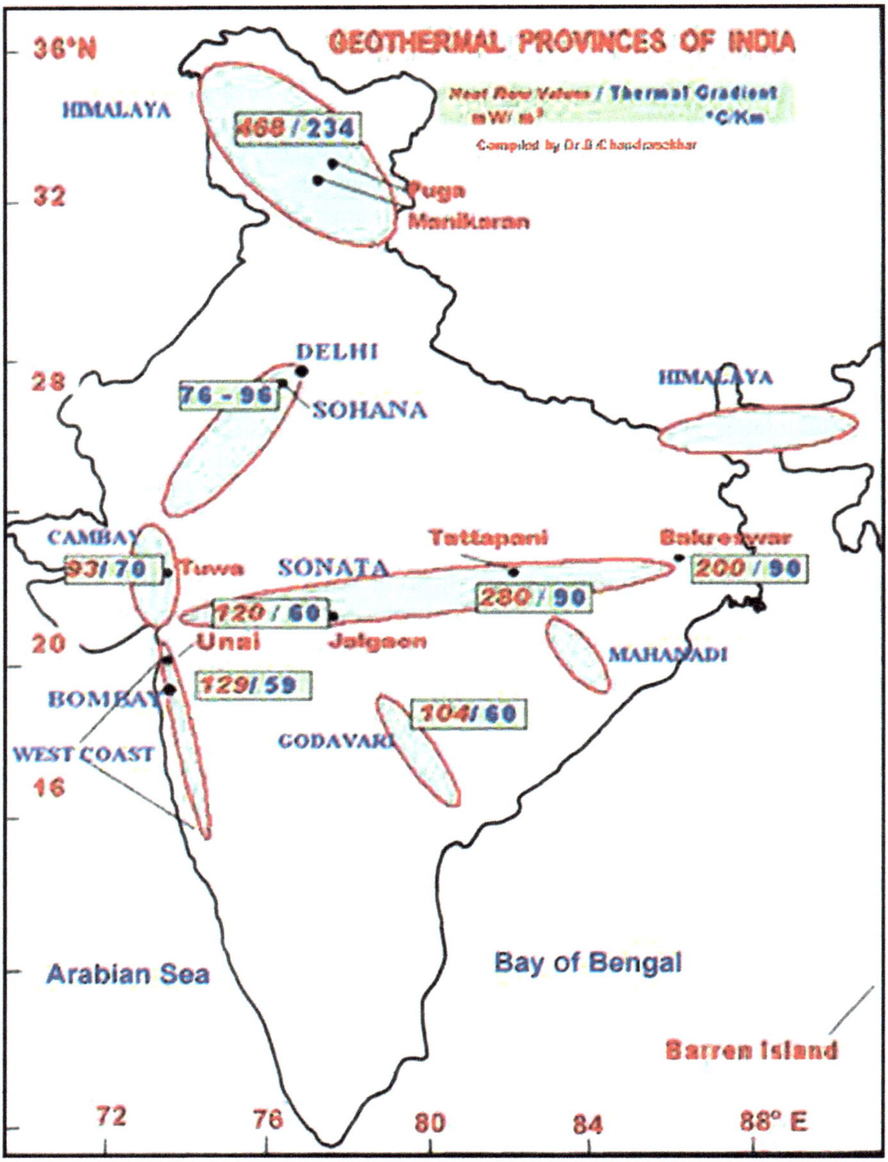

Chandrashekharam [9] has examined the characteristics of these provinces. He finds that the Himalayan Province is one of the most promising provinces in the coldest part of the country and it contains about 100 thermal springs with surface temperatures as high as 90 °C discharging more than 190 tons h^{-1} of thermal water. Its reservoir temperature is around 260 °C. Post-tertiary granite intrusives are responsible for the high-temperature gradient more than 100 °C/km and heat flow more than 468 MW/m^2. Cambay Province is another very important region of geothermal energy with more than 500 m of post-Cretaceous sedimentary formation overlying the well-known Deccan flood basalts. This province is having more than 15 thermal discharge sites with surface temperatures varying from 40 to 90 °C. Reservoir temperatures range between 150 and 175 °C and heat flow is 80–93 MW/m^2.

The West Coast Province is located within the world's famous Deccan flood basalts of Cretaceous age. In this province the thermal discharges are saline having Cl content which varies from 800 ppm to little over 1500 ppm [10]. Hence, geothermometers may not indicate the true reservoir temperatures. The saline component has been estimated in these thermal discharges to be about 1%. The reservoir temperatures are between 102 and 137 °C [11], whereas the surface temperature varies between 46 and 72 °C and heat flow is 75–129 MW/m^2.

Son-Narmada-Tapi, i.e. SONATA province, extends from Cambay in the west to Bakreswar in the east. It is an area with very high heat flow between 120 and 290 MW m^{-2} and geothermal gradient. The well-known Tattapani geothermal province spreading over an area of about 80,000 sq m. is located in this province. The Tattapani province has 23 thermal discharge sites with surface temperatures varying between 60 and 95 °C. The reservoir temperatures are between 105 and 217 °C, whereas the surface temperature varies between 60 and 95 °C.

Bakreswar Province is located in Bengal and Bihar districts and marks the junction between SONATA and Singhbhum shear zone. High helium gas is encountered in all the thermal discharges of this province. Therefore, it is planned to instal a pilot plant to recover helium from the thermal manifestation of this province.

Godavari Province is located in Andhra Pradesh. It is a northwest-southeast trending graben filled with Gondwana sedimentary formations. The lower Gondwana group of rocks consists of sandstone, shale and clays, and it includes 13 thermal discharges with surface temperature varying from 50 to 60 °C, whereas its surface temperatures varies between 50 and 60 °C. Two important thermal springs, namely, Bugga and Manuguru, also lie in this province. The reservoir of this province is at a depth of 2.5 km and its temperatures range between 175 and 215 °C.

The Barren Island is located 116 km ENE of Port Blair the Andaman and Nicobar island chain in the Bay of Bengal. It is considered that the volcanic activity in 1991 resulted in the formation of high temperature steaming and thermal discharges. In this province, fumarolic discharge is found to be in the temperatures range of 100–500 °C.

1.6 Potential Sites of Geothermal Energy in India

Some of the prominent geothermal energy sites in India are Puga Valley in Jammu and Kashmir, Chhumathang in Jammu and Kashmir, Bakreshwar in West Bengal, Unai and Jalgaon in Maharashtra, Manikaran in Himachal Pradesh and Tapovan in Uttrakhand. A new location of geothermal power has also been found in Tattapani in Chhattisgarh on the Narmada in Central India. In addition to it, Gujarat is also trying to generate geothermal electricity by using the resources available in Cambay between Narmada and Tapi River. Ottlakan [8] reports that India is all set to establish its first geothermal plants ranging from 3 MW to 5 MW.

Puga is another potential site of geothermal energy in India and is located in Jammu and Kashmir at a distance of about 180 km from Leh in Ladakh region across the great Himalayan range. In Puga Valley, hot spring temperatures are varying from 30 °C to 84 °C (boiling point at Puga), and discharge ranging is present up to 300 l min^{-1}. In order to capture the geothermal energy in the Puga Valley, a total of 34 boreholes ranging at depths from 28.5 to 384.7 m have been drilled. Thermal manifestations come in the form of hot springs, hot pools and sulphur condensates; borax evaporates with an aerial extent of 4 km. The temperature of the hottest thermal spring is 84 °C, and the maximum discharge from a single spring is 5 l/s.

Chhumathang spring is considered to be another geothermal area located about 40 km north of Puga. The thermal water coming out of the Chhumathang is quite similar to the thermal water at Puga difference that its water has relatively higher pH and sulphate. Another potential site of geothermal energy in India is Manikaran located on the right bank of Parvati River. Geothermal energy at Manikaran comes out in the form of hot spring over a distance of about temperature range of 34–96 °C, whereas on the left bank over a distance of about 450 m with temperature ranges 28–37 °C. Paterson in Uttarakhand is also a potential site of geothermal energy where the highest temperature recorded is 65 °C. The discharge from this spring varies between 0.83 and 9.2 l s^{-1}.

Tattapani is another most resource of promising geothermal energy at Balrampur district of Madhya Pradesh. Thermal manifestation at Tattapani is very intense in an area of 0.05 sq km with several hotspots, hot water pools and a marshy land. Recently, the Chhattisgarh Government has decided to set up the first geothermal power plant of the country at Tattapani. India's Renewable Energy Development Agency (IREDA) and National Thermal Power Corporation (NTPC) are working jointly towards the implementation of the project. This geothermal power project would use underground hot water springs to convert it into steam and then generate electricity by using special technology. The surface manifestations show occurrence of white to dirty deposits identified as silica. In addition to it, there are about 60 thermal water springs at 18 localities in the West Coast hot spring belt. HIMURJA, Himachal Pradesh, has also decided to select some geothermal energy areas in Beas Valley, Parvati Valley, Satluj Valley and Spiti Valley in the state for deep drilling up to 2 km for capturing the geothermal energy.

Razdan et al. [12] report that the deepest exploratory boreholes have been drilled in Puga (385 m), Chhumathang (220 m), Manikaran (700 m), Tapoban (728 m), Tattapani (620 m) and West Coast (500 m). Thermal discharges are at temperatures of 90–140 °C.

Presently, a number of Indian organisations are working in the field of geothermal energy which includes Central Electricity Authority; Geological Survey of India; Indian Institutes of Technology, Mumbai; Regional Research Laboratory, Jammu; National Geophysical Research Institute, Hyderabad; Oil and Natural Gas Corporation, Dehradun; National Thermal Power Corporation; etc. Magnetotelluric investigations in Tattapani geothermal area in Madhya Pradesh and magnetotelluric investigations in Puga geothermal area in Ladakh region, Jammu and Kashmir are some of the ongoing major geothermal power projects in India. It is estimated that one geothermal power project has a capacity of 25 MW.

1.7 Future Dimensions of Geothermal Energy in India

After decades of studies and field-level surveys by different agencies, India has finally decided to focus on tapping geothermal energy for various purposes. Obviously, the geothermal energy has great potential as a clean, green and naturally occurring renewable source of energy. Geothermal hot water can be used for many applications that require direct heating. For example, geothermal heat can be used for heating buildings, growing plants in greenhouses, drying crops, heating water at fish farms and several other industrial processes. It can also be used for generating electricity as well. It is, therefore, necessary to explore the possibilities of setting up of more geothermal power plants and exploit the full potential of such a naturally occurring renewable source of energy. India is set to have an estimated 10,600 MW of potential in the geothermal provinces. It is found that the Tattapani geothermal field is the most promising geothermal resource in Central India. Ministry of New and Renewable Energy (MNRE) of government of India has recently drafted a National Policy on Geothermal Energy (NPGE) [6] which aims at making India a global leader in this sector. It targets generating 1000 MW geothermal energy in the first phase by the end of 2020. The power generated through geothermal sources will be used to electrify rural parts of the country.

Some of the Indian states like Gujarat, Chhattisgarh, Andhra Pradesh and West Bengal are working towards setting up of geothermal plants with power capacity ranging from 3 to 5 MW. In addition to it, the NPGE also aims at guiding and commissioning geothermal energy-specific projects in the country. For promoting and popularising the use of geothermal energy, the related companies are giving emphasis on research and development also. However, there is a need to focus on research studies in certain priority areas including life cycle analysis of thermal power generation, sustainable production, cost reduction, sustainable use, expansion of use in new geographical regions, exploration of new areas and better and effective application of the geothermal energy. There are certain specific areas such as commercial development of enhanced geothermal system (EGS), development of better exploration and management tools, development of deep geothermal resources and cost-effective transportation of the geothermal energy, which need to be taken up on priority for research and development activities. Recently, the Times of India reported that the UK has experimented to heat the houses using green Scandinavian-style technology that takes heat from the nearby rivers and canals and pumps it into the homes. This type of experiments can be tried out in India also to transport the geothermal energy to the homes. An evaluation study on maintenance of geothermal plants by Atlason et al. [13] reveals that there is a need for more such studies focusing on planning maintenance conditions, shortening documentation time and risk analysis for effective and efficient functioning of the geothermal plants. More such studies are required focusing on India-specific problems.

The National Policy on Geothermal Energy (NPGE) [6] of India has envisioned to establish India as a global leader in geothermal power by deployment of geothermal energy capacity of 1000 MW in the initial phase till 2022. The Indian government focuses on assessing the potential of geothermal resources in the country and to promote the geothermal power production. Various resource assessments carried out by different agencies have established the geothermal energy potential of India as 10,600 MW spread over 340 hot springs across seven provinces in 11 states. MNRE, Government of India, contemplates initiatives in research and development of geothermal technology specifically for the purpose of cooling, drying, space heating, greenhouse cultivation, industrial processes, cold storages, poultry and fish farming, mushroom farming, horticulture, etc. It has a very ambitious plan in collaboration with the Bureau of Energy Efficiency (BEE) to increase the efficiency of conventional Heating, Ventilation and Air Conditioning (HVAC) systems by more than 50%. For this purpose, it is planned to do retrofitting in the system or replace the cooling towers by energy star qualified geothermal heat pumps. The major thrust of India is on innovation, research and development and demonstration in the field of geothermal power generation, geo-exchange pumps and hybrid models. For this purpose, financial and technical support is extended to all those organisations, universities and other third-level institutes for undertaking fundamental research and industry-led projects for developing and testing tools and devices to capture geothermal energy.

Conclusively, it can be said that though there is huge potential of using geothermal energy in India as a renewable source of energy, there are a number of challenges and limitations also to make it popular source of energy. Bhardwaj and Tiwari [4] point out that inadequate government policies, available technology, construction cost and funding for research and development are some of the additional hurdles in exploiting the potential of geothermal energy. It is, therefore, important that in view of the long-term interest of the country, the government should take proactive measures to promote the use of geothermal energy and utilise its potential for the development of the nation.

References

1. Fridleifsson B. Ingvar, geothermal energy for benefit of the people. https://pangea.stanford.edu/ERE/pdf/.../ISS/.../fridleifsson_benefit.pdf
2. Lund JW, Freeston DH, Boyd TL (2010) Direct utilization of geothermal energy 2010 worldwide review. In: Proceedings of world geothermal congress 2010, Bali, 25–29 Apr 2010
3. Li M, Lior N (2014) Comparative analysis of power plant options for enhanced geothermal system (EGS). Energies 7:8427–8445
4. Bhardwaj KN, Tiwari SC (2008) Geothermal energy resource utilization: perspectives of Uttarakhand Himalaya. Curr Sci 95(7):846–850
5. Lund JW, Freeston DH (2001) Worldwide direct use of geothermal energy 2000. Geothermics 30:29–68
6. National Policy on Geo-Thermal Energy (2017) (Draft), Ministry of New and Renewable Energy, Govt. of India, New Delhi
7. Lund JW (2010) Direct utilization of geothermal energy. Energies 3:1443–1471
8. Ottlakan, EA (2013) India to establish first geothermal energy plants. http://www.renewable-energy-technology.net/geothermal-bioenergy-news/india-establish-first-geothermal-energy-plants
9. Chandrasekharam D (2000) Geothermal energy resources of India-Country update. In: Iglesias E, Blackwell D, Hunt T, Lund J, Tamanyu S, Kimbara K (eds) Proceedings of world geothermal congress 2000, pp 133–145
10. Ramanathan R, Ganesh LS (1993) A multiobjective programming approach to energy resource allocation problems. Int J Energy Res 17(2):105–119
11. Chandrasekharam D (2005) Geothermal energy resources of India: past and present, Proceedings of World Geothermal Congrss-2005, Antalya
12. Razdan PN, Agrawal RK, Rajan S (2008) Geothermal resources and its potential in India. e-J Earth Sci India I(I):30–42
13. Reynir SA, Oddson GV, Unnthorsson R (2014) Geothermal power plant maintenance: evaluating maintenance system needs using quantitative kano analysis. Energies 7:4167–4184

Chapter 5
Biogas and Fuel Cell as Vehicular Fuel in India

Pankaj Kalita, Munu Borah, Rupam Kataki, Dipti Yadav, Dipam Patowary, and Rupam Patowary

Abstract There is a worldwide concern due to the fact that the rate of consumption of fossil fuels by far exceeds the rate of their formation. This has stimulated researchers to look into economically viable and more environmentally friendly alternatives such as biogas, fuel cells, etc. Development of technologies for utilisation of such fuels for the transport sector is in rapid progress worldwide. Utilisation of biogas is quite popular as cooking fuel in rural India and China. Various efforts were also made on utilisation of biogas as vehicular fuel in India. In this chapter, a review on various factors influencing the performance of biogas digester and functionalities of fuel cells and its application as vehicle fuel has been discussed.

Keywords Biogas • Vehicular fuel • Fuel cell • PEMFC • Fuel cell electric vehicle • Hydrogen technology

1 Introduction

Biomass is one of the premier sources of energy known to mankind. Even today, it is the largest source of renewable energy, accounting for almost 10% of world total primary energy supply. Further, it can make a noteworthy impact in supplying the increasing energy demand in a sustainable way [1, 2]. One of the technologies of harnessing the biomass energy is anaerobic digestion which results in biogas. Biogas is a flexible renewable energy source obtained from the anaerobic digestion

P. Kalita (✉) • D. Yadav
Centre for Energy, Indian Institute of Technology Guwahati, Guwahati 781 039, Assam, India
e-mail: pankaj.kalita@gmail.com; y.dipti@iitg.ernet.in

M. Borah
DST Women Scientist, National Physical Laboratory, New Delhi 110012, India
e-mail: scorpio06mb@gmail.com

R. Kataki • D. Patowary • R. Patowary
Department of Energy, Tezpur University, Tezpur 784 028, Assam, India
e-mail: rupamkataki@gmail.com; dipam189@gmail.com; rupam289@gmail.com

A.K. Chandel, R.K. Sukumaran (eds.), *Sustainable Biofuels Development in India*,
DOI 10.1007/978-3-319-50219-9_5

of manure, wastes, residues, energy crops and aquatic weeds. The process of anaerobic digestion has the potential of converting biodegradable organics into biogas which comprises methane (55–75%) and carbon dioxide (25–45%) [3] with a calorific value of 20 MJ/m^3 [4].

Anaerobic digestion (AD) process is carried out by different anaerobic species of bacteria that participate in different stages of the process of converting biomass consisting of organic compounds classified into carbohydrates, lipids and proteins to biogas. As a result, specific optimal parameters are also required for the proper functioning of each species of bacteria at different stages [5, 6]. The operating parameters in an AD process must be controlled so as to boost the microbial activity and thereby increase the digestion efficiency of the process. The production of biogas is governed by certain operational parameters, viz. temperature, pH, pretreatment, particle size, agitation, rate of organic load and retention time. Any swift change in these parameters can adversely affect the digestion process, hence the production of biogas [6].

Biogas is potential to provide the energy demand in a cost-effective and environmentally benign manner [1]. It finds its application ranging from cooking fuel to producing electricity and fuel vehicles [7]. In addition, besides the primary energy, the digested slurry (by-product) can be considered as a secondary energy source which is utilised in preparing briquettes for burning in stoves or as fertiliser for agricultural applications [8, 9]. Thus, there are distinct advantages of biogas technology over other bioenergy technologies (producer gas, pyrolysis, etc.) from users' perspective.

Application of biogas in automotive industries can be one of the possible ways to replace a significant amount of conventional vehicular fuels. However, its calorific value is deeply affected due to the presence of gases like carbon dioxide (CO_2), hydrogen sulphide (H_2S) and water vapour. Therefore, a prerequisite step of removing the unwanted gases so as to enable the full potential of biogas as a vehicular fuel becomes inevitable. A great deal of progress in purification technologies of biogas has been achieved in the European countries, and this has resulted in the large-scale feasible biogas purification for its utilisation as a vehicular fuel. According to the Natural and bio Gas Vehicle Association (NGVA), Sweden, Germany and Austria have been able to scale up its biogas-powered vehicles as compared to other European countries. Sweden have utilised more than half of its gas production in transportation vehicles, whereas Germany and Austria have utilised about 20% biogas production in natural gas-fuelled vehicles [10]. However, few developing countries have the majority of natural gas-fuelled vehicles instead of biogas-fuelled ones. The sole reason is that these countries have efficient natural gas grids. In most of these countries, upgradation and bottling of biogas are still in laboratory scale. As a result of which, commercialisation of biogas a vehicle fuel is not viable.

Biogas technology is an attractive option to generate clean fuel from organic wastes. The technology has been adopted in many countries specially the developing ones. Biogas technology has a long history in India. The technology here found attention in the 1950s, but the actual development started in the 1960s.

The first use of biogas has been reported in 1897 for lighting at the Matunga Leper Asylum, Bombay (now Mumbai). The development of the first biogas plant was done by Dr. N.V. Joshi in 1946 at IARI, New Delhi. The initial attempts in biogas technology were aimed at better understanding of the anaerobic digestion process. The early efforts in plant design resulted in costly and inefficient designs which were not easy to disseminate in rural areas. Problems of low gas output and plants being prone to burst were also a hindrance in wider use of the technology.

In 1961, Mr. Jashbhai Patel developed a floating drum biogas plant model named Grama Laxmi III which was more productive, requiring less maintenance with a longer life. The Khadi and Village Industries Commission (KVIC) promoted this model, and the model became famous as the KVIC model. The main advantage of this model was its constant pressure gas output because of its floating drum.

The constant research in plant design across the country led to development of 'Janata' fixed dome plant at the Planning Research and Action Division (PRAD) in the state of Uttar Pradesh. Based on a modified Chinese biogas plant design, it was 30% cheaper than the KVIC model of same capacity. In 1984, an NGO, AFPRO, developed a biogas plant model called Deenbandhu meaning 'friend of poor'. It was an improvement of the Janata type model and soon became popular because of its low cost. Early attempts in biogas technology were focused on manure generation, but later this focus shifted on biogas production.

2 Overview of Anaerobic Digestion Process

The subject of anaerobic digestion for production of biogas has received attention from several scientists and farming communities during the last century. AD is a process wherein complex organic compounds are broken down to simpler compounds to produce biogas (consisting of methane and carbon dioxide). It takes place in the absence of air and in the presence of suitable population of microorganisms. Digestion here signifies the different reactions that take place among the inputs, viz. methanogens and substrates that are fed into the digester. Animal excreta, plant biomass, industrial wastewater and sewage sludge which are mostly used as feedstock for AD consist mainly of carbohydrates, lipid, proteins and inorganic material. In the first stage known as hydrolysis, acidogenic bacteria degrade the above material to form alcohols, organic acid and long-chain fatty acid; while in the second stage, the acetogenic bacteria break down the fatty acid and produce acetate, CO_2, hydrogen, organic acids, alcohols and some organic nitrogen and sulphur compounds [6]. The most vital acid here is acetic acid, the principal organic acid used for the methane-forming organisms [11]. In the third stage known as acetogenesis stage, the other transitional products and acids other than acetate that were formed in the earlier stages are converted to acetic acid as well as carbon dioxide and hydrogen by different anaerobic oxidation reactions [12]. The principal acids produced are acetic acid (CH_3COOH), propionic acid (CH_3CH_2COOH), butyric acid ($CH_3CH_2CH_2COOH$) and ethanol (C_2H_5OH). In the last stage, methanogenic bacteria convert the acetic acid, carbon dioxide and hydrogen into methane and carbon dioxide [5]. In order to complete all the stages, the anaerobic bacteria should get a suitable environment for the above-mentioned processes. The reaction that takes place in the process of methane production is expressed by the following equations [13, 14]:

$$\begin{aligned} &\underset{\text{(Acetic acid)}}{CH_3COOH} \rightarrow \underset{\text{(Methane)}}{CH_4} + CO_2 \\ &2\underset{\text{(Ethanol)}}{CH_3CH_2OH} + CO_2 \rightarrow \underset{\text{(Methane)}}{CH_4} + 2CH_3COOH \\ &CO_2 + 4H_2 \rightarrow \underset{\text{(Methane)}}{CH_4} + 2H_2O \end{aligned} \tag{5.1}$$

2.1 *Biogas Digester*

Over the years a number of biogas plant designs have been developed. These can be classified in several ways depending upon the design and mode of operation. One common way to classify them is either movable drum type and fixed dome type. These two models in India are popularly known as the KVIC model (or the conventional type) and the Janata model (or Deenbandhu model), shown in Figs. 5.1, 5.2 and 5.3, respectively. Both these types have their merits and demerits. Biogas plants can also be classified as either batch-fed or continuous-fed. The movable drum (KVIC model) type basically comprises an underground brick masonry digester connected with an inlet and outlet and covered by a movable steel gasholder for the gas collection. Gasholder that moves up and down is guided by a central guide pipe depending upon accumulation and discharge of the gas. The movable gasholder made of mild steel alone accounts for approximately 40% of the total plant cost, and therefore these biogas plants are much more expensive than the fixed dome type. Maintenance costs of movable drum-type plants are also high due to the need to paint the gasholder every year to prevent corrosion [15]. Hence, this model is not found to be much effective in view of the raw material cost (mainly mild steel holder) and frequent maintenance requirements, even though this model provides consistent gas pressure. The fixed dome model such as Janata and Deenbandhu have been installed in different parts of India, but the performance of these plants depends on the location of their installation. The fixed dome model consists of an inlet or mixing tank, an outlet tank and a fermentation chamber or digester pit. In this design, there is no separate gasholder, and the upper portion of the digester pit itself acts as a gasholder. Displaced level slurry provides requisite pressure for the release of gas for its subsequent use [16]. A comparison of both models is provided in Table 5.1. Mahanta et al. [17] has modified the Deenbandhu model to duplex digester, which has been successfully implemented in north-eastern region of India. This duplex digester is constructed in such a way that it will provide 5 m^3 of biogas production per day in comparison to 3 m^3 Deenbandhu biogas plant. The schematic design of duplex digester model is shown in Fig. 5.4.

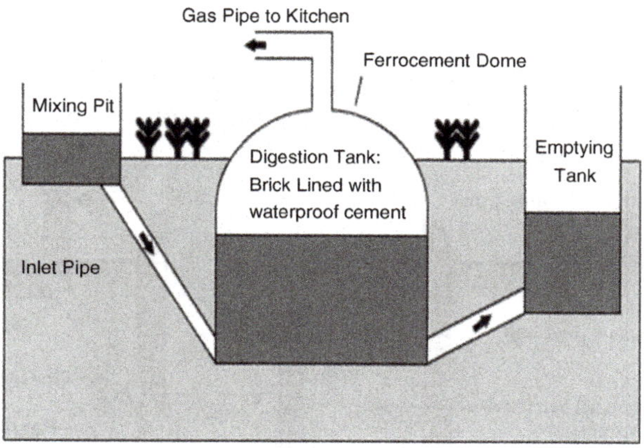

Fig. 5.1 Schematic diagram of fixed dome type digester

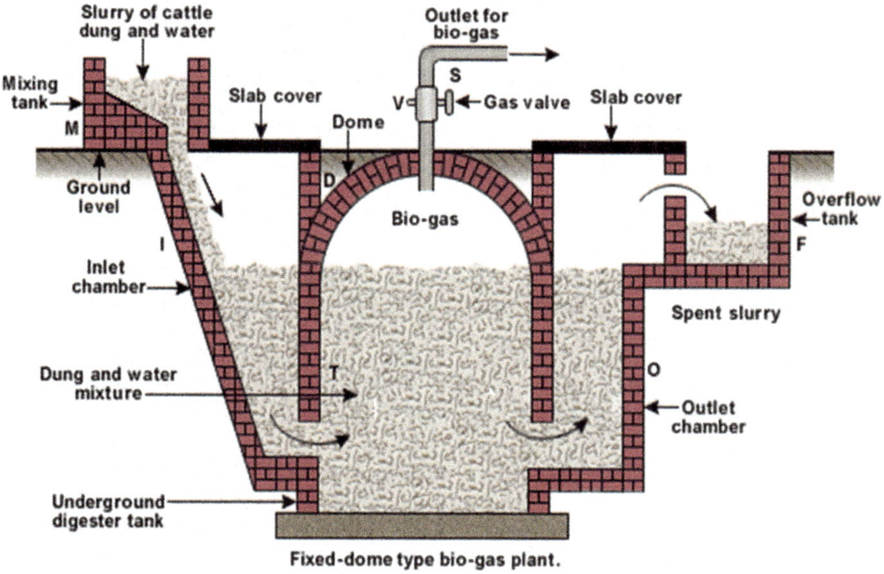

Fig. 5.2 Schematic diagram of fixed dome type

There are a number of parameters that affect the process of methanogenesis. These are carbon-to-nitrogen (C/N) ratio, temperature, pH, volatile solid content, biological oxygen demand (BOD) and chemical oxygen demand (COD). The influences of these parameters are discussed in section 2.3. The ambient temperature is an important parameter as the rate of biogas depends upon it. If the ambient temperature falls below 15 °C or exceeds 45 °C, the production of biogas ceases. The fluctuation of temperature is significantly high in north-eastern India. During winter, the temperature of this region falls below 15 °C, and hence biogas production becomes very small and digestion period increases.

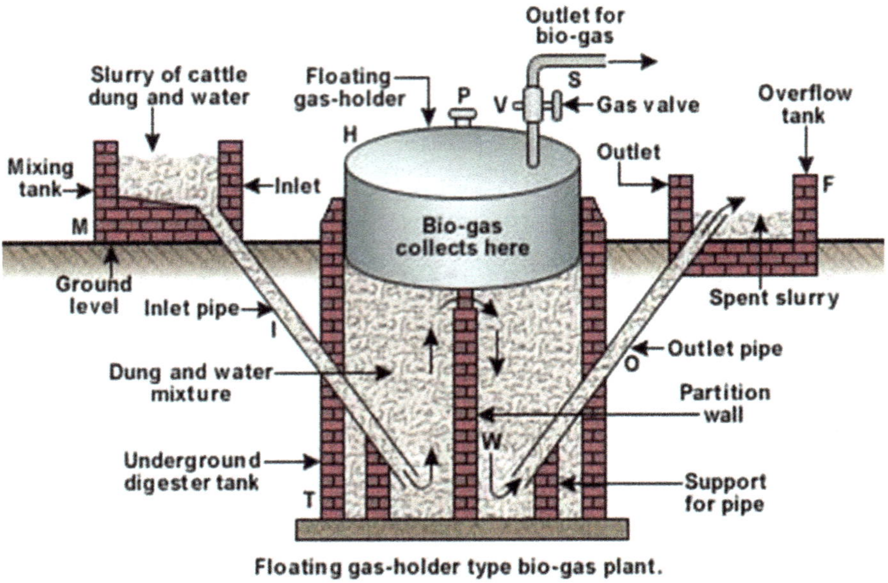

Fig. 5.3 Schematic diagram of a biogas digester with floating gas holder

Table 5.1 Comparison of fixed dome and movable drum types of biogas plants [17]

Comparison factor	Fixed domed	Movable drum
System	Regular filling and irregular discharge	Regular filling and regular discharge through overflow
Material	Concrete blocks, bricks, sand, quarry stones, lime, cement and steel	Concrete blocks, bricks, sand, quarry stones, lime, cement and steel
Main cost factor	Cement and bricks	Steel, cement and bricks
Construction aspects	Suitable to construct locally but skill required	Possible to construct locally but gas-holder is to be made in workshop
Thermal insulation aspects	Since constructed underground, it ensures uniformity of temperature and heat insulation	Heat is lost through gasholder which is difficult to insulate; it is, therefore, less suitable to use in colder region
Gas tightness	Gas storage dome is required to be given special treatment for gas tightness and painted periodically	No such problem
Feedstock	Agriculture waste and other organic matter	Animal and human fasces, chopped agriculture wastes as additive
Cost consideration	Since construction is more labour-intensive, hence it is relatively inexpensive	Due to greater use of steel, it is relatively more expensive

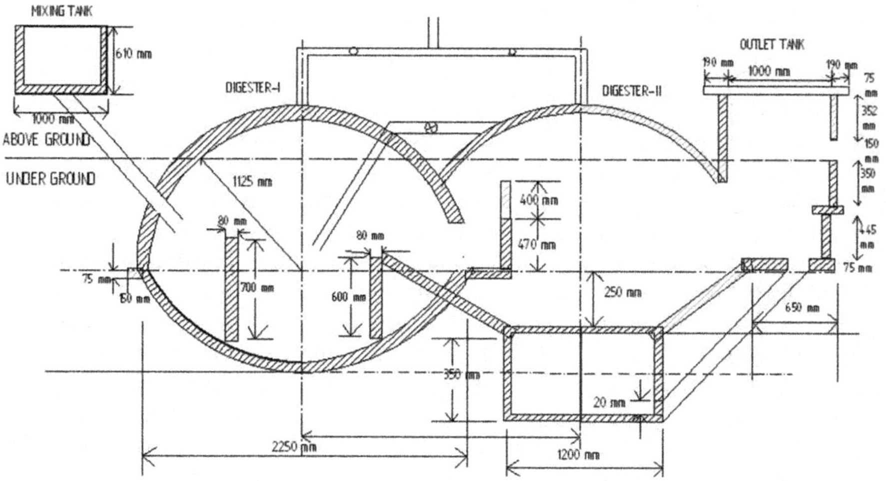

Fig. 5.4 Schematic diagram of duplex digester

Table 5.2 Properties and characteristics of biogas

Particulars	Usual	Range
Calorific value	21.5 kJ/l	20.1–25.9
Effective molecular weight	27.35	24–29
Density	1.0994 kg/m^3	0.96–1.17
Specific gravity	0.94	0.82–1.00
Viscosity	1.297×10^{-5} kg/s/m	
Optimum air-to-fuel ratio	5.5: 1 (15% biogas)	
Flammability	9–17% biogas in air	
Burning velocity	0.25/s in air	

2.2 *Properties and Characteristics of Biogas*

The major components of biogas are methane and carbon dioxide, and their percentages by volume are in the range of 55–70% and 45–30%, respectively, at standard pressure and temperature. The properties of biogas are presented in Table 5.2.

2.3 *Different Parameters Affecting Biogas Production*

In order to complete all the above-mentioned four stages, anaerobic bacteria should get a suitable environment. The key parameters that affect the reactions involved during methanogenesis are C/N ratio, temperature, pH, volatile solid content, BOD and COD. The effects of these parameters on biogas production are discussed below.

2.3.1 Carbon/Nitrogen Ratio

Carbon/nitrogen ratio is a measure of ratio of the amounts of carbon and nitrogen present in organic materials. A suitable ratio of carbon and nitrogen is essential for the production of bacteria as it is required to carry out the degradation process. Feedstocks for biogas production can be categorised as carbon rich or nitrogen rich depending upon the abundance of carbon and nitrogen contents. It is reported that carbon content in a feedstock for biogas production should be 25–30 times more than nitrogen as microorganisms, during digestion, utilise carbon at the rate of 25–30 times faster than nitrogen [1, 18–20].

To meet this requirement, the constituents of the feedstock are kept in a manner so as to ensure a C/N ratio of 25 to 30:1 and concentration of dry matter as 7–10%. According to experiments carried out in China [21] for equivalent gas yield, when C/N ratio lies in the vicinity of 13:1, it is possible to keep the range of concentration of ammoniac nitrogen between 400 and 500 ppm; when C/N ratio is 25:1, this range is maintained between 300 and 400 ppm; and when the C/N ratio is around 30:1, this range is kept between 100 and 200 ppm. Even in situations where the C/N ratio is close to 30:1, it will undergo efficient anaerobic fermentation only if waste materials are also biodegradable at the same time [15, 22]. Table 5.3 presents the C/N ratios of different organic wastes.

2.3.2 Temperature

The average minimum ambient temperature in north-eastern region of India falls below 13.5 °C during the winter season of the year, which lasts for 4–5 months. Further, temperatures during the day and night differ significantly in this region. Some fluctuations in the temperature and monthly average temperature for Guwahati city during the period from November to March are given in Table 5.4. The bacteria cannot survive under such extreme temperature variations as well as at low temperatures. Further, anaerobic digestion demands oxygen-free water for enhancing their population and activity.

Most digesters installed in field lack mechanism for the removal of the dissolved oxygen and temperature control. Hence, efficiency of these digesters is reported to be low, particularly during winter months. There are different temperature ranges during which mesophilic and thermophilic bacteria are most active causing maximum gas yield. It is generally found that mesophilic bacteria are most active in the temperature range 35–40 °C, and the thermophilic bacteria are active in the range of 50–60 °C. Choice between the mesophilic and thermophilic fermentation is governed by the natural climatic conditions in which the plant is located. However, it is possible to create conditions for thermophilic fermentation by external heat, but such a course is generally uneconomical. The length of the fermentation period is linked with the digester temperature.

Table 5.3 C/N ratios in different organic wastes [1, 15, 18–20]

Waste material	Nitrogen percentage on dry weight basis	C/N
Animal, human faeces		
Cattle wastes	0.29	25
Cow dung/buffalo dung	0.28	24
Poultry dropping	6.30	7.3
Sheep wastes	0.55	29
Pig wastes	0.60	13
Horse wastes	0.42	25
Urine	15–18	0.8
Blood	10.14	3
Slaughter house wastes	7–10	2
Night soil	0.85	29
Plant waste		
Wheat straw	0.3	128
Rice straw	0.63	67
Corn stalks	0.75	53
Bean stalks	1.30	32
Peanut stalk and leaves	0.59	19
Rotted sawdust	0.1	200–500
Purslane	4.5	8
Amaranthus	3.6	11
Cockstoot	2.6	19
Lucerne	2.8	18
Seaweed	1.9	79
Household wastes	2.2	25
Raw garbage	2.1	25
Bread	1.5	26
Potato tops	3.6	12
Cabbage Tomato	2.3	128

Table 5.4 Temperature data for Guwahati city during the period 1955–1990

Month	Maximum temp (°C)	Minimum temp (°C)	Average temp (°C)	Temperature difference (°C)
November	27.4	16.4	21.9	11
December	24.6	11.5	18.05	13.1
January	23.6	9.8	16.7	13.8
February	26.4	11.5	18.95	14.9
March	30.2	15.5	22.85	14.7
Average minimum of the month	13.5			

According to studies in China [23], when the digester operates at a temperature of 15 °C, it takes nearly a year for the digestion cycle to complete, whereas if the temperature is around 35 °C, the cycle can be easily completed in approximately less than a month's time. When the digester temperature is maintained at 25 °C, it takes about 50 days for the digestion of cattle waste, whereas if the temperature ranges between 32 and 38 °C, digestion is generally complete within 28 days. Neelakanthan et al. [24] carried out experiments at the National Dairy Research Institute, Karnal, to analyse the effect of temperature variation on the anaerobic fermentation of cattle wastes. Kopiske and Eggersgliib [25] suggested that at low temperature, biogas plants with some design modifications can also function quite effectively as in warm climate.

The methanogens are inactive in the extreme high and extreme low temperatures. The optimum temperature for their good activity is 35 °C [13]. When the ambient temperature goes down to 10 °C, the gas production virtually stops. Satisfactory gas production takes place in the mesophilic range between 30 and 40 °C. Proper insulation of the digester helps to increase the gas production in the cold season [13]. Ramchandra [14] observed a significant seasonal variation in the biogas utilisation rate and found that the maximum gas utilisation was 1.5 m^3/day in the month of August in north India that is the hottest month of the year, whereas the lowest of 0.8 m^3/day was observed during the peak winter in the month of January. Figure 5.5 shows the biogas yield at different temperatures. Table 5.5 presents the biogas production from a 4 m^3 fixed dome Janata model with maximum and minimum temperatures in different states of India [26, 27]. Mahanta et al. [28] reported the dependence of temperature on biogas production and the effect of different temperatures on biogas production rate which are shown in Figs. 5.6, 5.7 and 5.8.

2.3.3 pH Value

pH is a measure that determines the acidity and alkalinity of a solution. A solution having pH equal to 7 is a neutral solution, acidic if pH < 7 and alkaline if pH > 7.

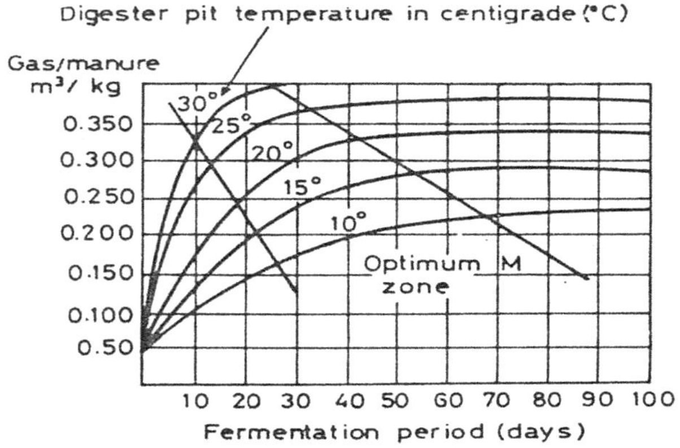

Fig. 5.5 Biogas yield at different temperatures

Table 5.5 Biogas production from cattle dung in different states of India [26, 27]

States	Monthly mean temperature		Rate of biogas production (m^3/tonne dung)
	Minimum	Maximum	
Andhra Pradesh	22.8	33.7	75
Assam	16.8	20.4	64
Bihar	17.0	31.7	68
Gujarat	20.7	31.6	71
Haryana	13.5	34.2	66
Himachal Pradesh	10.7	28.7	52
Jammu and Kashmir	3.2	21.0	23
Karnataka	21.9	29.4	69
Kerala	26.6	28.8	74
Madhya Pradesh	17.7	33.3	67
Nagaland	16.8	29.4	64
Orissa	20.6	31.3	72
Punjab	13.5	33.2	66
Rajasthan	15.8	33.7	69
Tamil Nadu	24.7	31.6	77
Uttar Pradesh	15.1	32.9	61
Delhi	13.5	34.2	66

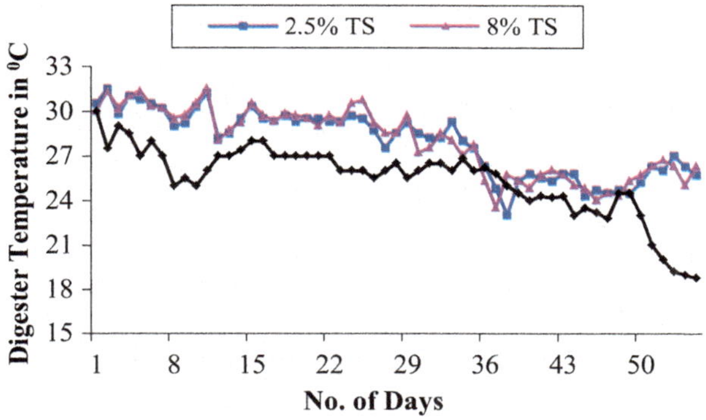

Fig. 5.6 Comparison of digester temperature

Augenstein et al. [29] proposed that microorganisms require a mild alkaline environment during anaerobic fermentation for efficient gas production. The optimum pH of feedstock should be between 6.25 and 7.50 for biogas production [30]. The pH inside a biogas digester is also a function of retention time. During the initial period of acidogenesis, the pH inside the digester decreases to a value of 5 as the acidogenic bacteria produces a large amount of organic acids. As a result of which, the digestion process is hindered. Methanogenic bacteria require a pH value below 6.5 to carry out the process of methanogenesis. Later, as the digestion process

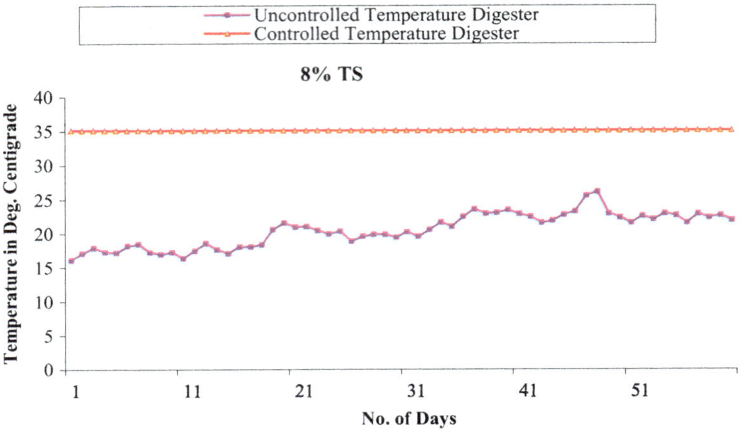

Fig. 5.7 Comparison of temperature at 8% total solids (TS)

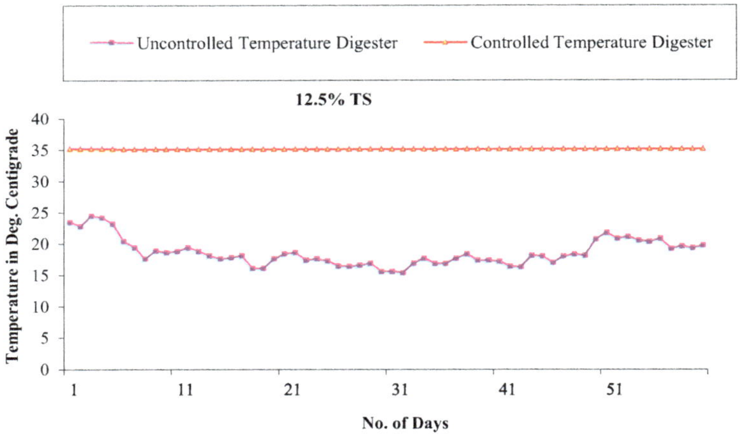

Fig. 5.8 Comparison of temperature at 12.5% total solids (TS)

continues, the pH value increases to 8 owing to the increase in concentration of NH_4 due to the digestion of N_2. The production of CH_4 is stabilised when the pH is between 7.2 and 8.2. Mahanta et al. [31] briefly described the effect of pH with respect to different solid content and the cumulative gas production rate shown in Fig. 5.9.

According to studies in China, during the period when ambient temperature ranges between 22 and 26 °C, it takes about six days for pH to acquire a stable value; similarly during the period when it ranges between 18 and 20 °C, it takes about 14–18 days for pH to attain a stable value [18]. The buffering provided by the carbon dioxide and bicarbonate during fermentation can be described by the following mathematical relationship [18]:

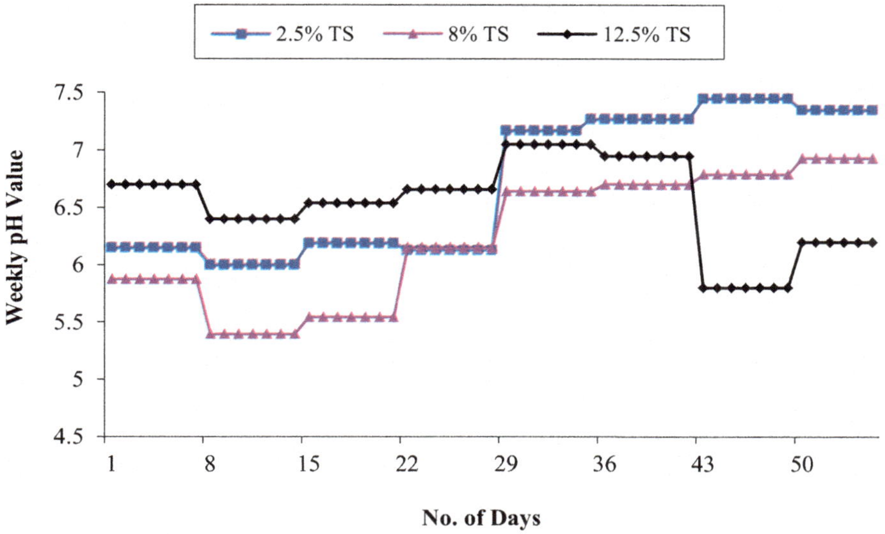

Fig. 5.9 Comparison of pH value

$$pH = 6.3 + \log[HCO_3/\text{dissolved } CO_2] \quad (5.2)$$

When biogas plant is fed with an improper mix of feedstock, there occurs an appreciable decline in the pH value following excessive formation of volatile fatty acids [18]. As a result, the proportion of carbon dioxide in biogas starts to increase leading to a further drop in pH value which adversely affects system's self-regulation. In this situation, it sometimes becomes necessary to bring the pH value to a desired range, which can be done by introducing additives. When pH value is low due to high acidity, limewater is preferred as an additive added to the digester contents. Lime is added in the form of lime supernatant in the requisite quantity only as any excess lime can start chemically reacting with carbon dioxide content of the biogas, thus adversely affecting the biogas yield. The following chemical reaction takes place [23]:

$$CO_2 + Ca(OH)_2 = CaCO_3 + H_2O \quad (5.3)$$

On the other hand, a requisite amount of HCl can be added to lower the pH inside the digester if it is too high. Figure 5.9 presents the variation of pH value at different time intervals during the fermentation of cow dung at an optimum concentration for gas production. From Fig. 5.10, it is seen that initially pH value is low, but when the fermentation period increases, the pH value falls in the optimum range for the biogas production.

According to Suhirman et al. [32], an addition of leguminous leaves into rice straw residue could increase the total gas production. The proportion of leaves and

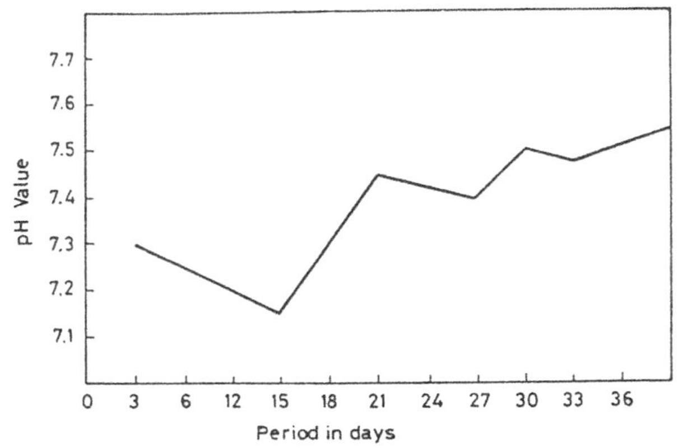

Fig. 5.10 Variation of pH value

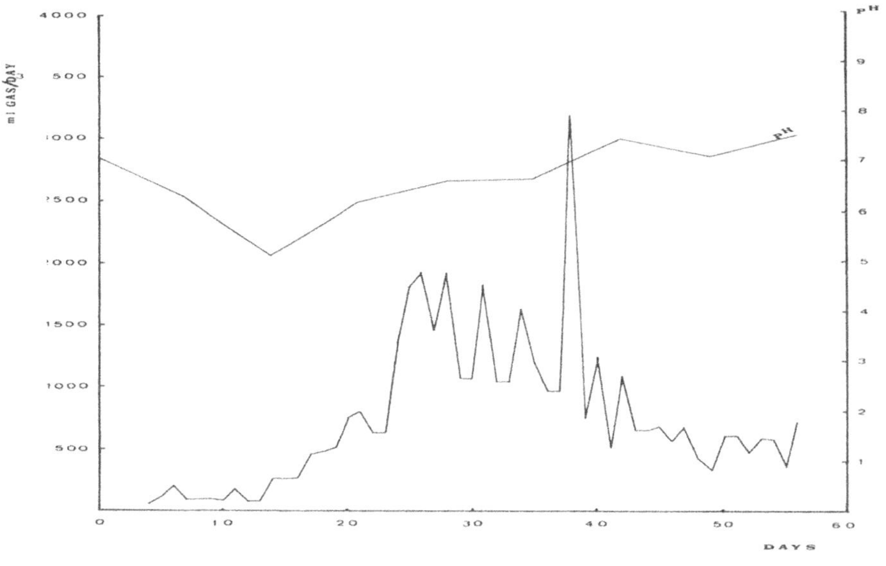

Fig. 5.11 Daily gas production and pH value

rice straw in a digester has to be considered in order to obtain a maximum gas production. The addition of leguminous leaves that resulted in a C/N ratio of residues between 20 and 30 produces more total biogas than the other treatments. The results of their experiment are plotted in Fig. 5.11, which shows the variation of pH value during 60 days of the digestion process and daily gas production. Initially the pH value drops during few days of digestion and then goes up to reach about a normal level. Further, from Fig. 5.11, it is seen that the daily gas production has not gone up before 40 days of digestion.

2.3.4 Dilution and Consistency of Input

Feedstocks fed into a biogas digester comprise both solid matter and water. The solid matter comprises of both volatile organic compounds and non-volatile or fixed solids. Only the volatile solid matters undergo digestion in anaerobic fermentation process. As per the findings of the Tata Energy Research Institute (TERI) [33], fresh cattle waste, i.e. cattle dung, comprises 20% of total solid (TS) matter and 80% of water. The solid matter can be further divided into 70% of volatile solids and 30% of fixed solids. According to [33] TS between 8 and 10% is desired in the feedstock for optimum biogas production. This range of TS is achieved by preparing a homogenous mixture (ratio of 1:1) of cattle dung and water. However for dry cattle dung, the amount of water should be increased accordingly to a desired consistency of the input, i.e. 1:1.25 or 1:1.2. The nature of the availability of dung (diluted or thick) can lower the gas production. Settling down of solid particles into the digester can be found in dilute cow dung, whereas for thick cow dung, the solid particles hinder the flow of gas produced at the bottom part of the digester [33]. Inert materials such as stones should be removed from the feedstock before feeding into the digester in order to prevent the loss of effective volume of the digester.

2.3.5 Loading Rate

Loading rate (LR) is defined as the amount of feedstock fed per m^3 of the digester capacity. LR is an important gas production parameter. Gas output is commonly expressed as m^3 of gas produced per kg of volatile solids destroyed. Overfeeding and underfeeding of feedstock hamper biogas production. For situations where feedstock is overfed, methane production will cease owing to accumulation of acids. Since micro-bacteria cannot survive in an acidic environment, gas production will be hampered. Similarly, for underfed situations, the gas production will be low because of alkaline solution which is not a positive state for the anaerobic bacteria [33].

Gore et al. [30] studied the effects of varying loading patterns on biogas yield. They basically analysed the impact of daily loading and alternate day loading on the biogas yield. According to their findings, a 50 kg of daily charge and 100 kg on alternate day basis produced 2.9043 m^3 and 2.9285 m^3 of gas, respectively. They also concluded that for a particular size of plant, there is an optimum feed charge rate that will produce maximum gas and beyond which further quantity of charge will not proportionately produce additional gas. According to the studies by Moharao [34], a daily LR of 16 kg of volatile solids per m^3 of the digester capacity produced 0.04–0.074 m^3 of gases per kg of raw dung fed. Kexin et al. [21] carried out experimental studies in China for analysing the effect of varying batch interval in batch-fed plants on biogas yield. Based on another study, Moharao [34, 35] recommended LR for plants working on night soil ranging from 1.04 to 2.23 kg of volatile solids per m^3 of the digester capacity. Higher LRs were recommended only when mean ambient temperatures are high.

Srivastava and Chynoweth [36] developed a mathematical model to describe the gas yield as a function of the organic loading rate corresponding to two different digester designs, namely, continuously stirred tank and non-mixed vertical flow reactor. Analysis of the digester operation with the help of the model indicates that an optimum gas yield can be achieved by selecting a digester design and an operating technique that will increase solid conversion through longer solids and microorganism retention. Pretreatment of the feed was identified as one of the contributing factors for increasing the biogas yield.

2.3.6 Hydraulic Retention Time

Hydraulic retention time (HRT) is the average period that a given quantity of the feedstock remains in the biogas digester. The retention time for a biogas digester using cattle dung is calculated by dividing the total volume of the digester by the volume of the feedstock added daily. According to Langrage [37], the inside temperature of the digester is essential for HRT. Higher temperature of the digester lowers the HRT. In tropical regions such as India, HRT for digesters is usually taken as 40–50 days, and in colder regions such as China, HRT is usually taken as 100 days [38]. Boodoo [38] studied the effects of different retention times employed in the anaerobic fermentation of slurry from cattle kept on slatted floors and fed mostly sugarcane and its by-products.

The HRT of a location is calculated by the following formula if the TS concentration and loading rate are known for a particular plant [33, 38]:

$$\mathrm{LR} = k\frac{\mathrm{TS}}{\mathrm{HRT}} \tag{5.4}$$

For the cattle dung, LR is expressed in units of $(\mathrm{kg} \cdot \mathrm{VS})/(\mathrm{day} \cdot \mathrm{m}^3)$, where TS is in percentage and HRT in days and k is found to have an approximate value of 7. Figure 5.12 presents the effects of the fermentation period on the gas yield at different temperatures.

Landine and co-workers [39] reported results from a study where anaerobic pretreatment facilities were located at a large potato processing plant in England. A horizontal-flow bulk volume fermenter (BVF) with a design loading rate of 0.5 kg $\mathrm{BOD/m^3/day}$ at 10 days' detention time was used. Figure 5.13 shows the temperature, detention time and loading on the bulk volume fermenter during the operational period of May 1982 to April 1983. COD, BOD and SS removals through the fermenter and methane produced are shown in Fig. 5.12 for the same operational period. Thus, it may be seen that the anaerobic portion of the plant, an ADI-BVF, provided greater than 90% removal of BOD, COD and SS while operating under higher loadings than assumed in the design and at temperature lower than 20 °C.

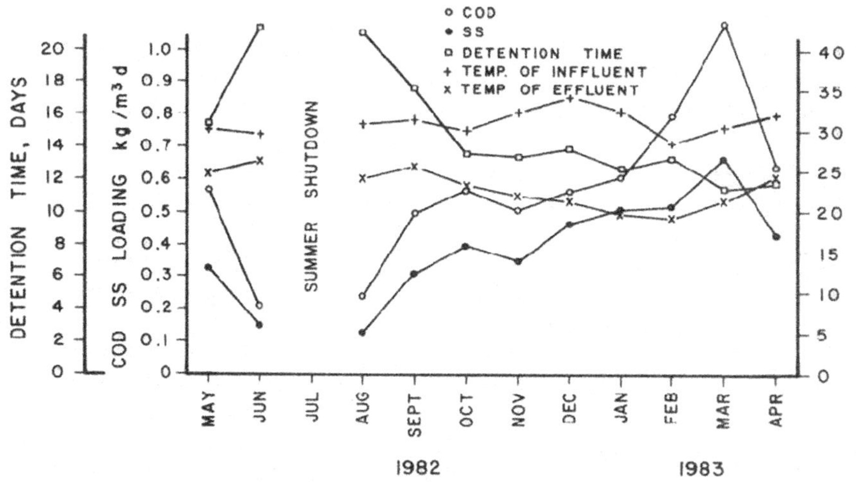

Fig. 5.12 Temperature, detention time and loading on ADI-BVF

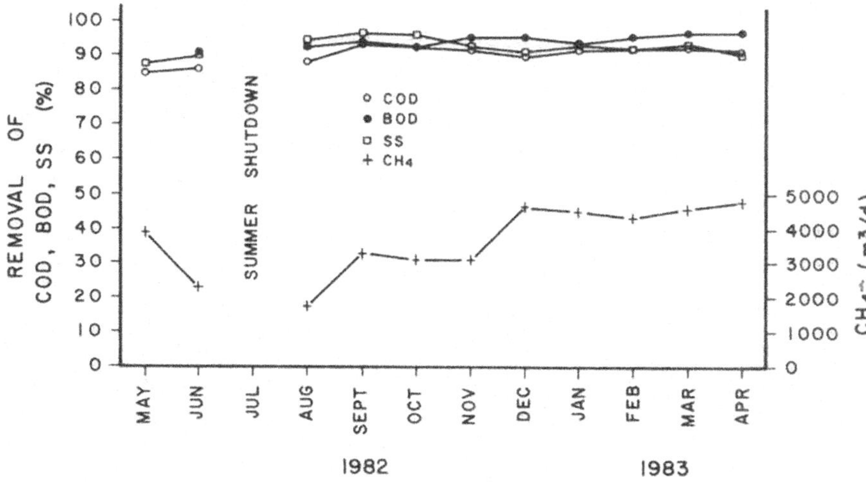

Fig. 5.13 Removals in ADI-BVF and methane production

The total nitrogen removal was found to be 80%; high nitrogen removal was attributed to the denitrification occurring in the aeration plant.

2.3.7 Toxicity

The presence of toxic materials, viz. mineral ions such as sodium, potassium, calcium, sulphur, magnesium and ammonium, heavy metal mineral ions, heavy

Table 5.6 Toxic levels of various inhibitors [41]

Inhibitors	Inhibiting concentration
Sulphate (SO_4^{-})	5000 ppm
Sodium chloride or Common salt (Nacl)	40,000 ppm
Nitrate	0.05 mg/ml
Copper (Cu^{++})	100 mg/l
Chromium (Cr^{+++})	200 mg/l
Nickel (Ni^{+++})	200–500 mg/l
Sodium (Na^{+})	3500–5500 mg/l
Potassium (K^{+})	2500–4500 mg/l
Calcium (Ca^{++})	2500–4500 mg/l
Magnesium (Mg^{++})	1000–1500 mg/l
Manganese (Mn^{++})	above 1500 mg/l

metals and detergents, in feedstock prevents the regular growth of pathogens in the digester. For instance, NH_4 composition ranging from 50 to 200 mg/l encourages the growth of anaerobic microbes, whereas its concentration beyond 1500 mg/l causes toxicity. Similarly, small presence of heavy metals such as copper, nickel, chromium, zinc, lead, etc. in feedstock is essential for bacterial growth, but their higher concentration forms toxic effects [40]. Presence of detergents such as soap, antibiotics and organic solvents impedes the activity of methanogenic bacteria [40]. Table 5.6 presents the toxic levels of various inhibitors.

2.3.8 Other Factors

Effect of Agitation on Biogas Yield

The rate of fermentation can be enhanced by close contact of microorganisms that can be done by significant mixing of feedstock in the digester contents. According to Coppinger [42], variation of degrees in mixing can improve biogas production. One of the major problems in biogas plants such as KVIC and Deenbandhu is scum formation appearing at the top of the digester. The scum blocks the flow of gas out from the upper portion of the digester, hence unavailability of biogas. It is suggested in [34] that the biogas recirculation can break scum formation and thereby improve biogas yield shown in Table 5.7.

Effect of Additives on Biogas Yield

Kumar et al. [43] analysed the effect of adding commercial charcoal to cattle dung slurry as feed on biogas yield. Addition of 5% of commercial charcoal to cow dung

Table 5.7 Effect of gas recirculation on biogas yield [27]

Experiment no.	Biogas yield Fortnightly m^3	Daily litres	Digester temperature (°C)	Ambient temperature (°C)
Without recirculation				
1	1.96	130	30.5	31.8
2	3.89	260	21.0	30.1
3	2.57	170	28.7	28.1
Mean	2.81	187	30.0	–
Without recirculation				
4	5.35	356	26.7	27.8
5	4.69	313	27.0	26.8
6	5.34	356	27.7	27.7
7	5.81	387	27.8	27.5
Mean	5.29	353	27.3	–

slurry on the dry weight basis raised the yield by 17 and 35% in batch and semi-continuous processes, respectively. Madanware and Mithal [44] carried out two sets of experiments, one at the controlled temperature of 38 °C and the other at the ambient temperature (15 °C), to study the impact of adding pectin to cattle dung slurry as feed on the biogas yield. Pectin not only enhances gas yield but also imparts process stability during periods of fluctuating temperatures. Geeta et al. [45] studied the impact of adding inert materials such as vermiculite, charcoal and lignite bovine excreta as feed on the biogas yield. These additives were found to increase biogas yield by 15–30%. Pebbles, glass marbles and plastic mesh when suspended in digester slurry reportedly led to an increase in the gas yield by 10–20%.

Prasad [46] studied the effect of adding bagasse, gulmohar leaves (*Delonix regia*), wheat straw, groundnut shells and leguminous plant leaves as additives to cow dung on the biogas yield, gas composition and extent of biodegradation. These additives were separately mixed with cow dung in the ratio of one part (oven dry) to ten parts of fresh dung containing 19.2% of total solids on the weight basis. Anaerobic fermentation was carried out under batch process in bottles in laboratory at ambient temperatures between 30 and 32 °C for 9 weeks. The volume of the biogas generated during 24 h was measured every day and the gas composition analysed periodically. Biogas yield, percent decomposition and methane and carbon dioxide contents in the gas from cow dung and individual additives comprising bagasse, gulmohar leaves, wheat straw and groundnut shells are given in Table 5.8. It indicates that it is advantageous to use these materials as additives for obtaining a high gas yield.

Table 5.8 Effects of additives on feed decomposition, biogas yield and gas composition [46]

Feed	Percent decomposition	Biogas yield (litre/g of feed)	Composition	
			CH_4	CO_2
Cow dung	26.1	196.9	57.6	40.2
Cow dung + bagasse	44.4	260.4	60.0	39.2
Cow dung + gulmohar leaves	45.1	240.9	60.0	39.2
Cow dung + wheat straw	44.5	265.4	51.2	48.0
Cow dung + groundnut shell	25.1	137.0	60.0	40.0

3 Government's Role in Dissemination of Biogas Technology

With the setting up of the Commission of Additional Sources of Energy under the Department of Science and Technology in 1981, the promotion of renewable energy technology started in India. An independent department of Non-conventional Energy Sources was established in 1982 which was later converted to the Ministry of Non-conventional Energy Sources in 1992. In 2006, the MNES got converted into present-day MNRE.

India started its National Project on Biogas Development (NPBD) in 1981 with three objectives to achieve. The objectives were to provide fuel for cooking and organic manure in rural areas through biogas plants, to lessen drudgery of women in villages and reduce burden on forests and to improve sanitation conditions in villages by linking biogas plants with toilets. The programme was renamed as the National Biogas and Manure Management Programme (NBMMP) in 2003–2004. During the 12th five-year plan (2012–2017), a target of establishing 6.50 lakh biogas plants was set up, and till March 2015, 2.86 lakh plants have been built across the country. The latest government data shows that the total number of biogas plants in India is 48.18 lakh against the theoretical potential of 120 lakh family-type biogas plants.

The NBMMP is being implemented by the State Nodal Departments/State Nodal Agencies and Khadi and Village Industries Commission (KVIC) and Biogas Development and Training Centres (BDTCs).

4 Applications of Biogas

Biogas covers a variety of area; it can be utilised for production of heat, cooking stove, electricity production/combined heat and power (CHP) production, industrial energy source for heat, injection into the gas grid, vehicular fuel, production of chemicals and combustible for fuel cells.

Table 5.9 Comparison of gaseous emissions for heavy vehicles (bus)

	g/km				
Fuel	CO	HC	NOx	CO_2	Particulates
Diesel	0.20	0.40	9.73	1053	0.100
Natural gas	0.40	0.60	1.10	524	0.022
Biogas	0.08	0.35	5.44	223	0.015

Biogas provides a clean fuel for cooking and lighting: biogas burner and lamp are structurally similar with their common components being a nozzle, an air inlet and a mixing chamber. Biogas lamp works on the principle that when the gas is burnt, the mantle of the lamp glows and emits light. To get a good combustible gas, the raw biogas is cooled, drained, dried and cleaned from H_2S because of its corrosive effect. The obtained gas can be either applied directly or upgraded to natural gas standard—biomethane (98% methane). In general, biogas has nearly about 60–65% CH_4 and about 35–40% CO_2, whereas natural gas has 85–98% CH_4 with slight percentages of ethane, butane and propane. The quality of biogas can be enhanced and utilised as a vehicular fuel if its methane content is enriched up to the level of natural gas. Biogas has the potential of exhibiting lower emission properties than natural gas and diesel as shown in Table 5.9 [47].

Biogas has been successfully used as vehicular fuel in countries such as Germany, Sweden, Austria, Italy and New Zealand. For its use as a vehicle fuel, biogas is first compressed in a three-stage compressor up to a pressure of 2500–3500 psi and stored in a high-pressure gas storage cascade. In addition, India's capital Delhi reportedly has the highest number of compressed natural gas-fuelled vehicles in the world. In addition to this, the Delhi government is planning to set up a biogas plant with production capacity up to 25,000 m^3 of compressed biogas per day to fuel around 120 buses on a daily basis [48].

Biomethane can be stored and pumped into the natural gas supply pipeline or can be transported as compressed biomethane (CBM), which is equivalent to compressed natural gas (CNG), or as liquefied biomethane (LBM), which is equivalent to liquefied natural gas (LNG). The methane content in the biomethane depends on the upgrading process, the quality of the biogas and the preconditioning of the biogas [49].

Biogas provides reasonably efficient fuel for both spark ignition and compression ignition engines: whereas diesel engines can run on biogas as a dual fuel comprising oil and gas, petrol engines can be made to run on biogas only. Biogas can be used to generate electricity: according to some studies, 1 kWh of electricity can be generated from 0.7 m^3 of gas, which can light as many as 15 electric bulbs of 60 W rating for 1 h [49].

The constituent of the biogas that can be used for energy production is methane. The biogas produced can be used directly for the heating purposes or used in an engine-driven generator to produce electricity [43]. The overall efficiency of the energy is about 80–90%. The combined heat to power couplings converts the biogas into mechanical power which in turn is converted to electricity by generators. The generated electricity powers the facilities and also fed to public power supply system. The profit earned by selling the generated electricity by the local power suppliers are an additional source of income for several operators of biogas plants. Biogas can also be supplied directly via pipelines into a biomass heating plant. By doing so, the user can avoid the implementation of an oversized boiler and the insulation of an additional heat source. Anaerobic fermentation produces biogas on renewable basis, eliminates foul smell and reduces harmful bacteria of organic wastes, improves nitrogen and phosphate contents of resulting sludge to yield highly enriched fertiliser and reduces BOD and COD [50].

5 Problems and Scopes

Biogas can be used as fuel in various applications without processing. Its use as fuel in vehicles directly is not suitable owing to the presence of CO_2. Presence of CO_2 in biogas, when used as a direct fuel, can lower the power output of the engine, reduce the range of vehicles by taking up storage spaces in cylinders and freeze valves and metering points where the compressed gas expands during mobility and refuelling. As a result of such discrepancies, CO_2 should be removed from the raw biogas in order to make it suitable for use as fuel for vehicles, in addition to the compression of the gas into high-pressure cylinders, carried by the vehicle.

The simplest and most economic method to remove CO_2 from biogas is by water scrubbing of biogas. This process can be conveniently integrated with compression, using a three- or four-stage compressor, and can easily be automated. Biogas with a composition of 100% CH_4 can be obtained using this method. However, the system produces 95% pure methane from raw biogas, originally containing 55% methane, which is pure enough for vehicle fuel. The scrubber also removes all corrosive sulphides.

It is convenient to modify vehicles to use methane as fuel in such a way that they can continue to use conventional fuel when outside the range of a gas refuelling station. Equipment designed for conversion of petrol engines to use natural gas or petrol is readily available from a number of manufacturers in Italy and the USA; and the same equipment can be used for conversion to methane or biogas. Since natural gas contains some higher alkanes (ethane, propane, butane, etc.) besides methane, giving the gas a higher energy than methane alone, larger diameter gas inlet supply lines and jets are needed for optimal running on methane. These modifications are especially important in the case of biogas containing less than 100% methane.

If the gas is introduced to the carburettor via a spacer and inlet pipe, fitted between it and the air cleaner, it is also essential that the hole in the spacer, through which the air flows, is of a suitable size and design to draw in the gas also, by the Venturi effect.

Even when the conversion is made correctly, there is likely to be some loss of power when the engines runs on methane or biogas instead of petrol, because of the compromise required adapting the engine to use both fuels interchangeably. This loss in power can be compensated for by increasing the compression ratio of the engine, to take advantage of the higher octane rating of methane, but then the engine can no longer run on petrol.

6 Biomethane as an Alternative Vehicle Fuel

Biomethane is produced through anaerobic digestion of organic waste or agricultural crops. Chemically, biomethane is practically identical to natural gas but produced from renewable resources. Using a combination of biomethane and natural gas is a way of guaranteeing the supply of methane gas used for transportation since the production of biomethane alone sometimes does not meet the demand. Several cities in Europe use biomethane as a transportation fuel for buses in the public transportation system. Examples of municipalities using refuse collection trucks, running on biomethane, also exist. However, there are important differences between the European countries in their use of biogas and biomethane. For example, in Sweden, 30 different upgrading plants exist for upgrading biogas into biomethane [51], whereas in France only one plant in Lille is in operation, and in most other European countries no upgrading plants exist at all. The use of biomethane as a vehicle fuel instead of conventional fossil fuels has several environmental benefits. Since the content of carbon in biomethane comes from nature's own photosynthesis, the combustion of biomethane does not contribute to the net addition of greenhouse gases [52]. Besides the reduction of the greenhouse gases, the emissions of nitrogen oxides (NOx), hydrocarbons and particles are less for biomethane compared with conventional fuels [53]. Biomethane has also the advantage of being able to be produced from a great variety of feedstocks that are available in Europe. This is an important feature in relation to Europe's ambition to reduce its dependency on imports of fossil fuels. Another advantage of using engines running on biomethane is that they are less noisy compared to engines running on diesel or petrol. This is a great advantage for buses and trucks serving urban areas. As the biogas is in a gaseous form and a low cetane fuel, it can be used in compression ignition (CI) engines in the dual fuel mode. Many researchers have studied the combustion, performance and the emission characteristics of engines fuelled with both gaseous and liquid fuels in dual fuel mode. In a dual fuel engine, after the compression of the charge, comprising the inducted fuel and air, a small amount of diesel, called the pilot fuel, is injected into the engine. This injected pilot fuel gets self-ignited and becomes the ignition source for the inducted fuel [54]. The main advantage of a dual fuel engine is that it can run with a wide variety of liquid and gaseous fuel without any major engine modifications [55].

Duc and Wattanavichien [56] gave an overview of several studies on the engine performance of diesel and dual fuel operation with biogas, which is not conclusive. The overview revealed that some research works indicated an increase and some a decrease and some reported no difference of performance in comparison with diesel operation. These seemingly contradictory results are the consequence of the combustion timing, which is affected by both the operating conditions and the fuel choice. Cacua [57] reported that the ignition delay is a critical parameter to control the performance and emissions of a biogas-fuelled dual fuel engine. Nathan et al. [58] reported that the biogas dual fuel engines generally exhibit low thermal efficiency, due to the presence of CO_2. At lower loads, the biogas addition results in a decrease (up to 10%) in the thermal efficiency, depending on the biogas quality (CH_4/CO_2 ratios), but it hardly matters at higher loads [59]. Moreover, as the compression ratio increases, the brake mean effective pressure and the brake thermal efficiency increase. At a higher compression ratio, the combustion temperature becomes higher, causing more NO emission [60]. Karim [61] claimed that the dual fuel operation results in a higher power output, better specific fuel consumption, superior emissions and quieter and smoother operation.

6.1 Upgraded Biogas

Biogas upgradation is the process of enriching of methane through purification, compression and bottling of the gas. Compared to natural gas that has between 75 and 98% methane with small percentages of ethane, butane and propane, the composition of raw biogas from a digester has approximately 60% methane and 40% carbon dioxide [62]. It also contains trace elements of hydrogen sulphide and moisture. This composition makes the quality of the gas not good enough to be used as fuel for to power machinery. But there is available and proven technology that can improve the methane content of biogas. It is currently possible to improve the quality of biogas by enriching its methane content to a level where it matches that of natural gas. With methane enrichment and compression of the gas, it can be used as fuel to power motor vehicle that runs on compressed natural gas (CNG). Biogas lowers emission levels compared to those of natural gas, and diesel makes it more desirable and climate friendly.

To produce biomethane from biogas, further purification is required. Biogas generally contains significantly more carbon dioxide than natural gas. In addition, natural gas also contains higher levels of hydrocarbons other than methane. The main objective of this upgrading process is to remove most of the CO_2 from the biogas to increase its energy density. To achieve the necessary specification for grid injection, further purification stages may be required including drying to a certain water dew point, depending on the pressure of the gas grid, into which the biomethane is injected. The proximity to the grid and the need to provide a physical injection point are additional practical challenges to be overcome.

A number of upgrading technologies are commercially available [63].

Pressure Swing Adsorption The CO_2 is separated by adsorption on a surface under elevated pressure. The usual adsorption material is activated carbon or zeolites.

Absorption The raw biogas meets a counterflow of liquid in a column which is filled with plastic packing. The liquid leaving the column will contain increased concentration of carbon dioxide as the latter is more soluble than methane.

Water Scrubbing It is the most common upgrading technique, and plants are commercially available from several suppliers in a broad range of capacities. In addition, organic physical scrubbing and chemical scrubbing based on amine solutions are applied technologies.

Membranes Dry membranes for biogas upgrading are made of materials that are permeable to carbon dioxide, water and ammonia. Usually membranes are in the form of hollow fibres bundled together. This is a common method for landfill gas upgrading.

Cryogenic Upgrading This is a relatively new technology. It makes use of the different boiling/sublimation points of the various gases particularly for the separation of carbon dioxide and methane.

The purification of biogas is done to augment the calorific value by reducing the carbon dioxide present in the raw biogas. This technology is an economically viable option for biogas produced at medium to large scales. The Ministry of New and Renewable Energy (MNRE) has developed a national master plan for waste-to-energy projects along with the National Biogas and Manure Management Programme and biogas fertiliser plants for biogas upgradation and bottling programme with assistance from UNDP/GEF. The Government of India is exploring anaerobic digestion potential from all sources: municipal solid waste, crop residue, sewage sludge, animal manure and industrial waste which includes distilleries, dairy plants, pulp and paper, poultry, slaughter houses and sugar industries excluding wastewater treatment plant—for two significant reasons: one being rocketing fuel pricing and the other stringent environmental regulations. The total potential of biogas from all sources has been estimated to be 48,382 million m^3/year. The equivalent upgraded amount estimated by Indian Petroleum and Natural Gas Statistics 2011–12, can contribute significantly both in transportation (~43.4% of total) and cooking (~41.7% of total) sectors.

In Europe especially Germany, Sweden, Switzerland, Italy and Denmark, the use of biogas and natural gas in both public and private transport sectors is a fast-growing phenomenon. Germany has 622 biogas/natural gas refuelling stations, while Italy has 521 stations. There are available passenger cars, buses and trucks that run on enriched biogas/natural gas from such manufacturers like Volvo, Mercedes, FIAT, MAN and Ford, among many others [61].

Since enriched biogas can be used to power motor vehicles and machinery. It is important to package the gas in bottles or provide refuelling infrastructure so that its handling and distribution are easy. To bottle the enriched gas, it has to be compressed first before being filled in high-pressure steel cylinders (available in the market for CNG storage) [63]. For example, to make enriched and compressed biogas (CBG) suitable for automobile application, the enriched biogas is compressed to about 20 MPa after moisture removal and filled in special high-pressure steel cylinders. To be used for motor transport, biogas has to be enriched to at least 95% methane, and then it can be used in vehicles originally modified to run on natural gas.

6.2 Scope of Using Duel Fuel Operation

By modifying the IC engine suitably, biogas can be used as fuel in duel fuel mode. The higher octane number of biogas with a relatively higher compression ratio makes suitable for engine applications which further results in increase in thermal efficiency [64]. In addition, the carbon content in biogas is relatively low in comparison to conventional diesel fuel, resulting in decrease in pollutants [65]. The most striking features that stand out in favour of using biogas in compression ignition (CI) engine is that there is no fearing of power which is evident in the case of spark ignition (SI) engines [66, 67]. The reason behind this is that SI engines are very sensitive to biogas composition leading to high cycle to cycle variations [68]. Biogas is used in the CI engines in duel fuel mode. In CI engines, the temperature of air at the end of compression stroke is more than 553 K. The auto-ignition temperature of biogas is 1087 K [55]. Hence, biogas mixture will not ignite by simply compressing the air. Therefore, a small amount of liquid fuel must be supplied along with the gaseous fuel to initiate the combustion process. The liquid fuel is called pilot fuel which acts as a source of ignition for the gaseous fuel. The gaseous fuel is called primary fuel on which the engine primarily runs. A CI engine can be easily modified to a dual fuel diesel engine (DFDE) by simply connecting a gas mixer at the inlet manifold and connecting a fuel control mechanism to control the supply of pilot fuel. It is seen that in a dual fuel engine, the combustion starts in a similar fashion of a CI engine. However, in the later part of combustion, the flame propagates in a manner similar to the SI engine. The power output of the engine is normally controlled by regulating the biogas flow. It is possible to achieve up to 85% diesel substitution when biogas is used as primary fuel [69]. The most promising feature of DFDE is the ability to switch over from dual fuel operation to diesel mode almost instantaneously in case of short age of primary fuel [70]. The idea of running CI engines on biogas is not of recent origin, but the process of refinement is still in progress.

Many researchers [65, 68] have tried mainly to examine the practical utility of biogas in CI engines in duel fuel mode. Studies on engine performance by using different qualities of simulated biogas have been found [65]. The study indicated the possibility of 60% gas-oil substitution of pilot fuel without knock. Yoon and Lee [66] carried out an experimental investigation to study the combustion and emission characteristics of biogas-run DFDE using soybean biodiesel. The study revealed that dual fuel combustion of biogas and biodiesel exhibited a superior performance in reduction of soot emissions due to the absence of aromatic, lower sulphur content and the lower need of air and oxygen content for biodiesel.

Walsh et al. [70] carried out a comprehensive study on utilisation of biogas in both SI and CI engines. Bari [71] experimentally studied the effect of CO_2 on the performance of a biogas-run diesel engine. The study indicated that the presence of CO_2 in biogas up to 40% did not deteriorate the engine performance and also explained the phenomena of dissociation of carbon dioxide to carbon monoxide and oxygen could improve the performance of the biogas-run DFDE. The results indicated that the maximum brake thermal efficiency was found to be 16.8% and 16.1% for diesel and *Jatropha* biodiesel, respectively, with dual fuel mode, whereas the same is found to be 20.9% when operated in diesel mode. Extensive studies on modification of diesel engines to run biogas in a dual fuel mode were carried out by Von Mitzlaff [57]. Chengqiu et al. [72] experimentally found that diesel substitution was more in the case of low-pressure biogas as compared to high-pressure biogas in DFDE. Moreover, the study indicated a stronger knocking produced by the high-pressure biogas-run engine as compared to low-pressure biogas-run engine.

6.3 Current Scenario of Biogas as Vehicle Fuel in India

Recent fuel development strategies to reduce oil dependency, mitigate greenhouse gas emissions and utilise domestic resources have generated interest in the search for alternative sources of fuel. Around 300 distilleries throughout India collectively have a potential of producing 1200 million Nm^3 of biogas along with 2000 tannery units capable of producing 787,500 Nm^3 of biogas [73]. The increasing number of poultry farms can also add to biogas productivity; as with the current population of 649 million birds, another 2173 million Nm^3 of biogas can be generated. Biogas has enormous potential as transport fuel, but due to its sophisticated long purifying process and cost, it is not so popular in India yet. There is no Indian government policy till now established for biogas as transportation fuel. Currently, more than 50 manufacturers worldwide offer a range of 250 models of commuter, light and heavy duty vehicles. Biogas vehicles have substantial advantages over vehicles run on petrol or diesel engines [63]. Carbon dioxide emission is reduced by more than 95%. Depending on how the electricity for upgrading and compressing of the gas is produced, the reduction might be as high as 99%. Many researchers in the country have been working on biogas upgradation to biomethane and try to utilise it in duel fuel engine. The *Government of India* reported that the nation could potentially produce 48,382 million m^3 a year; hence, there is huge potential for the use of biogas in the country.

7 Introduction and Basic Principle of Fuel Cell

Sir William Grove, in the UK, has invented a gaseous voltaic battery in 1839. He has found out that this voltaic battery can directly convert the chemical energy of hydrogen and oxygen into DC electricity at a platinum black anode and cathode immersed in sulphuric acid. On the basis of this voltaic battery, Mond and Langer had developed a cell in the year 1889 [74] that eventually becomes recognisable as the core of today's fuel cell. Mond and Langer cell contains two electrodes made from invariant high surface area materials that have been made separated by porous retaining medium in a flat structure like a capacitor. After a century, it is observed that fuel cell becomes a clean and efficient source of energy application. This is because of the potential of fuel cell to reduce the environmental impact and geopolitical consequences of the use of the fossil fuels. Hence, fuel cells have emerged as tantalising alternatives to the conventional use of the conventional combustion engines. Like a combustion engine, fuel cell also uses some sort of chemical fuel as its energy source; but the chemical energy directly gets converted to electrical energy without any inefficient combustion step. Other considerable advantages of fuel cell are its high efficiency, low emission, modular and distributive nature and zero noise pollution. Based on these properties, it is believed that fuel cell can play an essential role in future hydrogen fuel economy [75].

The basic design of fuel cell involves two electrodes on either side of the electrolyte. Hydrogen-rich fuel is supplied to the anode of the fuel cell, while oxygen is supplied to the cathode. The hydrogen gets split into electron and proton through chemical reaction. These electrons take the path other than the electrolyte and can produce electricity for a given load through the outer circuit. The protons pass through electrolyte and reunite at the cathode. The proton, electron and oxygen combine to form water as a by-product from fuel cell operation. The working principle of fuel cell has been shown in Fig. 5.14.

The key components of a fuel cell are an electrolyte for the conduction of ion, a cathode and an anode, as shown schematically in Fig. 5.14. Together these components are known as the membrane electrode assembly (MEA) or unit cell fuel cell. Hydrogen is brought into the anode compartment and an oxidant, typically oxygen, into the cathode compartment. The electrolyte serves as a barrier to gas diffusion and let the ions to migrate across it. The half-cell reactions occur at the anode and cathode, producing ions, which can traverse the electrolyte. For example, if the electrolyte conducts oxide ions, oxygen will be electro-reduced at the cathode to produce O-ions and consume electrons, whereas oxide ions, after migrating across the electrolyte, will react at the cathode with hydrogen and release electrons:

$$
\begin{aligned}
&\text{Cathode}: \frac{1}{2}O_2 + 2e^- \rightarrow O^{2-}\\
&\text{Anode}: H_2 + O^{2-} \rightarrow H_2O + 2e^-\\
&\text{Overall}: \frac{1}{2}O_2 + H_2 \rightarrow H_2O
\end{aligned}
\tag{5.5}
$$

The flow of ionic charge through the electrolyte balanced by the flow of electronic charge through an outside circuit, and it produces electric power.

8 Fuel Cell Plant: Its Description

As shown in Fig. 5.14, the fuel cell combines hydrogen from the supplied fuel and oxygen from the air to produce dc power, water and heat. A system should be built around the fuel cells so as to supply air and clean fuel, convert the power to usable form such as grid quality ac power and remove the depleted reactants and heat that are produced by the reactions in the cells. Figure 5.15 [76] shows a schematic diagram of fuel cell power plant. First a conventional fuel, e.g. natural gas, other gaseous hydrocarbons, methanol, naphtha or coal, is cleaned and converted to gas containing hydrogen. Energy conversion occurs when dc electricity is generated by means of individual fuel cells combined in stacks or bundles. For a particular type of application, a varying number of cells or stacks can be matched and get assembled. Finally, power conditioning converts the electric power from dc into regulated dc or ac for consumer use. Some important parameters are discussed below:

8.1 Pure Hydrogen

Fuel cell systems are fuelled with pure hydrogen gas as discussed above. Hydrogen gas is stored as compressed gas on board for particular application. Due to the low

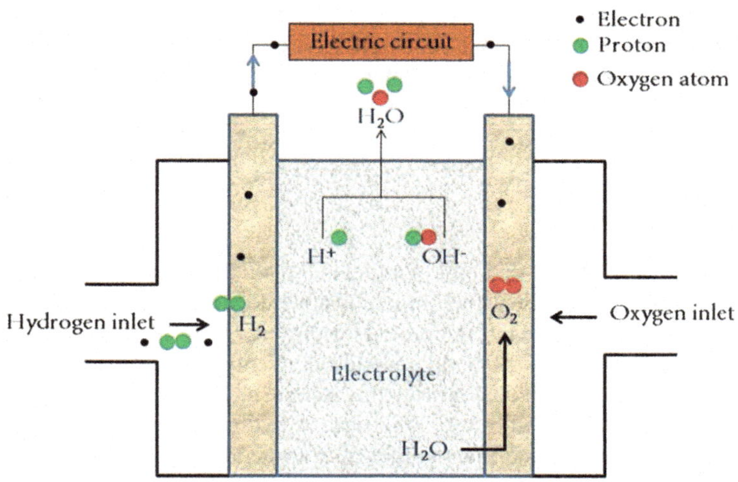

Fig. 5.14 Basic fuel cell operation

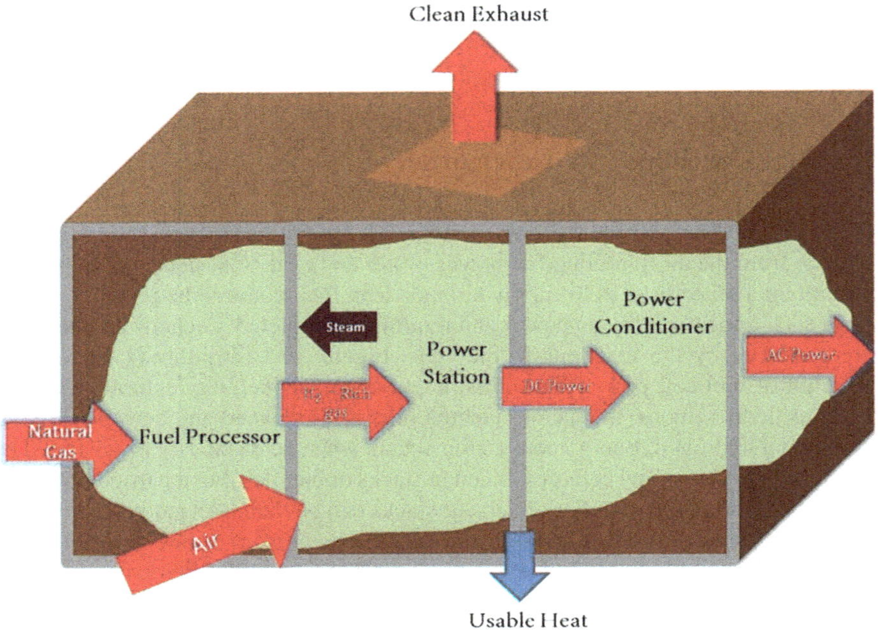

Fig. 5.15 Fuel cell power plant major processes [76]

energy density of hydrogen gas, it is found difficult to store enough hydrogen to generate power as that of conventional fuel gasoline. High-pressure tank and other technologies are being developed to store large amount of hydrogen for fuel cell application. Some other hydrogen-rich fuels that are most commonly used are methanol, natural gas, gasoline or gasified coal. High-temperature fuel cell systems can reform fuels within the fuel cell itself by a process called internal reforming, removing the need for on-board reformers and their associated costs.

8.2 Fuel Processor

As shown in Fig. 5.15, the first part of a fuel cell plant is the fuel processor. The fuel processor converts supplied fuel into a usable form by the fuel cell. If pure hydrogen is fed into the plant, the processor only needs to filter the impurities out of the hydrogen gas only. A hydrogen-rich conventional fuel such as methanol, gasoline, diesel or gasified coal powers the system, and a reformer is typically used to convert hydrocarbons into a gas mixture of hydrogen and carbon compounds called 'reformate'. The reformats is then sent to another reactor to remove impurities, such as carbon oxides or sulphur, before it is sent to the fuel cell stack. This prevents impurities in the gas from binding with the fuel cell catalysts. This binding process is also called 'poisoning' as it can reduce the efficiency and lifetime of the fuel cell. However, some fuel cells, such as molten carbonate (MCFCs) and solid oxide fuel cells (SOFCs), that operate at high temperatures can reformate the fuel itself. This kind of reforming is known as internal reforming of fuel cell and needs traps to remove impurities from the unreformed fuel before it reaches the fuel cell.

8.3 Current Inverters and Conditioners

The purpose of current inverters and conditioners is to adapt the electric current from the fuel cell to suit the electrical needs of the application, whether it is a simple electrical motor or a complex utility power grid. Fuel cell produces electricity in the form of direct current (DC), and in a direct circuit the electricity flows only in one direction. So for the use of fuel cell in AC power equipment, the direct current will have to convert into alternating current. Both AC and DC power must be conditioned. The power conditioning includes controlling current flow (amperes), voltage, frequency and other characteristics of the electric current to meet the needs of the application. These conversion and conditioning reduce system efficiency slightly around 2–6%.

8.4 Heat Recovery System

Significant amounts of heat are generated by fuel cell systems that operate at high temperatures such as molten carbonate (MCFCs), solid oxide fuel cells (SOFCs), etc. These excess amounts of heat energy can be used to produce steam or hot water or converted to electricity via a gas turbine or other technologies. Though fuel cell systems are not primarily used to generate heat, this kind of conversion of energy can increase the overall energy efficiency of the high-temperature fuel cell systems.

9 Types of Fuel Cell

Fuel cells are generally categorised by their electrolyte, the material sandwiched between the two electrodes. This material characterised the kind of chemical reactions that takes place in the cell, the kind of catalysts required for the optimal operating temperature in which the cell operates, the fuel required and other factors. These characteristics, in turn, affect the applications for which these cells are most suitable. There are several types of fuel cells currently under development, each with its own advantages, limitations and potential applications. The comparisons between different types of fuel cell investigated so far are given in Table 5.10.

Table 5.10 Comparison between different types of fuel cell

Fuel cell name	Electrolyte	Qualified power (W)	Working (°C) temperature	Electrical efficiency	Status
Metal hydride fuel cell	Aqueous alkaline solution (e.g. KOH)	–	Above 20–50	–	Commercial/ research
Electrogalvanic fuel cell	Aqueous alkaline solution (e.g. KOH)	–	Under 40	–	Commercial/ research
Direct formic acid fuel cell	Polymer membrane (ionomer)	To 50 W	Under 40	–	Commercial/ research
Zinc air battery	Aqueous alkaline solution (e.g. KOH)	–	Under 40	–	Mass production
Microbial fuel cell	Polymer membrane or humic acid	–	Under 40	–	Research
Reversible fuel cell	Polymer membrane (ionomer)	–	Under 40	–	Commercial/ research
Direct borohydride fuel cell	Aqueous alkaline solution (e.g. NaOH)	–	Under 50	–	Commercial
Alkaline fuel cell	Aqueous alkaline solution (e.g. KOH)	10 kW–100 kW	70	Cell 60–70% System 62%	Commercial/ research
Direct methanol fuel cell	Polymer membrane (ionomer)	100 kW–1 mW	Under 80	Cell 20–30% System 10–20%	Commercial/ research
Reformed methanol fuel cell	Polymer membrane (ionomer)	5 W–100 kW	90–120	Cell 50–60% System 25–40%	Commercial/ research
Direct ethanol fuel cell	Polymer membrane (ionomer)	Up to 140 mW/cm^2	Reformer 250–300 PBI 125–200	–	Research
Formic acid fuel cell	Polymer membrane (ionomer)	–	Above 25 90–120	–	Research
Proton exchange membrane fuel cell	Polymer membrane (ionomer), e.g. Nafion or polybenzimidazole fibre	100 w–500 kW	Nafion 70–100	Cell 50–70% System 30–50%	Commercial/ research
RFC redox	Liquid electrolytes with redox shuttle and polymer membrane (ionomer)	1 kW–10 mW	PBI 125–220	–	Commercial/ research
Phosphoric acid fuel cell	Molten phosphoric acid (H_3PO_4)	Up to 10 mW	–	Cell 55% System 40% Co-gen 90%	Commercial/ research

(continued)

Table 5.10 (continued)

Fuel cell name	Electrolyte	Qualified power (W)	Working (°C) temperature	Electrical efficiency	Status
Molten carbonate fuel cell	Molten alkaline carbonate (e.g. sodium bicarbonate $NaHCO_3$)	100 mW	150–200	Cell 55% System 47%	Commercial/ research
Protonic ceramic fuel cell	H^+-conducting ceramic oxide	–	700	–	Research
Solid oxide fuel cell	O^{2-}-conducting ceramic (e.g. zirconium dioxide, ZrO_2)	Up to 100 mW	700–1000	Cell 60–65% System 55–60%	Commercial/ research

10 Application of Fuel Cell

Fuel cell applications are categorised into three broad areas:

- Portable power
- Stationary power and
- Transportation

10.1 Portable Power Generation

Portable fuel cell application includes generation of power for products that are designed to be moved. Portable applications may include military applications (portable soldier power, skid-mounted fuel cell generators, etc.), auxiliary power units (APU) (e.g. for the leisure and trucking industries), portable products (torches, vine trimmers, etc.), small personal electronics (mp3 players, cameras, etc.), large personal electronics (laptops, printers, radios, etc.) and education kits and toys. To power these range of products, portable fuel cells are being developed in a wide range of sizes ranging from less than 5 W up to 500 kW (www.fuelcelltoday.com). The main drivers for fuel cells in portable applications are off-grid operation, longer run times compared with batteries, rapid recharging, significant weight reduction potential, convenience, reliability and lower operating costs.

10.2 Stationary Power Generation

Fuel cell can also be used in stationary units that can provide electricity and heat. These include combined heat and power (CHP), uninterruptible power systems

(UPS) and primary power units. CHP units are of sizes between 0.5 and 10 kW [76] and generally use either polymer electrolyte membrane fuel cell or solid oxide fuel cell technology.

10.3 Power for Transportation

Fuel cell can provide power to a vehicle, directly or indirectly. Because of their potential impact on the environment due to its control of emission of the greenhouse gases (GHG), applications of fuel cell focus on the transportation sector. A variety of fuel cell vehicles (FCV) have been developed and demonstrated till now, for example, GM HydroGen 1, Ford Demo IIa (Focus), DaimlerChrysler NeCar4a, Honda FCX-V3, Toyota FCHV, Nissan Xterra FCV, VW Bora HyMotion, Hyundai Santa Fe FCV, etc., as shown in Fig. 5.16. Auto makers such as Toyota, Honda, Hyundai, Daimler and General Motors (GM) have announced plans of commercialising their fuel cell vehicles by 2015 [77]. Distributed fuel cell system primarily focused on small scale like 50–250 kW for decentralised use or <10 kW for households [78]. The high cost of fuel cells remains a major issue that prohibits their widespread applications in various areas.

Fig. 5.16 Fuel cell vehicles by different automakers [79]

11 Polymer Electrolyte Membrane Fuel Cell

The polymer electrolyte membrane fuel cell (PEMFC), also known as proton exchange membrane fuel cell, is a very promising power source for residential and automotive applications due to its attractive features such as high power density, relatively low operating temperature, convenient fuel supply, longer lifetime and modular shape. PEM fuel cells use a solid polymer (Nafion membrane) as an electrolyte and porous carbon paper containing catalysts as an electrode. Pure hydrogen from fuel tank or on-board reformer is fuelled from the anode side, while oxygen from air is injected from the cathode side with the help of bipolar plate in both sides. At the anode, the hydrogen molecule is split into hydrogen ions (protons) and electrons. The hydrogen ions permeate across the electrolyte to the cathode, while the electrons are forced out of the anode and produce electric current that flows to the anode through the external load and produce electric power. Oxygen, usually in the form of air, is supplied to the cathode and combines with the electrons and the hydrogen ions to produce water.

The PEMFC is one of the most widely researched fuel cell technologies because it offers several advantages for transport and a number of other applications (Fig. 5.17). The possibility of the use of PEMFC in automobile application is

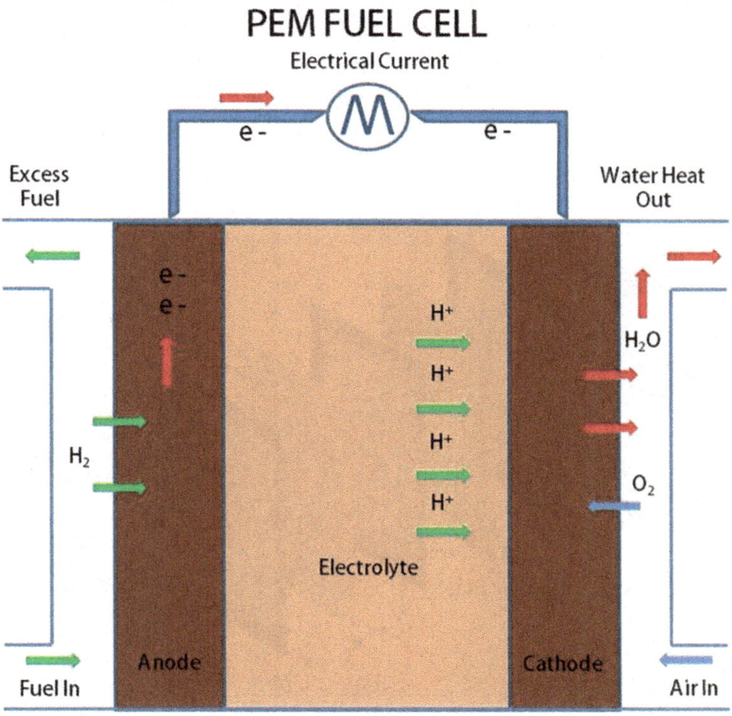

Fig. 5.17 Working principle of PEMFC

based on the low-temperature operation, high power density, fast start-up, system robustness and low emission. Hence, majority of motor manufacturers are actively pursuing PEMFC research and development. Other advantages result from the electrolyte being a solid material, compared to a liquid. The sealing of the anode and cathode gases is simpler with a solid electrolyte and, therefore, less expensive to manufacture. The solid electrolyte is also more immune to difficulties with orientation and has fewer problems with corrosion, compared to many of the other electrolytes, thus leading to a longer cell and stack life. These traits and the ability to rapidly change power output are some of the characteristics that make the PEMFC the top candidate for automotive power and space applications.

For the commercialisation of fuel cell to meet the extensive possibility of its use in various sectors, there is a need to resolve some issues [80]. One of the major issues of concern is the cost of fuel cell that should be competitive with today's internal combustion engine. The present cost of fuel cell is estimated to be about $200/kW [81], while the ultimate goal for replacing the internal combustion engine is $25–50/kW [82]. The US Department of Energy (USDOE) targeted the cost of fuel cell to be $30/kW by 2010. Researchers working in Japan have shown the possibility of reduction of fuel cell up to the acceptable level, which is as low as that of the present internal combustion engine, and the key factor is the mass production process of the bipolar plate [83].

The PEM fuel cell stack hardware consists of the membrane electrode assembly (MEA), the bipolar plate, seal, end plate, etc., as shown in Fig. 5.18 [84]. The bipolar plate arrangement is for current collection from each of the anode and cathode plates, respectively. When the reactant flow channels are formed on the

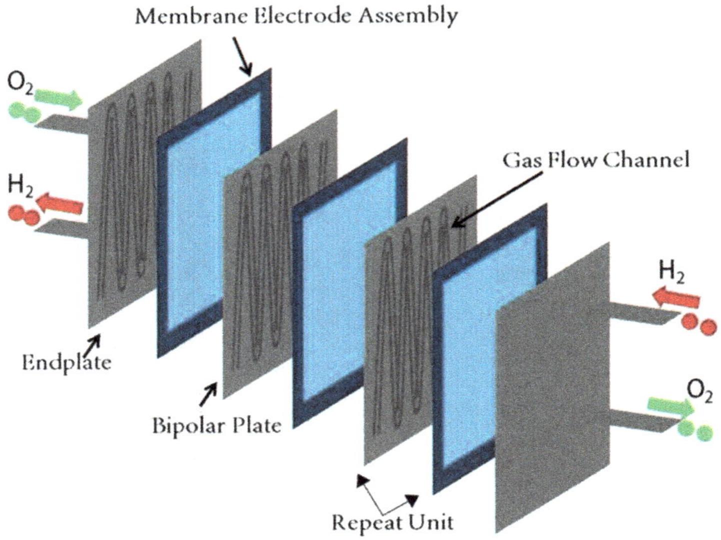

Fig. 5.18 Stack component of PEMFC

anode and cathode plates, the plates are normally referred to as fluid flow field plates. When the flow channels are formed on both sides of the same plate, one side serves as the anode plate and the other side as the cathode plate to the adjacent cell, and the plate is called bipolar (separator) plate, as shown in Fig. 5.18.

The MEA consists of two dispersed catalyst layers and two gas diffusion layers (GDL). The membrane separates the half reactions allowing protons to pass through, to complete the overall reaction. The electron created on the anode side is forced to flow through an external circuit thereby creating current. The GDL allows direct and uniform access of the fuel and oxidant to the catalyst layer, which stimulates each half reaction [85]. Among all the components of PEMFC, the bipolar plate is considered to be one of the most costly and challenging of the fuel cell stacks. To meet the cost constraint of fuel cell, the search for suitable, low-cost bipolar plate materials becomes a key element of stack development of PEMFCs.

12 Bipolar Plate

The main purpose of bipolar plate to fulfil in PEMFC stack is to supply fuel (hydrogen) and oxygen to the cell and also to manage produced heat and water. It is also used as a backing medium for stacking of individual fuel cells.

12.1 Functions of Bipolar Plate

Bipolar plate is an important component of the low-temperature fuel cell stack, such as PEMFCs. The biggest challenge for the development of PEMFC for automotive application is the reduction of cost and weight of the bipolar plate. The major cost of fuel cell comes from catalyst (platinum) and the bipolar plates, which rank second in the cost (up to 50%) depending on the material used and leads up to 80% of the total weight. Thus, widespread applications of PEMFCs heavily depend on both cost and weight reduction of bipolar plate. The bipolar plate is a multifunctional component and must have performed a number of functions simultaneously in order to achieve good stack performance within PEMFC stack. The functions that BPs need to perform are:

1. Distribute the fuel and oxidant within the cell.
2. Facilitate water management within the cell.
3. Separate individual cells in the stack.
4. Carry current away from the cell.
5. Facilitate heat management.

Related to each of these functions, the materials that BPs are made must have different physical and chemical properties. Some important physical properties of BP materials are:

- Coefficient of thermal expansion
- Density
- Hydrophobic

For uniform distribution of reactant gases, the bipolar plates should have less porosity with tight tolerance in channel dimensions. So as to maintain the current conduction, the bipolar plates should have higher electronic conductivity and lower contact resistance. Heat removal requires thermally conducting plates with preferably integrated cooling channels. Moreover it should be corrosion resisting to the PEM fuel cell environment. As bipolar plates operate in constant contact with the acidic water (pH $\approx$ 5) that is generated under the operating condition of the stack, high chemical stability and corrosion resistance are required. Oxide formed during corrosion can not only migrate and poison the catalyst but also increase the electrical resistivity of the plates and therefore result in reduced fuel cell performance. Due to these properties, the overall efficiency of the fuel cell depends on the performance of the bipolar plates in the fuel cell stack. In addition to that, low weight and low volume are essential for transport-based application of fuel cells [86]. Some other considerable criteria of the bipolar plate material are the ability to resist a temperature of 80 °C or more, high humidity and an electrical potential [87].

The PEMFC bipolar plate technical design criteria or major constraints are given below [88]:

1. Low cost (<$2/plate)
2. Easy for gas flow
3. High electrical conductivity (>100 S/cm)
4. Low permeability to gases
5. High manufacturability
6. Reasonable strength
7. Low weight
8. Low volume
9. High chemical stability and corrosion resistance (<16 μA/cm)
10. Low thermal resistance

The Department of Energy (DOE) target for bipolar plates in PEM fuel cell application is presented in Table 5.11. The possible bipolar plate materials should meet the requirements of all these design constraints, and in the long run, bipolar plate materials should be inexpensive and readily available for the purpose of mass production [89].

Table 5.11 The US Department of eEnergy (DOE) target for bipolar plates in PEM fuel cell application

Characteristics	Units	Status 2004	Status 2010	2015
Cost	$/kW	10	5	3
Weight	kg/kW	0.36	<0.4	<0.4
H_2 permeation flux	$Cm^3\ sec^{-1}\ cm^{-2}$ @80 °C, 3 atm (equivalent to < 0.1 mA/cm^2	$<2 \times 10^{-6}$	$<2 \times 10^{-6}$	$<2 \times 10^{-6}$
Corrosion	$\mu A/cm^2$	<1	<1	<1
Electrical conductivity	S/cm	>600	>100	>100
Resistivity	Ohm cm^2	<0.02	0.01	0.01
Flexural strength	MPa	>34	>25	>25
Flexibility	% Deflection at mid-span	1.5–3.5	3–5	3–5

12.2 *Bipolar Plate Material and Their Properties*

The materials investigated so far based on bipolar plate application can be broadly classified as:

1. Non-metal: non-porous graphite/electro graphite
2. Metals: non-coated and coated
3. Composites: polymer-carbon and polymer-metal

The detailed classification of BP materials is shown in Table 5.12 [89].

A carbon-polymer composite bipolar plate is a promising alternative to graphite and has the advantages of low cost, good corrosion resistance and low weight. Production of the carbon- polymer composites involves the hot moulding of carbon or graphite filler in a thermosetting (epoxy resin, phenolic resin, furan resin, vinyl esters) or a thermoplastic (polyvinylidene fluoride, polypropylene, polyethylene) matrix. Typical carbon contents range between 50 and 80% by weight. The carbon content used in preparing the composite bipolar plate ultimately determines the electrical conductivity and mechanical properties of the composites, whereas binder content determines the ultimate strength of the composite. Majority of PEM fuel cell stack producers utilise bipolar plates based on graphite. Nowadays more attention has been paid to composite or metals. Still further investigations of bipolar plates are needed to speed up the process of commercialisation. The advantages and disadvantages of different types of materials are discussed as in Table 5.13.

Table 5.12 Classification of materials for BPs used in PEM fuel cells (Hermann [89])

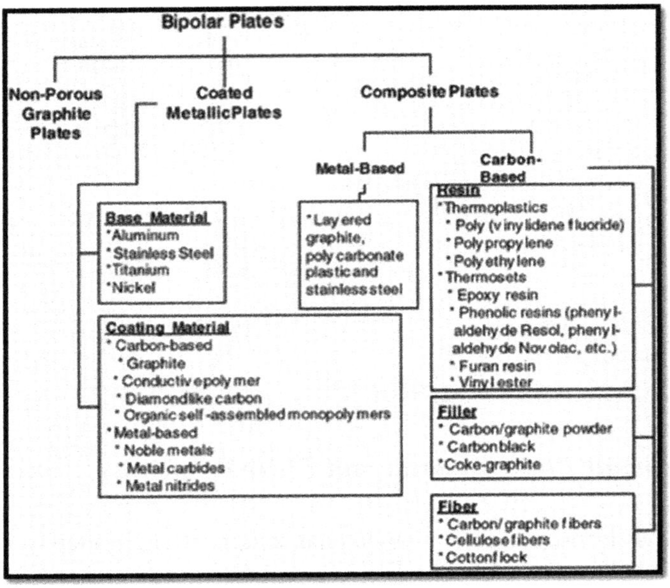

Table 5.13 Bipolar plate material and their advantages and disadvantages

Material	Advantages	Disadvantages
Graphite	Excellent corrosion resistance Low bulk resistivity Low contact resistance	Poor mechanical properties Porosity High weight and volume High processing cost
Carbon–carbon composites	Low density High corrosion resistance Low contact resistance	Low mechanical strength Low bulk electrical conductivity High price
Carbon–polymer composite	Low cost Good corrosion resistance Low weight No machining process Commercial availability of the row material	Low mechanical strength Low electrical conductivity
Metal	Good electrical conductivity High thermal conductivity Low cost Excellent mechanical properties Ease of fabrication Small volume	Severe corrosion (membrane poisoning and formation of insulating surface oxide)

13 Current Status: National and International

13.1 International

Although PEMFC has been demonstrated in automobiles and portable power units up to 250 kW all over the world since the 1990s, there are still many commercialisation issues such as high manufacturing cost and inadequate durability characteristics that need to be resolved. The PEMFCs are expected to be fully commercialised in the next coming years. Ballard Power Systems, Inc. is generally considered to be the world leader in PEMFCs. Ballard products are currently being tested by many major automobile and oil and gas companies. XCELLSIS Fuel Cell Engines is a combined effort between Ford, DaimlerChrysler and Ballard for developing automobile fuel cell unit based on the Ballard systems. Ballard also supplies fuel cells to Nissan, while Toyota, GM and Honda have their own fuel cell system engines. Interestingly, the major automakers like Toyota, Honda and Nissan have announced ambition plans to market their brand image in the fuel cell systems for transportation applications. Oil and gas companies in Japan are collaborating with fuel cell developers, as a part of government initiative to develop and demonstrate fuel cells (1 kW systems) for household stationary applications. These represent the determination and confidence level that have been achieved in the international R&D sector in bringing the fuel cell system on the industrial front. A major hurdle in the route of commercialisation of fuel cells still is the significantly high system and production cost.

Irrespective of the issues such as cost and durability, a significant outsurge in the confidence level of industry has been achieved in recent years to market PEM fuel cell systems' application of few watts to few kilowatts. The government-industry-academic tie-ups have been working effectively in the USA, Japan and the European Union more than ever before to facilitate immediate launch of useful and economically feasible systems for daily life applications. The US Department of Energy is trying to work with other laboratories, universities and industry partners to overcome critical technical barriers of fuel cell commercialisation. In Japan, METI is working closely with oil and gas companies, fuel cell developers, national laboratories and universities to facilitate fuel cell mass penetration in domestic household sector. In Korea, the Ministry of Knowledge Economy established a national R&D organisation for hydrogen and fuel cells, which will expedite the commercialisation of fuel cells in coordination with nearly 300 industries for stationary and transportation applications. International agencies such as the US Department of Energy (DOE), the New Energy and Industrial Technology Development Organization (NEDO) and the Japan and European Hydrogen and Fuel Cell Technology Platform (HEP) have established targets for commercialising fuel cells for transportation and stationary applications. Table 5.14 summarises the specific target levels set by these agencies [90].

Development of economical bipolar plates is another striving area. Bipolar plates represent one third of the overall cost of the fuel cell system. For a working

Table 5.14 The target levels, set by the dominant international agencies, for various fuel cell applications

International agency	Country	Performance requirements	
		Transportation	Stationary
DOE	USA	Lifetime ≥5000 h (~150,000 miles) Condition External temp. −40 °C to +40 °C Cost 50 $/kW	Lifetime >40,000 h Condition −35 °C to +40 °C Cost 750 $/kW
NEDO	Japan	Lifetime >5000 h Vehicle efficiency 60% (based on LHV) Condition Operating temp. >90–100 °C External temp. −40 °C and higher Cost ~10,000 Yen/kW	Lifetime >10 years (~90,000 h)
HEP	EU	Lifetime 5000 h (cars) 10,000 h (buses) Vehicle efficiency >40% Cost 100 Euros/kW	Lifetime >12,000 h Electrical efficiency 34–40% Cost 6000 Euros for system

fuel cell at the cost of $200/kW, the bipolar plate itself accounts $60–70/kW [91]. Machining of channels on the graphite plate is costly and time consuming. Due to this reason, a variety of metallic plates have been examined as potential systems to replace graphite bipolar plates. However, being highly susceptible towards corrosion, the metal plates have to be coated with efficient protective layer. Coated bipolar plates, however, tend to crack due to unequal coefficient of expansion of the metal and coating. A variety of polymer composite materials also have been investigated as replacement for the graphite bipolar plates. Metal-carbon-polymer composites have been recently identified as better materials both in terms of material and manufacturing cost. Thermoplastics such as polyethylene, polypropylene and polyvinylidene fluoride and thermoset resins such as phenolics, epoxies and vinyl esters are potential polymer candidates. Gas diffusion layers and accessories like sealant materials also required for fabricating the stack are other parts where innovative actions are advisable for bringing down the cost of the system for commercial mass penetration. In order to achieve consumer acceptance and mass penetrability into volume market, the fuel cell current cost (US $2500–3000) has to be brought down to a significant extent (nearly 10 times) by proper component design, process optimisation and system integration.

13.2 National

In India kW-sized PEMFC and PAFC have been developed and demonstrated for power generation and transportation sector. Moreover, a fuel cell battery hybrid vehicle with PEMFC stacks of 10kW undergone field performance evalution.

The main activities related to research and developments of fuel cell are:

- Materials, process and fabrication techniques
- Technology and infrastructure
- Demonstration of technology application
- Performance evaluation
- Training and awareness

Three main thrust areas for India to work on fuel cell are (1) production of cost-effective fuel cell; (2) research on materials, technology development and upgradation of the performance of fuel cell; and (3) expansion of infrastructure for the support of production and application of fuel cell.

Tata Motors, India's largest automaker, and the Indian Space Research Organisation (ISRO) have been joined in a fuel cell bus development MoU since 2006–2013, and it was reported that the first fuel cell bus from this effort was undergoing testing at ISRO facility.

There are many public and private organisations which have proven R&D skills for fuel cell system developments. National institutes like NCL, CECRI and NPL are associating closely in developing fuel cell systems for stationary applications. NMRL develops PAFCs and PEMFCs. The Green Initiative for Future Transport and the Green Initiative for Power Generation are two important steps taken by the government to effectively address the future energy and environmental requirements. These initiatives will be driven through public-private partnership, industry-driven planning process and network and cooperation between government, research organisations and academia. The Centre for Fuel Cell Technology in Chennai is established by the government with a specific objective of demonstrating PEMFCs for commercial applications. Companies such as Tata Motors, Mahindra & Mahindra and Ashok Leyland are associating with the government in its programmes for development of fuel cell systems. A list of prominent organisations which are involved in fuel cell development is given in Table 5.15.

Table 5.15 List of the national organisations/companies interested in the fuel cell technology

Organisation	Purpose/interest
NMRL	Working on PAFC and PEMFC
BHEL	Working on PAFC and MCFC
IIT-Delhi	Basic research; DMFC and electrode development
IIT-Chennai	Development of DMFC and PEMFC
IIT-Mumbai	DMFC and PEMFC
Spic Science Foundation	Fuel cell and components
CFCT	PEMFC, stack development
IISc	Polymer electrolytes

14 Conclusions

In this chapter various parameters influencing the performance of biogas digesters and functionalities of fuel cells have been reviewed and discussed. Potential of both the fuels to provide motive power to the vehicle has also been explored. From the literature it has been observed that biogas and fuel cell are getting more and more popularity across India and worldwide for various applications. In order to use raw biogas as vehicular fuel, the raw biogas has to be upgraded to biomethane, which is free from CO_2 and H_2S. There is also additional benefit in the use of enriched biogas as it reduces the emission level significantly. However, a common legislation still does not exist about the production, the quality and the distribution of the biogas. That is why biogas as vehicular fuel will be largely decided by future policies of the country and depends much on the sensitivity and concerns of the government. Fuel cell vehicles have been successfully demonstrated over the world, but it is yet to be deployed owing to its high manufacturing cost and poor durability. However, with dynamic manufacturing companies (viz. Ballard Power Systems, Inc., DaimlerChrysler, etc.), fuel cell in vehicles is expected to be commercialised in the next decades. It is worth mentioning that with combined efforts between the government, industry and academia, the hurdles and technical barriers pertaining to fuel cell commercialisation in vehicles can be overcome.

References

1. IPCC (2013) Summary for policymakers. In: IPCC special report on renewable energy sources and climate change mitigation
2. IEA (2010) Energy technologies perspectives: scenarios and strategies to 2050. International Energy Agency, Paris, p. 710
3. Myles RM (1985) Practical guide to Janata biogas plant technology. AFPRO Action Food Production, New Delhi
4. Steffen R (2000) Feed stock for anaerobic digestion. Making energy and solving modern waste problem. www.adnett.org
5. Yadvika (2004) Enhancement of biogas production from solid substrates using different technique. Bioresour Technol 95:1–10
6. Gerardi M (2003) The microbiology of anaerobic digesters. Waste water microbiology series. Wiley, New York
7. Kapdi SS (2004) Biogas scrubbing, compression and storage: perspective and prospectus in Indian context. Renew Energy 4:1–8
8. Fehrenbach H (2008) Criteria for a sustainable use of bioenergy on a global scale. UBA-Forschungsbericht 206:41–112
9. Kratzeisen M (2010) Applicability of biogas digestate as solid fuel. Fuel 89:2544–2548
10. Public consultation on the implementation of the renewed strategy to reduce CO_2 missions from passenger cars and light-commercial vehicles. European Natural Gas Vehicle Response 15 July 2007
11. Ostrem KM (2004) Combining anaerobic digestion and waste-to energy. 12th North America waste to energy Conference

12. Jarvis A (2004) Biogas: renewable energy from organic waste. Swedish Biogas Association, Stockholm. www.sbgf.org
13. Mata-alvarez J, Mace S, Liabres O (2000) Anaerobic digestion of organic solid wastes an overview of research achievements and perspectives. Bioresour Technol 74:3–16
14. Ramchandra P, Sameer M, Shishupal S, Phool SB et al (2006) SB field evaluation of biogas technology in India. Energy Technology Division, Tata Energy Research Institute, New Delhi
15. NAS (1977) Methane generation from human, animal and agricultural wastes, Report of an Ad. Hoc. panel of the advisory committee on technology innovation. National Academy of Science, Washington, DC
16. Bahadur S, Singh KK (1978) Janata biogas plants. Planning research and action division. State Planning Institute, Lucknow
17. Mahanta P, Saha UK, Dewan A, Kalita P, Buragohain B et al (2005) Biogas digester: a discussion on factors affecting biogas production and field investigation of a novel duplex digester. SESI J 15:1–12
18. Barnett A (1978) Biogas technology in the third world: a multidisciplinary review. IDRC, Ottawa, p. 51
19. Fry LJ, Merrill R (1973) Methane digesters for fuel gas and fertilizer, Newsletter No. 3. New Alchemy Institute, Santa Cruz
20. Gotass HB (1965) Composting. World Health Organisation, Geneva
21. Kexin L (1979) Fermentative technology of Chinese rural digester. Bio-Energy Laboratory, Chengdu Institute of Biology, Acadeemia Sinica
22. Homan E (1979) Biogas from manure, special circular 260, college of agriculture extension service. Pennsylvania State University, State College, PA
23. SPOBD (1979) Biogas technology and utilization, Chengdu seminar, Sichuan provincial office of biogas development, Sichuan
24. Neelakanthan S (1976) Effect of temperature, inoculum and agitation on biogas production from cattle dung. Indian J Dairy Sci 29:226–229
25. Kopiske G, Eggersgliib H, Strub A, Charter P, Schleser G et al (1982) Energy from biomass. In: Proceeding EC conference. Applied Science Publishers, London, pp 591–594
26. Bansal NK (1985) A study of greenhouse concept on conventional biogas system for enhancing biogas yield in winter months. Energy Res 9:119–128
27. Bansal NK (1988) A techno-economic assessment of solar-assisted biogas systems. Energy Res 10:216
28. Mahanta P, Dewan A, Saha UK, Kalita P (2004) Effect of temperature and agitation on the performance of biogas digesters. 2nd BSME- ASME International Conference on Thermal Engineering 2–4 Jan 2004, Dhaka
29. Augenstein DC, Wise DL, Wentworth RL, Cooney CL et al (1976) Fuel gas recovery from controlled landfill of municipal wastes. Resour Recover Conserv 2:103
30. Gore JA et al (1981) More gas from gobar gas plants, Indian farming, March
31. Mahanta P, Dewan A, Saha UK, Kalita P et al (2004) The influence of temperature and total solids concentration on the gas production rate of biogas digester. J Energy Southern Africa 15:112–117
32. Suhirman, RM (2000) Biogas production from rice straw and clitoria tiernatea. In: Wise DL (ed) Global bioconversions, vol IV. CRC Press, Boca Raton, FL, pp 177–188
33. TERI (1987) Fixed dome biogas plants: a design, construction and operation manual. TERI, New Delhi
34. Moharao GJ (1975) Aspects of night soil digestion, sewage farming and fish culture, A working paper, All India Institute of Hygiene and Public Health
35. Moharao GJ (1974) Scientific aspects of cow dung digestion. Khadi Gramodyog 20(7): 340–347
36. Srivastava VJ, Chynoweth DP (1987) Kinetic analysis of bio-gasification of biomass, waste blend and its engineering, significant energy from biomass and wastes. Elesevier, pp 1021–1034

37. Langrage B (1979) Biomethane 2: principles—technique utilization". EDISUD, La Calade, 13100 Aix-en-Provence, France
38. UN (1980) Guidebook on biogas development: 99. United Nation, New York
39. Boodoo A, Delaitre C, Preston TR (1979) The effect of retention time on biogas production from slurry produced by cattle fed sugarcane, Mauritius
40. Ladine RC, Cocci AA, Viraraghavan T, Brown GJ (1982) Anaerobic pretreatment of potato processing wastewater—a case history. In: Bell JM (ed) Proceedings of the 36th industrial wastes conference. Ann Arbor Science, Ann Arbor, MI, p 233
41. The Biogas technology in China (1989) Chengdu biogas research institute, Chengdu
42. Coppinger ER (1979) The operation of a 50,000 gallon anaerobic digester at the monroe state dairy farm, Ecotope Group, 2332. East Madison, Settle, WA
43. Kumar S (1987) Note on stimulation of biogas production from cattle dung by addition of charcoal. Biol Wastes 20:209–215
44. Madamwar DB, Mithal BM (1986) Effect of pectin on anaerobic digestion of cattle dung. Biotechnol Bioeng 28(4):624–626
45. Geeta GS, Raghavendra S, Reddy TKR et al (1986) Increase of biogas production from Bovine Excreta by addition of various inert materials. Agric Wastes 17:153–156
46. Prasad CR (1985) Utilization of organic wastes in biogas plant. Khadi Gramodyog:514–518
47. Glaub JC, and Digz LF, (1981) Biogas purification processes. Biogas and alcohol fuels production, vol II. Biocycle Journal of Waste Recycling. JP Press, Emmous
48. Biogas Forum India (2012) BiGFIN, E-Newsletter Vol III, 3, June 2012.
49. Biogas and bio syngas upgrading, Danish technical institute, December 2012 (Report)
50. Mital KM (1997) Biogas systems: policies, progress and prospects. New Age International (P) Ltd, New York
51. HEL, why anaerobic digestion of ferm wastes? Hamworthy Engineering Ltd, Fleet Corner, Poole, Dorset BH 17 7LA, UK
52. Jonsson O, Persson M (2003) Biogas as transportation fuel. FVS Fachtagung. http://www.fvee.de/fileadmin/publikationen/Workshopbaende/ws2003-2/ws2003-2_02_04.pdf
53. Owe J (2006) Market development for biogas as vehicle fuel in Europe—status 2006.
54. Pål B, Berglund M (2006) Environmental systems analysis of biogas systems—part I: fuel-cycle emissions. Biomass Bioenerg 30:469–485
55. Von Mitzlaff K (1988) Engines for biogas, a publication of Deutsches Zentrum fur Entwicklungstecknologien, GATE, p 132, Isbn 3528020326
56. Duc PM, Wattanavichien K et al (2007) Study on biogas premixed charge diesel dual fuelled engine. Energy Convers Manage 48:2286–2308
57. Cacua K, Amell A, Cadavid F et al (2012) Effects of oxygen enriched air on the operation and performance of a diesel, biogas dual fuel engine. Biomass Bioenerg 45:159–167
58. Nathan SS, Mallikarjuna JM, Ramesh A et al (2010) An experimental study of the bio-gasediesel HCCI mode of engine operation. Energy Convers Manage 51:1347–1353
59. Demirbas A (2006) Biogas production from the organic fraction of municipal solid waste. Energy Sources Part A 28:1127–1134
60. Karim GA (1983) The dual fuel engine of the compression ignition type e prospects, problems and solutions: a review. SAE paper 831073
61. Karim GA (2003) Combustion in gas fueled compression ignition engines of the dual fuel type. ASME J Eng Gas Turbines Power 125:827–836
62. How to use biogas to power your vehicle. http://energyzedworld.com/how-to-use-biogas-to-power-your-vehicle/
63. Papacz W (2011) Biogas as vehicle fuel. J Kones Powertrain Transp 18:1
64. Svensén B (2008) Biomethane as vehicle fuel: experiences from the biogas west project: Business Region Göteborg
65. Henham A, Makkar M (1998) Combustion of simulated biogas in a dual-fuel diesel engine. Energy Convers Manage 39:2001–2009
66. Yoon SH, Lee CS (2011) Experimental investigation on the combustion and exhaust emission characteristics of biogas – biodiesel dual-fuel combustion in a CI engine. Fuel Process Technol 92:992–1000

67. Walsh JL, Ross CC, Smith MS, Harper SR et al (1989) Utilization of biogas. Biomass 20: 277–290
68. Bari S (1996) Effect of carbon dioxide on the performance of biogas/diesel dual-fuel engine. Renew Energy 9:1007–1010
69. Bedoya ID, Saxena S, Cadavid FJ, Dibble RJ et al (2012) Exploring strategies for reducing high intake temperature requirements and allowing optimal operational conditions in a biogas fuelled HCCI engine for power generation. ASME J Eng Gas Turbines Power 134:1–9
70. Sahoo, BB (2011) Clean development mechanism potential of compression ignition diesel engines using gaseous fuels in dual fuel mode. PhD thesis. Centre for Energy, IIT Guwahati, India
71. Ganeshan V (2012) Internal combustion engines, 4th edn. Tata McGraw Hill Education Private Limited, New Delhi
72. Chengqiu J, Tianwei L, Jianli Z et al (1989) A study on compressed biogas and its application to the compression ignition dual-fuel engine. Biomass 20:53–59
73. Upgraded bottled biogas a green and low-cost fuel for automobiles in India. http://mnre.gov.in/file-manager/akshay-urja/march-april-2014/EN/20-23.pdf
74. Appleby AJ (1996) Fuel cell technology: status and future prospects. Energy 21(7):521–653
75. Haile SM (2003) Fuel cell materials and components. Acta Mater 51(19):5981–6000
76. Williams MC (2000) Fuel cell handbook. US Department of Energy, Morgantown
77. Vielstich W, Lamm A, Gasteiger HA (2003) Handbook of fuel cells: fundamentals, technology, applications. Wiley, West Sussex
78. Wang Y, Chen KS, Mishler J, Cho SC, Adroher XC (2011) A review of polymer electrolyte membrane fuel cells: technology, applications, and needs on fundamental research. Appl Energy 88(4):981–1007
79. Wind J, Spдh R, Kaiser W, Bцhm G (2002) Metallic bipolar plate for PEM fuel cell. J Power Sources 105:256–260
80. Bar-On I, Kirchain R, Roth R (2002) Technical cost analysis for PEM fuel cells. J Power Sources 109:71–75
81. Scholta J, Rohland B, Trapp V, Focken U (1999) Investigation on novel low-cost graphite composite bipolar plate. J Power Source 84:231–234
82. Tsuchiya H, Kobayashi O (2004) Mass production cost of PEM fuel cell by learning curve. Int J Hydrog Energy 29:985–990
83. Li X, Sabir I (2005) Review of bipolar plates in PEM fuel cells: flow-field designs. Int J Hydrog Energy 30:359–371
84. Mehta V, Cooper SJ (2003) Review and analysis of PEM fuel cell design and manufacturing, J. Power Sources 114(1):32–53
85. Busick D, Wilson M (2000) Development of composite materials for PEFC bipolar plates. Mat Res Soc Symp Proc 575:247–251
86. Heinzel A, Mahlendorf F, Niemzig O, Kreuz C (2004) Injection moulded low cost bipolar plates for PEM fuel cells. J Power Sources 131:35–40
87. Yuan XZ, Wang H, Zhang J, Wilkinson DP (2005) Bipolar plates for PEM fuel cell—from material to processing. J Mat Electr Sys 8:257–267
88. Borup RL, Vanderborgh NE (1995) Design and testing criteria for bipolar plate materials for PEM fuel cell application. Mat Res Soc Symp Proc 393:151–155
89. Hermann A, Chaudhuria T, Spagnolb P (2005) Bipolar plates for PEM fuel cells: a review. Int J Hydrog Energy 30:1297–1302
90. Borup R et al (2007) Scientific aspects of polymer electrolyte fuel cell durability and degradation. Chem Rev 107:3904–3951
91. Cunningham B, Baird DG (2006) The development of economical bipolar plates for fuel cells. J Mater Chem 16:4385–4388

Chapter 6
Cogeneration of Heat and Electricity from Biomass in India: Current Status and Future Challenges

Rupam Kataki, Moommoon Hiloidhari, Dipal Baruah, Debashis Sut, and Neonjyoti Bordoloi

Abstract Energy generation from waste or residual biomass has gained significant importance in light of the ongoing "*biomass carbon neutrality*" debate. Seasonal agro-residue biomass by-products meet the standard of carbon neutrality because emission burden from such by-products is very minimal or almost zero from a life cycle prospective. Development of advanced and efficient technology such as cogeneration or combined heat and power (CHP) to tap the energy potential of residual biomass is crucial. Cogeneration is a process in which energy is recovered from by-products for simultaneous production of process heat and power to generate electricity. Studies around the world have demonstrated sugar mill cogeneration as an attractive low-cost power option. India is the world 2nd largest producer of sugarcane, and many sugar mills in the country are traditionally using sugarcane bagasse cogeneration technology for heat and power. Moreover, many Indian states have surplus biomass and bio-wastes especially in the form of agricultural residues and are capable to implement biomass-based cogeneration. However, a majority of them have failed to properly utilize the resources for cogeneration and, therefore, actual generation is far below than potential. The present study explores the biomass and bio-wastes potential for power generation in India and the technologies adopted with particular focus on cogeneration. We observed high biomass potential in India, but variation in the tariff of biomass-based power among the states is a major hindrance for the Indian biomass power sector. The government programmes/policies have a major role to play in enhancing the growth of biomass-based technologies in the states with high biomass resources potential. We recommend a precise, up-to-date biomass database down to the scale of village level and

R. Kataki (✉) • D. Sut • N. Bordoloi
Department of Energy, Tezpur University, Tezpur 784028, Assam, India
e-mail: rupamkataki@gmail.com

M. Hiloidhari
School of Environmental Sciences, Jawaharlal Nehru University, New Delhi 110067, India

D. Baruah
Department of Mechanical Engineering, Girijananda Choudhury Institute of Management and Technology, Tezpur 784501, Assam, India

A.K. Chandel, R.K. Sukumaran (eds.), *Sustainable Biofuels Development in India*,
DOI 10.1007/978-3-319-50219-9_6

complete knowledge of the existing and state-of-the-art technologies for the success of biomass power in India. Selection of the appropriate technology to suit a particular biomass fuel(s) for a given geographical location is also important.

Keywords Cogeneration • Biomass • Bio-wastes • Combined heat and power (CHP)

1 Introduction

India's energy sector has been facing serious challenges of inadequate reserves of domestic energy to meet the growing energy demand, mostly due to improving living standard, rapid urbanization, industrial expansions, and growing population. The basic energy needs of India's thousands of its citizens are yet to be fulfilled. India has put forth a number of ambitious goals to generate cleaner energy, curb GHG emissions, and secure energy provision for all. The target of 175 GW renewable energy by 2022, electricity for all by 2019, injecting 40% renewable energy in its total electricity mix by 2030, 20% blending of biofuels by 2017, and reduce GHG emissions by 33–35% of 2005 level by 2030 are some of such initiatives. The country needs to increase its power generation capacity to over 750 GW by 2030 from its present capacity of 301 GW (as of June 2016) to meet the basic energy needs of its citizens [1]. This is important because the country needs massive input of energy to sustain its economic growth and to reduce environmental footprint of fossil fuels. Government of India's new initiatives such as "Make in India" and "24×7 power for all" to propel development will put additional pressure on the India's energy sector. Further, the transition from nonrenewable to renewable energy is crucial to achieve about 40% cumulative electric power installed capacity from nonfossil fuel-based energy resources by 2030 which is one of the Intended Nationally Determined Contribution (INDC) of the country [2].

Contrary to the demand, domestic energy reserves of India are not adequate and, therefore, the country is fairly dependent on foreign imports of oil. For instance, India imports nearly 80% of crude oil per year to sustain its energy sector. Similarly, coal and LPG are also imported. Major portion of coal produced in India is consumed in the power sector for electricity generation. About 55% of India's electricity generation are based on coal fired power plants. Other energy options like large hydro and nuclear power projects are handicapped by public protest due to environmental and safety issues. Insufficient and poor quality of electricity supply has forced almost all the sectors—industrial, commercial, institutional—to heavily rely on costly diesel fuel as a backup system. Lack of uninterrupted electricity supply has led to large-scale use of kerosene in the rural areas. It is reported that 78 million people in India mainly depend on kerosene for lighting [2]. Thus, there is a serious gap in energy supply–demand in the country. The growing energy demand, limited conventional fuel options, and environmental obligations for cleaner energy generation have prompted India to explore renewable and sustainable energy resources. Biomass, solar, wind, and small hydro are the major areas of renewable energy development in India. The estimated renewable energy potential from biomass, wind, solar, and small hydro in the country is about 900 GW, with Wind—102 GW (at 80 m mast height), Small Hydro—20 GW, Bioenergy—25 GW, and 750 GW solar powers [3].

Among different renewable resources, biomass energy constitutes about 18% of total primary energy use in the country, and more than 70% of the country's population depends on it. However, it is currently used in an inefficient manner with high levels of indoor pollution [2]. Wood and wood wastes, agricultural crops and their residues, municipal solid waste, animal waste, waste from food processing, and aquatic plants including algae are the main sources of biomass [4]. Biomass possesses the flexibility of being used in solid, liquid, or gaseous form [5, 6]. Solid form of biomass finds application in household and industrial purposes; both solid and gaseous forms find application in electricity and heat production whereas in the liquid form biomass finds application in the transportation sector [7]. Another unique feature of biomass is its storability which helps in ensuring a predicted output at a generation site, thus eliminating the dependency on weather.

In India, due to nonscientific and non-centralized collection and utilization practices, the full potential of bioenergy source has not been explored. However, India has a great potential in terms of surplus rice residue and bagasse for bioenergy applications. Rice residue has higher number of competing uses in comparison to sugarcane which makes bagasse the most potential bioenergy source in India. The rapid rate of urbanization in India makes bio-waste (Municipal Solid Waste (MSW), Sewage Sludge, etc.) another prospective contributor to bioenergy.

Flexibility of conversion to several other forms of energy is another added advantage of biomass. This enables the use of biomass in combined heat and power (CHP) technologies or combined cooling heat and power (CCHP) [8]. These technologies have the advantage of better efficiencies [9], lower consumption [10], and lower CO_2 emissions [11] than heat and electricity production individually. CHP plants based on biomass have been reported to have low operation and maintenance costs, high conversion efficiencies, low noise and vibration, and low levels of emission [11]. Integration of heat pumps with the CHP plants help in relocating the excess heat for process heating or to a storage facility [12]. CHP applications have been reported to greatly improve the efficiency of a power plant due to the easy harnessing of the thermal energy compared to the electrical energy. However, the heat demand source should be located close enough to reduce the difficulty in transportation and distribution of thermal energy which is generally faced.

The following couple of sections of this chapter aim at reviewing the present potential of biomass and bio-wastes in the country and also the technologies available for biomass-based electricity generation with or without heat recovery. A case study on biomass-based energy generation in India including the prospective of bioenergy technology is also presented.

2 Biomass and Bio-waste Potential in India, Zone-wise, and Biomass Type-wise

Biomass is one of the prime sources of renewable energy with widespread availability and lower environmental impact than fossil fuels. Carbon neutrality, although debatable, is another aspect which makes biomass an attractive source for providing firm energy. Today, biomass power has been attracting investments over INR 600 crores per year, generating more than 5000 million units of electricity and providing yearly employment of more than 10 million man-days in the rural areas [13]. At present, in India, biomass power shares 4946 MW of grid connected and 994 MW of off-grid power. Table 6.1 lists out the grid interactive and off-grid share of biomass-based power in India.

Bagasse-based cogeneration has a major share in the total grid connected capacity, and about 115 MW of energy is generated from waste to energy power plants. In the off-grid scenario, 652 MW of energy comes from non-bagasse-based cogeneration in captive power plants, about 18 MW from biomass gasifier systems installed in rural areas. Furthermore, about 164 MW equivalent biomass gasifier systems have been deployed in industries for thermal applications. In India, different feedstocks are used for power generation and some of them are bagasse, rice husk, straw, cotton stalk, coconut shells, soya husk, de-oiled cakes, coffee waste, jute wastes, groundnut shells, saw dust, etc.

In a recent study [13] it was estimated that around 686 MT gross agro-residue is available in India. Cereals, oilseed, pulses, and sugarcane crop groups were estimated to contribute around 545 MT while 61 MT came from horticultural crops like coconut, banana, and areca nut and 80 MT by others (cotton and jute). Among the crop groups, cereal residues were estimated to contribute about 54% and sugarcane

Table 6.1. Grid interactive and off-grid share of biomass-based power in India [4]

Sector	Installed capacity (MW)
I. Grid interactive power (MW)	
Biomass power (Combustion, gasification, and bagasse cogeneration)	4831.33
Waste to power	115.08
Subtotal grid interactive	4946.41
II. Off-grid/captive power (MW_e)	
Biomass (non-bagasse) cogeneration	651.91
Biomass gasifiers	
Rural	18.15
Industrial	164.24
Waste to energy	160.16
Subtotal off-grid	994.46
Total biomass based power	5940.87

residue (bagasse) about 16%. At an individual crop level, rice residue was estimated to be the highest contributor (154 MT) followed by wheat (131 MT).

Estimation of the surplus residue revealed that India generates 234 MT (34% of the gross residue) agro-residue annually [13]. Cereals group was estimated to be the highest contributor (89 MT) to the biomass pool followed by sugarcane (56 MT). Individual crop-wise, sugarcane is the highest contributor (56 MT) followed by cotton (47 MT) and rice (43 MT). It may be noted that although rice crop produces the highest gross residue among all the crops, its surplus residue production is less than sugarcane. More competing uses (cattle feed, animal feed, packing material, heating and cooking fuel) of rice residues (husk and straw) in comparison to sugarcane (bagasse, leaves and tops) lead to the fall in the surplus residue potential of rice residues. Banana and coconut (horticultural crops) were also estimated to have significant surplus residue potential in India with banana residue capable of providing 12 MT and coconut 10 MT [13].

Surplus portion of residue available in the country has been estimated to have a bioenergy potential of around 4.15EJ per annum, which is equivalent to 17% of the primary energy consumption in India. Crop group wise, cereals will be the highest contributor (1.49EJ) followed by sugarcane (1.11EJ). At an individual level, sugarcane will be the highest contributor providing around 1.11EJ [13]. At this point, it is worth mentioning that bagasse-based cogeneration in the country's 550 sugar mills is capable of generating 7000 MW additional power by deploying technically and economically optimal levels of cogeneration [4].

At the state level, the bioenergy potential from surplus residue has been estimated to vary between 0.84 PJ in Mizoram and 743.15 PJ in Uttar Pradesh [13]. The states with the most bioenergy potential were estimated to be Maharashtra (563 PJ), Punjab (467 PJ), Gujarat (426 PJ), Tamil Nadu (317 PJ), Rajasthan (200 PJ), Haryana (292 PJ), and Madhya Pradesh (207 PJ). Assam was estimated to have the highest bioenergy potential among the North-eastern states of India [13].

Table 6.2 provides agro-residue type-wise gross and surplus residue potential in India. On the other hand, Fig. 6.1 provides state-wise surplus agro-residue potential in India.

Overall, bioenergy available from surplus agro-residue in India amounts to be 4148 PJ per annum, of which maximum contribution comes from cereals residues followed by sugar residue as shown in Table 6.3.

Biomass from waste is another potential area for bioenergy generation in India. However, there is no consolidated information on availability of biomass from wastes in India. Although there are a few reported studies, they are limited to specific geographical area. Thus, it is very difficult to make any generalized comments about the national bioenergy potential from bio-waste in India. One of the reports [14] published by Central Pollution Control Board (CPCB), Ministry of Environment and Forests, Government of India, with the assistance of National Environmental Engineering Research Institute (NEERI), indicates a survey of solid waste management in 59 cities (35 metro cities and 24 state capitals of the country). Another report [15] indicates that among the four geographical regions in India, Northern India generates the highest amount of MSW (14.8 million tonnes per year)

Table 6.2 Agro-residue type-wise gross and surplus residue potential (in million tons, MT) in India [13]

Crop group	Crop	Gross residue potential, MT	Surplus residue potential, MT
Cereals	Rice	154.0	43.5
	Wheat	131.1	28.4
	Maize	35.8	9.0
	Bajra	24.3	5.1
	Barley	1.6	0.2
	Small millet	0.6	0.1
	Ragi	2.7	0.3
	Jowar	17.6	3.5
Oilseeds	Mustard and rapeseed	12.7	4.9
	Sesame	0.8	0.1
	Linseed	0.3	0.0
	Niger	0.1	0.0
	Safflower	0.6	0.5
	Soybean	13.5	4.6
	Groundnut	17.0	3.0
	Sunflower	3.8	0.6
Pulses	Tur (arhar)	7.2	1.4
	Gaur	2.6	1.8
	Gram	6.4	1.6
	Lentil	1.7	0.3
Sugarcane	Sugarcane	110.6	55.7
Horticulture	Banana	41.9	12.3
	Coconut	18.0	9.7
	Arecanut	1.5	0.5
Others	Cotton	75.9	46.9
	Jute	3.9	0.4
Total, MT		686.0	234.5

constituting around 30% of all MSW in India. Among the all Indian states, Maharashtra with an annual production of 8.1 million tonnes of MSW tops the list. The report also estimated the total available biodegradable fraction of waste to be around 25.54 million tonnes per year. Sewage sludge also has bioenergy potential in the country, but a central data base is not available in this case also. Although CPCB reports indicate the total sludge generated from several major cities and towns of India, still a complete picture of the sewage generation in the country is lacking. However, an extrapolated study [15] estimates the total sewage sludge generation in India to be around 1420.8126 tons per day with a bioenergy potential of 226.94 MW. The state-wise bioenergy potential from MSW and sewage sludge is summarized in Table 6.4.

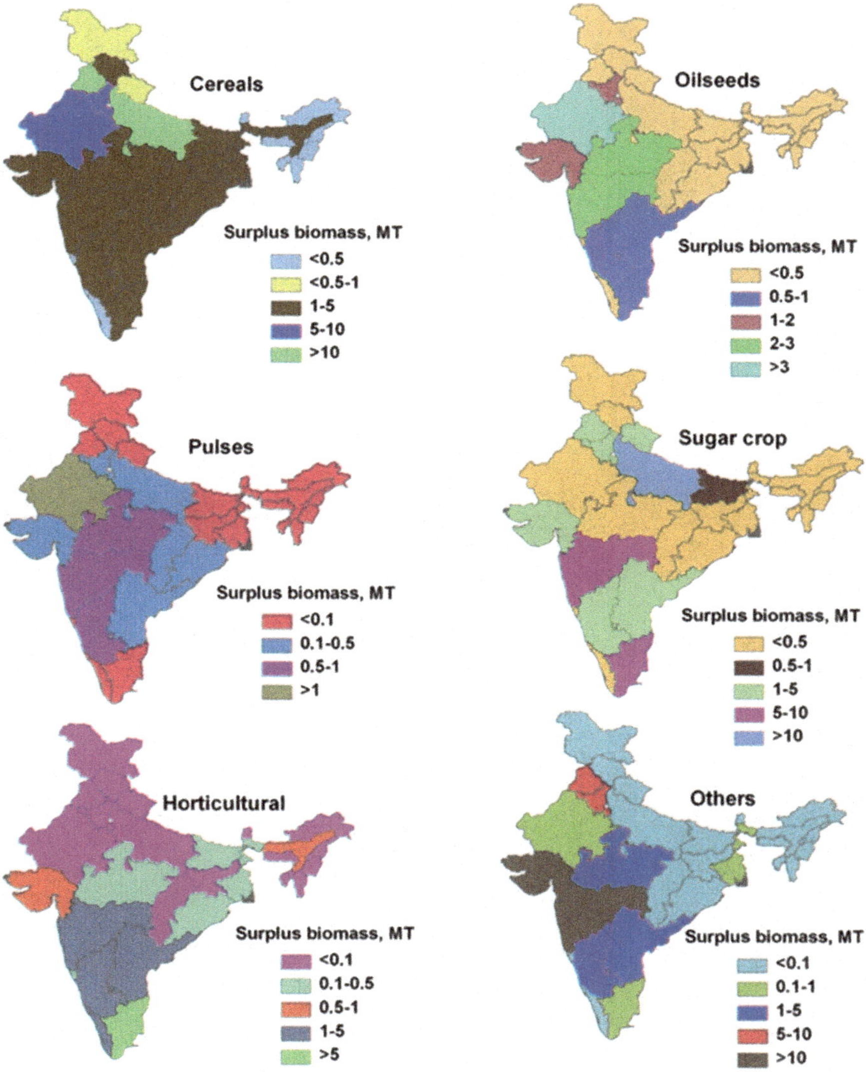

Fig. 6.1 State-wise surplus agro-residue biomass potential in India [13]

3 Review of Technologies for Cogeneration of Heat and Electricity

As stated earlier, biomass is a potential alternative energy source with less environmental impacts than fossil fuels. However, bulkiness and inconvenient form are the most common problems associated with biomass resources. Further, associated

Table 6.3 State-wise and surplus agro-residue bioenergy potential (in peta joule, PJ) in India [14]

Sl. No.	States	Surplus agro-residue bioenergy potential, PJ						
		Cereals	Oil	Pulses	Sugar	Horticultural	Other	Total
1	Andhra Pradesh	76.995	12.28	2.604	70.011	27.162	75.313	264.36
2	Arunachal Pradesh	1.678	0.076	0.005	0.063	0.18	0	2
3	Assam	16.956	0.398	0.098	3.693	11.607	0.471	33.22
4	Bihar	60.856	0.464	1.08	16.819	2.345	0.951	82.52
5	Chhattisgarh	27.756	0.751	1.597	0.071	0.166	0.015	30.35
6	Goa	0.439	0.041	0	0.221	2.191	0	2.89
7	Gujarat	36.117	19.618	7.515	56.368	35.558	270.666	425.84
8	Haryana	137.421	19.694	7.746	34.113	0	92.785	291.76
9	Himachal Pradesh	17.879	0.046	0.014	0.195	1.708	0.003	19.85
10	Jammu and Kashmir	10.051	0.445	0.001	0.002	0	0	10.5
11	Jharkhand	19.169	1.812	1.374	0.551	0	0.004	22.91
12	Karnataka	71.422	8.544	5.845	79.257	11.681	19.654	196.4
13	Kerala	2.955	0.001	0	1.908	93.746	0.028	98.64
14	Madhya Pradesh	78.07	41.113	10.65	9.648	6.122	61.728	207.33
15	Maharashtra	79.894	45.137	15.456	187.993	56.871	177.868	563.22
16	Manipur	1.954	0.003	0	0.083	0.889	0	2.93
17	Meghalaya	1.139	0.025	0.009	0.001	1.03	0.23	2.43
18	Mizoram	0.585	0.023	0.002	0.015	0.17	0.041	0.84
19	Nagaland	2.515	0.451	0.093	0.746	0.116	0.043	3.96
20	Orissa	34.651	0.683	1.568	3.851	7.86	0.86	49.47
21	Punjab	333.979	1.152	0.607	22.783	0	108.661	467.18
22	Rajasthan	117.152	56.931	18.899	1.779	0	5.659	200.42
23	Sikkim	0.816	0.052	0	0	0	0	0.87
24	Tamil Nadu	35.498	8.781	0.184	120.554	147.195	5.193	317.4
25	Tripura	4.151	0.021	0.01	0.187	1.227	0.076	5.67

26	Uttar Pradesh	258.723	4.752	7.069	471.515	0.853	0.24	743.15
27	Uttarakhand	13.288	0.211	0.022	26.053	0	0	39.57
28	West Bengal	44.644	2.4	0.335	4.773	4.276	6.038	62.47
	India total	1486.753	225.905	82.783	1113.253	412.953	826.527	4148.15

Table 6.4 Total MSW and sewage sludge generated and bioenergy potential in Indian states [14, 15]

State	Bio-waste potential		Energy generation potential	
	MSW (biodegradable fraction in tonnes per day)	Total sewage sludge (tonnes per day)	Out of MSW (MW)	Out of sewage sludge (MW)
Andhra Pradesh	1676	98.91	19.274	15.8
Arunachal Pradesh	18	NA	0.207	NA
Assam	246	19.33	2.83	3.09
Chhattisgarh	8.342	19.56	8.34	3.12
Goa	36.45	1.18	0.42	0.19
Gujarat	1592.72	95.42	18.31	15.24
Haryana	1367.28	33.51	15.72	5.35
Himachal Pradesh	31.89	1.45	0.37	0.23
Karnataka	865.2	101.19	9.95	16.16
Kerala	229.36	40.32	2.64	6.44
Madhya Pradesh	1027.8	68.98	11.82	11.02
Maharashtra	10,002.15	510	115.02	81.46
Manipur	72	1.34	0.83	0.21
Meghalaya	137	1.6	1.57	0.26
Mizoram	86	0.29	0.99	0.05
Nagaland	20	0.75	0.23	0.12
Punjab	1010.88	84.28	11.63	13.46
Rajasthan	2234.7	76.51	25.7	12.22
Sikkim	19	NA	0.22	NA
Tamil Nadu	1862	63.09	21.41	10.08
Tripura	114	1.2	1.31	0.19
Uttar Pradesh	1691.13	192.59	19.45	30.76
Uttarakhand	67.29	9.3	0.77	1.49

complex chain of logistics (harvest, collection, transport, storage) also creates difficulties in decision-making process. Energy density of biomass fuels is lower than fossil fuels. For example, energy content of air dried woody biomass is around 12–15 GJ/t whereas for sub-bituminous coal it is around 20–25 GJ/t (low heat values) [16]. Thus, for successful utilization of biomass, it is necessary to improve its properties for easy handling, storage, transportation, and conversion. One such method to overcome the difficulties associated with biomass is to convert solid biomass into liquid or gaseous fuels. This can be done via biological conversion, chemical conversion, or thermochemical conversion. In biological conversion/bio-digestion, microorganisms are used to produce gas from biomass. The method is suitable for moist biomass. Production of biogas from animal manure or plant biomass materials is one of the best examples of using biochemical conversion process to generate energy. Chemical conversion involves the use of enzymes to produce biofuels such as ethanol or other chemical products [17]. Thermochemical conversion of biomass is based on the application of heat and pressure to produce heat and electricity [18, 19]. As such, the discussion has been focused on thermo-chemical conversion mechanisms.

Technologies for heat and electricity production from woody biomass can be categorized as primary conversion technologies and secondary conversion technologies [20]. Primary conversion technologies are basically direct combustion or gasification based, whereas secondary conversion technologies are used in conjunction with primary conversion technologies for electricity and heat generation. Direct combustion involves burning of biomass in combustors in presence of oxygen that converts the chemical energy stored within the biomass into thermal energy. This thermal energy is later harnessed by using steam turbine, steam engine, or organic Rankine cycle.

On the other hand, biomass gasification is the partial oxidation of solid biomass, in presence of heat to produce gaseous or liquid fuels. The chemical energy of the gaseous fuel, also known as producer gas, can be utilized by means of gas turbines or internal combustion engines. The producer gas is mostly composed of carbon monoxide, nitrogen, and hydrogen. It is worth mentioning that all the conversion paths mentioned above can be used for either electricity production or combined heat and power (CHP) generation, depending upon whether or not the excess heat available after electricity generation is being exploited. However, some CHP layouts involve the combination of two different secondary technologies, for example, gas turbine is used for electricity generation and steam turbine for heat retrieval. Biomass cogeneration or CHP generation is a very effective energy technology, which provides energy saving in comparison with other technological solutions. Combined heat and power (CHP)—an integrated system that simultaneously generates electricity and useful thermal energy (e.g., steam) from a single fuel—is a versatile technology that can generate useful energy more efficiently and thereby significantly improve energy efficiency and deliver substantial benefits for end-user facilities, utilities, and communities (Fig. 6.2) [21]. The electricity from CHP systems can be used on-site or sold back to the grid. The thermal energy can be used, typically on-site, for a variety of purposes such as steam production, refrigeration, and space heating or cooling. In fact, the most efficient CHP systems are those that are able to productively utilize the available thermal energy with minimal waste [21]. Cogeneration technologies can operate within a range of power-to-heat output ratios, allowing units to adapt to specific energy demand requirements over time. The addition of energy storage capacity to cogeneration plants can also provide an added level of flexibility to regulate electricity and heat outputs while minimizing energy losses. The small-scale power generation technologies most favorable for CHP are reciprocating engines, diesel engines, natural gas engines, steam turbines, gas turbines, micro-turbines, and fuel cells [22, 23]. Selecting a CHP technology for a specific application depends on many factors, including the amount of power needed, the duty cycle, space constraints, thermal needs, emission

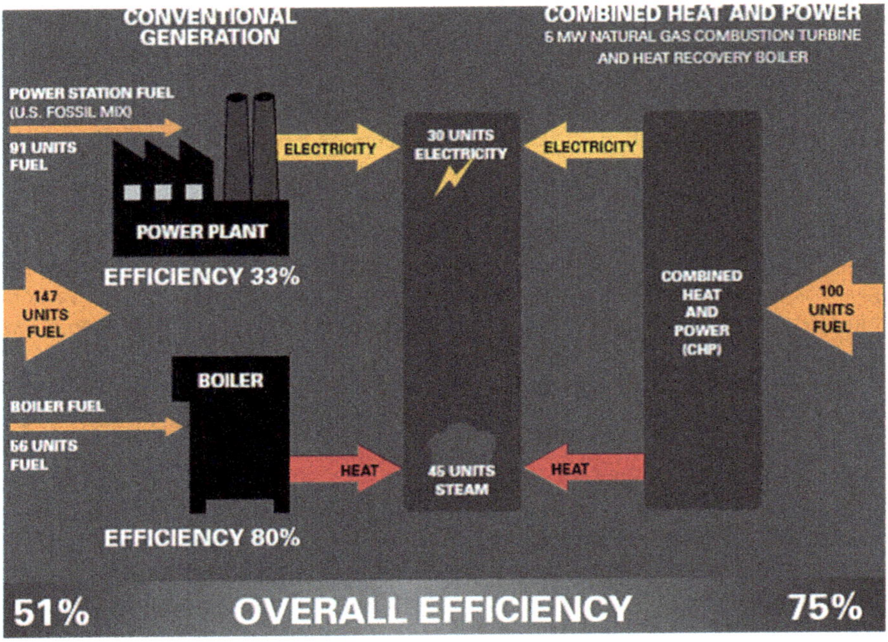

Fig. 6.2 CHP systems have higher overall efficiency than separate electric and thermal systems (adopted from [21]).

regulations, fuel availability, utility prices, and interconnection issues [22]. Table 6.5 provides a comparison of different CHP technologies.

The various conversion methodologies at different stages of development with commercial viability and maturity are discussed below:

3.1 Primary Conversion Technologies

Direct combustion and gasification are the most widely used primary conversion technologies for electricity and heat generation using low moisture biomass. Different methods of direct combustion and gasification for electricity and heat generation are discussed in the following sections.

3.1.1 Direct Combustion

Direct combustion is a thermochemical process involving complete oxidation of the biomass in an aerobic environment resulting in the generation of heat in the range of 800–1000 °C. Although direct combustion-based electricity generation accounts

Table 6.5 Comparison of different CHP technologies [22]

Parameters	Diesel engine	Natural gas engine	Steam turbine	Gas turbine	Micro-turbine	Fuel cells
Electric efficiency (LHV)	30–45%	25–45%	30–42%	25–40% (simple), 40–60% (combined)	20–30%	40–70%
Size (MW)	0.05–5	0.05–5	Any	3–200	0.025–0.25	0.2–2
Footprint (sqft/kW)	0.22	0.22–0.31	<0.1	0.02–0.61	0.15–1.5	0.6–4
CHP installed cost ($/kW)	800–1500	800–1500	800–1000	700–900	500–1300	>3000
O&M cost ($/kWh)	0.005–0.008	0.007–0.015	0.004	0.002–0.008	0.002–0.01	0.003–0.015
Availability	90–95%	92–97%	Near 100%	90–98%	90–98%	>95%
Hours between overhauls	25,000–30,000	24,000–60,000	>50,000	30,000–50,000	5000–40,000	10,000–40,000
Start-up time	10 s	10 s	1 h–1 day	10 min–1 h	60 s	3 h–2 days
Fuel pressure (psi)	<5	1–45	n/a	120–500 (may require compressor)	40–100 (may require compressor)	0.5–45
Fuels	Diesel and residual oil	Natural gas, biogas, propane	All	Natural gas, biogas, propane, distillate oil	Natural gas, biogas, propane, distillate oil	Hydrogen, natural gas, propane
Noise	Moderate to high (requires building enclosure)	Moderate to high (requires building enclosure)	Moderate to high (requires building enclosure)	Moderate (enclosure supplied with unit)	Moderate (enclosure supplied with unit)	Low (no enclosure required)
NOx emission (lb/MWhr)	Mar-33	2.2–28	1.8	0.3–4	0.4–2.2	<0.02

(continued)

Table 6.5 (continued)

Parameters	Diesel engine	Natural gas engine	Steam turbine	Gas turbine	Micro-turbine	Fuel cells
Uses for heat recovery	Hot water, LP steam, district heating	Hot water, LP steam, district heating	LP-HP steam, district heating	Direct heat, hot water, LP-HP steam, district heating	Direct heat, hot water, LP steam	Hot water, LP-HP steam
CHP output (Btu/kWh)	3400	1000–5000	n/a	3400–12,000	4000–15,000	500–3700
Useable Temp for CHP (F)	180–900	300–500	n/a	500–1100	400–650	140–700

for more than 90% of the total biomass-based electricity production, there is an inherent problem of high emission always associated with this technology [24].

There are several technologies to harness the heat released in the combustion process, but steam production for electricity generation using steam engine is the most common conversion route [25].

Different types of combustors, viz., fixed grate, moving grates, bubbling or circulating fluidized beds, auger floor, and suspension burners, are used for direct combustion. Selection of a particular type of combustor basically depends upon the capacity of the plant and characteristic of the feedstock. For example, stoker grates are suitable for small-scale plants (<6 MW_{th}) while fluidized beds are more appropriate for large-scale layouts (>10 MW_{th}) with higher moisture content [26].

In direct combustors, the fuel is heated, dried, and combusted in a single compartment. Combustors are basically classified depending upon the manner in which fuel is fed into and through the combustor. Some common varieties of combustors are discussed below.

- Fixed-grate: In fixed-grate combustors, the fuel is generally placed on a static sloping grate, and combustion air is provided both above and below the grate. As the fuel burns, the ash falls down through the perforated grate, where it is collected and removed. Fixed-grate combustors are advantageous as they are capable of handling variable-size fuel. One major disadvantage of fixed grate combustors is their inability to control the amount of combustion air precisely as compared to some other configurations.
- Moving-grate: Although similar in construction to fixed grate burners, moving-grate combustors include a moving grate capable of sliding or shaking. The fuel moves through the combustor on the moving grate. In another configuration, a series of grates are arranged in the form of a staircase so that the fuel falls from one grate to the other. This configuration helps in maintaining an optimum delivery of combustion air if a uniform size of fuel is maintained.
- Fluidized-bed: In fluidized-bed combustor, fuel is combusted by adding it to a fluidized bed of sand like material. The fluidized bed is produced by passing air though the bed material using distributors. This arrangement ensures even heating of the fuel, and high efficiency of combustion is achievable. A wide variety of fuels can be handled in this combustor.
- Auger floor: This type of combustor utilizes a screw auger which feeds the fuel into the combustor by rotation of the auger shaft. The rotation of the auger can be varied to maintain an optimum feed rate for different types and sizes of feedstock.
- Suspension/entrained combustion: In suspended fuel systems, very small particles of fuel, such as sawdust, are blown into the combustor where it is oxidized in a ball of fire.

3.1.2 Gasification

Gasification technology is based on the partial oxidation of feedstock in an anaerobic environment, within a closed reactor, into gaseous or liquid fuels which can be further processed for energy generation or for production of value-added chemicals [27]. Gasifiers are basically classified depending upon nature of contact between the gas and fuel. Gasifiers are categorized as (1) fixed bed (also known as moving bed), (2) fluidized bed, and (3) entrained flow. Gasifiers are also classified into atmospheric and pressurized reactors based on the operating pressure of the reactor. Depending upon the nature of heating, gasifiers are classified as allothermal or indirectly heated if the heat is provided externally and auto-thermal or directly heated if the heat is provided by the partial combustion of feedstock within the gasifier. The overall gasification process is divided into four stages, viz., drying, pyrolysis, oxidation (combustion), and reduction (char gasification). The processes are temperature dependent with drying occurring below 150 °C followed by pyrolysis, oxidation, and reduction which occur in the range of 150–700 °C, 700–1500 °C, and 800–1100 °C, respectively [28]. Drying involves the evaporation of moisture in the feedstock releasing steam. Pyrolysis results in the vaporization of the volatile components in the feedstock resulting in the production of a mixture containing hydrogen, carbon monoxide, carbon dioxide, methane, hydrocarbon gases, tar, and water vapor [29]. Steam is also produced by the oxidation of hydrogen in the fuel. The reduction reactions are facilitated by the oxidation or combustion of char. There are four major reduction reactions, viz., (1) Water–gas reaction, (2) Boudouard reaction, (3) Shift conversion, and (4) Methanation occurring inside the gasifier results in the production of a mixture of combustible gases primarily containing hydrogen, carbon monoxide, carbon dioxide, and methane [30]. Although the major components of producer gas obtained from gasification and biogas obtained from anaerobic digestion are the same, the production processes and their conversion efficiencies are entirely different. Electricity production via gasification occurs at efficiencies of 30–35% if dry biomass is used. As the moisture content of the biomass increases, the efficiency comes down to about 15% if the moisture content in the biomass is around 70% by weight. On the other hand, electricity production from anaerobic digestion occurs at around 15% efficiency, and moisture content has little influence in the process [31].

The most important parameters influencing the gasification process are gasification temperature, residence time, and feedstock size. In general, it is observed that higher particle sizes along with higher residence time result in higher rate of gasification whereas higher temperature results in higher hydrogen content in the producer gas leading to higher gas yield and lower methane content leading to lower heat value of the producer gas [31]. The advantageous aspects of versatility and flexibility of the gasification technology make it a viable option to be used in conjunction with different secondary conversion technologies [32]. Additionally, flexibility of using fuels at a wide range of moisture content makes it more advantageous in comparison to direct combustion technologies.

The following section discusses the different types of gasifiers coming under the classification based on the gas and fuel interface, which is the most common method of classification. Fixed bed, fluidized bed, and entrained bed type of gasifiers are summarized below.

Fixed Bed Gasifiers

In fixed bed gasifiers, the fuel remains in an almost static position while the gasification agent (air, oxygen, steam, or mixture of them) flows through it. The location of the different reaction zones is dependent upon the direction of gas flow through the fuel. Accordingly, the most widely used subtypes of fixed bed gasifiers are updraft and downdraft.

In downdraft gasifiers, the flow direction of both fuel and gas is downwards. Since the pyrolysis gases pass through the bed of hot char, most of the tar is cracked into noncondensable gases and water [27]. The gasifier is provided with air intake nozzles which admits limited quantity of air or oxygen enriched air into the feedstock bed resulting in the pyrolysis of the feedstock. Thermal energy required for drying, pyrolysis, and gasification is provided by the reactions occurring above the gasification zone. The charcoal produced during pyrolysis consumes the gas products of flaming pyrolysis and are reduced to product gas. The gas produced in downdraft gasifiers has been found to be suitable in internal combustion engines, gas burners, gas turbines, or as product gas distributed through pipelines. Advantages of downdraft gasifiers are quick response to load change along with high char conversion, lower tar and ash carry over, and simple construction. These factors make downdraft gasifiers one of the most widely used type of gasifiers. However, downdraft gasifiers have higher gas outlet temperatures, are difficult of scale up, face problem of ash fusion at high grate temperatures, and have fuel moisture limitations [33].

In updraft gasifiers, the feedstock is provided from the top, and product gas also comes out from the top. The gasifying agent is passed through the gasifier from the bottom and ascends through the bed of biomass. Thus, the gasifying agent first comes in contact with non-converted char and ash at the bottom of the bed. The pyrolysis of the biomass takes place above the gasification area. Reactions at different zones of the gasifier is facilitated by the residual heat in the updraft of hot air. Pyrolysis results in the production of a mixture of noncondensable gases, condensable gases, and char. The gases rise up and the char falls down along with the other solids. This output gas is a mixture of products of gasification and pyrolysis.

Fluidized Bed Gasifiers

In fluidized bed gasifiers, use of a fluidizing agent aids in enhancing the uniformity and adjustability of the temperature distribution [19, 34] by dropping down the slagging of the reaction significantly [35]. This helps in obtaining biomass conversion rate upto 100% [36]. The fluidizing gas is fed from the bottom of the reactor which enables it to act as the gasifying medium. The feedstock is fed from the top upon the whole of the fluidized bed. The fluidized bed helps in enabling rapid gasification reactions as the newly arrived particles rapidly come upto the bed temperature due to their contact with the hot bed of solids. Depending upon the velocity of the fluidizing gas, fluidized gasifiers are classified as bubbling fluidized bed (BFB) and circulating fluidized bed (CFB). Even after the thorough mixing of the solid bed, the fluidizing gas is continuously fed. As the solids are continuously mixed, full conversion of char is not achievable. The char particles entrained from the fluidized bed results in efficiency losses of the gasifier. Very low propagation of oxygen from the bubbles to the emulsion phase takes place in BFB. This results in the occurrence of combustion reactions in the fluidized phase leading to a decrease in the gasification efficiency. In CFB gasifiers, high residence times and thorough mixing are ensured by circulating the solids in a loop. The gas is prevented from bypassing the bed due to the absence of bubbles.

In order to prevent accumulation of ash, fluidized bed gasifiers are typically operated at temperatures of 800–1000 °C. This enables the use of high ash content fuels such as lignite, MSW, and biomass. Also, these gasifiers are flexible to use different types of fuels and are therefore preferred for larger plants [28].

Entrained Bed Gasifiers

Entrained bed gasifiers involve the use of powdered fuel which is entrained through the gasifying medium. These types of gasifiers operate at temperatures of around 1400 °C and with the pressures between 20 and 70 bar and are generally preferred in integrated gasification combined cycle (IGCC) plants. The powdered fuel (<75 μm) along with the gasifying agent is directly injected into the reactor chamber. In order to ease the feeding of the fuel in pressurized reactors, fuel is mixed with water to form a paste. The entrainment of the fuel particles is ensured by the passing the gas at an optimum speed. For gasifiers using a paste of water and fuel, a larger reactor volume is required in order to evaporate the water used in the mixture. The oxygen consumption of these types of reactors is usually 20% more than dry feed systems due to higher draft demands. The volatile materials and the char readily react with the incoming oxygen releasing heat. As a result, the temperature rises above the ash melting point, and the tars are destructed more rapidly. Thus, high carbon conversion efficiency is obtained [28].

3.2 *Secondary Conversion Technologies*

There are a number of secondary conversion technologies available for heat and power generation from biomass. While some technologies integrate better with direct combustion technologies, some are more compatible with gasification technologies. The type of technology and the output scale greatly influences the efficiencies of these technologies [37]. It is, however, observed that higher efficiencies are achievable if the output scale is higher, regardless of the technology. The following section discusses some of the most widely used secondary conversion technologies used in biomass-based cogeneration.

3.2.1 Steam Turbine Cogeneration System

In an ideal steam turbine cogeneration plant, the condenser is replaced with a process heater. Thus, no waste heat is produced in this type of plant. All the energy transferred to the steam in the boiler is utilized in producing power and for process heating. Steam is extracted from the exit of the turbine. Depending upon the thermal load, the steam exits the turbine at pressures higher than or at least equal to the atmospheric pressure. Thus, this configuration is also known as a back pressure turbine. Although the ideal steam turbine cogeneration plant seems to be the most efficient, it has its limitations. It cannot adjust to the variations in power and process-heat loads. In a more practical configuration, a condenser along with a process heater is included. When both power and process heat are required, some amount of steam is extracted from the turbine at the required process-heating pressure, and the rest expands to the condenser pressure and is then cooled. Depending upon the pressure and temperature requirement, the steam is extracted from one or more intermediate stages. This configuration is also known as an extraction condensing steam turbine. The heat rejected in the condenser goes out as the waste heat. When the process-heating requirement is high, all the steam is routed to the process-heating units and none to the condenser. The waste heat becomes practically zero. If still the process-heating requirement is not satisfied, an expansion or pressure-reducing valve is used to throttle some of the steam leaving the boiler to the process-heating unit. The power produced in this case becomes practically zero. When the process heating requirement is nil, all the steam passes through the turbine and the condenser, and the cogeneration plant operates as an ordinary steam power plant [38].

It may be observed that in steam turbine plants, initially combustion takes place in a boiler, and the heat is then transferred through a heat exchanger in order to evaporate the working fluid. This makes steam turbines versatile in accepting different kinds of fuels. For biomass-based operation, generally, bark, sawdust, woodchips, and pellets are used [39]. However, it is advisable to pre-dry the biomass fuel in order to increase the efficiency [40].

3.2.2 Gas Turbine Cogeneration System

The gas turbine-based cogeneration system works on the basic principle of Brayton cycle. The gas turbine system consists of a compressor, combustion chamber, turbine, and generator. In the combustion chamber, air from the compressor is drawn and mixed with the fuel. The fuel air mixture is then ignited. The high temperature flue gases coming from the combustor are used to drive a turbine which in turn drives the electric generator and the air compressor. A part of the developed mechanical power is used to run the compressor and the rest is converted into electric power. The Brayton cycle can be of two types, viz., open cycle Brayton cycle and closed cycle Brayton cycle and accordingly the gas turbine cogeneration systems are classified as open cycle gas turbine cogeneration system and closed cycle gas turbine cogeneration system [38].

In the open cycle cogeneration system, the gas turbine exhaust flue gases, typically at a temperature of 480–540 °C, are used as a heat source. This heat is then recovered by using it to generate steam or as hot air for process heating. Although the industrial gas turbine-based power plants operate at 25–35% thermal efficiency only, depending upon the type and size of gas turbine, a cogeneration system deriving energy from the high temperature flue gases helps in increasing the overall plant efficiency to around 85–90%. In the closed-cycle system, helium or air acts as the working fluid which circulates in a closed circuit. Before entering the turbine, a heat exchanger is used to heat the working fluid, and it is then cooled down after the exit from the turbine releasing useful heat. The working fluid when used in this manner remains clean, and it does not cause corrosion or erosion [38].

Integration of a gasification unit, a heat recovery steam generator (HRSG), and a gas cleaning unit with the gas turbine results in a configuration known as the biomass integrated gasification combined cycle (BIGCC) [41–43]. Integration with a gas engine [44] is also possible, but gas turbine is generally used due to its high exhaust temperatures [45]. Depending upon the gasification technology and whether or not the HRSG is used, various configurations of BIGCC is possible. However, all of these conversion pathways operate on a gaseous fuel. For large-scale applications, BIGCC emerges as a highly efficient process [46]. The main drawback in the implementation of BIGCC technology is that major modifications of the feedstock handling system are required in the existing natural gas-based technology. This is due to higher mass flow requirement of producer gas in comparison to natural gas. This is due to the lower heating value of producer gas in comparison to natural gas. Moreover, applicability of gas turbines is limited to large-scale applications (>1 MWe).

3.2.3 Reciprocating Engine Cogeneration System

Reciprocating engines are used extensively in industries as they start quickly, follow load well, have good part-load efficiencies, and generally have high reliabilities. Also, the fuel-related operating costs of reciprocating engines are lower than that of gas turbines of comparable sizes because of their high electrical efficiencies. Reciprocating engine has four possible sources of usable heat, viz., from engine jacket-cooling water, exhaust gas, lube oil cooling water, and turbo-charger cooling system. Recovery of heat from these sources is usually in the form of low-pressure steam or hot water. Although there is possibility of generating medium pressure steam from high temperature exhaust, the hot exhaust gas contributes to only about 50% of the available thermal energy from a reciprocating engine.

Reciprocating engines exhibit better performance with smooth consumption profiles [47]. Storage systems are, however, required in order to smoothen the consumption profile [48]. At small and microscale levels of application, reciprocating engines are capable of ensuring better returns on investment [46] due to good part-load performance and low upfront costs [26, 49].

3.2.4 Steam Engine

Steam engine utilizes steam generated from thermal evaporation of water or other working fluids for driving the engine. The mode of operation of the steam turbine enables the use of a variety of fuels for heat generation. Although steam engines are based on matured and well-proven technologies, they lack applicability due to their relatively lower efficiencies and inability to harness the excess heat [49].

3.2.5 Organic Rankine Cycle (ORC)

ORC technology is essentially similar to steam turbines in which the working fluid is some "organic" fluid. ORCs are categorized into high-temperature and low-temperature ORCs depending upon the temperature of the working fluid. *n*-Pentane or toluene are used as working fluids for high-temperature ORCs with capacities more than 200 kW_e resulting in high efficiencies along with the capability of producing heat. On the other hand, low-temperature ORCs with capacities less than 200–250 kW_e utilize hydrocarbons as the working fluid. It is worth mentioning that low-temperature ORCs have lower efficiencies and are unable to be utilized for CHP layouts [50, 51]. The use of organic fluids enables the use of low heating value feedstock like biomass without compromising the efficiency [52–54]. This is ensured due to the lower temperature of vaporization of the organic fluids which in turn aids in the setting up of Rankine cycles with lower temperatures than that compared to the conventional ones. Capability of handling solid fuels along with associated benefits of low temperature operation makes ORCs suitable for solar and geothermal applications also [54, 55].

Other advantages of this technology include reduced blade damage [56], good part-load operation [57], and non-requirement of a preheating stage [54]. This is attributed to the lower temperature of vaporization of organic fluids in comparison to water.

Cogeneration systems are also classified according to the arrangement of energy use and the operating patterns adopted. Accordingly, cogeneration systems are classified either as topping or bottoming cycle cogeneration systems. In a topping cycle, the fuel is first used to produce power and then the residual or waste energy is used to generate thermal energy, which is used to supply heat for process heating or for other thermal requirements. In a bottoming cycle, firstly high temperature thermal energy is produced from the primary fuel and then the rejected heat is utilized in a recovery boiler and turbine to generate power. Manufacturing processes which operate at very high temperatures and also reject heat at very high temperatures, such as furnaces and kilns, are suitable for the application of bottoming cycle cogeneration systems. However, topping cycle cogeneration is more widely used in comparison to bottoming cycle plants [38].

4 Potential of Heat and Electricity Production via Cogeneration

The potential for power generation from agricultural and agro-industrial residues is estimated at about 18,000 MW [4]. On the other hand, the potential of surplus power generation through bagasse cogeneration in sugar mills is estimated at 7000 MW. Thus, the total estimated biomass power potential in India is about 25,000 MW. The potential for bagasse cogeneration lies mainly in the sugar producing states like Maharashtra and Uttar Pradesh. As per the MNRE (Ministry of New and Renewable Energy, GoI), over 300 biomass power and cogeneration projects aggregating to about 4760 MW capacity have been installed in the country up to December 2015 for feeding power to the grid. In addition, around 30 biomass power projects aggregating to about 350 MW are under various stages of implementation. The highest numbers of biomass power and bagasse cogeneration projects in terms of installed capacity have been developed in Maharashtra followed by Uttar Pradesh and Karnataka as shown in Fig. 6.3. In India, biomass cogeneration projects mostly rely on sugarcane bagasse. India is the world's second largest sugarcane producer after Brazil; hence, bagasse-based energy generation potential in the country is also very high. State-wise bagasse generation potential in the country is presented in Table 6.6. There are 211 sugar mills in India having cogeneration facilities based on sugarcane bagasse. State-wise list of sugar mills with bagasse cogeneration facilities in India are given in Table 6.7. Uttar Pradesh has the highest numbers of sugar mills (54) having cogeneration facilities followed by Maharashtra (45) and Karnataka (39).

5 Case Study on Biomass-Based Bioenergy Generation in India

Baking and drying form the most important operations in a food processing industry. These units are mostly dependent on thermal energy for their operation. These industries are mostly dependent upon conventional sources of energy such as

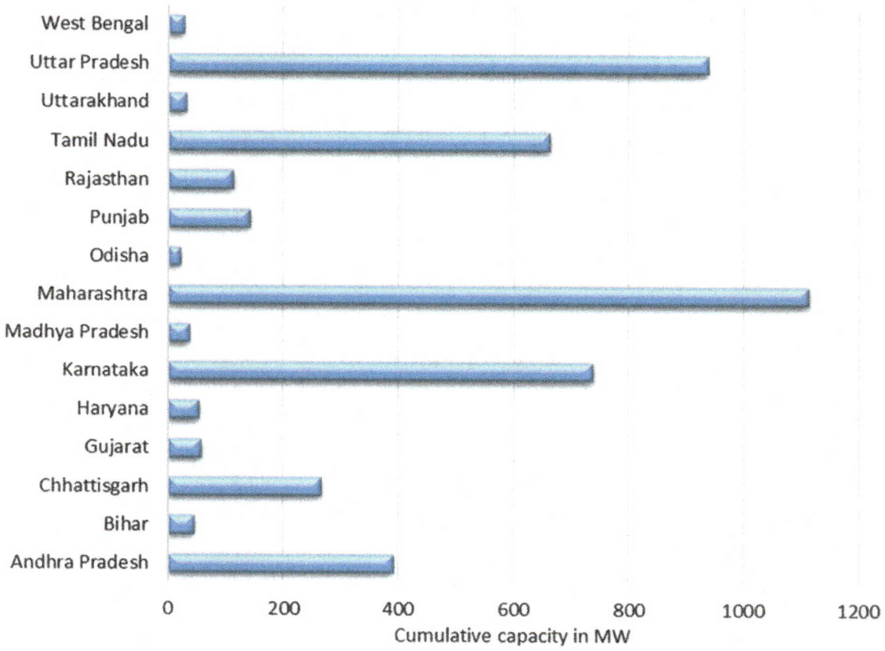

Fig. 6.3 State-wise cumulative commissioned biomass power and bagasse cogeneration projects [4]

coal and diesel to meet their thermal energy requirements. Biomass-based heat and energy generation has immense potential to substitute the conventional sources of fuel in these industries. Thinking in a similar line and considering the availability of surplus rice husk in the area, a biscuit producing factory located in the Sonitpur district of Assam had installed a biomass gasifier in order to substitute a part of their fossil fuel consumption.

NEBISCO Industries Pvt. Ltd. is a biscuit producing factory located at Tezpur in the district of Sonitpur, Assam. The factory carries out large-scale biscuit baking in tunnel shaped ovens. To meet the thermal energy demand, the industry operates using two ovens with five burners each. The ovens were using high speed diesel (HSD) as fuel until 2011 when one of the two ovens was modified to use producer gas as fuel from a biomass gasifier. Since then it has been using rice husk for thermal energy while the other oven still operates on diesel.

The gasifier is a fluidized bed gasifier with gas flow rate of 875 Nm^3/h, thermal output of 903,000 kCal/h, consuming rice husk at upto 440 kg/h, and with an auxiliary power consumption of 40 kWh. The gasifier operates for 8 h a day for 300 days in a year. The industry produces on an average 750 kg of biscuits per hour.

Utilization of the biomass-based gasifier was reported to account for fuel saving of about $200 per day amounting to a saving of $0.27 per day per kg of produced

Table 6.6 State-wise sugarcane bagasse production potential in India [58]

States	Sugarcane farming area, million ha	Sugarcane production, million tonne	Productivity, tonnes/ha	Gross bagasse production, million tonne	Net bagasse production available for energy, million tonne
Andhra Pradesh	0.204	16.73	82.01	5.02	3.76
Gujarat	0.202	14.18	70.20	4.25	3.19
Karnataka	0.43	38.81	90.26	11.64	8.73
Madhya Pradesh	0.069	2.68	38.84	0.80	0.60
Maharashtra	1.02	81.86	80.25	24.56	18.42
Tamil Nadu	0.38	39.28	103.37	11.78	8.84
Bihar	0.23	12.07	52.48	3.62	2.72
Haryana	0.1	6.96	69.60	2.09	1.57
Punjab	0.08	4.67	58.38	1.40	1.05
Uttar Pradesh	2.17	128.82	59.36	38.65	28.98
Uttarakhand	0.108	6.6	61.11	1.98	1.49
West Bengal	0.016	1.18	73.75	0.35	0.27
Others	0.086	3.83	44.53	1.15	0.86
All India	5.095	357.67	70.20	107.30	80.48

Table 6.7 State-wise sugar mills in India having sugarcane bagasse-based cogeneration facilities [58]

Sl. No.	State	No. of sugar mills with cogeneration facility	Exportable Energy (capacity in MW)	
			Season	Off-season
1	Bihar	3	27.00	21.00
2	Uttar Pradesh	54	960.16	159.50
3	Uttaranchal	1	27.50	–
4	Haryana	6	32.60	–
5	Punjab	13	165.00	–
6	Andhra Pradesh	23	317.75	37.00
7	Tamil Nadu	27	514.70	89.00
8	Maharashtra	45	623.20	7.00
9	Karnataka	39	553.50	418.50
India total		211	3221.41	732

biscuit. The operation of the gasifier is, however, dependent upon the continuous supply of husk to the gasifier site. The supply chain involves the collection and transportation of rice husk from rice mills in the vicinity of the factory. Absence of competitive uses of the rice husk in the region has made avenues for revenue generation for the mill owners who otherwise had to burn or dump the husk as waste. Employment generation in respect of transportation and handling of the rice husk at the user end was also observed. The local farmers of the region could also find a new avenue of income by selling the husk themselves directly to the industry or to a collection center that may operate as a medium for collection of the husk from the farmers and delivering to the industry. Thus, it becomes evident that deployment of similar units in various other industries in the region could increase the revenue generation of the farmers, rice processing units, and local manpower. One problem faced by the industry is the ash generation. In the absence of any competitive use of the ash, the industry is bearing additional cost for ash disposal. The ash generated can, however, be utilized in other applications such as concrete mixture.

Sintering, agglomeration, deposition, erosion, and corrosion are some technological issues related with the viable application of biomass gasification [59]. Gas handling equipment faces the problems of deposition, corrosion, and erosion due to the presence of ash in the producer gas. Thus, gas cleaning and periodic maintenance of the equipment demand top priority to prevent the eventuality of a breakdown. The industry under study was found to adhere to strict maintenance schedules. However, problems of ash deposition on the reactor walls and tar deposition on pipes carrying gas were still reported. The ash-related problems of the system can be reduced to a great extent by using pretreatment methods such as leaching and fractionation [60]. Although the type of biomass influences the effectiveness of water leaching, fractionation has been reported to reduce the ash content by almost 50% [61].

Utilization of the biomass gasifier in the industry has brought down the fuel cost by an estimated amount of $200 per day per kg of product and also resulted in an estimated 77% reduction in the CO_2 emission. Taking into account the technological and economical merits, it may be commented that rice husk-based gasification technology is suitable for similar industrial applications in the region. However, development of a systematic network of collection of feedstock is required in order to ensure comprehensive benefits for the local people. Also, technological intervention is required in order to reduce the drudgery of maintenance.

6 Future Prospects

There exists a huge potential of biomass for dedicated use in electricity and heat production in the country. Agriculture being the most important contributor to the national economy, crop residue and bagasse become the highest contributor to biomass resource in the country. This crop residue is available mainly as a by-product of crop production and agro-industries. However, at present there has been very little exploitation of these resources. It is evident from Fig. 6.3 that only five states, viz., Tamil Nadu, Karnataka, Andhra Pradesh, Maharashtra, and Uttar Pradesh, have been able to properly harness the potential of these resources. It is also observed that the states of Punjab and Haryana, who are among the top crop producing states, have not been able to harness the potential of biomass resources in their respective states. This has been attributed to the fact that there has been an increasing trend of burning the crop residue (husk and straw) by the farmers in Haryana and Punjab. This indicates a general lack of awareness amongst a major section of the country's population regarding the prospective use of biomass resources. Thus, there is a pressing need to educate the various sections of the society about the huge prospects of generation of renewable energy in general and bioenergy in particular from the agro-residues. There also exists a vast potential for energy generation from bio-waste in India. Municipal solid waste, sewage sludge, bio-wastes from textiles, leather, food, and fruit processing industries are significant sources of bio-waste in the country. However, adequate data bases of resource assessment of these sources are still lacking. In general, there is an urgent need to have a comprehensive biomass and bio-waste resource assessment in the country.

In the technological prospective, many new and proven technologies have not been able to make inroads into the national and international market. However, there exists a sense of reluctance among the users in deploying these technologies. In this regard, the Government has a major role to play in demonstrating the capability of these technologies by setting up pilot plants in various regions of the country. Once the robustness and economic viability of these technologies are made known to the general population, biomass-based energy generation will have a flourishing market in the country which will in turn help in uplifting the economic well-being of its population.

7 Conclusion

Biomass has always been a significant energy source for India considering the benefits and potentials it offers. Biomass has the capability of acting as a reliable source of nonconventional and renewable energy due to its widespread availability and storage ability. However, unavailability of a reliable supply chain and a well-organized market structure has hindered the progress of biomass based technologies. Also, wide variations in physical properties and uncertainty in availability round the year, which depends on the cropping pattern and weather, is another source of concern. For example, biomass from agriculture residue is only available for a short period of 2–3 months in a year after harvesting season, if not properly stored. Similar problems are also there with other sources of biomass. Therefore, there is the need to evolve a robust and an organized biomass market based on innovative business models. The success of such business models depends upon the level of motivation of the rural entrepreneurs who have the most crucial responsibility of maintaining the supply to the biomass processing facilities. Fiscal and financial incentives from Government both at Central and state and involvement of commercial institutions and nongovernmental organizations are also prerequisites for successful surplus biomass utilization for cogeneration for power and heat.

Development and exploitation of energy plantations which take up plantation of energy crops on marginal and degraded lands is another important aspect.

Biomass-based cogeneration of heat and power has been taken by many Indian states with Uttar Pradesh, Maharashtra, Karnataka, Andhra Pradesh, Tamil Nadu, and Chhattisgarh leading the table. However, it is also observed that many states endowed with very good biomass potential have not been able to properly utilize their resources and figure low in biomass power achievements. Uttar Pradesh has been the most successful state in utilizing its biomass potential through cogeneration system.

It is also observed that there is a wide variation in the tariff of biomass-based power in the different states. It may be commented that Government initiatives/policies has a major role in enhancing the growth of biomass-based technologies in the country in states with high biomass power potential in particular. Fruitful exploitation of biomass for CHP plants requires an in-depth biomass resource assessment in order to formulate a biomass inventory. Such a database will help in formulating comprehensive roadmaps for deployment of biomass-based plants. Furthermore, a comprehensive idea on the existing and state-of-the-art technologies for biomass-based CHPs is also essential to select the appropriate technology suitable for a specific type of biomass fuel for a given geographical location.

References

1. Renewable energy in India: growth and targets (2015) Ministry of new and renewable energy. May13
2. http://www4.unfccc.int/submissions/INDC/Published%20Documents/India/1/INDIA%20INDC%20TO%20UNFCCC.pdf. Accessed 20 May 2016
3. Ailawadi VS, Bhattacharyya SC (2006) Access to energy services by the poor in India: current situation and need for alternative strategies. Nat Resour Forum 30:2–14
4. Ministry of new and renewable energy. Annual report 2015–2016. http://www.mnre.gov.in/
5. Demirbas A (2008) Biofuels sources, biofuel policy, biofuel economy and global biofuel projections. Energy Convers Manage 49:2106–2116
6. Panwar NL, Kothari R, Tyagi VV (2012) Thermochemical conversion of biomass – eco friendly energy routes. Renew Sustain Energy Rev 16:1801–1816
7. Balat M, Balat M, Kırtay E, Balat H (2009) Main routes for the thermo-conversion of biomass into fuels and chemicals. Part 2: gasification systems. Energy Convers Manage 50 (12):3158–3168
8. Hossain AK, Davies PA (2013) Pyrolysis liquids and gases as alternative fuels in internal combustion engines – a review. Renew Sustain Energy Rev 21:165–189
9. Gustavsson L (1994) Biomass and district-heating systems. Renew Energy 5:838–540
10. Raj NT, Iniyan S, Goic R (2011) A review of renewable energy based cogeneration technologies. Renew Sustain Energy Rev 15(8):3640–3648
11. Kuhn V, Klemeš J, Bulatov I (2008) MicroCHP: overview of selected technologies, products and field test results. Appl Therm Eng 28(16):2039–2048
12. Blarke MB, Lund H (2008) The effectiveness of storage and relocation options in renewable energy systems. Renew Energy 33(7):1499–1507
13. Hiloidhari M, Das D, Baruah DC (2014) Bioenergy potential from crop residue biomass in India. Renew Sustain Energy Rev 32:504–512

14. Central Pollution Control Board report: Ministry of Environment and Forests, Government of India. 2012
15. Annepu RK (2012) Sustainable solid waste management in India. Columbia University, New York
16. Sahyog report: http://cordis.europa.eu/result/rcn/144096_en.html
17. International Energy Agency (2007) Good practice guidelines: bioenergy project development and biomass supply. OECD/IEA, Paris
18. Küçük M, Demirbaş A (1997) Biomass conversion processes. Energy Convers Manage 38 (2):151–165
19. Janajreh I, Al Shrah M (2013) Numerical and experimental investigation of downdraft gasification of woodchips. Energy Convers Manage 65:783–792
20. Zhang K, Chang J, Guan Y, Chen H, Yang Y, Jiang J (2013) Lignocellulosic biomass gasification technology in China. Renew Energy 49:175–184
21. Gowrishnkar V, Angelides C, Druckeniller H (2013) Combined heat and power systems: improving the energy efficiency of our manufacturing plants, buildings, and other facilities. NRDC Issue Paper, April
22. Review of Combined Heat and Power Technologies (1999) Office of industrial technologies, US department of energy, Washington, DC
23. United States Environmental Protection Agency. An engineering-economic analysis of combined heat and power technologies in a μGrid application, March 2002
24. González A, J-Roger R, Puig R, Navarro P (2015) Review of micro- and small-scale technologies to produce electricity and heat from Mediterranean forests' wood chips. Renew Sustain Energy Rev 43:143–155
25. Kirkels AF, Verbong GPJ (2011) Biomass gasification: still promising? A 30-year global overview. Renew Sustain Energy Rev 15:471–481
26. Bain RL, Overend RP, Craig KR (1998) Biomass-fired power generation. Fuel Process Technol 54:1–16
27. Obernberger I (1998) Decentralized biomass combustion: state of the art and future development. Biomass Bioenerg 14:33–56
28. Basu P (2010) Biomass gasification and pyrolysis: practical design and theory. Elsevier, New York
29. Basu P (2006) Combustion and gasification in fluidized beds. CRC Press, London
30. Kivisaaria T, Björnboma P, Sylwana C, Jacquinotb B, Jansenc D, Grootc AD (2004) The feasibility of a coal gasifier combined with a high-temperature fuel cell. Chem Eng J 100:167–180
31. Duman G, Uddin Md A, Yanik J (2014) The effect of char properties on gasification reactivity. Fuel Process Technol 118:75–81
32. Yoshida Y, Dowaki K, Matsumura Y, Matsuhashi R, Li D, Ishitani H (2003) Comprehensive comparison of efficiency and CO_2 emissions between biomass energy conversion technologies—position of super critical water gasification in biomass technologies. Biomass Bioenerg 25:257–272
33. Passey R, Spooner T, MacGill I, Watt M, Syngellakis K (2011) The potential impacts of grid-connected distributed generation and how to address them: a review of technical and non-technical factors. Energ Policy 39:6280–6290
34. Dogru M (2013) Experimental results of olive pits gasification in a fixed bed downdraft gasifier system. Int J Green Energy 10:348–361
35. Siewert A, Niemelä K, Vilokki H (2004) Initial operating experience of three new high-efficiency biomass plants in Germany. Power Generation and Europe Conference
36. Osowski S, Fahlenkamp H (2006) Regenerative energy production using energy crops. Ind Crop Prod 24:196–203
37. Bocci E, Sisinni M, Moneti M, Vecchione L, DiCarlo A, Villarini M (2014) State of art of small scale biomass gasification power systems: a review of the different typologies. Energy Procedia 45:247–256

38. Cengel YA, Boles MA (2006) Thermodynamics: an engineering approach, 5th edn. McGraw Hill, Boston
39. Wahlund B, Yan J, Westermark M (2000) Comparisons of CO_2-reducing alternatives for heat and power generation: CO_2-capture and fuel shift to biomass. In: Durie RA, McMullan P, Paulson CAJ, Smith AY, Williams DJ (eds) 5th International Conference on Greehouse Gas Technology. CSIRO Publishing, Cairns, pp 229–234
40. Deshmukh R, Jacobson A, Chamberlin C, Kammen D (2013) Thermal gasification or direct combustion? Comparison of advanced cogeneration systems in the sugarcane industry. Biomass Bioenerg 55:163–174
41. Corti A, Lombardi L (2004) Biomass integrated gasification combined cycle with reduced CO_2 emissions: performance analysis and life cycle assessment (LCA). Energy 29:2109–2124
42. Jurado F, Cano A, Carpio J (2003) Modelling of combined cycle power plants using biomass. Renew Energy 28:743–753
43. Larson ED, Williams RH, Leal MRLV (2001) A review of biomass integrated- gasifier/gas turbine combined cycle technology and its application in sugarcane industries, with an analysis for Cuba. Energy Sustain Dev 5(1):54–76
44. Zhou Z, Yin X, Xu J, Ma L (2012) The development situation of biomass gasification power generation in China. Energy Policy 51:52–57
45. Arena U, Di Gregorio F, Santonastasi M (2010) A techno-economic comparison between two design configurations for a small scale, biomass-to-energy gasification based system. Chem Eng J 162:580–590
46. Gustavsson L, Johansson B (1994) Cogeneration: one way to use biomass efficiently. Heat Recover Syst CHP 14:117–127
47. Alanne K, Saari A (2004) Sustainable small-scale CHP technologies for buildings: the basis for multi-perspective decision-making. Renew Sustain Energy Rev 8(5):401–431
48. Onovwiona HI, Ugursal VI, Fung AS (2007) Modeling of internal combustion engine based cogeneration systems for residential applications. Appl Therm Eng 27(5-6):848–861
49. Salomón M, Savola T, Martin A, Fogelholm CJ, Fransson T (2011) Small-scale biomass CHP plants in Sweden and Finland. Renew Sustain Energy Rev 15(9):4451–4465
50. Maraver D, Sin A, Royo J, Sebastián F (2013) Assessment of CCHP systems based on biomass combustion for small-scale applications through a review of the technology and analysis of energy efficiency parameters. Appl Energy 102:1303–1313
51. Maraver D, Sin A, Sebastián F, Royo J (2013) Environmental assessment of CCHP (combined cooling heating and power) systems based on biomass combustion in comparison to conventional generation. Energy 57:17–23
52. Dong L, Liu H, Riffat S (2009) Development of small-scale and micro-scale biomass-fuelled CHP systems—a literature review. Appl Therm Eng 29(11-12):2119–2126
53. Liu H, Shao Y, Li J (2011) A biomass-fired micro-scale CHP system with organic rankine cycle (ORC)—thermodynamic modelling studies. Biomass Bioenerg 35(9):3985–3994
54. Tchanche BF, Lambrinos G, Frangoudakis A, Papadakis G (2011) Low-grade heat conversion into power using organic rankine cycles—a review of various applications. Renew Sustain Energ Rev 15(8):3963–3979
55. Invernizzi C, Iora P, Silva P (2007) Bottoming micro-Rankine cycles for micro-gas turbines. Appl Therm Eng 27(1):100–110
56. Obernberger I, Hammerschmid A (2001) Biomass fired CHP plant based on an ORC cycle–Project ORC-STIA-Admont. Final Report, Bios-Energy System. 4. http://www.bios-bioenergy.at/uploads/media/Report-ORCAdmont-Thermie-2001-03-26.pdf
57. Obernberger I, Thonhofer P, Reisenhofer E (2002) Description and evaluation of the new 1000 kWel organic rankine cycle process integrated in the biomass CHP plant in Lienz. Austria Euroheat Power 10:1–17
58. Indian Sugar Mills Association. Premier association of the sugar industry in India. http://www.indiansugar.com/Cogeneration.aspx

59. Baxter LL (1993) Ash deposition during biomass and coal combustion: a mechanistic approach. Biomass Bioenerg 4:85–105
60. Arvelakis S, Koukios EG (2002) Physicochemical upgrading of agro-residues as feedstocks for energy production via thermochemical conversion methods. Biomass Bioenerg 22:331–248
61. Arvelakis S, Gehrmann H, Beckman M, Koukios EG (2002) Effect of leaching on the ash behavior of olive residue during fluidized bed gasification. Biomass Bioenerg 22(1):55–69

Part II
Raw Material, Feedstock, Compositional Analysis and By-products and Application

Chapter 7
Advancement in Development of Biodiesel Production in the Last Two Decades: An Indian Overview on Raw Materials, Synthesis, By-products, and Application

Mahboob Alam, Santosh Kumar Yedla, Sandup Tshering Bhutia, Vaibhav V. Goud, and Nageswara Rao Peela

Abstract The fossil fuel resources are depleting at a faster rate, and environmental concerns are increasing worldwide. Therefore, there is a need to explore renewable energy resources to meet the increasing energy demands and to reduce the environmental pollution. Biodiesel is one of such renewable resources. This chapter reviews the advancements taking place in the development of biodiesel processes in India, particularly for raw materials, synthesis, by-products, and applications. India has huge potential to produce biodiesel from nonedible feedstocks (such as *Jatropha curcas*, *Pongamia pinnata*, Neem, Castor, Karanja, and Rubber seed) which are widely available. Algal oil has also been explored as a biodiesel feedstock. The USA and European Union are using edible oil as feedstock for the production of biodiesel. However, India cannot afford edible oil as the needs of the country are met by imports up to 50%. Several processes (such as co-solvent, enzymatic, microwave, and ultrasound) have been explored for the production of biodiesel. Use of biodiesel as a transportation fuel reduces emission of harmful pollutants such as CO_2 (global warming), NO_x (photochemical smog), and SO_x (acid rain). It also helps in rural upliftment by increasing employment in agriculture sector. The main by-product of biodiesel production is the crude glycerol, and in this chapter, various applications of this by-product have been discussed.

Keywords Biodiesel • Edible and Nonedible oil feedstocks • Algal oil • Jatropha • Pongamia • Karanja • Transesterification • Crude glycerol

---

M. Alam • S.K. Yedla • S.T. Bhutia • V.V. Goud • N.R. Peela (✉)
Department of Chemical Engineering, Indian Institute of Technology Guwahati, Guwahati 781039, Assam, India
e-mail: peelanr@iitg.ernet.in

A.K. Chandel, R.K. Sukumaran (eds.), *Sustainable Biofuels Development in India*,
DOI 10.1007/978-3-319-50219-9_7

1 Introduction

The demand for new and cleaner fuel from nonconventional sources has been the prime focus of the twenty-first century. Nonconventional sources of energy have been able to provide partial energy supplements to cater to increasing requirements for fossil fuels. India is one of the fastest growing economies in the world. It is very critical to sustain this economy via sustainable and non-sustainable developments. It is very important for a nation like India to have energy security for socioeconomic developments. According to Greenpeace (2009), renewable energy sources will be able to meet 35% of India's energy demands by 2030. India's domestic oil production is able to meet 30% of its energy requirements. The gradual decrease in fossil fuels and increasing environmental degradation raise serious issues with nonconventional sources (fossil fuels) of energy. This ever-increasing hunger for energy has to be supplemented with new and alternate sources of energy. The search for new alternate energy sources has given rise to a new sustainable and environment-friendly fuel known as "BIODIESEL.". The oil is a derivative of chemical interactions of vegetable oils (edible and nonedible), animal fats, and algae with an alcohol that can be used directly or as a blend in diesel engines. Biodiesel has gained significant prominence because of the following characteristics: (1) renewable and biodegradable, (2) higher cetane value, (3) low sulfur and aromatic content, and (4) lower emissions of CO, particulate matters, and unburnt hydrocarbon. However, there are certain drawbacks which need to be addressed as well: (1) higher NO_x emissions, (2) lower oxidative and storage ability, and (3) higher viscosity [1]. Biodiesel development in India mainly focuses on the cultivation and processing of Jatropha plant seeds which are very rich in oil content (40%). Many developed countries use edible oil seed crops such as groundnut, soybean, rape seed, sunflower, etc., for the production of biodiesel. India being a developing country cannot afford to use edible oils as feedstocks for biodiesel production [2]. The Indian perspective in biodiesel exploration and production is solely based on nonedible (nonfood) feedstocks to be cultivated on lands that are not suited for agriculture (wastelands), thus avoiding a conflict of fuel vs. food security [3]. In India, various feedstocks are being considered as possible sources of biodiesel, and hence, extensive research and development projects are being carried out in consultations with concerned government agencies and other institutions. Indian Legislation intends the use of B5 (5% blend) and successively B20 (20% blend) to become mandatory all over India by 2017 without engine modifications.

2 Raw Materials for Biodiesel Production

Various sources have been recognized as possible feedstocks for biodiesel production. The biodiesel sources can broadly be categorized into three types: first-generation feedstocks (edible oils), second-generation feedstocks [nonedible oils, animal fat, waste material (cooking oil)], and thirst-generation feedstocks (algal oils) [4]. Currently, more than 95% of the world's biodiesel production comes from first-generation feedstocks (edible oils). The use of edible oil for biodiesel production started way back in the nineteenth century, where the first compression ignition engine running on peanut oil was demonstrated. The various feedstocks used for biodiesel production are listed below.

2.1 *Edible Oils (First-Generation Feedstocks)*

2.1.1 *Elaeis guineensis* (Palm)

Edible oils such as palm oil, which is one of the cheapest vegetable oil, provide a promising alternative as a feedstock for biodiesel (FAME) production with conversions in the range 98–99% [5] and yields up to 87% [6]. The palm fruit contains about 50% oil by weight (mesocap and kernel). Physicochemical properties of palm oil (Table 7.1) are in agreement with the ASTM standards [7]. India produces about 70,000 tons of palm oil per year (Kerala alone 20,000 tons) [8]. India has the potential to produce approximately 435,000 Mt of biodiesel from PFAD (Palm Fatty Acid Distillates) annually. Oxidation stability of palm oil biodiesel is good compared to other biodiesel feedstocks such as Jatropha, karanja, Castor, and Kusum because of higher saturated fatty acid concentration in its composition [9].

2.1.2 *Linum usitatissimum* (Linseed)

Linseed oil (*Linum usitatissimum*), a colorless to yellowish oil derived from the dried seeds of the plant, contains about 30–35 wt. % oil. India accounts for about 5×10^5 tons of seed production with over 70% contribution from Madhya Pradesh and Uttar Pradesh [10, 11].

2.1.3 *Cocos nucifera* (Coconut)

Kerala accounts for almost 65% of coconut production in India. The dried kernel has oil content varying from 65 to 75% [8]. Several other edible oil feedstocks such as sunflower oil, coconut oil [8, 12], canola oil, rapeseed oil, mustard oil [13], soybean oil [14], and groundnut oil have successfully been used as biodiesel feedstocks.

Table 7.1 Properties of biodiesel obtained from various edible and nonedible sources

S NO	Property	Unit	Diesel	ASTM D 6751	JOME	Palm oil	POME	Neem biodiesel	Rubber seed oil	Castor Biodiesel			Punnakka methyl ester
										B_{10}	B_{20}	B_{100}	
1	Color	–	–	–	Light yellow	–	Yellowish red	Light to dark brown	Dark brown	Light yellow			Greenish yellow
2	Odor	–			Aromatic	–	Disagreeable	Strong	–	Mild odor			Disagreeable
3	Density at 15 °C	kg/m^3	850	875–900	880	–	924	820–940	860	–	–	–	910
4	Kinematic viscosity at 40 °C	mm^2/s	2.60	1.9–6.0	4.84	4.50	40.2	3.2–8.7	–	4.54	4.97	15.98	38.17
5	Flash point	°C	70	>130	162	135	225	120	72	85.3	88.7	190.7	224
6	Pour point	°C	−20		−6	–	−3	2	12	−26	−30	−45	–
7	Water content	%	0.02	<0.03	Nil	0.01		0.036	–	–			–
8	Ash content	%	0.01	<0.02	Nil	0.01	0.07	–	–	–			–
9	Carbon residue	%	0.17	–	0.025	–		–	–	0.009	0.007	0.036	–
10	Sulfur content	%	–	0.05	–	0.002		–	–				–
11	Acid value	mg KOH/g	0.35	<0.05	0.24	0.24	5.40	–	–	–	–	–	–
12	Iodine value	–	–		104	–	87	–	–	–	–		–
13	Saponification value	–	–	–	194	–	184	–	–	–	–		203
14	Calorific value	MJ/kg	42	–	37.2	–	8742	39.6–40.2	44	44.43	44.78	37.90	32.5
15	Cetane number	–	46	–	51.6	54.6	42	48–53	45	–	–	–	–
16	Induction point	–	–	3		–		–	–	−5	−7	−23	–

JOME Jatropha methyl ester; *POME* Pongamia methyl ester

2.1.4 Waste Cooking Oil

Other potential feedstocks are waste vegetable oils and animal wastes. Waste cooking oil and animal waste create serious disposal problems and possible pollution of land and water bodies. However, conversion to biodiesel via transesterification process tackles the problem in an environment-friendly as well as in an economical way. Waste cooking oil (WCO) residue from various domestic and industrial sources, e.g., restaurants, food industries, domestic household, etc., can be used as a raw material for biodiesel production. The use of these raw materials not only reduces the environmental hazards but also lowers production cost significantly. Biodiesel generated from waste cooking oil can nearly replace 1.9% of diesel oil in India [15]. India being an agricultural country is bestowed with different varieties of livestock. The livestock industry generates a large amount of oils and fats. Animal fats are highly viscous and readily solidify at ambient temperature because of their high content of saturated fatty acids. More than half of animal by-product is unsuitable for consumption. The oil derived from animal waste is cleaner and burns efficiently in diesel engines as it has high cetane number. However, the oil has high cloud point and tends to crystallize at much higher temperatures than biodiesel derived from vegetable oils.

2.2 Nonedible Oils (Second-Generation Feedstocks)

India is the fourth largest edible oil market, and much of the demand (~50%) is met by imports. Since both the edible oil and biodiesel markets are expected to rise in the near future, it becomes necessary to rely on sustainable nonedible feedstocks for biodiesel. Extensive use of edible oils causes significant food vs. fuel issues which may lead to serious concerns in the long term. In order to overcome this conflict, focus has been shifted to nonedible oil feedstocks (second and third generation). Nonedible feedstocks such as jatropha [16], pongamia [17], mahua [18], and neem [19] have been extensively investigated, and jatropha/pongamia are found to be promising feedstocks. Availability of nonedible feedstocks in various places in India is shown in Fig. 7.1.

2.2.1 *Jatropha curcas*

Jatropha is a multipurpose, drought-resistant plant found in most tropical and subtropical regions of the world. Jatropha *curcas L* occurs in various regions like South East Asia, Africa, South America, and India. It possesses enormous source of hydrocarbons which in turn results in its use as commercial source of fuels. Profile of fatty acids present in Jatropha oil is shown in Table 7.2. The yield of biodiesel

Fig. 7.1 Occurrence of nonedible oil for biodiesel feedstock

from Jatropha oil is reported to be 97% [16]. Properties of Jatropha Oil Methyl Ester as a fuel are comparable to those of a standard diesel fuel [17].

Disadvantages Jatropha seed oil contains phorbol ester or PMA (phorbol-12-myristate 13-acetate) which is toxic in nature; that is why nutritional utilization is not possible [20]. PMA could activate cPLA2α gene expression in NSCLC (non-small cell lung cancer) resulting in contribution to tumor progression [21]. Another major disadvantage of jatropha curcas oil is high viscosity which results in carbon deposits and incomplete fuel combustion, thereby reducing engine life [22].

Pongamia pinnata

Pongamia generally occurs in India, Myanmar, and Australia. Annual rainfall of 500–2500 mm and humid and subtropical environment are useful to grow pongamia [16]. Therefore, we can see these plants in coastal areas and river banks. It produces seeds with oil concentration in the range of 27–39%. Due to high concentration of oil, it has been receiving considerable attention as feedstock for biodiesel production. Various methods like micro-emulsion, pyrolysis, and transesterification are used to produce biodiesel from pongamia. Potential availability of oil from

Table 7.2 Fatty acid and oil content of various nonedible oils

	Nonedible oil									
Fatty acid	Castor	Jatropha	Karanja	Mahua	Sal	Polonga	Neem	Kusum	Rubber	Jojoba
% of oil in seed or kernel	45–50	25–30	27–39	35–42	14–20	43–52	40–50	28–34	45–52	45–50
Lauric C12	–	–	–	–	0.58	–	–		–	–
Myristic C14	–	–	–	–	–	0.09	0.2–2.60	1	–	–
Palmitic C16	0.5–1	17.57	3.7–7.9	16–28.2	3.69	12.01–14.6	13.6–16.2	5–8	8.7–10.6	–
Stearic C18	0.5–1	7.49	2.4–8.9	20–25.1	47.04	12.95–19.96	14.4–24	2–6	8–12	0.55–0.77
Oleic C18:1	2–6	32.46	44.5–71.3	41–51	43.98	34.09–37.57	49–62	40–67	17–20	–
Linoleic C18:3	1–5	41.25	10.8–18.3	8.9–18.3	1.2	26.33–38.26	–	–	21–26	–
Linolenic C18:2	0.5–1	–	–	14.74	–	0.27–0.3	2.3–15.8	2.5–5.2	33–39	28–31
Palmitoleic	–	1.209	–	–	–	2.5	0.8–3.4	20.31	–	0.25
Arachidic C20	–	–	2.2–4.1	0.0–0.3	3.5	0.94	–	–	–	–
Margaric	–	–	–	–	–	–	–	–	–	–
Petroselenic	–	–	–	–	–	–	–	–	–	–
Gondoic	–	–	–	–	–	–	–	–	–	–
Behenic	–	–	4.2–5.3	–	–	–	–	–	–	14.2
Lignoceric	–	–	–	–	–	–	–	1.5–3.5	–	–
Ceric	–	–	–	–	–	–	–	–	–	–
Ricinoleic	85–95	–		–	–		–	–	–	–
Saturated	–	–	–	–	54.8	–	–	–	–	–
Unsaturated	–	–	–	–	45.2	–	–	–	–	–
Other	0.2–0.5	–	–	–	–	–	–	–	–	–
Seed production (10^6 tonnes/year)	0.25	0.20	0.06	0.20	0.20	–	0.10	0.06	–	–
Oil (tonnes/ha/year)	0.5–1	2–3	2–4	1–4	1–2	–	2–3	–	–	–

Collected from the website: http://www.chempro.in/fattyacid.htm, accessed on 01st of June 2016 pongamia is estimated to be 5500 tons/year [17]. Fatty acid profile of Pongamia is shown in Table 7.2. The pongamia tree is helpful in various ways such as shade, seeds, plant fodder, and green manure. It also has medicinal properties [16]. Physicochemical properties of pongamia oil are listed in Table 7.1.

***Azadirachta indica* (Neem)**
Neem is native of India and Burma and also occurs in Asia, Africa, and Central and South America. It grows in a range of weather and soil conditions: tropical and subtropical climate, semiarid and subhumid conditions, and any type of soil such as clay, alkaline, and saline are suitable for the growth of neem plants. These plants can survive in an annual mean temperature range of 21–32 °C. All parts of neem (leaves, bark, flower) are useful to cure many diseases such as malaria, ophthalmia, and tuberculosis. Extract of neem leaves is used to treat diabetes, scrofula, and erysipelas [17].

On an approximate basis, 20 million neem trees are found in India. Indian neem trees have the potential to provide 1 million tons fruits and 0.1 tons kernel. Each mature tree may produce 30–50 kg fruits per year, and this process will continue up to 150–200 years. Calorific value of neem oil is same as diesel, and therefore, it can be blended with diesel up to a concentration of 30%. It has been suggested that no major engine modification is needed while using neem oil [19]. Physical and chemical properties of neem methyl ester and conventional diesel are reported in Table 7.1.

***Hevea brasiliensis* (natural rubber)**
Natural rubber comes from various countries: Thailand (35%), Indonesia (23%), Malaysia (12%) India (9%), and China (7%). In India, annual rubber seed production is about 150 kg/ha. Kerala and Tamil Nadu are known for seed production and processing [23].

***Schleichera Oleosa* Merr. (Kusum)**
It is a medium to large size, evergreen dense tree growing to 35–45 ft in height. It is found in India (Himalayas, central and southern region), Java, Timor, and Ceylon. Annual production of Kusum oil is approximately 66000 tons in India, out of which 4000–5000 tons are collected and used for medicinal purposes [24, 25]. Seeds are brown and irregular elliptical in shape and contain up to 40.3% of oil which is bitter in taste and yellowish in color [26].

***Ricinus communis* (Castor)**
Castor plant is believed to be originated in Abyssinia now known as Ethiopia, distributed throughout tropical, subtropical, and temperate region [27]. This plant is able to stand long period of drought. Major castor producing countries are India, China, Brazil, Thailand, and the USSR. Annual production of castor oil in India is approximately 0.73 Mt, which makes it a leading producer in the world. The major use of castor oil in India is in textile, print industry, and synthesis of high-grade lubricants. Major component of castor oil is ricinoleic acid (89.5%) which is unsaturated hydrocarbon and solubilizes in most organic solvents. Other properties of castor oil such as low Sulfur content (0.04%), ash content (0.02%), heating

value (39.5 GJ/t), iodine value about 80, and flash point of 260 °C and also good carbon trading potential are listed [28]. Properties of castor oil biodiesel match with conventional diesel (Table 7.1).

Calophyllum Inophyllum **Linn (punnakka)**
Punnakka is medium to large, evergreen tree with shining leaves and golden seed. Average height of punnakka is around 8–20 m with irregular branch [29]. It is found in many countries like India (coastal area), East Africa, South East Asia, Australia, and South Pacific [7]. It is a best source as a feedstock for biodiesel production as its kernel contains up to 70% of oil. Oil availability depends upon productivity of the tree and efficiency of extraction process [17].

Madhuca indica **(Mahua)**
Mahua tree contains 30–40% fatty oil called Mahua oil and is also referred to as butter tree by locals since the oil fat is used for skin care, soaps, and detergent manufacture. The tree starts bearing seeds after a period of 7 years of planting, and harvesting starts from the tenth year. The seed yield depends on its growth and development and ranges from 20 to 200 kg per tree every year [18]. India has an estimated annual production potential of 181,000 Mt/year [30].

Manilkara zapota **(Sapodilla)**
Manilkara zapota, popularly known as sapodilla, an evergreen forest tree is found in abundance throughout Karnataka, Maharashtra, Gujarat, Tamil Nadu, West Bengal, and Andhra Pradesh. The local name varies as chikoo (chiku) in Northern India and sapota in southern parts of India. According to the National Horticulture Board, the sapota plantation covers around 160,000 ha. The fruit appears brownish with seeds of brown or black in color. The seeds have an oil content of 23–30% weight of the seed [1].

Cashew Nut Shell Liquid
Thermally cracked cashew nut shell liquid (CNSL) is a valuable raw material for producing biodiesel. The liquid is a by-product of the industrial processing of cashew nut. The production potential in India is approximately 2×10^5 tons. The raw nut shell contains over 20% CNSL. A ton of cashew shells yield up to 100 kg of good quality of CNSL [31].

Terminalia Bellirica **(Bahera)**
Terminalia Billerica, known as "Bahera," is a deciduous tree mostly found in Madhya Pradesh, Uttar Pradesh, Punjab, and Maharashtra. The tree bears seeds (bedda nuts) which have an oil content of 40%. The tree provides fodder for cattle as well as timber for furniture.

2.3 Algal Oil (Third-Generation Feedstocks)

The primary feedstocks for first-generation biodiesel (rapeseed, soybeans, and palm) are all food-based crops. In order to shift this focus from food-based feedstocks, second- and third-generation feedstocks are introduced which reduces dependence on food sources. Third-generation biofuel mainly focuses on nonfood sources. One such source is utilization of algal oils for biodiesel production. Algal biomass generally contains proteins, carbohydrates, and lipids/natural oils. Microalgae grow quickly compared to terrestrial crops, and hence, there is a focus on the biofuel scenario. Microalgae make use of sunlight and CO_2 for their growth and give higher oil productivity than terrestrial oil seed crops. Since CO_2 is utilized for algal growth, this could serve as a boon in reducing CO_2 emissions from the atmosphere. Microalgae contribute about 40–50% oxygen in the atmosphere, thereby supporting biological life on our planet. Microalgae have the potential to produce 1,36,900 l, while Jatropha can produce 1892 l of oil per acre [32, 33]. Variation in amount of oil depends upon different species of microalgae (Table 7.3). Algae can be classified into 12 classes including the green algae (Chlorophyceae), diatoms (Bacillariophyceae), yellow-green algae (Xanthophyceae), golden algae (Chrysophyceae), red algae (Rhodophyceae), brown algae (Phaeophyceae), and picoplankton (Prasinophyceae and Eustigmatophyceae). Chlorophyceae (green algae) and Bacillariophyceae (diatoms) are best studied for biofuel production due to their lipid content [34]. Microalgae are also a potential source of the lipids and fatty acids required for the food and fuel. The lipid or oily part of the algae biomass can be extracted and converted into biodiesel with similar process applied for other feedstocks. The carbohydrate content of algae can also be fermented into bioethanol and butanol fuel [35].

Table 7.3 Oil content of microalgae [33]

Algal oil (third-generation feedstocks)	
Microalgae	Oil content (dry wt. %)
Botryococcus braunii	25–75
Chlorella sp.	28–32
Crypthecodinium cohnii	20
Cylindrotheca sp.	16–37
Dunaliella primolecta	23
Isochrysis sp.	25–33
Monallanthussalina	>20
Nannochloris sp.	20–35
Nannochloropsis sp.	31–68
Neochloris oleoabundans	35–54
Nitzschia sp.	45–47
Phaeodactylum tricornutum	20–30
Schizochytrium sp.	50–77
Tetraselmis suecica	15–23
B. braunii	25–75

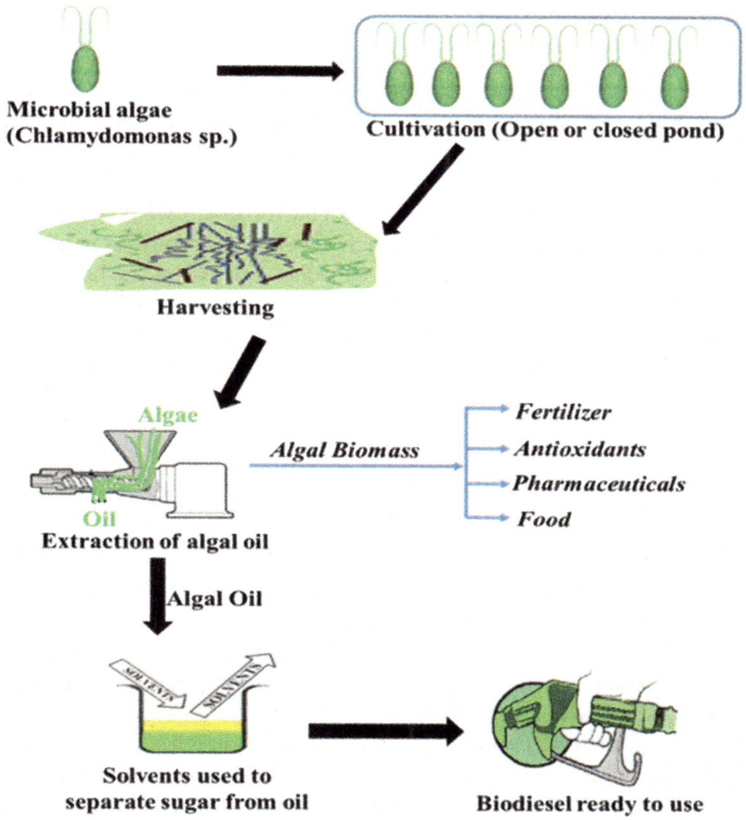

Fig. 7.2 Microalgae biodiesel production route [36]

Biodiesel production from microalgae is represented below as Fig 7.2 [36]. The fatty acid composition changes with ambient condition such as temperature, sunlight, heavy metals, nitrogen, and phosphorus concentration.

Advantages of using microalgae as biodiesel feedstock

- Removal of CO_2 by bio-fixation and reducing greenhouse gas emissions
- Wastewater treatment by removal of NH_4^+, NO_3^-, and PO_4^{3-} using wastewater as nutrient for algal growth.
- Processing of algal biomass into other useful products (ethanol, methane, livestock feed, organic fertilizer (due to high N, P ratio), and energy).
- Ability to grow in extreme conditions or areas unsuitable for agricultural purposes

3 Biodiesel Production

Biodiesel is generally produced by transesterification reaction. As the name suggests, one mole of triglyceride reacts with an alcohol, generally methanol, and forms three moles of fatty acid methyl ester (FAME) and one mole of glycerol [35]. Various methods such as catalytic [35], enzymatic [37], ultrasound irradiation [38], and microwave [39] have been developed for biodiesel production. Cost of biodiesel production is a major factor which depends upon various parameters such as quality and type of feedstock, plant capacity, processing technology, and its storage. It has been found that nonedible oil is considered as low-cost biodiesel feedstock. In addition, two other parameters, i.e., transesterification process and purification of by-product (glycerol), are also responsible in deciding the overall cost [40].

3.1 *Catalytic Transesterification*

In the first step of the process, the oil from feedstock is extracted using extractor. The by-product that comes from the extractor can be used for supplying the heat required for the main process. In the second step of the overall process, catalyst (sodium hydroxide) and methanol are mixed in appropriate proportion to form sodium methoxide (which is a strong base) and then sent to the reactor containing vegetable oil. Methoxide part of the sodium methoxide reacts with triglyceride to form FAME (biodiesel) and glycerol (by-product) [41]. Neutralization of the base catalyst is carried out using an acid (HCI or H_3PO_4). Glycerol and FAME are separated by decanting. The separated FAME must be washed with water to remove several impurities such as trace catalyst, salt, methanol, soap, and glycerol. These impurities and residues may be detrimental to any combustion system [42]. The schematic of catalytic transesterification for biodiesel production is given in Fig. 7.3 [43].

3.2 *Co-solvent Transesterification*

The co-solvent transesterification of vegetable oil provides a new way of producing the biodiesel [37]. In this method, solvent is subjected to temperature and pressure conditions above its critical point, resulting in supercritical conditions. Under such conditions, there is no separate phase exit, favoring high reactivity [44]. Various co-solvents such as carbon dioxide, propane, and hexane are used in the transesterification process [37]. With the use of co-solvents, better yields of biodiesel can be obtained under same operating conditions. The main use of solvent is to enhance the miscibility of reactants and the solubility of alcohol in oil phase [45].

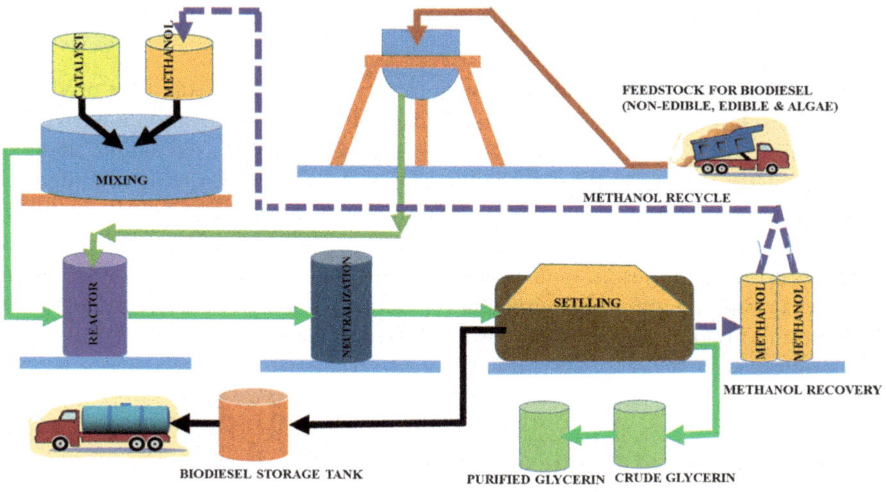

Fig. 7.3 Schematic diagram of transesterification process [43]

3.3 *Enzymatic Transesterification*

Enzymatic transesterification is another method for the production of biodiesel from vegetable oil [35]. A group of enzymes, called lipases, which have the ability to moderate the transestefication reactions are used to produce biodiesel [35, 37]. Generally, extracellular lipases and intracellular lipases are used in the enzymatic transesterification. Extracellular lipases are recovered from the live producing microorganism, while intracellular lipases are employed in the whole cell form maintaining the enzyme immobilized [37]. Enzyme-catalyzed reactions possess several advantages such as low energy intensive, less side product formation, and eco-friendly.

3.4 *Microwave Transesterification*

Microwave transesterification is another method that is being implemented for the production of biodiesel. In this method, microwave irradiation is used to improve the heat transfer between the phases [35]. Heat is directly delivered to the reaction media, making the process well controlled which in turn results in better yields [39]. Due to its advantages such as less reaction time, better yields, and eco-friendly nature, the microwave transesterification is becoming a favorable choice for many researchers to produce biodiesel [35, 39].

3.5 *Ultrasound Transesterification*

Ultrasound transesterification has proven to be an effective method for producing the biodiesel. In this method, ultrasound irradiation transesterification improves the miscibility between the phases by the generation of cavitation bubbles [35, 38]. Ultrasonication provides the mechanical energy for mixing and required activation energy for initiating the transesterification reaction and thereby increases the reaction rate and yield of product [46]. One of the advantages of cavitation bubbles is that they cause a local increase in temperature and the formation of micro jets. As a result, there is no requirement for external heating and intense mechanical agitation in the production of biodiesel production [38].

4 Application of Biodiesel and Its Side Products

4.1 *Application of Biodiesel*

Social and economic development of any country depends upon energy. This energy comes from fossil fuels and petroleum products, but resource of fossil fuels is limited and deteriorating environment. According to US Energy Information Administration (2011), fuel consumption increased from 86 million barrel per day to 112 million barrel per day by 2035. The limited reserves cannot afford this usage [47, 48]. Fossil fuels produce several gases such as polycyclic aromatic hydrocarbon (PAH), nitrated PAH (carcinogens) in nature, and SO_x (acid rain). Hence biodiesel is recognized to be an alternative fuel. Advantages of biodiesel are numerous and include biodegradable, nontoxic, low sulfur content (0.001%), and low CO, PAH, and nitrated PAH emissions. It is also easy to synthesize compared to gasoline and diesel [48].

One disadvantage of biodiesel is emission of NO_x. NO_x emissions are harmful and may destroy O_3 layer. Various researchers have tried to find out ways to reduce NO_x emissions during and after combustion process. Efficiency of engine depends upon the composition of reactant and contaminants. For example, methanol reduces flash point, density, and viscosity [49]. Quantity of water affects pitting in piston and blockage of filter and reduces heat during combustion in engine. Fouling of injection depends on the concentration of glycerol, acrolein, and aldehyde. Formation of glyceride is responsible for the carbon deposition in piston, valve, and nozzle. A series of precautions should be taken for using biodiesel in engine because it may create problems such as filter lugging, injection coking, piston ring sticking and breaking, seal swelling, hardening, cracking, and lubricant degradation. A suitable additive is needed to improve the oxidation stability of biodiesel. Advantages and disadvantages of Biodiesel are listed in Table 7.4.

Table 7.4 Advantages and disadvantages of biodiesel [40, 47, 50, 51]

S. No	Advantage	Disadvantage
1	Biodiesel possesses excellent combustion properties due to high oxygen content (10–11%). Reduces CO_2 and smoke due to free soot, higher cetane number reduces ignition delay	Biodiesel has about 12% less energy content compared to petroleum diesel
2	Renewable, nontoxic, nonflammable, easily available, biodegradable, eco-friendly: free from sulfur and polyaromatic hydrocarbons (cancer-causing agent)	The fuel has higher viscosity and lower volatility (requires higher injector pressure), higher cloud, and pour points
3	It creates rural employment	It causes corrosion in fuel tank, pipe, and injector due to low oxidation stability
4	Excellent lubricity due to high viscosity which decreases engine wear and tear and increases engine performance	It creates imbalance in food supply and demand market if edible oil is used as its feedstock
5	Easy transportation, distribution, handling, and storage because of higher flash point (<100–170 °C) than petroleum diesel (60–80 °C)	It causes excessive carbon deposition, gum formation (polymerization), filter clogging, seal failure, and higher NO_x emissions
6	For blends up to B20, engine modification is not needed	Transesterification process is expensive and also poses problems to environment such as waste disposal and soap formation
7	Reduce environmental effect of waste, e.g., cooking oil and lards	

Easy to Use No vehicle modification is needed up to B20. For B100, minor modification is needed [48].

Power, Performance and Economy Biodiesel shows horsepower, torque, and fuel mileage similar to those of petroleum fuels. Performance of engine reduces when 100% biodiesel (with no blending) is used as fuel because it contains 5–10% less energy content as compared to petroleum fuels [50, 52]. High viscosity of biodiesel possesses excellent lubricity compared to petroleum-derived diesel [53].

Emission and Greenhouse Gas Reduction By using Biodiesel, it reduces harmful substances like CO, particulate matter, and hydrocarbons (HCs) as compared to conventional fuel. This benefit occurs because biodiesel contains 11% O_2 by weight; presence of O_2 allows the fuel to burn completely resulting in fewer emissions. This has been approved by Environmental Protection Act. Conventional fuels also produce many diseases like cancer, nausea, and chronic respiratory disease as compared to biodiesel confirmed by Clean Air Act [53]. Cloud of diesel and black smoke (which are generally related with diesel vehicle) can easily be reduced by using biodiesel [53].

Toxicity, Biodegradability, Safety, and Recycling Biodiesel is a nontoxic, biodegradable, and renewable fuel [54]. Biodiesel from various feedstocks is easily broken down into harmless components according to EPA standard (1982) and high biodegradation rate in aquatic environment [54]. The maximum degradation (mineralization) rate for biodiesel (rape ethyl ester, REE) is examined by CO_2 evolution 25 mg/l/day, while diesel is 12.5 mg/l/day that is half of REE (rapeseed ethyl ester) [55]. Biodiesel is easy to handle and transport from one place to another place due to its high flash point (>160 °C). Flash point is the lowest temperature at which the vapor above a combustible liquid can be made to ignite.

4.2 *Application of Crude Glycerol*

Crude glycerol (50–60%) comes as by-product during biodiesel production (Fig. 7.3). Crude glycerol treated with inorganic acid (H_2SO_4 or HCl) and neutralized with base followed by distillation for removal of methanol and water results in a solution containing 80–85% of glycerol (termed as "raw glycerol"). Raw glycerol can be converted into other useful products with the help of catalyst.

4.2.1 Chemicals Produced Through Catalytic Conversion

Value-added opportunities for glycerol are discussed below based on the following procedures such as catalytic conversion (esterification, hydrogenolysis, etherification, oxidation, and reforming, etc.) and biological conversion (fermentation).

Glycerol Esters Production of monoglycerides, diglycerides, triglycerides, and their derivatives by esterification reaction was reported by Behr et al. [56] and Guerrero et al. [57]. They are widely used as emulsifier in food, cosmetic, and pharmaceutical industries. Monoglycerides can be obtained by esterification of glycerol with oleic acid and lauric acid in the presence of ordered mesoporous catalyst which has $ROSO_3H$ group [58, 59]. Mono- and dilaurins are formed in the presence of multivalent metal salts when glycerol reacted with lauric acid as reported by Nakamura et al. [60]. They found that chloride catalyst ($ZrOCl_2{\cdot}8H_2O$ and $AlCl_3{\cdot}6H_2O$) and sulfate catalysts ($(FeSO_4){\cdot}H_2O$ and $ZrSO_4{\cdot}H_2O$) are active for the formation of monolaurin and dilaurins, respectively. These compounds are generally used in pharmaceutical industry. Monolaurin is responsible for the suppression of HIV virus [61].

Glycerol Ethers At high temperatures, glycerol polymerizes and oxidizes partially to acrolein which is toxic in nature and cannot be mixed to fuel. By selective etherification, it can be converted into fuel additive [62]. Glycerol reacts with tert-butanol or isobutene and forms glycerol tert-butyl ethers (GTBEs) [43]. GTBEs are generally added to fuels [63] to reduce particulate matter, CO, hydrocarbon, and unregulated aldehyde in emission [62]. Cloud point of biodiesel reduces when these ethers are added [64]. On etherification of glycerol with t-butanol in the presence of Amber-15 catalyst, GTBEs will be produced [65, 66]. Long-chain ethers (mainly

C_8) can be produced by telomerization reaction (glycerol with butadiene) and are widely used as surfactants and detergents [67].

Propanediols Propanediol (propylene glycol) can be obtained by hydrogenolysis of glycerol in the presence of Ni, Pd, Pt Cu, and copper chromite at 200 °C and pressure below 14 bar hydrogen [68]. It is a valuable compound generally utilized as pharmaceutical, cosmetics, flavoring agent and solvent for food color and also acts as a moisture retaining agent in medicine. In recent years, 1,2-propanediol is being used as antifreeze instead of ethylene glycol due to its nontoxic nature [69]. Another propanediol with various applications is 1,3-propanediol. It is utilized in the synthesis of polymers (polytrimethylene and terephthalate), beautifiers, medicine, and heterocyclic compounds [70]. Conversion of glycerol to 1,3 Propanediols by hydrogenolysis reaction can be carried out by Pt/WO_3 catalyst supported on ZrO_2 [71].

Epichlorohydrin The accessibility of substantial measure of unrefined glycerol has energized the utilization of glycerol as a raw material in the generation of epoxides. Epichlorhydrin is majorly utilized as a part of the generation of epoxy resins [72]. Epichlorhydrin is also used in the production of pharmaceutical, paper sizing agents, dyes, and textile conditioners [56, 72]. Epichlorohydrin has been produced from crude glycerol by DOW Chemical Company using GTE process [73]. This process consists of hydrochlorination of crude glycerol at high temperature and pressure in the presence of carboxylic acid to produce mixture of 1,3 dichloropropanol and 2,3 dichloropropanol. This step is followed by the addition of base to form epichlorohydrin [73]. Epichlorohydrin has also been synthesized by Solvay chemical company with commercial name Epicerol [56].

Acrolein Dehydration of crude glycerol results in acrolein. It is utilized for the synthesis of acrylic acid, esters, super absorber polymer, detergents, and a retention agent in the synthesis of paper [74]. Ott et al. obtained a maximum selectivity of acrolein (75 moles %) at 50% conversion of glycerol when experiment was carried out at 350 °C, 25 MPa, and 70 ppm of electrolyte ($ZnSO_4$) by using critical water as reaction medium. They also reported an improved yield and activity with ZSM-5 as catalyst and higher temperatures [75].

Dihydroxyacetone It is a value-added compound used as an active ingredient in cosmetics and all sunless tanning skin care preparation and as a building block for several fine chemicals [76, 77]. Dihydroxyacetone can be obtained by using selective oxidation of 2° hydroxyl group of glycerol [62]. Glyceric acid is formed as major product by chemoselective oxidation of glycerol with air on Pd/C or Pt/C catalyst on the basis of experiment conducted. However, addition of Bi on Pt enhances the selectivity toward the secondary hydroxyl group of glycerol to produce dihydroxyacetone with a selectivity of 50% at 70% conversion [78].

Glycerol Carbonate Glycerol carbonate is known as cyclic ester formed with glycerol and carbonic acid. It is a value-added chemical with numerous applications such as a protic solvent, ingredient for cosmetics, filling plasticizers, solvent for battery electrolyte, lubricants, a substitute for ethylene carbonate, propylene carbonate, cyclocarbonate derivatives, coatings, paints, and surfactants [79]. Glycidol is another polymeric material produced from glycerol carbonate, used as a feedstock for polyurethane and polycarbonates [57, 62, 69, 80]. Glycerol carbonate can be formed by direct carboxylation reaction (glycerol with CO_2) in the presence of zeolite or ion exchange resins [81].

Biosurfactants Biosurfactants are biodegradable, less toxic, and eco-friendly and have unique surface-active properties. Biologically derived surfactant can be an excellent surfactant compared to petroleum-derived surfactant [82]. Recently, depth of research have been done in the field of biosurfactant production from glycerol [83, 84]. By using *Pseudomonas aeruginosa* DS10-129, rhamnolipid surfactant of 1.77 g/l was obtained from glycerol. Zhang et al. [85] found rhamnolipids (concentration 15.4 g/l) using *Pseudomonas aeruginosa* cultured on basal mineral medium having glycerol. Rhamnolipids have been used for synthesizing fine chemicals, surface coating, surface characterization, biological control agent, and additive for environmental remediation [85].

5 Conclusions

Biodiesel is highly potential fuel for future transportation needs of India. Nonedible feedstocks (such as such as *Jatropha curcas*, *Pongamia pinnata*, Neem, Castor, Karanja, and Rubber seed) are attractive as they are widely available in India. Nonedible oils are renewable and inexhaustible source of energy with an energetic content close to diesel fuel. Use of biodiesel as transportation fuel reduces emission of various harmful gases such as CO_2, SO_x, and NO_x in comparison with petroleum diesel. Properties of biooil such as cetane number, calorific value, iodine number, and acid value can be improved by transesterification reaction.

References

1. Kumar RS, Sureshkumar K, Velraj R (2015) Optimization of biodiesel production from Manilkara zapota (L.) seed oil using Taguchi method. Fuel 140:90–96
2. Panwar NL, Kaushik SC, Kothari S (2011) Role of renewable energy sources in environmental protection: a review. Renew Sustain Energy Rev 15(3):1513–1524
3. Biresselioglu ME, Zengin Karaibrahimoglu Y (2012) The government orientation and use of renewable energy: case of Europe. Renew Energy 47:29–37
4. No S-Y (2011) Inedible vegetable oils and their derivatives for alternative diesel fuels in CI engines: a review. Renew Sustain Energy Rev 15(1):131–149
5. Habibullah M, Rizwanul Fattah IM, Masjuki HH, Kalam MA (2015) Effects of palm-coconut biodiesel blends on the performance and emission of a single-cylinder diesel engine. Energy Fuels 29(2):734–743

6. Faizal MH, Latiff ZA, Iskandar MAM (2015) Investigation on charecteristics of Pome blended diesel engine. Teknologi 8(75):1–5
7. Chavan SB, Kumbhar RR, Deshmukh RB (2013) Callophyllum Inophyllum Linn ('honne') oil, A source for biodiesel production. Res J Chem Sci 3(11):24–31
8. Satyanarayana M, Muraleedharan C (2011) Comparative studies of biodiesel production from rubber seed oil, coconut oil, and palm oil including thermogravimetric analysis. Energy Sources Part A Recover Util Environ Eff 33(10):925–937
9. Sarin R, Sharma M, Sinharay S, Malhotra RK (2007) Jatropha-Palm biodiesel blends: an optimum mix for Asia. Fuel 86(10–11):1365–1371
10. Dixit S, Kanakraj S, Rehman A (2012) Linseed oil as a potential resource for bio-diesel: a review. Renew Sustain Energy Rev 16(7):4415–4421
11. Kanakraj S, Dixit S, Rehman A (2013) Optimization of degummed Linum usitatissimum methyl ester from methanolysis as a potential resource for diesel engines. Electron J Energy Environ 1(2):9–20
12. Kumar D, Kumar G, Poonam Singh CP (2010) Fast, easy ethanolysis of coconut oil for biodiesel production assisted by ultrasonication. Ultrason Sonochem 17(3):555–559
13. Chakraborty R, Bepari S (2012) Biodiesel synthesis from mustard oil (brassica nigra) using calcium oxide-calcium aluminate catalyst developed from duck eggshell. Int J Adv Sci Res Technol 3(2):242–256
14. Narkhede N, Patel A (2013) Biodiesel production by esterification of oleic acid and transesterification of soybean oil using a new solid acid catalyst comprising 12-tungstosilicic acid and zeolite Hβ. Ind Eng Chem Res 52:13637–11364
15. Pukale DD, Maddikeri GL, Gogate PR, Pandit AB, Pratap AP (2015) Ultrasound assisted transesterification of waste cooking oil using heterogeneous solid catalyst. Ultrason Sonochem 22:278–286
16. Kumar A, Sharma S (2011) Potential non-edible oil resources as biodiesel feedstock: an Indian perspective. Renew Sustain Energy Rev 15(4):1791–1800
17. Patel NK, Nagar PS, Shah SN (2013) Identification of non-edible seeds as potential feedstock for the production and application of bio-diesel. Energy Power 3(4):67–78
18. Shikha K, Rita CYJ (2012) Biodiesel production from non edible-oils: a review. J Chem Pharm Res 4(9):4219–4230
19. Karmakar A, Karmakar S (2012) Biodiesel production from neem towards feedstock diversification: Indian perspective. Renew Sustain Energy Rev 16(1):1050–1060
20. Ahmed WA, Salimon J (2009) Phorbol ester as toxic constituents of tropical Jatropha curcas seed oil. Eur J Sci Res 31(3):429–436
21. Tsou J, Chang K, Wang W, Tseng JT, Su W, Hung L, Chang W, Chen B (2008) Nucleolin regulates c-Jun/Sp1-dependent transcriptional activation of cPLA 2 a in phorbol ester-treated non-small cell lung cancer A549 cells. Nucleic Acids Res 36(1):217–227
22. Pramanik K (2003) Properties and use of Jatropha curcas oil and diesel fuel blends in compression ignition engine. Fuel Energy Abstr 44(3):170
23. Kumar SS, Purushothaman K (2012) High FFA Rubber seed oil as an alternative fuel for diesel engine—an overview. Int J Eng Sci 1(10):16–24
24. Sharma YC, Singh B (2010) An ideal feedstock, kusum (Schleichera triguga) for preparation of biodiesel: optimization of parameters. Fuel 89(7):1470–1474
25. Singh RK (2015) Kusum—the best lac host tree for Kusmi lac production. Pop Kheti 3 (2):55–58
26. Naveen Kumar HSP (2012) "Kusum oil as a potential fuel oil for CI engene. Conference Paper
27. Leonor C, Forero B, De Paula F (2001) Biodiesel from castor oil: a promising fuel for cold weather. Power 12:4
28. Sattanathan R (2015) Production of biodiesel from castor oil with its performance and emission test. Int J Sci Res 4(1):273–279
29. Rao SBH, Rao VT, Reddy HCK (2013) Palm oil and Calophyllum inophyllum oil are potential feed stocks for future biodiesel in compression ignition engines: a review. Int J Mech Eng Technol 4(5):301–312

30. Borugadda VB, Goud VV (2012) Biodiesel production from renewable feedstocks: status and opportunities. Renew Sustain Energy Rev 16(7):4763–4784
31. Velmurugan A, Loganathan M, Gunasekaran EJ (2014) Experimental investigations on combustion, performance and emission characteristics of thermal cracked cashew nut shell liquid (TC-CNSL)—diesel blends in a diesel engine. Fuel 132:236–245
32. Kumar M, Sharma M (2015) Assessment of potential of oils for biodiesel production. Renew Sustain Energy Rev 44:814–823
33. Khan SA, Hussain MZ, Prasad S, Banerjee UC (2009) Prospects of biodiesel production from microalgae in India. Renew Sustain Energy Review 13:2361–2372
34. Veillette M, Chamoumi M, Nikiema J, Faucheux N, Heitz M (2012) Production of biodiesel from microalgae. Adv Chem Eng 2:245–268
35. Vyas AP, Verma JL, Subrahmanyam N (2010) A review on FAME production processes. Fuel 89:1–9
36. Kumar M, Sharma MP (2014) Status of biofuel production from microalgae in India. J Inter Sci Technol 2(2):72–75
37. Abbaszaadeh A, Ghobadian B, Omidkhah MR, Najafi G (2012) Current biodiesel production technologies: a comparative review. Energy Convers Manag 63:138–148
38. Vyas AP, Verma JL, Subrahmanyam N (2011) Effects of molar ratio, alkali catalyst concentration and temperature on transesterification of Jatropha oil with methanol under ultrasonic irradiation. Adv Chem Eng Sci 1(2):45–50
39. Motasemi F, Ani FN (2012) A review on microwave-assisted production of biodiesel. Renew Sustain Energy Rev 16(7):4719–4733
40. Silitonga AS, Atabani AE, Mahlia TMI, Masjuki HH, Anjum I, Mekhilef S (2011) A review on prospect of Jatropha curcas for biodiesel in Indonesia. Renew Sustain Energy Rev 15 (8):3733–3756
41. Lotero E, Liu Y, Lopez DE, Suwannakarn K, Bruce DA, Goodwin JG (2005) Synthesis of biodiesel via acid catalysis. Ind Eng Chem Res 44:5353–5363
42. Glisic S, Skala D (2016) Design of three step purification. In: 2nd International congress on green process engineering
43. Aalam CS, Saravanan C (2015) Biodiesel production techniques: a review. Int J Res Appl Sci Eng Technol 3(6):41–45
44. Han H, Cao W, Zhang J (2005) Preparation of biodiesel from soybean oil using supercritical methanol and CO_2 as co-solvent. Process Biochem 40:3148–3151
45. Gerpen JV (2005) Biodiesel processing and production. Fuel Process Technol 86:1097–1107
46. Mahamuni NN, Adewuyi YG (2009) Optimization of the synthesis of biodiesel via ultrasound-enhanced base-catalyzed transesterification of soybean oil using a multifrequency ultrasonic reactor. Energy Fuels 51(2):2757–2766
47. Demirbas A (2007) Importance of biodiesel as transportation fuel. Energy Policy 35:4661–4670
48. Murugesan A, Umarani C, Chinnusamy TR, Krishnan M, Subramanian R, Neduzchezhain N (2009) Production and analysis of bio-diesel from non-edible oils—a review. Renew Sustain Energy Rev 13(4):825–834
49. Balat M (2011) Potential alternatives to edible oils for biodiesel production—a review of current work. Energy Convers Manag 52(2):1479–1492
50. Atadashi IM, Aroua MK, Aziz AA (2010) High quality biodiesel and its diesel engine application: a review. Renew Sustain Energy Rev 14(7):1999–2008
51. Bozbas K (2008) Biodiesel as an alternative motor fuel: production and policies in the European Union. Renew Sustain Energy Rev 12(2):542–552
52. Demirbas A (2009) Progress and recent trends in biodiesel fuels. Energy Convers Manag 50 (1):14–34
53. Bowman M, Hillgoss D, Rasmussen S (2006) Biodiesel: a renewable and Reprinted from: Biodiesel: a renewable and biodegradable fuel New US specification ensures product identity and quality for biodiesel. Hydrocarbon Process 85:103–106

54. Pasqualino J, Montane D, Salvado J (2006) Synergic effects of biodiesel in the biodegradability of fossil-derived fuels. Biomass Bioenergy 30(10):874–879
55. Zhang X, Peterson C, Reece D, Moller G, Haws R (1998) Biodegradability of biodiesel in the aquatic environment. Trans ASAE 41(5):1423–1430
56. Behr A, Eilting J, Irawadi K, Leschinski J, Lindner F (2008) Improved utilisation of renewable resources: new important derivatives of glycerol. Green Chem 10(1):13
57. Guerrero-Perez M, Rosas J, Bedia J, Rodriguez-Mirasol J, Cordero T (2009) Recent inventions in glycerol transformations and processing. Recent Pat Chem Eng 2(1):11–21
58. Diaz I, Marquez-Alvarez C, Mohino F, Perez-Pariente J, Sastre EJ (2000) Combined alkyl and sulfonic acid functionalization of MCM-41-type silica part 2. Esterification of glycerol with fatty acids. J Catal 193(2):295–302
59. Pérez-Pariente J, Díaz I, Mohino F, Sastre E (2003) Selective synthesis of fatty monoglycerides by using functionalised mesoporous catalysts. Appl Catal A Gen 254(2):173–188
60. Nakamura R, Komura K, Sugi Y (2008) The esterification of glycerine with lauric acid catalyzed by multi-valent metal salts. Selective formation of mono- and dilaurins. Catal Commun 9:511–515
61. Kosmider A, Leja K, Czaczyk K (2011) Improved utilization of crude glycerol by-product from biodiesel production in chapter 19. In: Montero G, Stoytchera M (eds) Biodiesel—quality, emissions and by-products. Intech Open Acces Publisher, Croatia, pp 341–365
62. Pagliaro M, Ciriminna R, Kimura H, Rossi M, Della Pina C (2007) From glycerol to value-added products. Angew Chem Int Ed Engl 46(24):4434–4440
63. Karinen RS, Krause AOI (2006) New biocomponents from glycerol. Appl Catal A Gen 306:128–133
64. Rahmat N, Abdullah AZ, Mohamed AR (2010) Recent progress on innovative and potential technologies for glycerol transformation into fuel additives: a critical review. Renew Sustain Energy Rev 14:987–1000
65. Klepá K, Mravec D, Hájeková E, Bajus M (2003) Etherification of glycerol. Pet Coal 45:54–57
66. Srinivas M, Sree R, Raveendra G (2014) Selective etherification of glycerol with tert- butanol over 12-tungstophosphoric acid catalysts supported on Y-zeolite. Indian J Chem 53:524–529
67. Behr A, Leschinski J, Awungacha C, Simic S, Knoth T (2009) Telomerization of butadiene with glycerol: reaction control through process engineering, solvents, and additives. ChemSusChem 2(1):71–76
68. Dasari MA, Kiatsimkul P-P, Sutterlin WR, Suppes GJ (2005) dasari 2005.pdf. Appl Catal A Gen 281:225–231
69. Johnson DT, Taconi KA (2007) The glycerin glut: options for the value-added conversion of crude glycerol resulting from biodiesel production. Environ Prog 26(4):338–348
70. Saxena RK, Anand P, Saran S, Isar J (2009) Microbial production of 1, 3-propanediol: recent developments and emerging opportunities. Biotechnol Adv 27(6):895–913
71. Kurosaka T, Maruyama H, Naribayashi I, Sasaki Y (2008) Production of 1,3-propanediol by hydrogenolysis of glycerol catalyzed by Pt/WO3/ZrO2. Catal Commun 9(6):1360–1363
72. Lewandowski G, Bartkowiak M, Milchert E (2008) Low-waste technology of glycerine epichlorohydrin production. Oxid Commun 31(1):108–115
73. Bell BM, Briggs JR, Campbell RM, Chambers SM, Gaarenstroom PD, Hippler JG, Hook BD, Kearns K, Kenney JM, Kruper WJ, James Schreck D, Theriault CN, Wolfe CP (2008) Glycerin as a renewable feedstock for epichlorohydrin production. The GTE process. Clean Soil Air Water 36(8):657–661
74. Corma A, Huber GW, Sauvanaud L, O'Connor P (2008) Biomass to chemicals: catalytic conversion of glycerol/water mixtures into acrolein, reaction network. J Catal 257(1):163–171
75. Ott L, Bicker M, Vogel H (2006) Catalytic dehydration of glycerol in sub- and supercritical water: a new chemical process for acrolein production. Green Chem 8(2):214–220
76. Nguyen B-C, Kochevar IE (2003) Factors influencing sunless tanning with dihydroxyacetone. Br J Dermatol 149(2):332–340

77. Enders D, Voith M, Lenzen A (2005) The dihydroxyacetone unit—a versatile C 3 building block in organic synthesis. Angew Chem Int Ed Engl 44(9):1304–1325
78. Garcia R, Besson M, Gallezot P (1995) Chemoselective catalytic oxidation of glycerol with air on platinum metals air on platinum metals. Science 127:165–176
79. Lanjekar K, Rathod VK (2013) Utilization of glycerol for the production of glycerol carbonate through greener route. J Environ Chem Eng 1(4):1231–1236
80. Rokicki G, Rakoczy P, Parzuchowski P, Sobiecki M (2005) Hyperbranched aliphatic polyethers obtained from environmentally benign monomer: glycerol carbonate. Green Chem 7(7):529
81. Vieville C, Yoo JW, Pelet S, Mouloungui Z (1998) Synthesis of glycerol carbonate by direct carbonatation of glycerol in supercritical CO_2 in the presence of zeolites and ion exchange resins. Catal Lett 56:245–247
82. Banat IM, Franzetti A, Gandolfi I, Bestetti G, Martinotti MG, Fracchia L, Smyth TJ, Marchant R (2010) Microbial biosurfactants production, applications and future potential. Appl Microbiol Biotechnol 87(2):427–444
83. da Silva GP, Mack M, Contiero J (2009) Glycerol: a promising and abundant carbon source for industrial microbiology. Biotechnol Adv 27(1):30–39
84. Rahman KSM, Rahman TJ, McClean S, Marchant R, Banat IM (2002) Rhamnolipid Biosurfactant production by strains of Pseudomonas aeruginosa using low-cost raw materials. Biotechnol Prog 18(6):1277–1281
85. Zhang G, Wu Y, Qian X, Meng Q (2005) Biodegradation of crude oil by Pseudomonas aeruginosa in the presence of rhamnolipids. J Zhejiang Univ Sci B 6(8):725–730

Chapter 8
Potential of Bioelectricity Production in India Through Thermochemical Conversion of Lignocellulosic Biomass

Rakesh Punia, R. R. Marlar, Sachin Kumar, and S. K. Tyagi

Abstract India is an agricultural nation and its economy mainly depends on the agriculture. About 500 million tons (MT) of crop residues are generated out of total agriculture produce per year, which are either used as animal feed, livestock bedding, packaging material, cooking in households, papermaking, etc. or burned in the farms and create the environment pollution along with deterioration of the farm fertility. However, there is a great potential if the surplus biomass could be utilized as a feedstock to produce bioelectricity through gasification and combustion. About 23% of India's land is also covered by forest. Forestry waste and other agro-wastes are readily available to use these wastes for the electricity production. In this chapter, the availability of biomass, gasification and combustion technologies and its potential in power generation have been described.

Keywords Biomass energy • Bioelectricity • Gasification • Combustion

1 Introduction

Nonconventional energy from biomass sources is generally obtained as a by-product of crop cultivation, wood industry and agro-processing. India has biggest issue to make the correlation between the developments of nonconventional bioenergy and the maintenance of ecology. The drastic growing Indian population creates a great hurdle to implement the programmes of biomass energy. Sustaining

R. Punia
Department of Chemical Engineering, Motilal Nehru National Institute of Technology Allahabad, Allahabad 211004, India

R.R. Marlar • S.K. Tyagi
Thermochemical Conversion Division, Sardar Swaran Singh National Institute of Bio-Energy, Kapurthala 144601, India

S. Kumar (✉)
Biochemical Conversion Division, Sardar Swaran Singh National Institute of Bio-Energy, Kapurthala 144601, India
e-mail: sachin.biotech@gmail.com

A.K. Chandel, R.K. Sukumaran (eds.), *Sustainable Biofuels Development in India*,
DOI 10.1007/978-3-319-50219-9_8

growth of energy utilization requires the proper implementation of bioenergy programmes and upgradation of strategy according to outcomes of implemented programmes for a small span of time. In rural areas, more than 85% of the population still have a heavy demand of energy resources to meet the daily energy requirement from traditional fuels. Comparatively, urban areas have about 35% demands of energy resources from traditional fuel. In most developing countries, this is one third of the total required energy and is reaching the levels as high as 75–90% for cooking and heating purposes [1–3].

Globally the biomass has 12% contribution to the generation of energy from other energy sources, but it is 18–49% in developing countries. In India, about 32% of the total energy demand might be achieved by the available biomass. Over 500 million tons of biomass is annually generated in India. The potential of India in biomass energy generation is about 1600 MT for domestic sector, excluding through cogeneration. Thermal conversion (gasification and combustion) processes are best suitable option for the processing of biomass feedstock for energy requirements. Biomass resource is the best alternative source of energy to the fossil fuels (gasoline, residue oil, coal, etc.) and to the decentralized electricity production from a capacity from 10 kW to more than a megawatt (MW) [4, 5].

The Ministry of New and Renewable Energy (MNRE) of the Indian government established four Action Research Centres (ARCs) in 1990 to promote and support the research and development activities of biomass energy through gasification technology in different areas of the nation. Installed capacity of power generation from biomass gasification is around 128 MW in India [6] and expected to change soon by various Indian government programmes for technology advancement. On a global scale, India ranked fourth for the most attractive country in renewable energy investment, just behind the USA, Germany and China. By the year 2022, MNRE has target about 10,000 villages to make them biomass energy efficient. Private sectors are largely promoted to invest in renewable power generation in India by a conducive, strong and investor-friendly environment policy.

2 Biomass Potential

Biomass is a biodegradable and renewable organic compound, originating from plants, microorganisms and animals. Biomass has been utilized as a source of energy by humans for heating, cooking and lighting since the Stone Age. It was not until about 400,000 years ago that reusable structures to contain fire began to be used by hunter-gatherers. As settled groups of humans began to form, the uses for biomass as an energy source expanded. It was still the primary heating and cooking source, but it also began to be used to process materials for building and tools.

Broadly biomass may categorize in two groups: (1) primary or virgin and (2) waste biomass. Biomass may also be divided into two broad groups: (a) virgin and (b) waste biomass. Direct from plants or animal biomass is called as primary biomass which is generally collected especially for the purpose of producing energy

Table 8.1 Range of biomass available for biofuel production

Virgin	Biomass from terrestrial	Forest grows, energy crops, cultivated crops, grasses
	Aquatic biowastes	Water plant, algae
Waste	Municipal wastes	Municipal solid biowaste, landfill biowastes, sewage
	Agricultural solid wastes	Livestock and manures, residue of agricultural crops
	Forest residues	Bark, floor residues, leaves
	Industrial biowastes	Black liquor, demolition wood, waste fats or oils

for shorter time. Waste biomass is a secondary biomass which is generated from agricultural field, municipal waste, forest residues, etc. Primary biomass comprise woody energy, herbaceous crops, aquatic crops and industrial crops likewise sugar cane, eucalyptus, poplars, cotton, sorghum, sunflowers, etc. These biomass sources are best for the process of combustion, pyrolysis and gasification to produce biofuels, synthesis gas and hydrogen gas. Annually, India has large quantity of agricultural solid residues generated which is big tussle to dispose off. Table 8.1 shows the broader fields of primary and waste biomass [7].

2.1 Potential for Biomass in India

India has great capability for biomass energy production. The majority of Indian rural areas use traditional biomass such as animal dung, wood stalks, forest and agricultural biomass wastes to get energy and light. Biomass is best suitable source to provide energy for Indian households and industry. Biomass is commonly utilized as domestic fuel apart from being good energy source for many small-scale industries and fuel for captive power plants. About 500 MT/year is the estimate production of biomass in India from agriculture, forest covers and agro-industry. Punjab, Uttar Pradesh, Maharashtra and Madhya Pradesh are the Indian states having highest potential of biomass production.

2.1.1 Forest and Tree Cover Biomass Potential

Energy from biomass can be produced by direct consuming it, and they are grossly available in nature. Renewable biomass resources for energy have about one third shares to total energy demand of developing countries.Three quarters of the population of India are in the rural areas, and the rural energy demand can be fulfilled by this available biomass resource up to three fourths of its total energy demand. Indian forest and tree cover has share of about 23.4% to the total geographical area of the country. Biomass has also an indirect use as a fuel for different energy utilizations such as transport fuel, electrical energy, thermal energy, specialty chemicals, etc. [8].

Table 8.2 Availability of forest-type biomass in India

Forest type	Total forest cover area in (ha)	Volume in (m^3/ha)	Timber wood in (Tons/ha)	Branches and foliages in (tons/ha)	Stumps and roots in (tons/ha)	Subtotal biomass in (tons/ha)	Total biomass in (MT)
Deciduous	12,684.23	316.06	161.316	38.716	51.621	252.653	3.19
Evergreen	3962.23	428.23	196.998	47.277	63.036	307.301	1.22
Secondary deciduous	2960.28	216.67	154.983	37.196	49.595	241.773	0.72
Southern thorn	6676.15	73.03	42.282	10.148	13.530	65.960	0.44
Euphorbia shrub	304.4	52.72	36.860	8.846	11.795	57.5	0.02
Total	26,587.80	1086.71	592.428	142.183	189.577	924.188	5.58

Indian surplus land could generate 62–310 MT of wood annually, after fulfilment of total conventional demand of biomass likewise industrial wood, swan wood and domestic fuelwood. The availability of forest biomass in India is shown in Table 8.2. The plantation biomass has annual potential of 930–4650 PJ, which could supply about 3–18 % of total energy consumption [9, 10]. During year 2011, the growing stock of India was 6.047 billion m^3 with corresponding comparison to forest cover of 4.499 billion m^3 and tree cover of 1.548 billion m^3. On an average, the forest cover growing stock per hectare was 59.79 $m^{3.}$ While, for the year 2005, the forest cover growing stock was 4.602 billion m^3 and 1.616 billion m^3 for tree cover. The percentage difference of growing stock of tree cover to the forest cover is 34.14% and 35.12% during 2003 and 2005, respectively [5, 11].

Biofuel sources have higher potential from plantation forest tree resources outside forest (TOF) of 5.07% share of total tree biomass. Short rotation tree plant of regular leaf shedding plantation forests have high carbon sequestration potential. The car and char briquette can be prepared from fast-growing conifers such as chir pine-produced litter (pine needles). The ideal plant for wasteland that is salt affected is *Eucalyptus sp.* to increase the biomass of such land [12].

The herbaceous and non-perennial woody biomass can be collected from Indian forests in huge quantity. The energy requirement for about 80,000 remote villages of the country needs strategic and proper implementation of plans. Therefore, 12.6 M kcal energy would be required in a year or 1.05 M kcal in a month. Only forest biomass is insufficient for this huge energy demand. India has agricultural strength and the crop residue of nation is abundantly available. The gap of biomass demand for energy production could be achieved by adding this available crop residue stock to the surplus forest biomass.

2.1.2 Crop Residue Biomass Potential

Indian economy is mainly supported by agriculture and this sector has strong potential for India's bioenergy strategic planning. The contribution of the agricultural sector is only 17% to the Indian GPD, and it is the livelihood source to 60% of its population. National agricultural practice performance acquires about 60% of available arable land [13]. The arable land of nation is 159 million hectare (Mha), which is 11.2% of total global arable land. In view to the agricultural production worldwide, India stands first in producing jute and is ranked second in rice, wheat, pulses, sugar cane, cotton and groundnut. This makes strong potential in the field of crop residue biomass for India [14–17].

Ministry of Agriculture (MoA), government of India, has crop statistics of India, and it is the major source of data for estimation of available crop residue biomass [18]. The five years (2003–2004 and 2007–2008) assessed crop residue data shows an average of 686 MT/annum of gross residue available in India from 39 types of residue of 26 crops [19]. Out of available gross crops residue, the contribution of cereals, pulses, oilseed and sugar canes has 545 MT; horticulture crops such as banana, areca nuts and coconut have 61 MT, and others (jut and cotton) have 80 MT. Crop-wise, the highest contribution of 368 MT (54%) is through cereals followed by 111 MT (16%) by sugar cane. Individually crop residue generation is highest by rice (154 MT) then by wheat (131 MT).

The surplus stock of biomass residues from some selected crops is about 34% available in India. The annual national potential of surplus residue is about 234 MT. Different crop-wise contributions in the total surplus residue are in the order of cereal (89 MT), sugar cane (56 MT), miscellaneous crops (47 MT), cotton (47 MT), rice (43 MT), horticulture (23 MT), oilseeds (14 MT) and legume crops (5 MT). Indian potential of gross and surplus residue of selected crops is presented in Table 8.3. Among all crops, the highest gross residue is produced by rice crop, but the surplus residue of sugar cane is high because rice husk/straw has other uses. From horticulture crops, coconut and banana also hold good contribution in surplus residue potential.

Generally, variation in the different data to other study could be due to the variation in crop selection, residue production ratio (RPR) and heating values and surplus residue fraction [20].

The potential of crop residue varies drastically among different states of India. The minimum gross potential 0.21 MT is of Mizoram and maximum 121 MT is of Uttar Pradesh. The most supported land of India for agriculture is of Uttar Pradesh, and about 90% of state agricultural residue is generated by sugar cane, wheat and rice. Sugar cane is widely cultivated in the state. Punjab is an advanced state of India in the agricultural field, and it contributes about 83 MT of gross residue.

Annual surplus portion of biomass energy potential from residue in India would be 4.15 EJ and contributes about 17% to the primary utilization of Indian energy in 2011 [21]. Cereals contribute 1.49 EJ followed by sugar cane (1.11 EJ), miscellaneous (0.83 EJ), horticultural (0.41 EJ), oilseed (0.23 EJ) and legumes (0.08 EJ) to

Table 8.3 Potential of gross and surplus crop residue biomass in India [19]

Crop category	Crop	Gross potential (MT)	Surplus potential (MT)
Cereals	Rice	154.0	43.5
	Wheat	131.1	28.4
	Maize	35.8	9.0
	Bajra	24.3	5.1
	Barley	1.6	0.2
	Small millet	0.6	0.1
	Rage	2.7	0.3
	Jawar	17.6	3.5
Oilseeds	Mustard and rapeseed	12.7	4.9
	Sesame	0.8	0.1
	Linseed	0.3	0.0
	Niger	0.1	0.0
	Safflower	0.6	0.5
	Soybean	13.5	4.6
	Groundnut	17.0	3.0
	Sunflower	3.8	0.6
Pulses	Tur (arhar)	7.2	1.4
	Gaur	2.6	1.8
	Gram	6.4	1.6
	Lentil	1.7	0.3
Sugar cane	Sugar cane	110.6	55.7
Horticulture	Banana	41.9	12.3
	Coconut	18.0	9.7
	Areca nut	1.5	0.5
Others	Cotton	75.9	46.9
	Jute	3.9	0.4
Total (MT)		686.0	234.5

the total primary energy consumption (24.91 EJ) of India. Sugar cane residue contributes highest at individual crop level.

3 Biomass Energy

Generation of thermal or electrical energy using renewable resource technology exists for large-scale or small-scale distribution systems. Over 44 million households worldwide rely on biomass-based fuels for cooking and lighting, and over 166 million use new high-efficiency biomass-burning stoves. Electricity is the most vital form of energy input demand for infrastructure development of the country in agriculture and industry. According to the repost of Ministry of Power, the total installed capacity of power production in 2014 is about 255.681 GW, in which

Table 8.4 Sector-wise, all India installed capacity GW as of December 2014

Sector	Thermal (GW)				Nuclear (GW)	Hydro (GW)	Renewable energy sources (GW)	Grand total (GW)
	Coal	Gas	Diesel	Total				
Central	46.525	7.429	0.00	53.954	4.780	10.691	0.00	69.425
State	55.891	6.974	.602	63.467	0.00	27.482	3.804	94.753
Private	51.755	8.568	.597	60.920	0.00	2.694	27.888	91.503
All India	154.171	22.971	1.199	178.341	4.780	40.867	31.692	255.681

Table 8.5 Break-up of renewable energy source

Mini hydroplant (GW)	Wind power (GW)	Biopower (GW)		Solar power (GW)	Total capacity (GW)
		Biomass/ cogeneration power	Waste to energy		
3.804	21.136	4.014	0.1066	2.632	31.692

generation through thermal source is 178.342 GW and renewable energy source is 31.692 GW. The thermal power contribution is about 69% followed by 16% of hydropower. The share of nuclear power is 2% (smallest), and the remaining 12% power generation through renewable sources (Tables 8.4 and 8.5) [22]. Renewable energy is derived from naturally replenishing resources such as the sun, wind, rain, tides and geothermal heat. About 405 EJ (81%) of total energy production is fossil fuel based. The wide variety of renewable energy resources will have regional access to some energy from these sources [23].

The next major technological advancement occurred with industrialization and the understanding of energy conversion procedures (Fig. 8.1). There are many advanced techniques for conversion of biomass to energy, which were began to be implemented, and many more were developed. There are many current technologies for harnessing the energy contained in biomass that run across a wide range of methods, products and uses.

Further understanding of thermodynamics and chemistry has allowed for the development of many different methods of harnessing the energy contained in biomass. Thermochemical conversions make up the majority of the current use of surplus biomass. Carbonaceous fuel thermochemical conversion techniques were being developed in the early 1800s. Coal and peat were being turned into gaseous fuels for town lighting and heat.

3.1 Thermochemical Conversion

Biomass energy is produced through different thermochemical processes; gasification, pyrolysis and direct combustion have multiple applications of energy to

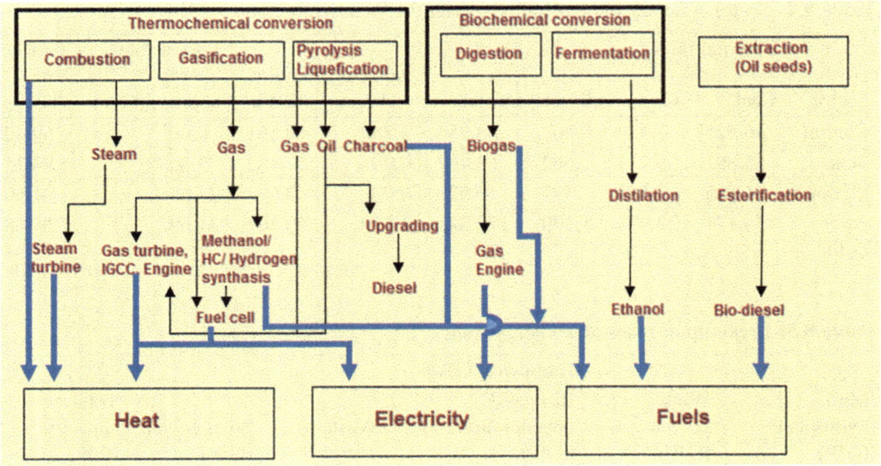

Fig. 8.1 Current biomass conversion method overview

essential chemicals. Among thermochemical conversion processes, gasification is the only one promising a more energy-efficient process than that of combustion [24, 25]. Thermochemical conversion of biomass is the only technique, which best suits to achieve present energy demand from available fuel resources. Combustion is widely applied on the domestic scale in both developing and developed nations for space heating and cooking. It is also extensively used on an industrial scale for process heat and electricity and also commercially applicable for the production of electricity and heat generation. However, for the fuel production of hydrogen and hydrocarbons, it is still currently in development phases [26]. Pyrolysis is the thermochemical biomass conversion in the absence of oxygen to get the products like charcoal, biocrude oil and tars. It has been commercially used for the production of charcoal for a long time.

4 Biomass Gasification

Gasification is the technique to obtain energy from solid fuel through its gaseous conversion. The conversion of solid biomass to gaseous fuel using gasification technology has been utilized in different forms since 200 years. The process was discovered in 1798 in France and England independently. Thermal conversion of solid biomass through gas-forming reactions is gasification (indirect combustion) to generate syngas/flue gas. Gasification is also defined as a partial combustion of solid organic materials in the presence of limiting oxygen as stoichiometric combustion requirement. Delignification process can be eliminated by the gasification and pyrolysis to produce syngas from lignocellulosic materials. Biochemical

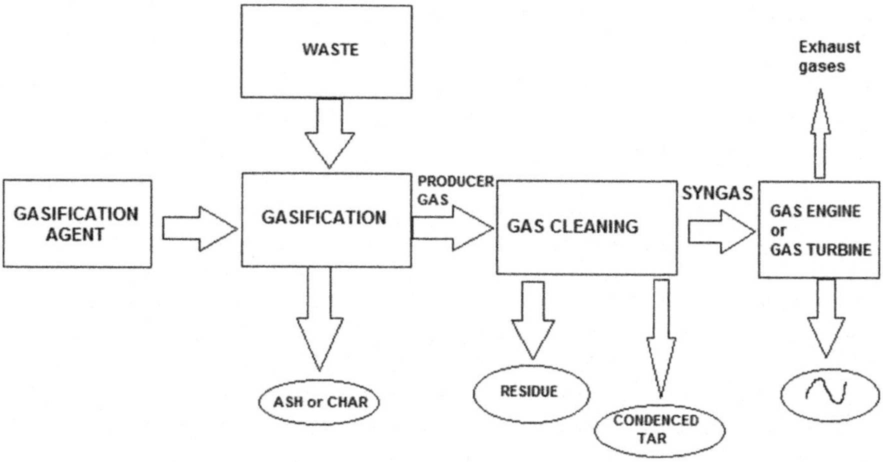

Fig. 8.2 Schematic diagram of the gasification-based power production

products are produced by using various reforming processes to convert this syngas into different alternative compounds of fossil chemicals. Syngas purification and separation process generates eco-friendly clean fuels such as hydrogen, methane, etc. Gasification technology has its unique ability to accept broad range of feedstock which is directly linked with strength of population. By the various unique abilities of gasification, it is best suitable process for generating biomass energy from solid waste.

Gasification converts biomass waste into syngas, a mixture of some heating value compounds (carbon monoxide, hydrogen, methane), along with undesired contaminants such as char particles, tar, chlorides, alkali metals and sulphide. Compositions of products depend upon the macro and micro characteristics of biomass feedstock. Gasification of biomass consists of some important chemical reactions such as Boudouard reaction, water gas or steam, hydrogasification, shift reaction, methanation and steam reforming. There are several different types of gasifiers that take advantage of different aspects of these reactions to achieve some desired effect on the product gas.

In terms of operating conditions, efficient utilization of syngas produced from gasification of solid biomass waste is far better than traditional combustion. Syngas from gasification can be utilized as gaseous fuel in boilers, gas turbine, furnaces and DG sets with efficient energy-saving and ecological preservations. The schematic of the gasification-based power production is shown in Fig. 8.2. Small-scale gasifiers can be operated in rural or remote areas with local biomass wastes. This technique has capability to resolve the problem of rural or remote area electricity supply. Gasifiers are commonly categories on the basis of flow pattern of feedstock and produced syngas. Some of the commonly used types of gasifiers are downdraft, fixed bed, entrained bed, fluidized bed, vertical shaft, moving grate furnace, plasma furnace and rotary kiln.

4.1 Indian Context of Energy from Biomass Gasification

The most successful utilization of biomass for energy is sugar cane bagasse in agriculture, manure in livestock residues and pulp and paper residues in forestry. Electric power generation potential of India from bagasse-based cogeneration capacity is of 3500 MW and 16,000 MW from surplus biomass. Present scenario of biomass gasification power plants in India is 537 MW commissioned and 536 MW is in under construction. India has a total installed biomass power plants of 288 with the capacity of about 2.7 GW. Biomass energy technology contributes about 34% to the total energy consumption of India with capacity of 18 GW electricity generation potential from biomass resources [27, 28]. The total contribution of India in providing biomass to the world demand is more than 1% (257 TWh per year). Depending on the technology, the direct heat utilized in industry through biomass and waste combustion is about 4.5 EJ, and heat of 2–3 EJ is used as combined heat and power (CHP) plants to residential sector. Generally, on findings of developing countries, these heat utilized exclude the traditional combustion of biomass. Bagasse from Indian 550 sugar mills has considerable contribution (5 GW) in the power production [29].

From the last six decades, the energy consumption of India increased by 16%, whereas the capacity of electricity generation by 83%, which positioned India to the fifth highest energy consumption country. At present, India faces the crisis of electricity supply, which is 15% shortfall in peak demand. The capacity of installed biomass-based power plant of India is doubled from 2.56 GW to 4.12 GW from 2010 to 2014, respectively [22, 29].

MNRE has been supporting the country for biomass gasification energy technique to promote and coordinate research activities in various fields by four Action Research Centres (ARCs).

Defined fields of various Action Research Centres are:

1. Indian Institute of Technology, Delhi: biomass characterization and developing process technology
2. Indian Institute of Technology, Bombay: developing product research, modification of technology, method standardization, process inventing, quality assurance criterion, cost optimization, testing methodology and instrumentation
3. Indian Institute of Science, Bangalore: R&D of nonwoody biomass gasification and scaling up wood-based systems
4. Madurai Kamaraj University: monitoring revalidation, testing training centre, field evaluation, development of procedures and implementation of application packages

Indian Ministry also focused on the electric power plants of biomass gasifiers using local available biomass waste like rice husk, wood stalks, cotton stalks, mustard stalks, etc. Mainly biomass gasifier programmes are involved to provide power for off-grid/remote areas, captive power plant industries and up to 2 MW capacities tail-end grid-connected power projects.

The focus of the biomass gasifier programme is to find the best alternates of fossil fuels such as coal, gasoline, residue oil, etc. Small-scale biomass power plants in different areas are nearer to the biomass feedstock, connecting them to the main power grid for multiple benefits such as electric supply to remote areas, reduction of transportation cost, distribution loss, etc. Grid and off-grid projects of about 150 MW capacity biomass gasification have been installed. For meeting captive power and thermal applications rice mills and other industries approximately 300 gasifier systems are installed. Furthermore, nearly seventy biomass gasifier plants are providing electricity power to 230 villages in the country.

4.2 *Statewise Biomass Gasifiers*

Biomass gasifiers are approximately installed in over all states of India. Gujarat ranked one with 24% and Tamil Nadu ranked second with 15% to the total installed gasifiers in India. Gujarat alone has the highest number of biomass gasifier installations with 101. From total installed gasifiers in the country, nearly 67% are accounted for heating utilization, while 33% utilized for electricity production. Most of the electricity generation applications are for captive or decentralized electricity generation and distribution. Some states with most potential for biomass production are Punjab (150 MW), Andhra Pradesh (200 MW), Bihar (200 MW), Gujarat (200 MW), Karnataka (300 MW), Tamil Nadu (350 MW), Maharashtra (1000 MW) and Uttar Pradesh (1000 MW). Gasifiers have numerous fields of utilization other than heating such as CO_2 generation, alternate chemicals, metal processing, food processing, ceramic industry, etc.

4.3 *Environmental Performance of Gasifiers*

The thermochemical conversion of solid biomass through gasification is best suitable energy generation technology for human being without disturbing ecological system. Gasification is clean and efficient technology to produce chemicals and energy fuels to control the air pollution along economic cost. The flue gas from gasification process is a better process for getting clean exhaust gas than incinerator to decompose biowaste. The basic healthy living and power requirement for rural areas being less than 1 kWh/day/person, small-scale downdraft gasifiers running generators could provide that power for 100 to nearly 1500 people. However, in rural areas of developing countries, the use of small-scale distributed energy production with regionally available renewable sources can be a cost-effective alternative to extending existing grids or constructing new plants. Bioenergy technology mainly depends upon the availability of biomass, and its supply from nearby areas will affect the performance of power generation and cost. This will make one nation independent to the foreign fossil fuel import. Important factors influencing the utilization of bioenergy in a rural or remote area generally depend upon agro-climate conditions.

5 Combustion

The largest portion of the world's power is derived from combustion of fuels. Combustion is the process of exothermic oxidation of fuel rapidly. Broadly, the union of carbon, hydrogen and sulphur with oxygen is termed as combustion. If the products of combustion are CO_2, water and SO_2, respectively, from the above three elements, the combustion is termed as complete combustion. If CO appears in the product gas, it is called incomplete or partial combustion because CO can further combine with oxygen to produce CO_2 [30]. Combustion means stored chemical energy in the fuel and is released to heat, light, infrared radiation and other forms of energy. Fuel, oxygen and high temperatures are the important requirements for combustion reaction [31]. Ignition of biomass materials requires at least 550 °C, so the starting part of combustion is bit difficult. After ignition with guaranteed air supply, the combustion will proceed continuously up to the complete combustion of fuel [32]. The schematic representation of combustion is shown in Fig. 8.3.

When heat is applied to the surface layer of biomass through flame from match or flint, the rays are focused on the biomass. The different stages of combustion are shown in Fig. 8.4. Combustion is the most direct process to utilize as a useful energy. When the temperature reaches the 100 °C, drying of biomass takes place [33]. Biomass is very complex in nature which contains moisture stored in the pores

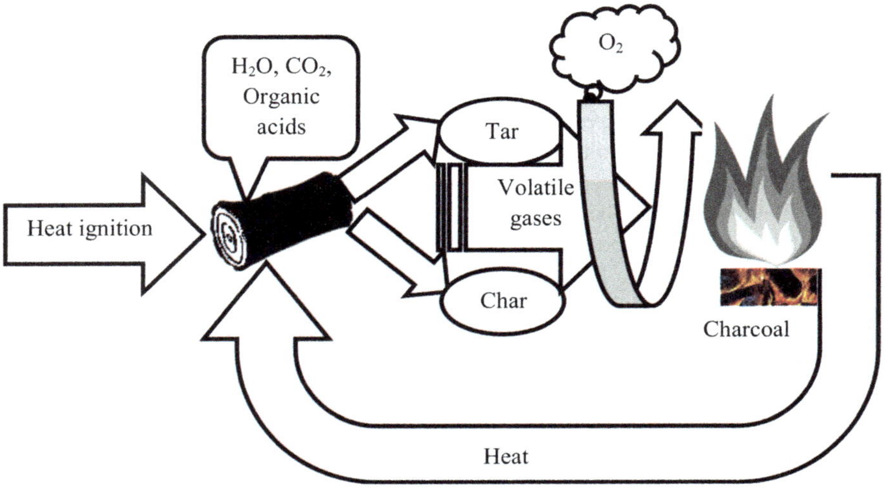

Fig. 8.3 Schematic representation for combustion

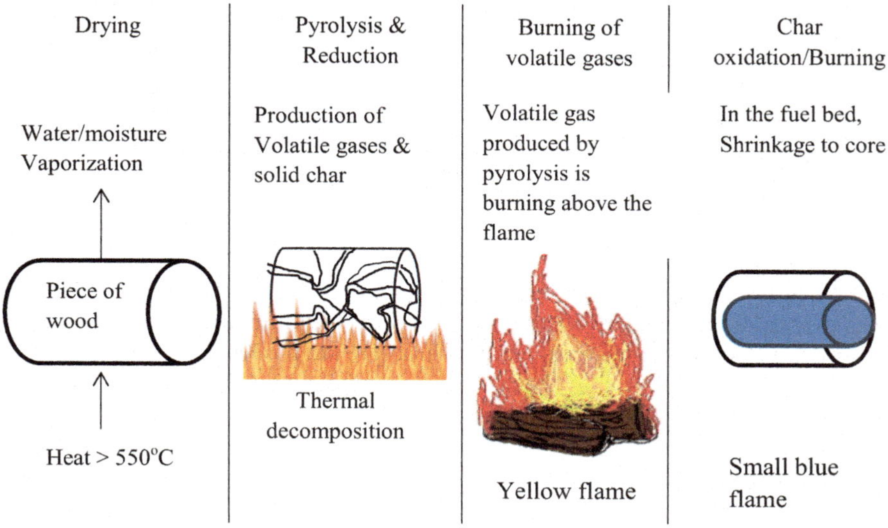

Fig. 8.4 Stages of combustion

due to capillary or intermolecular forces in cellulosic structure. Subsequently, the biomass gives off water, CO_2 and some organic acids [31]. The CO_2 and the water vapour cover the biomass surface results and prevent the oxygen supply and production of smoke.

$$H_2O_{(l)} \rightarrow H_2O_{(g)} \tag{8.1}$$

As the temperature increases, the surface of the biomass chars gets burned into charcoal. Meanwhile, heat travels inwards to the biomass, which causes evaporation of moisture or water into water vapour from the inner part of the biomass [32]. Instantly pyrolysis occurs and the tar above the biomass surface is converted into volatile gases. Then gases and tar spread over the surface of the biomass and react with char to form volatile gases which is called pyrolysis. The volatiles are mainly reacting with CO, CO_2, CH_4, H_2, higher hydrocarbons (C_xH_y), tar (CH_mO_n) and other miscellaneous compounds. Simply we can describe the combustion as biomass into char and volatiles. These volatile gases react with oxygen above the fuel bed and produce flame and give heat. After ignition of the gases, burn with a luminous flame heat that is continuous to produce more gases from the biomass until only carbon monoxide and hydrogen are given out.

$$\text{Biomass} \rightarrow \text{Volatiles} + \text{Char} \tag{8.2}$$

The charcoal in the fuel bed merely glows with little small blue flame. Products of combustion are CO_2, water and soot (carbon). The char oxidation is as follows [34]:

$$C + \frac{1}{\Phi}O_2 \rightarrow 2\left(1 - \frac{1}{\Phi}\right)CO + \left(\frac{2}{\Phi} - 1\right)CO_2. \tag{8.3}$$

where

$\Phi$is the stoichiometric ratio.

When the burning is slow, combustion produces lots of tar, highly flammable gases and lots of heat. If burning is slow with the little air, then more carbon dioxide, water vapour, charcoal and low heat rate are produced.

6 Combustors

The combustion process occurring in the reactor is called combustors. The applications of combustions are significant from small-scale to large-scale systems. The combustion applications are open fires, cookstoves and power generation. The power output from combustion is a wide range of scales from stoves to furnaces and boilers. The largest usage of biomass goes for cooking, lighting and heating applications mostly in developing countries [35].

6.1 Domestic Level Combustors

Biomass combustion for domestic level such as heating, lighting and cooking is concentrating considerably on targeting to be clean and energy efficient. Design and development of clean combustion chamber for cooking is significant research activity globally [32]. Pollutant emissions also cause chronic diseases that concern major health issues and GHG emissions also [36]. The recent technological approach in the cookstove research has multiple choices for clean combustion like preprocessed fuels (pellets), forced air, TEG-attached cookstoves and so on. Pellet cookstoves reduced remarkable amount of emissions from 15 to 25 mg/MJ [35].

6.2 Power and Heat Generation Combustors

The power generation from biomass combustion throughout the world is 50,000 MWe and included small- and large-scale systems. The power generation through combustion is by generating steam from the heat which is generated from combustion process; simultaneously that steam operates the steam turbine to produce electricity.

The schematic Rankine cycle for power generation is explained in Fig. 8.5. Medium-type efficiency needed pressure is 6–10 MPa and needed temperature is 540 °C [35]. Multiple types of blade are also available in industry to increase the efficiency. Steam turbine engine can operate with reciprocating or screw-type turbines also. Combustion products exhausted should be clean, and the steam is getting condensed to water feed pump rotationally. The efficiencies of biomass-based boiler systems have less than the fossil fuel-based power systems due to various parameter moisture content, lower steam temperatures and pressure at combustion. Biomass combustion boilers integrated with gasifiers achieved 35% electrical efficiency [35]. Power-generating combustors are using three types of combustor principles to produce the electricity, which are grate burners, suspension burners and fluidized beds [37]. Grate burners are supplying towards the direction to burners. Suspension burners allow the whole coarse particles in the combustion chambers. Fluidization occurs by adjusting airflow and at the same time to target complete combustion also. Circulating fluidized bed collects the unburned particles and returned to the fluidized bed. Bubbling fluidized bed runs without reinjecting the unburned particles [37].

6.3 Cofiring

Dual or more fuels used for combustion process simultaneously is the principle of cofiring. It is the attractive system to reduce GHG emissions using combination of

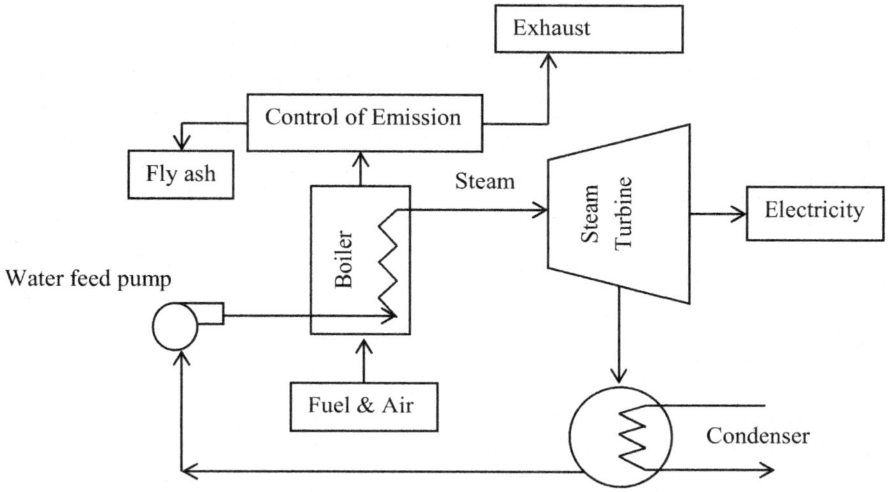

Fig. 8.5 Schematic representation for Rankine cycle

coal and biomass in the power plants. Biomass can be inserted as a premixed fuel with coal. Pulverized combustion from coal and combustion has the high efficiencies compare to the biomass combustion boilers. This combustion process involves powder-sized fuel particles. Pulverized coal combustion should be less than 75 μm [37].

6.4 Alternative Combustors

Steam Rankine cycle is the basic principle for the power and heat generation from solid fuels. Alternative combustors support the supercritical Rankine cycles with the higher temperatures (647 K) and pressures (22.1 MPa), above the critical point of water and coal or fossil fuels as fuels are sometimes cofiring. Due to high temperatures, we need to concentrate on corrosion factors and ash fouling also. Conversion of biomass to fuel by gasification is other alternative combustors [35].

7 Conclusion and Future Prospects

India has huge potential to generate an additional 20 GW of electricity from residues of biomass. In order to find the efficient potential, various fiscal incentives are being given by Indian Government. To make the project economically attractive, the following are the key incentives like capital subsidy, renewable energy certificates and Clean Development Mechanism (CDM) which can be implemented effectively. The government provides a onetime capital subsidy based on the installed capacity of the project. New investment is the most potent indicator of growth of energy sector. As per an estimate in 2009, the total financial investment in green energy in India was at INR 135 billion. Along this, Indian Renewable Energy Development Agency (IREDA) and other public sector agencies are also actively involved in renewable energy projects [38]. Biomass energy will be able to meet economic and environmental challenges in rural areas of India which poised this country to become self-dependent in power generation from biomass in the near future.

References

1. Keyhani A, Ghasemi-Varnamkhasti M, Khanali M, Abbaszadeh R (2010) An assessment of wind energy potential as a power generation source in the capital of Iran, Tehran. Energy 35:188–201
2. Ramachandra TV, Kamakshi G, Shruthi BV (2004) Bioresource status in Karnataka. Renew Sustain Energy Rev 8:1–47

3. D'Sa A, Murthy N (2004) LPG as a cooking fuel option for India. Energy Sustain Dev 8:91–106
4. Demirbas A (2001) Biomass resource facilities and biomass conversion processing for fuels and chemicals. Energy Convers Manage 42:1357–1378
5. Forest Survey of India, Ministry of environment and forests government of India, Dehra Dun, India (2008) State of Forest Report, 2005
6. Ministry of New and Renewable Energy, Annual report 2010–2011
7. Bhavanam A, Sastry RC (2011) Biomass gasification processes in downdraft fixed bed reactors: a review. Int J Chem Eng App 2:425–433
8. McKendry P (2002) Energy production from biomass (part 3): gasification technologies. Bioresour Technol 83:55–63
9. Ramachandran A, Jayakumar S, Haroon RM, Bhaskaran A, Arockiasamy DI (2007) Carbon sequestration: estimation of carbon stock in natural forests using geospatial technology in the Eastern Ghats of Tamil Nadu, India. Curr Sci 92:323–331
10. Sudha P, Somashekhar HI, Rao S, Ravindranath NH (2003) Sustainable biomass production for energy in India. Biomass Bioenergy 25:501–515
11. Forest Survey of India, Ministry of environment and forests government of India, Dehra Dun, India (2005) State of Forest Report, 2003
12. Raizada A, Parandiyal AK, Ghosh BN (2003) Estimation of carbon flux through litter fall in forest plantations of India. Indian Forester 129:881–894
13. The World Bank Data (2013) http://data.worldbank.org/indicator/AG.LND.AGRI.ZS/countries. Accessed 1 Nov 2013
14. Baruah DC, Hiloidhari M (2009) Adequacy of crop residue biomass as renewable energy for tea drying in Assam: a spatial assessment. J Crop Eng 46:43–50
15. Singh J, Panesar BS, Sharma SK (2008) Energy potential through crop biomass using geographical information system—a case study of Punjab. Biomass Bioenergy 32:301–307
16. Das S, Jash T (2009) District-level biomass resource assessment: a case study of an Indian State West Bengal. Biomass Bioenergy 33:137–143
17. Chauhan S (2010) Biomass resources assessment for power generation: a case study from Haryana state, India. Biomass Bioenergy 34:1300–1308
18. Directorate of Economics and Statistics, Ministry of Agriculture, Government of India. Agriculture Statistics at a glance 2010. http://agricoop.nic.in/Agristatistics.htm. Accessed on Aug 2011
19. Hiloidhari M, Das D, Baruah DC (2014) Bioenergy potential from crop residue biomass in India. Renew Sustain Energy Rev 32:504–512
20. Sukumaran RK, Surender VJ, Sindhu R et al (2010) Lignocellulosic ethanol in India: prospects, challenges and feedstock availability. Bioresour Technol 101:4826–4833
21. http://www.eia.gov/project. Accessed 12 Nov 2013
22. Ministry of Power Central Electricity Authority, India (2014) Executive Summary Power Sector Dec 2014
23. Brown A (2010) Survey of energy resources. Technical report, World Energy Council
24. Bridgwater AV, Toft AJ, Brammer JG (2002) A techno-economic comparison of power production by biomass fast pyrolysis with gasification and combustion. Renew Sustain Energy Rev 6:181–246
25. Devi L, Ptasinski KJ, Janssen FJJG (2003) A review of the primary measures for tar elimination in biomass gasification processes. Biomass Bioenergy 24:125–140
26. Demirbas A (2010) Biorefineries. Springer, London
27. Singh R, Setiawan AD (2013) Biomass energy policies and strategies: harvesting potential in India and Indonesia. Renew Sustain Energy Rev 22:332–345
28. U S Energy Information Administration (2013) India is the fourth largest energy consumer in the world after the United States, China, and Russia, full report, and overview. Accessed 18 Mar 2013

29. Radhey A (2012) Gain report (Number-IN2081), biofuels annual. http://mnre.gov.in/progbiomasspower.html
30. Bhatt BI, Vora SM (2009) Stoichiometry, 4th edn. Chemical Engineering Series, McGraw-Hill
31. The Tata Energy Institute (1992) Consolidation of information. Cooking Stoves Handbook Pilot Edition. United Nations Educational Scientific and Cultural Organization UNISIST. PGI-82/WS/20 Paris
32. Quaak P, Knof H, Stassen H (1999) Energy from biomass: a review of combustion and gasification technologies. The World Bank Technical Paper No. 422. Energy Series, Washington, DC
33. Regional Wood Energy Development Program in Asia (1993) Improved solid biomass burning cookstoves: a development manual Bangkok pp 4–5
34. Kausley SB, Pandit AB (2010) Modeling of solid fuel stoves. Fuel 89:782–791
35. Brown RC (2011) Thermochemical processing of biomass conversion into fuels, chemicals and power. Iowa State University, Wiley Series in Renewable Resources
36. Kshirsagar MP, Kalamkar VR (2014) A comprehensive review on biomass cookstoves and a systematic approach for modern cookstove design. Renew Sustain Energy Rev 30:580–603
37. Turns SR (2000) An introduction to combustion—concepts and applications, 2nd edn. McGraw-Hill, New York
38. Ministry of New and Renewable Energy, Government of India. Renewable energy in India: Progress, Vision and Strategy

Chapter 9
Feedstock Transportation, Agricultural Processing, Logistic from Farm to Bio-refinery: Recent Developments, Mechanization, and Cost Analysis

B. S. Pathak and Anuj K. Chandel

Abstract The ever increasing energy demand, high energy costs, predicted scarcity of fossil fuel, and their severe implication on environment have substantially fueled the search for sustainable alternatives of crude petroleum sources. Biofuels that are derived from plant biomass are renewable and have best answers to replace petroleum products. First-generation biofuels have proved their scale and success to compete with gasoline but are restricted to attain their goals because of alarming socioeconomic consequences, climate change mitigation, food/feed concerns, and use of arable agricultural lands. These issues have called for the development of second generation biofuels, which do not compete directly with food and feed crops. Crop residues are primary feedstock for biofuel production in near future. However, second generation biofuel are in initial stage of technology development and have several challenges for large-scale production as their production costs are substantially higher compared to first-generation biofuels. Crop residue management and transportation of crop residues from agricultural farm to factory (processing site) are two very important cost contribution factors in the overall economics of biomass conversion into biofuels. Second-generation biofuels cannot be cost competitive without significant reformation in production technologies, feedstock cultivars, and most importantly, biomass feedstock supply logistics. In Indian context, there is a lot to be done in the biomass collection followed by cost-competitive storage and supply system logistics. This chapter summarizes the

B.S. Pathak (✉)
Sardar Patel Renewable Energy Research Institute, Post Box No-2, Vallabh Vidyanagar, Gujarat 388120, India

KC—5, Kavi Nagar, Ghaziabad 201002, Uttar Pradesh, India
e-mail: bspathakprof@gmail.com

A.K. Chandel (✉)
Centro de Tecnologia Canavieira (CTC), Fazenda Santo Antonio s/n-Caixa Postal 162, 13400-970, Piracicaba, Sao Paulo, Brazil
e-mail: anuj.kumar.chandel@gmail.com

A.K. Chandel, R.K. Sukumaran (eds.), *Sustainable Biofuels Development in India*,
DOI 10.1007/978-3-319-50219-9_9

recent advances, mechanization, and economic analysis with regard to crop residues transportation, agricultural processing, and logistic from farm to bio-refinery.

Keywords Agricultural residues • Biomass feedstock logistics • Biofuels • Bio-refinery • Economic analysis • Fossil fuel • First-generation biofuels • Second-generation biofuels

1 Introduction

The need of energy demand worldwide is continuously increasing due to the rapid civilization, industrialization, and ever expanding economy. Currently, fossil fuels are the primary source of energy globally. Because of the predicted shortage of fossil fuels along with excessive use and increased global warming, attention is focused on the use of renewable energy sources such as solar, wind, hydro, tidal and biomass worldwide [1–3]. Amongst different renewable energy sources, biofuels produced from biomass are of the immense interest that are anticipated to play a significant role in economy the near future [4]. Biologically produced fuels in solid, liquid, or gaseous fuels derived from bio-organic matter are normally called biofuels [5].

In general, biofuels are mainly categorized into three groups: (1) Natural biofuels, (2) Primary biofuels, and (3) Secondary biofuels. The natural biofuels are usually produced from organic sources, such as landfill gas, animal waste, and vegetables. While primary biofuels include fuelwoods, which are utilized mainly for heating, cooking, electricity production, or brick furnace, secondary biofuels are made up of biodiesel and bioethanol and are synthesized or produced via biomass processing [6]. The first-generation biofuels (bioethanol and biodiesel) are mainly synthesized from food crops as well as oil seeds. They are limited due to socioeconomic consequences (competition of feedstock with food commodities), climate change mitigation, and using arable agricultural lands [7, 8]. Owing to these concerns, interest is renewed for producing second-generation biofuels from nonfood feedstocks such as wood processing residues, agricultural residues, and nonedible constituents of beet, sugarcane, corn, etc. because of their non-competency with human food chain [8]. Given that, various agro-processing residues such as rice husk, groundnut shell, and sugarcane bagasse are already being used as preferable source to produce heat and electrical power in boilers; crop residues are anticipated to turn out to be the main feedstock for biofuel production in near future.

Biomass feedstock logistics involves from taking away the biomass from the agricultural fields directly to the bio-refinery. Complete logistics package usually includes the biomass harvesting, collection, pre-processing, storage, and transportation. The Biomass Research and Development Board's Feedstock Logistics Interagency Working Group identified following five primary barriers related to commercialization of biofuels [9, 10].

1. *Low bulk and energy densities of biomass from agricultural and forest resources*: The low densities of feedstocks make transport, handling, and storage ineffective. For this, economic densification and preprocessing machineries are required to attain higher bulk and/or energy densities for the effective transportation, storage, and other logistics operations.
2. *High moisture content of biomass at the time of harvest:* High moisture content of biomass cause aerobic instability during storage and thus decrease the efficiency of transportation and preprocessing operations. Competent strategies and equipment are required to deal with high-moisture biomass.
3. *Incompetent feedstock logistics equipment:* Current logistic operations are inadequate for economically harvesting, store, and transport feedstocks for biofuels. Innovative approaches of integrating system components are required with industrial alliance for higher efficiency with reduced costs.
4. *High variation in biomass quality:* Biomass qualities vary with feedstock source and season, causing inefficiencies in handling and conversion systems. Because of this, logistics operations that enhance consistency of delivered feedstock should be improved.
5. *Overburden transportation networks:* Trucking is generally utilized for biomass transportation. This is expensive as well as damages the roadways and eventually leads to increased traffic on roads. More concerted efforts are required to understand the trucking regulations, payload limits, prices regulations, and control traffic effects. Finding the alternatives of trucking for transportation of bioass by rail or other source of transport is necessary.

Improvement in feedstock logistics and fast establishment of new systems and technologies are important to bring down the cost of biofuel production. Though significant progress has been made, more effort is still required.

Second-generation biofuels are still immature at industrial scale. The biofuel production processes further require improvements and investment to establish competent and reproducible production technologies at commercial scale. Significant improvement in production technologies, feedstock cultivars, and most importantly, biomass feedstock supply logistics are the need of hour to make cost-competitive biofuels [9, 10].

2 Feedstock

Biofuels are derived from biomass which is organic matter of recent origin. Crops and animals, agro-processing industries, forests, agro-forestry, road sides, and waste land generate by-products and wastes which are potential sources of biomass to produce biofuels. Countries with large resources of unused cultivable land grow energy crops for the purpose of manufacturing biofuels. Pressure on land does not permit large-scale cultivation of energy crops in India. Crops and agro-processing residues and forest waste, estimated in 2001 at 523.44 Mt and 157.18 Mt, respectively, have been reported as the two potential suppliers of biomass which, being surplus to their traditional uses, can be made available for production of biofuels [11]. Because of easier accessibility and concentrated availability, surplus crop and agro-processing residues, expected to exceed 200 Mt in 2015, have been projected as the future feedstock for generating power and producing biofuels. Since many agro-processing residues like bagasse, rice husk, and groundnut shell are already being used as fuels to generate heat and electrical power, crop residues are likely to become the main feedstock for biofuel production. In 2008, the availability of surplus crop residues was estimated at about 125 Mt [12].

Food grain crops (rice and wheat), oil crops (mustard, soybean, and groundnut), sugarcane, and cotton produce the bulk of crop residues in India. The geographic distribution of these seven crops showing major producer states is given in Fig. 9.1

Fig. 9.1 Major producer states of seven selected crops

[13]. The residues of these crops are traditionally used as fodder and domestic fuel and for making ropes, thatching, etc. [14].

The surplus residues of these crops, estimated at 78 Mt in 2005–2006, are burnt after harvest which results in loss of soil organic matter and serious air pollution Figs. 9.2, 9.3, 9.4, 9.5.

Utilization of surplus crop residues as feed stalk for biofuel production will make these otherwise wasted sources of biomass a marketable commodity and add to the income of the farmers.

Fig. 9.2 Uncontrolled burning of rice straw in Punjab

Fig. 9.3 Smoke screen in Punjab

Fig. 9.4 Charred field in Punjab

Fig. 9.5 Cotton stalks on fire (Gujarat)

3 Characteristics of Residues

Unlike wood, crop residues (CR) vary widely in their characteristics which influence the selection of raw material and the conversion route to obtain the desired end product/biofuel. Characteristics of the residue of a crop differ according to variety and environmental conditions under which it is grown. In order to develop CR to biofuel processes and technology, we need information on:

1. Physical properties like bulk density, angle of repose, moisture content, and equilibrium moisture content of CR.
2. Thermal properties like higher and lower heating values and ash softening/fusion temperature.
3. Proximate analysis to determine fixed carbon, volatile matter, and ash content.

4. Elemental analysis, also called ultimate analysis, to determine the chemical composition of the residue in terms of different elements like carbon, hydrogen, nitrogen, alkali metals, silica, etc.
5. Summative analysis, also called chemical analysis, to estimate cellulose, hemicellulose, lignin, and benzene extract content.
6. Thermo-gravimetric analysis to obtain information on temperature-dependent weight loss rate.

Proper understanding of the physical, thermal, and chemical properties of biomass used as feedstock to produce biofuels is a basic need, and characterization of the raw material should be a part of the work in any major project to develop biofuels technology (Tables 9.1, 9.2, and 9.3).

Even in a small sample consisting three crop residues and three processing residues, bulk density varies from 30 kg/m^3 to160 kg/m^3, ash from 3.3% to 19.2%, N from 0.2% to 1.6%, and lignin from 12.1% to 31.3%.

Table 9.1 Some characteristics of crop and agro-processing residues [15]

Sr. No.	Crop residues	Bulk density (kg/m^3)	EMC at 80% RH	Higher c.v. (MJ/Kg)	Proximate analysis Fixed carbon (%)	Volatile matter (%)	Ash (%)
1.	Rice straw	30	36.70	15.00	11.10	69.70	19.20
2.	Rice husk	105	29.40	15.50	12.50	71.00	16.50
3.	Wheat straw	60	34.00	17.20	17.93	73.60	8.47
4.	Groundnut shell	100		20.01	11.67	83.90	4.43
B.	Cotton stalks	160	27.05	17.40	15.30	81.40	3.30
6.	Sugarcane bagasse	70	34.86	20.00	15.86	79.20	4.94

Table 9.2 Elemental analysis of crop residues [15]

Sr. No	Crop residues	Elemental analysis (%) C	H	N	Na	K	P	Mg	Ca	SiO_2	C:N ratio
1	Rice straw	36.80	5.00	1.00	0.09	2.50	0.06	0.53	0.08	15.60	37:1
2	Rice husk	37.80	5.00	0.30	0.02	0.30	0.03	0.17	0.10	16.77	126:1
3	Wheat straw	43.80	5.40	1.00	0.06	0.78	0.04	0.35	0.10	7.08	44:1
4.	Groundnut shell	41.10	4.80	1.60	0.05	1.20	0.12	0.40	0.10	2.52	26:1
5.	Cotton stalks	51.00	4.90	1.00	0.09	0.61	0.08	0.43	0.12	1.33	51:1
6.	Sugarcane bagasse	48.20	6.10	0.20	0.06	0.51	0.04	0.36	0.14	1.30	241:1

Table 9.3 Chemical analysis of crop residues [15]

		Chemical analysis (%)			
Sr. No	Crop residues	Cellulose	Hemi-cellulose	Lignin	Bz extract
1	Rice straw	41.40	20.40	12.10	5.60
2	Rice husk	44.10	17.80	17.20	3.40
3	Wheat straw	39.60	24.10	17.00	7.30
4.	Groundnut shell	36.50	13.90	31.30	12.20
5.	Cotton stalks	41.90	19.00	27.20	9.30
6.	Sugarcane bagasse	40.00	22.60	14.80	15.90

4 Recovery of Crop Residues

The logistics of supply of CR in large quantity for energy generation has always been questionable (Fig. 9.6). Residues are thinly spread over very large areas. For example, the 14–15 million tonnes of rice straw burnt each year in Punjab has to be collected from an area of 2.7 million hectares. Rice is combine-harvested and the operation is completed in 20 days. The time available for mowing the standing stubble (Fig. 9.7), windrowing the straw, and baling and transporting it to storage sites is 25–30 days only. Recovering and transporting 500,000 tonnes of rice straw (3,500,000 cubic meter volume) per day will require a huge fleet of trucks; tractors; trailers; and large number of mowers, liners, and balers. Large-scale utilization of crop residues for producing biofuels will not be feasible without appropriate mechanization of logistic operations. The machinery needed for this purpose is manufactured/can be manufactured in India. Normally, larger an operation, more economical it is. But in the case of crop residues, greater the quantity to be collected, higher the cost/tone of residue because of the longer distance over which it is to be transported. A reliable estimate of recoverable residue from a production area can be made by using the approach suggested below.

$$Q = \mathrm{GA} \times \mathrm{NAUC} \times \mathrm{AUC} \times Y \times R \times L(\text{tonnes})$$

where

Q = quantity of residue (CR) recoverable
GA = geographic area (km^2)
NAUC = net area under cultivation (% of GA)
AUC = area under a crop (% of NAUC)
Y = yield (tonnes/km^2)
R = CR: grain ratio
L = CR recovery (%)

This approach also makes it easy to arrive at the radius of collection area from which the required amount of residue can be recovered.

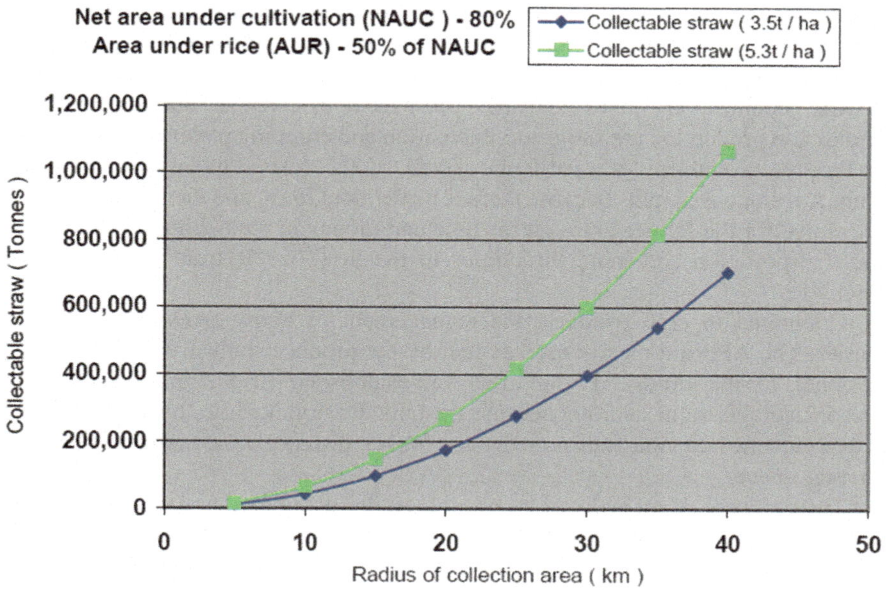

Fig. 9.6 Straw from increasing radius of collection

Fig. 9.7 Combine harvested field

5 Collection, Transport, and Storage of Residues

The characteristics of crop residues make their collection and storage in large quantity rather difficult. They are thinly spread over millions of hectares, their harvest periods are short requiring collection and removal within few weeks, and availability is seasonal requiring storage till the next harvest. The bulk density of

CR is low creating difficulty in transport and resulting in the demand for large space for storage. The moisture content of CR is generally high at the time of harvest which can result in decomposition in storage. Because of their bulkiness, handling of CR becomes very labor intensive. Most residues carry serious fire risk. Crop yields and production are subject to fluctuation and cropping patterns change which makes the availability of a particular residue in the desired quantity rather uncertain. A residue is surplus because there is no demand for it, and therefore there is no incentive for the farmer to invest his time and money in recovering it. Regulations on transportation add more difficulties to the flow of CR from the field to the user site.

Compared to crop residues, the management of agro-processing residues is easier. The APR are mostly used as fuel by the producer industry. The marketing channels for the surplus APR are often well established. Briquetting of APR results in considerable value addition because the bulk density increases by a factor of 5–6. The briquettes can be transported over longer distances and require much less storage space.

The problems in recovery of crop residues, discussed earlier, can be solved, and the movement of the surplus CR from the farm to the power/biofuel production plant can be economically achieved by introducing a suitable package of mechanization hardware and practices for the following operations for leafy residues:

- Residue harvest
- Windrowing
- Baling
- Transportation
- Satellite storage
- Transportation to user site
- Storage at user site

The operations for woody residues are likely to be:

- Collection
- Shredding
- Satellite storage
- Briquetting
- Transportation to user site
- Storage at user site

Crop-wise equipment package recommended for mechanizing the recovery, densification, and transport of the residues of the seven crops are listed in Table 9.4.

These recommendations take into account the size of tractors commonly available in India. The management of large quantities of residues will require bigger equipment and more powerful tractors. Figures 9.8, 9.9, and 9.10 show such equipments used for baling and transporting.

Table 9.4 Equipment for mechanizing recovery, densification, and transport of crop residues

Sr. no	Crop	Residue	Operation	Machinery	Power required
1	Rice (combine harvested)	Straw	(1) Mowing (2) Windrowing (3) Baling (4) Transport	Reciprocating or disc mower Liner Baler Tipping trailer	35–50 hp tractor 35–50 hp tractor 50 hp tractor 35–50 hp tractor
2	Wheat (combine harvested)	Straw	(1) Mowing (2) Windrowing (3) Baling (4) Transport or (1) Harvesting (2) Transport or (1) Harvesting (2) Transport	Reciprocating or disc mower Liner Baler Tipping trailer Reciprocating mower Straw combine Tipping trailer Flail Harvester Tipping trailer	35–50 hp tractor 35–50 hp tractor 50 hp tractor 35–50 hp tractor 35–50 hp tractor 35–50 hp tractor 35–50 hp tractor 45–60 hp tractor 35–50 hp tractor
3	Cotton (hand picked)	Stalks	(1) Uprooting and windrowing (2) Shredding (3) Transport (4) Grinding (5) Briquetting	Tractor blade Shredder Tipping trailer Hammer mill Briquetting machine	35–50 hp tractor 35 hp tractor 35–50 hp tractor 25 hp motor 25 hp motor
4(a)	Soybean (hand harvested)	Haulms	(1) Transport	Tipping Trailer	35–50 hp tractor
4 (b)	Soybean (combine harvested)	Haulms	(1) Windrowing (2) Transport (3) Grinding (4) Briquetting	Liner Tipping trailer Hammer mill Briquetting machine	35–50 hp tractor 35–50 hp tractor 25 hp motor 25 hp motor

(continued)

Table 9.4 (continued)

Sr. no	Crop	Residue	Operation	Machinery	Power required
5	Groundnut (stationery Thresher)	Haulms	(1) Transport (2) Grinding (3)Briquetting	Tipping trailer Hammer mill Briquetting machine	35–50 hp tractor 25 hp motor 25 hp motor
6	Rapeseed and Mustard	Stalks	(1) Harvesting (2) Transport (3) Grinding (4) Briquetting	Flail harvester Tipping trailer Hammer mill Briquetting machine	60 hp tractor 35–50 hp tractor 15–25 hp motor 25 hp motor
7	Sugarcane	Trash	(1) Windrowing (2) Baling (3) Transport	Liner Baler Tipping trailer	35–50 hp tractor 50 hp tractor 35–50 hp tractor

Fig. 9.8 Round baler

Fig. 9.9 Roto-cut baler

Fig. 9.10 Bale transport

6 Cost of Crop Residues

Under a study done at Sardar Patel Renewable Energy Research Institute(SPRERI), in 2008, the cost of rice straw in Punjab delivered at factory site was estimated to be Rs. 1095.90/t. The per tonne cost break up was

Payment to farmer	Rs. 94.34 (Rs. 500/ha)
Recovery and baling	Rs. 548.80
Transport, handling, and storage	Rs. 221.10
Losses and profit margin	Rs. 231.66
Total	Rs. 1095.90

The above estimate was for collection 66,628 tonnes rice straw from a geographical area of 10 km radius under conditions of high NAUC and at 5.3 t/ha recovery rate. In view of the steep rise in costs and prices during the last few years, the cost of rice straw would now be higher than Rs. 1095.90/t. On the other hand, introduction of lager and more efficient equipment for baling, bale handling and transport will bring about a reduction in the cost.

7 Conclusion

Biofuels are emerged as sustainable, renewable, and eco-friendly alternatives to conventional petroleum-based fuels. Biomass residues including agricultural residues are expected to become the potential feedstock for biofuel production in near future. The technology routes for the production of nonfood feedstocks (such as crop residues)-based biofuel are still not mature for the commercialization due to their high production costs and ineffective conversion technologies compared to first-generation biofuels. Biomass logistics are a critical part of the biofuel production chain as their prices are reliant on many factors such as region, biomass type, and bio-refinery specification. The logistic prices can be even more than half the price of transported feedstock. Therefore, improving the logistics system efficiency may significantly decrease the price through accelerated establishment of new systems and technologies eventually adding price competitiveness with conventional fossil fuels.

References

1. Goldemberg J, Guardabassi P (2009) Are biofuels a feasible option? Energy Policy 37:10–14. doi:10.1016/j.enpol.2008.08.031
2. Dragone G, Fernandes B, Vicente AA, Teixeira JA (2010) Third generation biofuels from microalgae. In: Mendez-Vilas A (ed) Current research, technology and education topics in applied microbiology and microbial biotechnology. Formatex, Madrid, pp 1315–1366

3. Rajkumar R, Yaakob Z, Takriff MS (2014) Potential of the micro and macro algae for biofuel production: a brief review. Bioresources 9:1606–1633. doi:10.15376/biores.9.1
4. Chen C-Y, Chang J-S, Chang H-Y, Chen T-Y, Wu J-H, Lee W-L (2013) Enhancing microalgal oil/lipid production from *Chlorella sorokiniana* CY1 using deep-sea water supplemented cultivation medium. Biochem Eng J 77:74–81
5. Sharma L, Singh AK (2015) Algal biofuel: a sustainable green energy. In: Bag N, Murgan R (eds) Biotechnology in India: initiatives and accomplishments. New India Publishing Agency, New Delhi, pp 205–222
6. Nigam PS, Singh A (2011) Production of liquid biofuels from renewable resources. Prog Energy Combust Sci 37:52–68
7. Chhetri AB, Tango MS, Budge SM, Watts KC, Islam MR (2008) Non-edible plant oils as new sources for biodiesel production. Int J Mol Sci 9:169–180
8. Alam F, Date A, Rasjidin R, Mobin S, Moria H, Baqui A (2012) Biofuel from algae—is it a viable alternative? Proc Eng 49:221–227
9. Biomass Research and Development Board (2010) Biofuel feedstock logistics: recommendations for research and commercialization—a report by the feedstock logistics Interagency Working Group. November, 2010. p 38 http://www.biomassboard.gov/pdfs/biomass_logistics_2011_web.pdf
10. Langholtz MH, Stokes BJ, Eaton LM (Leads), ORNL/TM-2016/160 (2016) U.S. Department of Energy. Billion-Ton Report: Advancing Domestic Resources for a Thriving Bioeconomy, vol 1: Economic availability of feedstocks. Oak Ridge National Laboratory, Oak Ridge, p 448
11. Pathak BS, Khan TA, Sharma P (2005) Biomass production, its utilization and surplus for energy generation in India. Proceedings of National Seminar on Biomass Management, SPRERI, ANAND pp 10–35
12. Pathak BS, Khan TA, Sharma P (2008) www.lab.cgpl.iisc.in. Accessed 04 Mar 2008
13. Pathak BS, Khan TA, Sharma P (2009) Management of surplus crop and agro-processing residues to harness their potential for power generation. Sardar Patel Renewable Energy Research Institute, Anand
14. Pathak BS, Khan TA, Sharma P http://punevis.nic.in/PDF/newsletters/bio uti.pdf
15. Pathak BS, Jain AK (1984) Characteristics of crop residues. Technical report, Energy in Agriculture Project. Punjab Agriculture University, Ludhiana

Chapter 10
First- and Second-Generation Ethanol in India: A Comprehensive Overview on Feedstock Availability, Composition, and Potential Conversion Yields

Rajeev K. Sukumaran, Anil K. Mathew, M. Kiran Kumar, Amith Abraham, Meera Chistopher, and Meena Sankar

Abstract With imports of crude oil touching 80% of the total requirement, India is a nation heavily dependent on foreign oil. There are serious efforts to replace part of the transportation fuels with renewable alternatives—notably biofuels. Bioethanol from biomass is probably the near-term solution, since the country is dependent on imports for edible oils, which in turn makes the possibility of near-term implementation of biodiesel as a renewable transportation fuel remote. Being an agrarian economy, the country has abundant feedstock for bioethanol production in the form of agricultural residues. The country generates a total of nearly 650 million metric tons (MMT) of lignocellulosic agro-residues annually, and though a major part of it is used for other purposes, there are still huge quantities of the feedstock available as surplus. Major technical bottlenecks in the conversion of lignocellulose to ethanol have been addressed worldwide, and companies claim to have started operation in the production of second-generation ethanol. While these international technologies can be implemented here directly or with modifications, the major challenge in India's context would be to ensure availability of quality raw material at the sites of production. Sustainability of such industries would depend on the availability of feedstock at affordable prices which inherently involve logistic issues. Understanding the type and amount of biomass available in the country and its geographical location is important in establishing bioethanol industries, as well as the potential of converting these feedstocks to ethanol. The chapter introduces the background of India's need for bioethanol and why first-generation bioethanol cannot fully cater to the gasoline blending demands of the nation and proceeds to describe the availability of agro-residues as potential feedstock for bioethanol in the country. Data on the generation of agro-residues which form

---

R.K. Sukumaran (✉) • A.K. Mathew • M. Kiran Kumar • A. Abraham • M. Chistopher • M. Sankar
Centre for Biofuels, CSIR-National Institute for Interdisciplinary Science and Technology, Thiruvananthapuram 695019, Kerala, India
e-mail: rajeevs@niist.res.in

A.K. Chandel, R.K. Sukumaran (eds.), *Sustainable Biofuels Development in India*,
DOI 10.1007/978-3-319-50219-9_10

major agricultural crops in the country is provided along with estimates for the consumption of these feedstocks for competing applications. It gives a fair estimate of the surplus biomass availability in the country and gives the possible realistic amounts of ethanol that can be generated from these biomass residues based on primary and secondary data about the compositions and conversion efficiencies. The study shows that there is a potential of producing ~30 billion liters of ethanol annually from surplus agro-residues in the country (estimated at ~175 MMT) even at very conservative calculations. The importance of logistics and collection models suited for the country is also highlighted

Keywords Bioethanol • Biomass • Agro-residues • Cane trash • Straw • Feedstock • Cellulose

1 Introduction

India is currently the world's fastest growing economy at 7.2% [1]. The growth is expected to accelerate to 7.9% in 2016–2017, and, apparently, this economic growth places tremendous demand on the energy requirement of the nation. India is the fourth largest consumer of petroleum in the world and also the fourth largest emitter of greenhouse gases (GHG). The country meets 95% of its transportation fuel needs through petroleum products and is increasingly dependent on imports to meet this demand [2]. Crude oil import of the country is constantly on the rise and has never declined over the past decade (Fig. 10.1). Import dependency of petroleum has also been consistently high at above 75% and has almost hit 80% in the financial year 2014–2015. The financial burden this places on the country is huge, with rupees 8.66 trillion spent on imports of crude oil in 2013–2014 and rupees 6.87 trillion in the year 2014–2015, the latter reduction only due to global drop in oil prices (Fig. 10.1). Petroleum imports alone contribute to an average 32% of the country's total imports [3]. This heavy dependency on imported oil is a serious

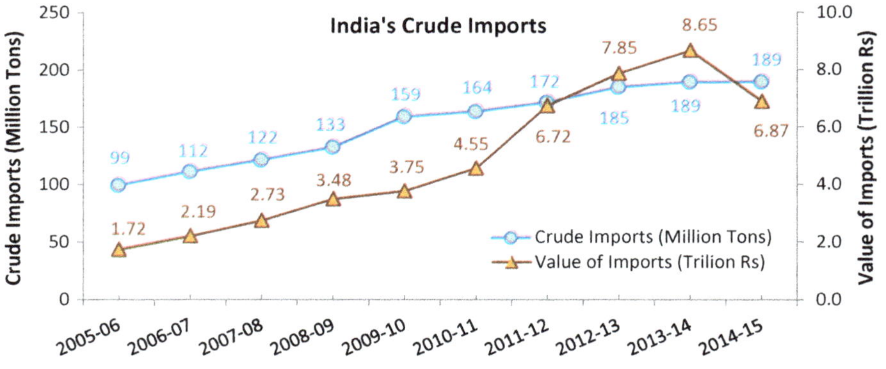

Fig. 10.1 India's crude oil imports: quantity and value. Data source: [3, 4]

threat to the country's energy security, and the country is adopting multipronged strategies to reduce the oil imports, with promotion of alternative and renewable transportation fuels a major step toward this.

Renewable transportation fuels, if substituting the petro-based, will play a major role in bringing down the import dependency, and the replacement of even a small percentage of the petro-based transportation fuels can result in the savings of billions of rupees and reduce significantly the GHG emissions. Alternative fuels developed based on indigenous renewable and sustainable feedstock are proposed as the solution to the overwhelming vulnerability of energy security, especially in the transportation fuel sector. Biofuels derived from plant biomass are therefore very promising candidates to supplement petro-based fuels, and since they are considered environment friendly, there is the added advantage of GHG emission reductions. India has a very progressive outlook toward biofuels and the nation endeavors to facilitate and bring about optimal development and utilization of indigenous biomass resources for production of biofuels. The most recent biofuel policy document brought out by the Ministry of New and Renewable Energy, Government of India, projects an indicative target of 20% blending of both biodiesel and bioethanol by 2017 [5]. While the blending targets proposed for biodiesel are intended to be recommendatory in near terms, those proposed for ethanol have been made mandatory—starting with 5% blending in 2008 and continued further in incremental increases leading to the indicated target. With the target date approaching soon, the actual amount of ethanol blended in gasoline has remained far less (<2%) with the entire ethanol coming from first-generation feedstock (almost exclusively molasses) or from imports.

India's take on biofuels is different from many of the international approaches, especially in that the feedstock to be used for biofuel generation should not have any (direct) conflict with food security. Thus, the policy specifies that it should be based solely on nonfood feedstocks to be raised on degraded or wastelands that are not suited for agriculture [5]. This makes sense, since India is set to overtake China as the most populous country in the world, and feeding this enormous population requires that the nation's agricultural land is not utilized for anything other than food production. This also has to be read in the context that ~51% of India's arable land is already under cultivation against a global average of only 11% [6], and any use other than food production to which this land is utilized is going to seriously affect the food security. The policy at the same time envisions a central role for biofuels in the energy and transportation sector of the country in coming years and is expected to bring accelerated development and promotion of the cultivation, production, and use of biofuels. Since the country is a net importer of edible oils, without any surplus oil to be diverted to biodiesel production, the policy advocated the use of nonedible crops like *Jatropha curcas* and *Pongamia pinnata* cultivated in marginal lands for its biodiesel program. Cultivation of these crops was encouraged in such lands, avoiding irrigated agricultural land for the purpose, which meant new cultivations to be established to meet the demand of oil. However, the productivities that may be expected from such suboptimal cultivation practices can be significantly lower. Consequently, the biodiesel programs based on these nonedible oil crops have not performed as per the expectations. This leaves bioethanol as the major biofuel with near-term potential to supplement petroleum-based liquid transportation fuels.

Currently, India's entire bioethanol comes from first-generation technologies and specifically from sugarcane molasses as the feedstock. Though India is the second largest producer of sugar in the world, the country is also the largest consumer and, depending on the production cycle, can often become a net importer of sugar. There are about 330 molasses-based distilleries in India with a total capacity of ~4 billion liters of rectified spirit per year [7]. There are 152 distilleries associated with the sugar industries that use molasses as the feedstock for ethanol production, with a total installed capacity of ~2.2 billion liters [8]. While there are grain-based distilleries, most of their output is potable alcohol in various forms of liquor and not available to the biofuel basket. Also, the production of molasses ethanol is not consistent due to the cyclic nature of sugarcane production surplus and deficit in the country, influenced by climatic and socioeconomic reasons. The production of molasses ethanol in the calendar year (CY) 2015 was high due to the high production (~350 MMT) of sugarcane, but this may come down in the coming years as a result of the reasons specified earlier. Even for the available molasses ethanol, a significant portion goes into the potable alcohol sector and other industrial uses, and only the leftover is available for blending in gasoline. The production of ethanol in CY 2015 was estimated at 2.22 billion liters, up from 2.03 L of the previous year and the imports amounted to 120 million liters. The total consumption was 2.35 billion liters of which only 800 million liters were estimated to cater fuel applications [7]. It may be noted that gasoline consumption of the country was 21.79 billion liters during April–December 2015 [4] which places the estimate for financial year 2015–2016 at ~29 billion liters and, consequently, the ethanol requirement for 5% blending at 1.45 billion liters. It may easily be understood that the 800 million liters of molasses ethanol diverted to fuel applications cannot even cater to the current 5% mandatory blending requirement. Besides, there is also a consistent demand for molasses ethanol from the potable alcohol and other industries, and the percentage availability of ethanol for biofuel applications cannot be expected to cross the current level of ~35% of the distillery output. This would also imply that India's target of 20% blending of ethanol in gasoline can be achieved only through increased imports of ethanol or by increase in the domestic production. The latter cannot be accomplished without compromising some way or another on the food security, since it means either diverting sugar juice to alcohol production or increasing the land area available for cultivation of sugarcane. Latest government policies do allow the use of surplus cane juice for direct ethanol production, but the implications it might have on sugar production is unpredictable. Apparently, bioethanol production in India has to look beyond molasses as the feedstock, and since grains and tubers are already out of question due to their food value, lignocellulosic biomass remains as the most promising option. This however is limited by the lack of mature technologies ready for implementation. While this aspect probably would be addressed in the near future through entry of multinationals with lignocellulosic ethanol technologies proven at least at pilot scale and in demonstration plants, the major impediment lies in the availability of feedstock in sufficient density in any geographical location for setting up large-scale ethanol production plants. The lack of data on the actual availability of biomass feedstock at a geographical location is still a major limitation in India despite several studies addressing the biomass production and availability issues [9–14]. It may be noted that unlike several other countries where forest residues are a major feedstock, India has peculiar situation of relatively low forest coverage and needs to protect it. Thus, it cannot be considered for a feedstock for bioethanol. India is an agrarian economy, and agro-residues form a formidable quantity of lignocellulosic biomass generated annually (read renewable). Though a major share of this is consumed for other applications like fodder, cogeneration, firewood, building material, etc., the surplus is more than sufficient to produce enough alcohol for the nation's ambitious blending program. Most of the above studies on biomass availability in the nation have relied on secondary data (based on crop production), which do not give an estimate on the current actual use of these important feedstocks. It is highly important to have reliable estimates of the country's biomass resource potential for any lignocellulosic ethanol program. Technology providers for bioethanol—either home grown or foreign—need to know the feedstock they are dealing with, to provide the most appropriate technology for their conversion to ethanol. Also it is important to understand the nuances of biomass variability and composition even within the same feedstock group, since these determine the digestibility of biomass and hence the possible yields of ethanol. Previously, through the NIIST–TIFAC study on biomass resource availability, we had brought out the data on Indian feedstocks, which analyzed primary data from surveys and the secondary data available from national and international agencies through calculations based on crop production data [11, 14]. However, the above reports covered only data up to 2008 and are now very old for practical purposes. The chapter therefore tries to provide a comprehensive overview of the availability, composition, conversion efficiencies, and potential ethanol yields from the major first- and second-generation feedstock in the country. The data presented here on biomass availability are more of secondary nature and sourced from national agencies, while those on composition are mostly primary data sourced from our own experimental analyses.

2 First-Generation (1G) Bioethanol in India: The Critical Role of Sugar Industry

2.1 Molasses Ethanol: Production Yields and Usage

Current production of fuel ethanol in India relies exclusively on sugarcane black strap molasses as the feedstock. Molasses is a by-product of the sugar manufacturing. Typically, during sugarcane production, the juice from cane crushing passes through different stages of sugar crystallization leaving behind the juice which is called molasses. Thus, after the initial extraction of about 77% of the available sucrose from the concentrated can juice syrup, the leftover is called A or A heavy molasses which still contain significant amounts of sugar. This is then boiled again, and an additional 12% of the sugar is crystallized to leave what is known as B or B heavy molasses; the process is continued one more time, and the final molasses is known as the C molasses or the black strap molasses which still contains about 30% of sugar which cannot be recovered economically [15]. The availability of this feedstock as can be perceived is dependent on the amount of cane diverted to sugar production. Sugar production in India is dependent on three major factors which are (1) area under production, (2) sugarcane yield per hectare, and (3) proportion of cane crushed by the sugar factories [16]. For the 10-year period from 2005–2006 to 2014–2015, the area under sugarcane cultivation has increased from 4.2 million hectares to 5.03 million hectares, while a close observation will show that the area has remained fairly consistent through 2011–2012 till the most recent estimate (Table 10.1). India's sugarcane production follows a cyclic trend with a 5–7-year cycle. Typically, the cane production has 3–4 years of increased production, which reaches high toward the end of this period. This in turn results in lower sugar prices, and due to the lower sugar price realization of sugar mills, there is some increased payment arrears to the farmers. As a result, there are lesser plantations in pursuing years resulting in lower production for the next 2–3 years, in turn, shooting up the prices. The new increased prices will again bring back sugarcane cultivation in more area for the next season and the cycle continues [16]. Since 1G fuel ethanol production in India is dependent on molasses, which in turn is dependent on sugar production, this cyclic nature of sugarcane production seriously affects the consistent supply of ethanol.

It is estimated that 85–100 kg of sugar is obtained per metric ton of sugarcane (8.5–10%), whereas the molasses yield is about 40 kg (4%). The recovery of

Table 10.1 Sugarcane and molasses ethanol production in India

Year	Production* (MMT)	Area under cultivation* (million hectares)	Maximum molasses yield** (MMT)	Theoretical maximum ethanol yield (MMT)	Theoretical maximum ethanol yield (billion liters)
2005–2006	281.17	4.20	8.4	2.0	2.51
2006–2007	355.52	5.15	10.7	2.5	3.18
2007–2008	348.19	5.06	10.4	2.5	3.11
2008–2009	285.03	4.42	8.6	2.0	2.55
2009–2010	292.30	4.17	8.8	2.1	2.61
2010–2011	342.38	4.88	10.3	2.4	3.06
2011–2012	361.04	5.04	10.8	2.5	3.22
2012–2013	341.20	5.00	10.2	2.4	3.05
2013–2014	350.02	5.01	10.5	2.5	3.13
2014–2015	366.80[a]	5.03[a]	12.5[a]	2.9	3.72

*Data till 2013–2014—DAC, Govt. of India agricultural statistics 2011–2014, **Estimated value
[a]Values from ISMA [19]

ethanol from molasses is about 22–25% as per Indian standards [17]. India also has the production of traditional sweeteners like *jaggery* and *khandsari*, and about 20–30% of the cane produced is used for the production of these alternative sweeteners and as seeds for plant propagation, which means the actual amount of cane available for sugar production is only about 70–80%. Calculations done on the availability of molasses based on these estimates (actual cane available for sugar production = 75% of total production, molasses = 4% of available cane, ethanol yield = 23.5% of molasses) and sugarcane production data show that there is an estimated annual availability of about 10.5 MMT of molasses which is nearly the same as the estimates by ISMA [18]. Assuming that this entire molasses is converted to ethanol, there is a potential to produce at least 3.0 billion liters of ethanol (Table 10.1).

The demand for molasses ethanol by potable alcohol industry is the largest at about 45%, while the demand by chemical industry is 40% which leaves only about 15% of other applications including its use as gasoline blends [16]. India has about 330 molasses-based distilleries with an installed capacity of 4.0 billion liters of ethanol annually [7]. Of this, about 152 distilleries are associated with sugar industry and have a capacity of producing ~2.2 billion liters of alcohol [8]. While there are no reliable data on the actual production of molasses ethanol by these distilleries, the analysis by the US Department of Agriculture's Foreign Agricultural Service (USDA-FAC) estimates the production for year 2015 at 2.219 billion liters [7]. By applying the historical demand estimates, this would imply that the potable industry consumed approximately 1.0 billion liters of molasses ethanol, while the chemical industry's share was ~0.89 billion liters for the year 2015. The consumption for fuel was estimated at 0.8 billion liters met partially through imports and the carried over surplus stocks of about 0.38 billion liters [7]. The ethanol production was indicated as coming from 115 distilleries with a total nameplate capacity of 2.0 billion liters which means the percentage capacity use was 111.

2.2 *Projected Demand for (1G) Molasses Ethanol for Blending*

India's biofuel policy is unique in that it allows only the use of nonfood feedstock to be used for fuel ethanol production. This also implies that the 1G ethanol production in India is heavily dependent on sugar production and is also subject to the price fluctuations affecting sugar industry due to the cyclic nature of sugarcane production in the country. With limited land resources, the country would be struggling to meet its blending demands from current molasses-based 1G ethanol production. The demand for petroleum products has grown at a CAGR of 7.2% over the period 2010–2011 to 2015–2016 which imposes a demand of 6.3 billion liters of ethanol at 20% blending target for the year 2016–2017 when this target is aimed to be achieved (Fig. 10.2).

Historical data on ethanol production over the past 4 years (2011–2015) where the area of land under sugarcane cultivation had assumed a more or less stationary value shows that the actual production of ethanol was more or less stagnant at 2.1 billion liters [7]. Assuming that the consumption of ethanol by potable alcohol industry and chemical industry together was ~80–85%, the surplus ethanol available for blending was only about 350 million liters. Even with a scenario of 100% utilization of potentially available molasses for ethanol production, the maximum ethanol yield could have been only 3.72 billion liters in the year 2014–2015 (Table 10.1) of which, after deducting for potable and industrial usage, the surplus was only about 650 million liters. This was against a projected demand of 1.29 billion liters at a target of 5% blending. The projections for gasoline demand by Purohit and Dhar [20] indicate that the country will require 33.5, 39.5, and 45.6 billion liters of the fuel in the years 2020–2021, 2025–2026, and 2030–2031, respectively. The corresponding demand on ethanol for 5, 10, and 20% blending targets would be 4.2, 5.9, and 9.2 billion liters in 2020–2021; 4.9, 6.9, and 10.8 billion liters in 2025–2026; and 25.7, 7.9, and 12.5 billion liters in 2030–2031, respectively. Needless to say, this demand is impossible to be met from molasses ethanol production in the country, since this would mean a drastic increase in the area under sugarcane cultivation or increase in productivity of the crop both of which seems unlikely. Even with a go-ahead to produce ethanol from B heavy molasses [21] in addition to the C molasses, this seems to be an impossible task. While there has been some growth in the ethanol production in the past decade [16], the same cannot be expected for the future years due to the reasons mentioned

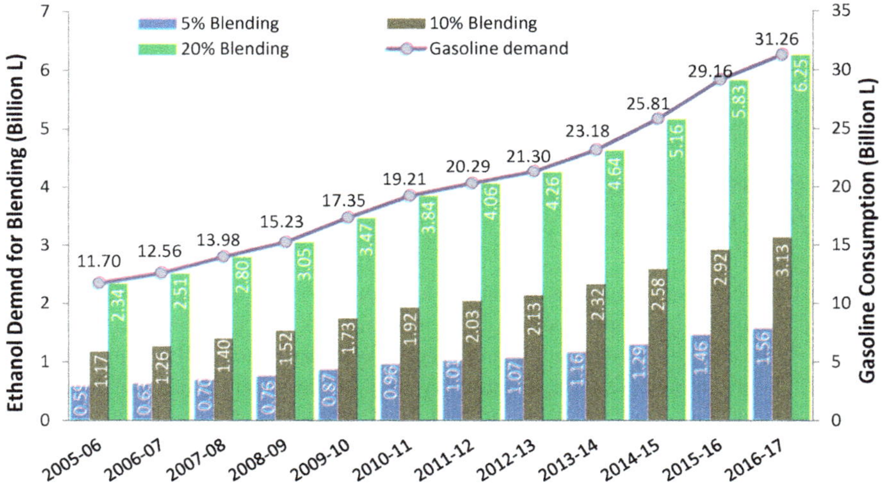

Fig. 10.2 Historical and current gasoline demand with the corresponding demand for ethanol at different blending targets. Data source: Gasoline Demand -[3, 4]. Ethanol Demand-Estimated

below. India's sugarcane productivity has been stagnating at 65–70 tons/ha, and an increase in the area under cultivation would be achieved only at the cost of diverting agricultural land from food crops. The latter would be against the country's policy on fuel ethanol production from nonfood feedstock. However, if ethanol production from cane juice (not molasses) is permitted in the country, sugar industries would produce more of ethanol and less sugar, as ethanol fetches more price. State would not be able to monitor the amount of juice that goes for sugar/ethanol, and if the balance tilts towards ethanol, there would be a sugar scarcity. Defining what can be termed surplus is difficult and there is no monitoring mechanism in place. As ethanol can fetch better prices compared to sugar, the production of sugar can be affected seriously posing the danger of ethanol production at the cost of sugar. Another alternative would be to import ethanol, but ethanol imports at the scales required would go against one of the prime objectives of the blending program—to reduce import dependency on transportation fuels—and would seriously affect the energy security of the nation.

Apparently, 1G ethanol cannot be the only solution to gasoline blending program, and alternative feedstocks for ethanol production have to be accessed to meet the nation's ambitious target of 20% ethanol blending by 2017. While home-grown mature technologies are lacking for the second-generation (2G) ethanol, it still seems to be the way forward since India has a huge resource of lignocellulosic agro-residues to turn to.

3 Second-Generation Ethanol: The Feedstock Challenge

Second-generation ethanol is bioethanol from nonfood feedstock and typically refers to ethanol produced from lignocellulosic biomass. Plant biomass contains about 35–50% cellulose, a polymer of glucose; 20–35% hemicellulose, which is a polymer of sugars, predominantly the five-carbon sugar xylose; and 15–25% of lignin, which is a phenyl propene (nonsugar) polymer [22]. Both cellulose and hemicellulose can be broken down to their component sugars, which then can be fermented to produce ethanol. This is typically accomplished through enzymatic hydrolysis employing cellulases and hemicellulases. Since plant biomass is highly recalcitrant, the hydrolysis of lignocellulosic biomass requires specific pretreatment steps, which are designed to enhance the accessibility of cellulases/hemicellulases to the substrate. The major challenges associated with lignocellulose conversion to ethanol are out of scope of this chapter, and details may be obtained from our previous report [14]. While India still does not have a mature home-grown technology for commercial-scale production of 2G ethanol from lignocellulosics, there are global technologies like Proesa™ from Beta Renewables, Italy, and formicobio™ from Chempolis, Finland, available, and the Indian company Praj Matrix is rapidly advancing toward an advanced 2G ethanol technology. While commercial viability of such technologies is still to be proven on a longer time scale, the fact remains that there are 2G ethanol technologies now available with support from the technology providers for implementation. This would imply that 2G bioethanol productions could start in India in a reasonable time, if not in the immediate future. It therefore becomes important to have an assessment of the biomass resource availability in the nation along with the geographical distribution and density of occurrence so that learned decisions can be made on feasibility and location of bioethanol plants. The important biomass residues available in the country for conversion to ethanol include forestry and agro-residues. While the cellulosic part of municipal solid wastes (e.g., leaves, straw, paper, etc.) can technically be converted to ethanol, the feasibility seems remote in the current scenario and is therefore not considered here. Also, from studies conducted previously by our group [11, 14] and others [9, 10, 12, 13, 20], it is evident that the only category of biomass available in sufficient densities, geographically favorable locations, and feasible for economic conversion to bioethanol is the agro-residues. While agro-residues do have other important uses in rural areas where they are generated, the most important ones are their use as fodder, as thatching material, and as cooking and heating fuels [12, 20], the quantities generated are sufficient to cater to these and still leave a surplus which can be used as feedstock for bioethanol. It may also be noted that it is the same surplus fraction that is targeted in many of the studies mentioned above, to be used as fuel for biomass-based power generation. Though these are competing uses, the percentage diversion of agro-residue feedstock to either of these applications cannot be computed accurately now due to the obvious reason that both are projections, where the actual demand for biomass is yet to be understood. Hence, in the following discussions, the surplus agricultural residue feedstock is treated as available entirely for bioethanol application. One of the major challenges in the feedstock resource mapping in India is the lack of reliable data on the usage of these feedstocks for competing applications. Often empirical estimates are made on the usage of feedstock for the competing applications, which has limitations. Our previous study [11, 14] calculated the proportion of residues used for other applications based on primary data collected through site visits and interviews with farmers and cooperatives, and hence the factors used to derive those estimates have been used here for a want of better methods. Though India's net cropped area has been relatively constant at 140 million hectares (Mha), from the 1970s, there has been an increase in gross cropped area (due to multiple cycles of cultivation in the same land), and in the production and yield of several crops, largely due to the increase in area under irrigation. Increase in human and cattle population has resulted in an increase in demand of the agro-residues for alternate applications, and therefore it is presumed that the conversion factors used for calculating surplus availability remain fairly accurate.

3.1 Crop Production, Conversion Factors, and Biomass Generation Estimates

India with a total area of ~329 Mha uses ~141 Mha for agriculture which at 42.8% is one of the highest against a global average of ~11%. The total (gross) cropped area (counting land usage more than once for cropping) was 195.25 in the year

2011–2012 [23]. The major crops in the nation are rice and wheat together accounting for ~37% of the cropped area in 2012–2013, whereas sugarcane is the largest single crop with an annual production of 341 MMT (Table 10.2).

The statistics by the Department of Agriculture and Cooperation (DAC), Government of India, shows that the area under cultivation of the major crops—rice, wheat, sugarcane, maize, and cotton—has not changed much during the past 9 years. Production of these crops has also remained almost similar except in the case of sugarcane (Fig. 10.3a and b). Plantation crops, vegetables, and fruits were not considered since reliable data is not available and their amounts are also minor compared to these principal crops. Nevertheless, it may be noted that these crops may assume a regional value, if distributed small-scale operations are to be considered.

Table 10.2 Area under cultivation, production, and yields of the principal crops in India (2012–2013)

SL. no	Group of crops	Crops	Area (Mha)	Production (MMT)	Yield (tons/ha)
1	I. Food grains	Rice	42.75	105.24	2462
2		Wheat	30	93.51	3117
3		Sorghum (jowar)	6.21	5.28	850
4		Pearl millet (bajra)	7.3	8.74	1198
5		Maize	8.67	22.26	2566
6		*Total coarse cereals*	*24.76*	*40.04*	*1617*
7		Chick pea (gram)	8.52	8.83	1036
8		Pigeon pea (Tur)	3.89	3.02	776
9		Lentil (masur)	1.42	1.13	797
10		*Total pulses*	*23.26*	*18.34*	*789*
11	II. Oilseeds	Groundnut	4.72	4.7	995
12		Rapeseed and mustard	6.36	8.03	1262
13		Soya bean	10.84	14.67	1353
14		Sunflower	0.83	0.54	655
15		Sesamum	NA	0.69	NA
16		Safflower	NA	0.11	NA
17		Castor	NA	1.95	NA
18		Linseed	NA	0.15	NA
19		*Nine major oilseeds*	*26.48*	*30.94*	*1168*
20	III. Other cash crops	Sugarcane	5.00	341.2	68,254
21		Cotton	11.98	34.22	486
22		Jute and mesta	0.86	10.93	2281
23		Chilies	0.79	1.28	NA

Source: MoA [23]

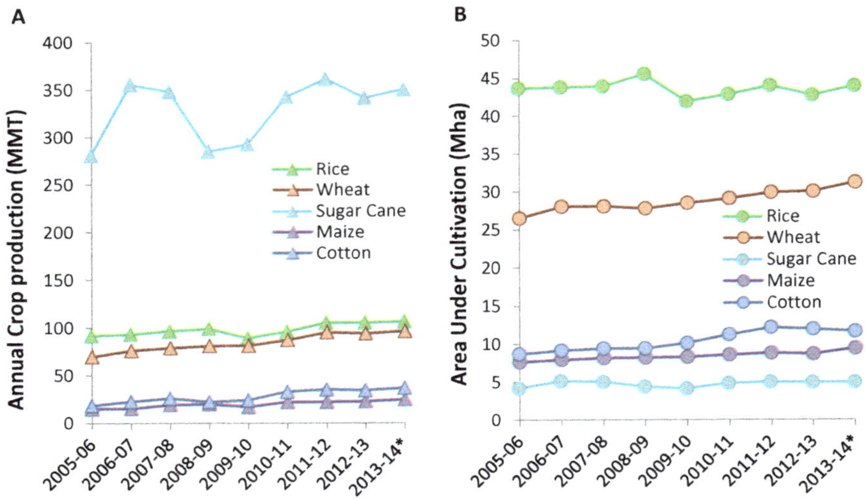

Fig. 10.3 Historical trends in the production of major crops and the area under their cultivation. Source: DAC, MoA [23]. *Asterisk* Advance Estimates by DAC

Primary data on the residue generation of agricultural crops is limited, and hence most of the studies addressing agro-residue generation in the country have estimated the residue potential based on the few studies which have brought out the residue to crop ratios (RCRs) like the one by Ravindranath et al. [12]. The relatively recent study by our group [11, 14] has calculated the RCR based on primary data and provided agro-residue generation data similar to the study by Ravindranath et al. [12]. These RCRs were used in the current study to estimate the potential of agro-residue generation in the country and are given in Table 10.3. Where RCR were not available from this study, the RCR was taken from Ravindranath et al. [12], and where RCR was not available in both the studies, it was assumed that the RCR was the same as that for the broader class to which the crop belongs (e.g., category "other pulses" assumed to have the same conversion yields as "pigeon pea" or "lentil," both of which had an RCR of 1.25). The analyses indicate that the total availability of agro-residues in the country is 657 MMT which, though is higher, is not very different from our previous report which indicates the availability of 623 MMT [11, 14]. However, the current report did not consider the residues—pine needles, bamboo, and water hyacinth biomass—as it deals exclusively with agro-residues. Apparently, the production of agro-residues can be considered as increasing, which correlates well with the increase in crop yield, the latter attributable probably to increase in area under irrigation from the time of the last report.

The analysis proves once again that the residues that are generated in highest quantities are rice straw (152 MMT), sugarcane residue (145 MMT), wheat (136 MMT), cotton stalks (68 MMT), maize (41 MMT), and other coarse cereals (29.01). Together, these accounted for almost 90% of the agro-residues generated.

Table 10.3 Generation of agricultural residues in India

Category	Crops	Area (Mha)	Production (Mt)	Types of residues	RCR [11]	Annual residue generation (MMT)
Sugar	Sugarcane	5.00	341.2	Trash	0.275	93.83
	Sugarcane			Bagasse (dry)	0.200	51.18
Food grain	Rice	42.75	105.24	Straw	1.2	126.29
	Rice			Husk	0.24	25.26
	Wheat	30	93.51	Straw	1.45	135.59
Cotton	Cotton	11.98	34.22	Stalk	2	68.44
Other coarse cereal	Maize	8.67	22.26	Stover	1.5	33.39
	Maize			Cobs	0.275	6.12
	Maize			Husk, silk, etc.	0.175	3.90
	Pearl millet (bajra)	7.3	8.74	Stover	1.25	10.93
	Sorghum (jowar)	6.21	5.28	Stover	2	10.56
	Other coarse cereals	2.53	3.76	Straw	2[b]	7.52
Oilseeds[a]	Soya bean	10.84	14.67	Stalk	2	29.34
	Rapeseed and mustard	6.36	8.03	Stalk	2	16.06
	Groundnut	4.72	4.70	Stalk	2	9.40
	Castor		1.95	Stalk	2	3.90
	Sesame		0.69	Stalk	2	1.38
	Sunflower	0.83	0.54	Stalk	2	1.08
	Linseed		0.15	Stalk	2	0.30
	Safflower		0.11	Stalk	2	0.22
	Other oilseeds		0.1	Stalk	2	0.20
Pulses	Chick pea (gram)	8.52	8.83	Stalk	1	8.83
	Pigeon pea (tur)	3.89	3.02	Stalk	1.25	3.78
	Lentil (masur)	1.42	1.13	Stalk	1.25	1.41
	Other pulses	9.71	5.36	Stalk	1.25	6.70
Spices	Chilies	0.79	1.30	Stalk	0.74	0.96
				Total		656.56

Crop data source: MoA [23]
RCR adopted from Pandey et al. [11]
[a]RCR assumed to be the same for oilseeds
[b]RCR adopted from Ravindranath et al. [12]

Other significant residues were oilseeds (62 MMT) and pulses (21 MMT). Soybean alone contributed nearly 50% of the projected residues from "oilseed" category.

3.2 *Current Uses and Surplus Availability*

Agro-residues find a variety of applications in India, and the major bulk of the residue is consumed at the source of generation itself or in the vicinity by the farmers as a fodder for their cattle. The usage of the different residues vary depending on their calorific value, lignin content, density, palatability by livestock, and nutritive value [20]. Thus, most of the cereal and pulse residues find their way into the animal fodder basket and are unavailable as feedstock for bioethanol applications. India has the world's highest cattle population at 512 million [23], and since the grazing lands in the country have lower productivity, agro-residues are the primary source of feed for the cattle. It was estimated that about half of the total agro-residue generated is used as fodder [11, 20]. A small percentage goes for other rural applications, such as their use as thatching material, as direct burning, as a cooking fuel or for heating, etc. About 20% of the residues are used for other commercial applications including paper manufacture, biomass power plants and gasifiers, boilers (rice mills, distilleries, etc.), and other applications like handicrafts, packaging, etc. The usage of agro-residues as fodder also depends on the geographical region and farmer preferences. For example, while rice straw is used as a fodder in southern India, it is not used typically in north and northwestern region, where wheat straw becomes the preferred fodder. Despite a large fraction diverted to different end uses, a large quantity of agro-residues still remains surplus and may be used as bioethanol feedstock. Our previous study had identified that more than 90% of the residues from major crops (rice, wheat, sugarcane, maize, and sorghum) are consumed for rural and commercial applications, and the actual availability of biomass from these crops is less than 10% of the generation. Also since the collection, transportation, and storage systems for the residues are highly underdeveloped and in some cases nonexistent, there is considerable loss in the material. The losses in collection, transportation, and storage were estimated at ~20% by Purohit and Dhar [20]. In our earlier study, the actual usage patterns of the residues from principal crops were estimated based on primary data obtained through surveys, and a study discovered that almost 75% of the total crop residues are consumed. Only less than 10% of residues generated in largest quantity (rice and wheat straws, sugarcane bagasse, and corn stover) is available as surplus.

Since any authentic recent study on the actual usage of agro-residues in India is not available, the consumption was calculated as a percentage of the generated residues based on crop residue consumption data from our previous study [11]. Table 10.5 shows the surplus residue availability calculated based on the residue generation data of 2012–2013, which is the most recent consolidated data on crop production available from the Department of Agriculture and Cooperation, Ministry of Agriculture, Government of India [23]. The estimations on the surplus availability of agro-residues indicate a similar trend as we had observed earlier with sugarcane trash (tops, leaves, etc., leftover in the field after harvesting) being the agro-residue available in the largest quantity in the country at 75.03 MMT. This is followed by cotton stalk (41.65 MMT), wheat straw (11.1 MT), and rice straw (9.47). Residues from oilseed crops and pulses were available in formidable amounts (18.55 and 6.25 MT, respectively). However, the latter two are estimates on the collective amount of different crops and will be geographically more distributable compared to the top four residues—cane trash, cotton stalk, rice, and wheat straws. Together these residues account for almost 90% of the surplus residues available in the country (Fig. 10.4). Total surplus agro-residue availability in the country is 175.11 MMT which is 27% of the total generation (Table 10.4).

It is apparent that if 2G bioethanol has to be produced in the country, sugarcane trash, cotton stalk, and the straws from rice and wheat would have to be the possible feedstock, since the percentage share of these residues in the total surplus biomass is 43, 24, 6.3, and 5.4, respectively.

Sugar Cane field in Tamil Nadu showing trash left over after harvesting. Cane trash is the agro-residue available in largest quantities in India

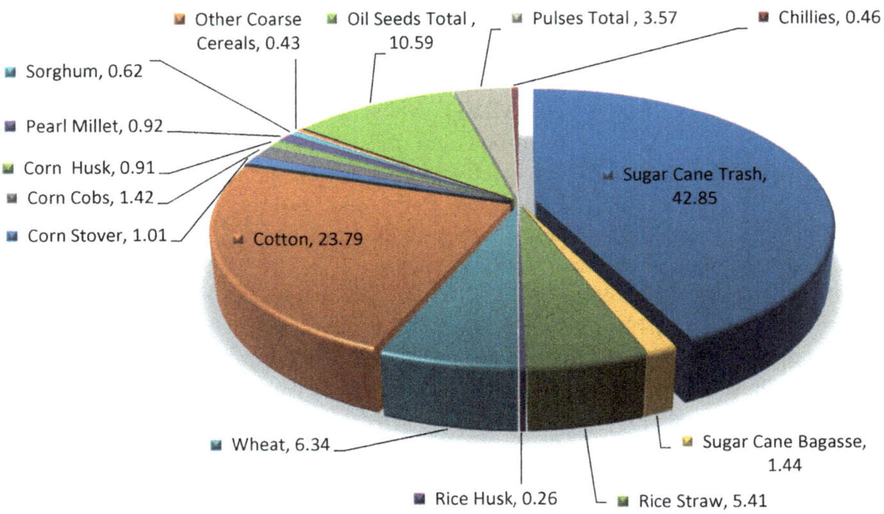

Fig. 10.4 Percentage share of different agro-residues in the total surplus biomass availability

3.3 *Feedstock Composition and Maximum Theoretical Ethanol Production*

If the entire 175.11 MMT of surplus agro-residues in India can be converted to ethanol, it can generate 35 billion liters of alcohol using the conversion factor 214 lge/t_{DM} (liters gasoline equivalent/ton dry matter) of IEA [6]. However, this conversion factor is a rather generic one and may not represent all biomass types, since the composition of biomass varies from residue to residue and also depending on geo-climatic and varietal differences. Hence, the bioethanol production potential from the major identified agro-residues was computed using the content of cellulose and hemicellulose in those biomass types. A representative chemical composition of each residue was calculated from our own experimental results, from reported literature, NREL biomass compositional database [24], and the Phyllis2 database [25] of Energy Research Centre of Netherlands. The yield of fermentable sugars from cellulose and hemicellulose in the biomass was calculated using the US-DOE's conversion factors (yield of glucose per unit quantity of cellulose is 1.11 and the yield of pentose sugar per unit quantity hemicellulose hydrolyzed is 1.12). The stoichiometric value for glucose or xylose to ethanol conversion by fermentation is 0.51, and the efficiency for hydrolyzing cellulose to glucose was assumed at 75%. Similarly, the efficiency for converting hemicellulose to pentose sugars was assumed to be 85% based on experimental results performed under typical conditions. Also the fermentation efficiencies were assumed to be 75% and 50%, respectively, for biomass-derived glucose and xylose. The theoretical ethanol production for each biomass residue was computed based on the above

Table 10.4 Generation and surplus availability of agro-residues in India

Residue type	Annual residue generation (MMT)	% surplus[a]	Surplus residue availability (MMT)
Sugarcane trash	93.83	79.96	75.03
Sugarcane bagasse	51.18	4.94	2.53
Rice straw	126.29	7.50	9.47
Rice husk	25.26	1.79	0.45
Wheat straw	135.59	8.19	11.10
Cotton stalk	68.44	60.85	41.65
Maize stover	33.39	5.29	1.77
Maize cobs	6.12	40.48	2.48
Maize husk, silk, etc.	3.90	40.74	1.59
Pearl millet (bajra) stalk	10.93	14.75	1.61
Sorghum (jowar) stover	10.56	10.26	1.08
Other coarse cereals—stalk/stover	7.52	[b]10.00	0.75
Soya bean stalk	29.34	–	–
Rapeseed and mustard stalk	16.06	–	–
Groundnut stalk/leaves	9.40	–	–
Castor stalk/leaves	3.90	–	–
Sesame stalk/leaves	1.38	–	–
Sunflower stalk/leaves	1.08	–	–
Linseed stalk/leaves	0.30	–	–
Safflower stalk/leaves	0.22	–	–
Other oilseed stalks/leaves	0.20	–	–
Total oilseed residues	61.88	29.98	18.55
Chick pea (gram) stalk/leaves	8.83	–	–
Pigeon pea (tur) stalk/leaves	3.78	–	–
Lentil (masur) stalk/leaves	1.41	–	–
Other pulse stalks/leaves	6.70	–	–
Total pulse residues	20.72	30.16	6.25
Chili stalk/leaves	0.96	83.33	0.80
	656.56		175.11

[a]Percentage of surplus residue calculated based on previous study Pandey et al. [11]
[b]Conservative estimate based on usage patterns of similar coarse cereals

assumptions. Wherever compositional data was unavailable, the conversion factor 214 lge/t_{DM} [6] was used for computing the potential yield of ethanol from the agro-residue. The potential yield of ethanol from the agro-residues identified as second-generation feedstock is given in Table 10.5.

The theoretical yield of ethanol estimated at ~39 billion liters is closer to the calculations made using IEA's [6] conversion factor. Assuming 10 or 20% losses of biomass during collection, transportation, and storage, the projected yields of ethanol are 35.06 billion and 31.6 billion liters, respectively. This amount is

Table 10.5 Bioethanol production potential using surplus agro-residues as feedstock

Crop residue	Surplus residue (MMT)	Cellulose (%)	Hemi cellulose (%)	Lignin (%)	Ethanol yield from glucans[a] (kg/ton)	Ethanol yield from xylans[b] (kg/ton)	Net ethanol yield (L/ton)	Gross ethanol yield at no collection loss (billion liters)	Gross ethanol yield at 10% collection loss (billion liters)	Gross ethanol yield at 20% collection loss (billion liters)
Sugarcane trash	75.03	37.0 ± 3.4	26.9 ± 3.8	24.43 ± 4.4	117.9	65.4	232.25	17.43	15.68	13.94
Cotton stalk	41.65	38.7 ± 4.5	18.6 ± 3.1	25.32 ± 3.1	123.3	45.2	213.55	8.89	8.00	7.12
Oilseed stalk—total[c]	18.55						201.00	3.73	3.36	2.98
Wheat straw	11.1	40.4 ± 7.9	25.7 ± 3.5	14.19 ± 4.2	128.7	62.5	242.29	2.69	2.42	2.15
Rice straw	9.47	36.2 ± 4.5	20.4 ± 1.6	20.18 ± 2.6	115.3	49.6	209.00	1.98	1.78	1.58
Pulse stalk—total[c]	6.25						201.00	1.26	1.13	1.01
Sugarcane bagasse	2.53	41.6 ± 4.3	25.0 ± 2.6	20.58 ± 2.8	132.5	60.8	244.97	0.62	0.56	0.50
Corn (maize) cobs	2.48	36.9 ± 3.4	35.6 ± 5.9	14.30 ± 2.4	117.6	86.5	258.65	0.64	0.58	0.51
Corn (maize) stover	1.77	37.6 ± 3.0	26.0 ± 4.9	18.85 ± 2.7	119.9	63.2	231.90	0.41	0.37	0.33
Pearl millet stover[c]	1.61						201.00	0.32	0.29	0.26

Corn (maize) husk[c]	1.59						201.00	0.32	0.29	0.26
Sorghum stover	1.08	39.5 ± 3.1	22.9 ± 4.7	19.1 ± 3.5	125.8	55.7	230.03	0.25	0.22	0.20
Chili stalk	0.8	39.9 ± 2.1	17.9 ± 1.8	25.3 ± 0.2	127.1	43.5	216.24	0.17	0.16	0.14
Other coarse cereal stalks[c]	0.75						201.00	0.15	0.14	0.12
Rice husk	0.45	32.9 ± 3.6	21.2 ± 7.1	19.37 ± 1.5	104.8	51.5	198.14	0.09	0.08	0.07
Total ethanol yield								38.95	35.06	31.16

[a]Assumes 0.75 % efficiency in hydrolysis and 0.75% efficiency in fermentation
[b]Assuming 0.85 % efficiency in hydrolysis ad 0.5 % efficiency in fermentation
[c]Residues for which reliable compositional data is not available: Ethanol yield = 214 lge/TDM Eisentraut [6]

sufficient to replace the county's entire current gasoline demand and can very well satisfy the projected 20% blending requirement (6.2 billion liters), which has to be implemented by 2017 according to the national biofuel policy. Purohit and Dhar [20] project the demand for fuel ethanol as 10.8 billion liters by the year 2025 and about 12.5 billion liters by 2030, if it has to be blended in gasoline at 20% level. They had projected the gasoline demand to be 45 billion liters by 2030. Second-generation ethanol from agro-residues can meet more than double these targets even at current crop production levels. Even if we assume that the crop production stagnates, with improvements in efficiency of hydrolysis and fermentation, there is a potential to dope the entire projected demand of ~45 billion liters of gasoline for 2030 with bioethanol at 20% level. The biggest advantage in these cases would be that it does not require additional land to obtain sufficient feedstock and does not compete with food applications. More importantly, any increase in land area for food production will automatically increase the agro-residue output also, making more feedstock available for 2G ethanol production.

4 Feedstock for 2G Ethanol in India: The Way Forward

India's hope on achieving 20% ethanol blending in gasoline is probably dependent exclusively on second-generation ethanol technology implementation, since the projected demand of ethanol for even near-term targets cannot be achieved using the surplus ethanol coming from molasses-based ethanol industry. It seems the time is ripe for implementing 2G ethanol technologies with international technology providers declaring readiness to sell their technologies and at least one Indian company almost ready with a home-grown ethanol technology. Technical challenges aside, the major impediment to biomass-based ethanol is the availability of biomass itself. 2G ethanol plants require a consistent and sustainable supply of quality raw material within a given radius of the plant (ensuring feasible costs for logistics) and in quantities required for year-round operation. This in turn requires a thorough assessment of the nation's biomass resource potential with geo-climatic distribution as well. India does have an annually updated database on crop production generated and maintained by the Ministry of Agriculture, and there have been notable studies (e.g., [12]), which have brought out reliable residue to crop ratios that allow the estimates on annual agricultural residue generation in the country. With limited land surface to spare for energy crops, the only feasible feedstock for fuel ethanol production in India seems to be agro-residue. Being an agrarian economy, India is a leading producer of several cereals, sugar, cotton, and plantation crops, all of which generate lignocellulosic residues that are potential feedstock for bioethanol. Though the total quantity of such residues is considerable, there is a major difference in the density and distribution of crops in comparison with countries like the USA, Brazil, China, or the European Union where the cropped areas are huge and concentrated geographically. This ensures sufficient feedstock for an ethanol production facility, and a 2G ethanol biorefinery can enjoy a single and uniform feedstock. Also the harvesting and bailing operations are mechanized, and the logistics and storage systems adequately addressed, allowing operational speed and reduction both on energy and cost of operation. Land holding for farmers is typically small in India, and due to these and the difference in crops cultivated in individual landholdings, it is difficult to obtain the same kind of crop residues year round for operation of 2G ethanol plants. The 2G ethanol technology for India should therefore address these unique issues and should definitely be capable of handling multiple feedstocks and of smaller scales. Mechanized harvesting, organized collection, and bailing of crop residues are now being implemented across the country, and there are agencies in the private sector now catering to supply of biomass in densified forms. This is a major development since our last review in 2009 when organized operations in this area were limited, if not nonexistent. While these companies cater mainly to biomass power generation companies now, the demand from a future 2G biorefinery will definitely lead to the existing and new companies catering to the biomass requirements of such biorefineries. The collection model required for biomass procurement at the farmer level is also expected to evolve, once such demands are in place. However, the increased demand can result in price escalations and competitions with other applications including their fodder use, which can have serious consequences. Therefore, it becomes the responsibility of the state to evolve policies and mechanisms for the regulation of biomass uses and assigning the types and quantities of biomass for different applications, which is easily said than done. This report would not make any recommendations on these aspects, since where the surplus biomass is going to end up is a question probably decided by market forces unless there is a regulatory intervention. Nevertheless, it is prudent to state that the 2G ethanol is not an option but a necessity for India, and feedstock resource assessment and scientific characterizations of the composition of feedstock are going to be highly important for implementing 2G ethanol technologies. Integration of the residue availability statistics with the existing annually updated crop database prepared and maintained by the Ministry of Agriculture, together with remote sensing data and geographic information systems (GIS) to have an interactive map of biomass resource availability, seems to be the need of the time and is strongly recommended. It should also include the varietal and compositional details of crop residues mapped to the geological coordinates, which will provide all necessary details for implementation of 2G ethanol or for that matter any national program that requires agro-residues as feedstock. The Biomass Resource Atlas commissioned by Ministry of New and Renewable Energy [10] was an important step in this direction, but the data reference for this resource was on 2004 and since then not updated. Moreover, the resource has limited coverage. This does not in any way underestimate the pioneering effort to integrate the abovementioned data but points to the necessity of making it or a new and comprehensive database, which is periodically updated. The new system should have GIS integration with biomass data acquired through primary and secondary surveys by national agencies, remote sensing, and compositional data through biochemical analysis of crop residues mapped to their locations.

5 Conclusions

It is now widely recognized that the near-term solution for having renewable and sustainable liquid transportation fuels lies in developing lignocellulosic ethanol. This is especially true in the case of India, where the limited availability of edible or nonedible plant-/animal-derived oils seriously limits the possibilities of biodiesel. With over 650 MMT of agricultural residues generated annually, this is inarguably the most abundant feedstock available in the nation for conversion into biofuels. Several international players have claimed mature 2G ethanol technologies and have even announced operations in India in the very near future, and this opens the question of feedstock availability. Unlike countries with land to spare for energy crops, India has to depend on agro-residues as the feedstock since the arable land utilization for energy crops is not a feasible option for the nation due to the need to produce food for a rapidly growing population. Assessment of the availability of agro-residues and their compositions in terms of sugar polymers, collection, densification, transportation, and storage of the biomass is an important issue to be addressed for any successful biofuel programs in the nation. Several studies have addressed the issue of biomass availability from agriculture, and quite often the estimates are in reasonable agreement with each other. Our earlier assessment which covered agro-residue production data till 2008, which relied on primary as well as secondary data collection, estimated the biomass generation at 623 MMT and had revealed that the agro-residue with largest surplus availability was sugarcane trash. The most updated information presented in this study indicated the total agro-residue generation at 657 MMT, and the surplus availability was highest for cane trash itself indicating no change in pattern of cultivation and consumption. It may be very well assumed that this pattern might remain fairly consistent unless and until there are changes in the land usage patterns and/or irrigation and introduction of high-yield varieties of the major cereal crops and sugarcane. Since these are longer-term possibilities, it may well be assumed that the near-term changes in the generation of agro-residues would not happen. However, with government-imposed policies and incentives, disposal of residues by burning has been reduced to a large extent, and the influx of these additional amounts of residues into the surplus availability pool can increase the amount of agro-residues actually available for biofuel applications. It seems that collection and preprocessing of agro-residues are slowly evolving as profitable industries which mean that there could be a better management of the surplus residues. It is still felt that there is a lot to be done in the compositional characterization of biomass to assess the bioethanol potential, its collection technologies, and the logistics of transportation and storage. All these areas have to develop rapidly now and ideally the information should be on public domain, just like the data from the Department of Agriculture on annual crop production. Also it is suggested that the data on biomass residue generation to be mapped to their geographical locations and made accessible through interactive knowledge portals which would make the decision-making on where to set up biofuel plants and their feasibility can be assessed. This is important since the type of land holdings in India and hence the pattern of agriculture are significantly different from the developed nations, and 2G bioethanol technologies for the nation has to address this issue. An integrated online portal with information on the type and amount of biomass generated, their compositions and potential conversion yields, and geographical location is probably an immediate need to accelerate developments in this area.

References

1. United Nations (2016) World economic situation and prospects, 2016. http://www.un.org/en/development/desa/policy/wesp/wesp_current/2016wesp_ch1_en.pdf
2. Gunatilake H, Holst DR, Sugiyarto G, Baka J (2011) Energy security and economics of Indian biofuel strategy in a global context, ADB economics working paper series no 269
3. MoPNG (2015) Indian petroleum and natural gas statistics 2014–2015, Government of India, ministry of petroleum and natural gas, economics and statistics division, New Delhi
4. PPAC (2016) Petroleum planning and analysis cell, ministry of petroleum and natural gas, government of India, Data on petroleum consumption till Jan 2016. http://ppac.org.in/content/147_1_ConsumptionPetroleum.aspx. Accessed 2 Feb 2016
5. GOI (2009) Government of India- ministry of new and renewable energy-national policy on biofuels
6. Eisentraut A (2010) Sustainable production of second—generation biofuels: potentials and perspectives in major economies and developing countries-information paper by international energy agency
7. Aradhey A (2015) India biofuels annual 2015, global agricultural information network report no IN 5079. http://gain.fas.usda.gov/Recent%20GAIN%20Publications/Biofuels%20Annual_New%20Delhi_India_7-1-2015.pdf. Accessed 1 Jul 2015
8. ISMA (2014) List of ethanol producers, Indian sugar mills association web pages. http://www.indiansugar.com/Ethanol.aspx. Accessed 19 Oct 2015
9. Joshi P, Sharma N, Sarma PM (2016) Assessment of biomass potential and current status of bio-fuels and bioenergy production in India. Curr Biochem Eng 3:4–15
10. MNRE (2010) Biomass resource atlas of India V2.0, a project of NRE executed by combustion, gasification and propulsion laboratory (CGPL), Indian Institute of Science. http://lab.cgpl.iisc.ernet.in/atlas/Default.aspx. Accessed 16 Dec 2016
11. Pandey A, Biswas S, Sukumaran RK, Kaushik N (2009) Study on availability of Indian biomass resources for exploitation: a report based on a nationwide survey. TIFAC, New Delhi
12. Ravindranath NH, Somashekar HI, Nagaraja MS, Sudha P, Sangeetha G, Bhattacharya SC, Abdul Salam P (2005) Assessment of sustainable non-plantation biomass resources potential for energy in India. Biomass Bioenergy 29(3):178–190
13. Sharma N, Tabacchioni S, Chiarini L, Sara PM (2013) Biomass and biowaste inventories report, SAHYOG project grant agreement no. 289615
14. Sukumaran RK, Surender V, Sindhu R, Binod P, Sajna KV, Janu KU, Rajasree KP (2010) Lignocellulosic ethanol in India: prospects, challenges and the feedstock Dilemma. Bioresour Technol 101(13):4826–4833
15. Pérez R (1995) Molasses. In: Tropical feeds and feeding systems, first FAO Electronic Conference 1995
16. Ray S, Miglani S, Goldar A (2011) Ethanol blending policy in India: demand and supply issues. ICRIER policy series no 9, December 2011

17. Raju SS, Parappurathu S, Chand R, Joshi PK, Kumar P, Msangi S (2012) Biofuels in India: potential, policy and emerging paradigms. National center for agricultural economics and policy research (NCAP)- Policy paper no 27, NCAP, New Delhi
18. Verma A (2014) The Indian ethanol program. Paper presented at the F.O. Licht's sugar and ethanol Asia, February 25–26, 2014. http://www.indiansugar.com/uploads/Bangkok-_26th_Feb_2014_1_%20(1).pdf. Accessed 20 Nov 2015
19. ISMA (2016) Statistics. http://www.indiansugar.com/Statics.aspx. Accessed 15 Feb 2015
20. Purohit P, Dhar S (2015) Promoting low carbon transport in India: biofuel roadmap for India, UNEP DTU Partnership, isbn 978-87-93130-66-1
21. GOI (2015) Incentives to sugar mills for supply of ethanol, press release -ID 119854-ministry of consumer affairs, food and public distribution. http://pib.nic.in/newsite/mbErel.aspx?relid=119854. Accessed 28 Apr 2015
22. Wyman CE (1994) Ethanol for lignocellulosic biomass: technology, economics and opportunities. Bioresour Technol 50:3–16
23. MoA (2014) Agricultural statistics at a glance 2014, government of India—ministry of agriculture, department of agriculture and co-operation, directorate of economics and statistics. Oxford University, New Delhi, ISBN 13: 978-0-19-945965-0
24. DOE (2016) US department of energy biomass program, theoretical ethanol yield calculator and biomass feedstock composition and property database. http://www.afdc.energy.gov/fuels/ethanol_feedstocks.html Accessed 2 Feb 2016
25. Phyllis 2 (2016) Database for biomass and waste, energy research centre of the Netherlands. https://www.ecn.nl/phyllis2. Accessed 2 Feb 2016

Chapter 11
India's Biodiesel Programme: Status, Prospects, and Shortcomings

Sanjib Pohit and Pradip Kumar Biswas

Abstract Even after launching biodiesel programme more than a decade ago, India is yet to achieve significant progress. This calls into question whether India has followed a flawed approach. This chapter argues that though basic policy approach is not flawed, India needs to tweak the policy environment for the success of the programme. The principal changes required are multi-feed feedstock approach, attractive incentive mechanism both at feedstock stage as well as biodiesel production stage, and R&D for increasing the yield from feedstock.

Keywords Biodiesel • India • Institution

1 Introduction

At the outset, energy is critical, directly or indirectly, in the entire process of evolution, growth, and survival of all living beings, and it plays a vital role in the socio-economic development and human welfare of a country. This is very much true in the case of Indian economy which has experienced unprecedented economic growth over the last decade. Today, India is the ninth largest economy in the world, driven by a real GDP growth of 8.7% in the last 5 years (7.5% over the last 10 years). No doubt, the high growth path implies high demand for energy resources. Unfortunately, India is an energy-deficit nation by global standards [1]. The energy deficit is all the more pronounced in the liquid transportation fuels sector which faces two basic challenges, i.e., rising energy demand in the face of limited reserves and higher dependence on increasingly costlier imported crude oil. The problem is more acute since more than 95% of India's surface transport is dependent on petroleum products [2].

S. Pohit (✉)
National Council of Applied Economic Research, New Delhi, India
e-mail: spohit@gmail.com; spohit@ncaer.org

P.K. Biswas
College of Vocational Studies, University of Delhi, New Delhi, India
e-mail: pkbiswas63@hotmail.com

A.K. Chandel, R.K. Sukumaran (eds.), *Sustainable Biofuels Development in India*,
DOI 10.1007/978-3-319-50219-9_11

With global demand and global energy prices likely to increase in the medium- to the long term, its impacts on balance of payment, could adversely affect the country's future development. This grim energy prospect for India has forced her policy-makers to intensify their efforts to search for alternate fuel options for transport sector. In this context, biofuel may offer alternative options for partially meeting India's demand in transport sector. Moreover, since India's transport sector is heavily dependent on diesel, the focus has mainly been on biodiesel as an alternative to fossil fuel for transport sector.

In this context, this chapter attempts to trace out the development in the biodiesel sector with a view to identify the potentials and shortcomings in the policy space so that policy corrections can be made and true potentials are realised in the coming years. The plan of the rest of the chapter is as follows. The next section provides a brief overview of India's biodiesel programme and the rational thereof. The role of various actors in the process is also highlighted. Following this, we focus on the ground realties in this sector. Subsequently, we stress on the lacunae on the policy diasporas with a view to identify the changes required in the policy space for efficient functioning of biodiesel programme. The following section summaries our findings.

2 Overview of India's Biodiesel Programme: Policy Space

India entered into the biofuel production arena way back after the second oil shock in the late 1970s [16]. However, policy push really happened about a decade back when oil price exhibited spikes and India began a 5% ethanol blending (E5) pilot programme in 2001. However, policy-makers soon realised that it is not prudent for India to follow the path of other countries like Brazil, USA, and Germany where food crops are used as feedstock for biofuel. This could lead to rise in food prices within India leading to economic hardship on the poorest section of the society. Nor it is advisable for India to divert prime farm land for growing of feedstock for biofuel since stagnant productivity growth in respect of food crops coupled with increasing population pressure implies that policy-makers do not have the choice to adopt this path lest it leads to food price rise. It should be also noted that India's biofuel programme emphasises on replacing diesel by biodiesel since the latter is the dominant fuel for transport sector in India. Thus, the focus is on increasing the production capacity of biodiesel and feedstock thereof.

Given these constraints, Indian policy-makers looked for oleic seed-bearing plants which can be grown in wastelands and/or low-quality lands. National Mission on biodiesel initially identified about 400 varieties of oilseed-bearing plants. Out of these, jatropha, a small shrub that grows on degraded land was selected as the main feedstock for biodiesel for the following reasons: (a) high oil content (40% by weight), (b) lower gestation period (2–3 years), and (c) harvesting time of its seeds does not coincide with agricultural harvesting time.

India's Jatropha programme has been driven primarily by central government [15]. In 2003, the government launched biodiesel mission. It was divided into two phases—Phase I, a research and demonstration phase, spanning the period 2003–2007 and Phase II, an implementation Phase, spanning the period from 2007 to 2012, with a tentative target of achieving 20% blending by 2012. Phase 1 is primarily funded by government and aimed to create critical momentum for promotion of jatropha as a valuable undertaking, leading to self-sustaining developments by all actors in Phase II. The Ministry of Environment and Forests in conjunction with the state forest department was instrumental to implement plantation programmes in degraded forests, while the department of rural development, along with other government bodies was asked to use existing funding mechanism to implement plantation through local councils and NGOs across the country.

The programme roped in several universities and research institutes to develop appropriate plant varieties. To encourage sector's growth, it fixed purchase price for biodiesel and feedstock and designated some twenty oil marketing companies (OMCs) to purchase and distribute the biofuel. A little more than 700,000 ha of land was brought under jatropha plantation by 2008. This was abysmally low level of plantation if India had to fulfil a 20% blending target of biodiesel. Contrary to expectation, yield of feedstock per acre was found to be half of what was postulated at the lab stage. Thus, programme was behest from the beginning with chronic feedstock shortage for biodiesel. By August 2008, the mission was planned to be shelved. An alternative policy with focus on wider set of plants producing non-edible oilseeds (instead of jatropha only) that could be grown on marginal, degraded, or wasteland was subsequently adopted. The National Policy on Biofuels was formally adopted by government in December 2009 [17]. In the policy, the blending target for biodiesel is fixed at 20% by 2017 (see Table 11.1). As Table 11.1 shows, the policy framework adopts a holistic approach encompassing various policy tools for promoting all round sector development from feedstock to production of biodiesel.

With regard to institutional mechanism, a National Biofuel Coordination Committee (NBCC) headed by the Prime Minister was set up to provide high-level coordination, policy guidance, and review on different aspects of biofuel development. At the state level, the respective government should work with their respective research institutions, forestry department, universities, etc., towards development and promotion of biofuel program. The role of different ministries in various aspects of biofuel development and promotion in the country has been laid out (see Table 11.2). It must be mentioned that the roles mentioned here are by and large that of the central ministries. They are supposed to work in tandem with the corresponding ministries at the state level.

It must be mentioned that not all state governments have shown interest in promoting biodiesel (feedstock cultivation, production and use) within their territories. On the other hand, some states like Chhattisgarh, Madhya Pradesh, and Andhra Pradesh have been quite proactive in promoting biodiesel as reflected in announcement of additional policy incentives over and above that of central government (Table 11.3).

Table 11.1 Salient features of national biofuel policy in respect of biodiesel

Feedstock	Focus on non-food feedstock	
Mode of production	• Promote plantations on wasteland, degraded land, forest land • Promote contract farming schemes concomitant with seed buyback programmes	
Policy mechanism to promote biodiesel production	Subsidies	• Feedstock cultivation activities may be given subsidies towards labour cost • Introduce statutory support prices (SMP) for oilseed procurement • Fix a minimum purchase price (MPP) for biodiesel
	Financing	• Provide finaincial incentives in respect of schemes for biofuel projects • Mobilise fund from multilateral/bilateral lending institutions under assistance for carbon financing opportunities • A National Biofuel Fund may be created to provide financial incentives towards existing biofuel-related activities, new- and second-generation feedstocks, and development of advanced technologies, etc.
	Fiscal incentives	• New- and second-generation biofuel production may be given subsidies/ grants for their growth • Existing concessional excise duties on bioethanol and biodiesel would continue • Custom and excise duties for plant and engine technologies may be reduced • Bringing biofuels under the ambit of "Declared Goods" by the GOI so as to ensure their unrestricted interstate and intrastate movement • Barring an excise duty of 16% on bioethanol, biodiesel is proposed to be free from other central duties
	Research and development	• To establish competitive domestic biofuel industries, promote government's role in research, development, and demonstration • Promote public–private partnerships
	International collaboration	• Encourage international partnerships for developing domestic biofuel industries and ancilliary activties in respect of technology transfer, joint research, and technology development • Biofuel technologies and projects would be allowed 100% foreign equity through automatic approval to attract foreign direct investment (FDI) in case of biofuel is for domestic use • Plantations of inedible oil-bearing plants would not be open for FDI participation

Table 11.2 Roles of different ministries

Ministry of	Role
New and renewable energy	Overall policy-maker, promoting the development of biofuels and research and technology development for its production
Petroleum and natural gas	Responsible for marketing biofuels as well as development and implementation of pricing and procurement policy
Agriculture	Promoting research and development for the production of biodiesel feedstock crops
Science and technology	Supports research in biodiesel crops, specifically in the area of biotechnology
Rural development	Promote Jatropha plantations on wastelands
Road transport and highway	Responsible for plantation along highway rights-of-way and use biodiesel-blended fuel. They need to coordinate with automobile manufacturers association in India for engine modification, emission norms
Railways	Undertake plantation of jatropha over wastelands along rail rights-of-way and trials of biodiesel-blended fuel on railroad locomotives
Environment and forest	Promote plantation of jatropha and tree-borne oilseeds in forest wastelands; get Central Pollution Control Board to monitor health and environmental effects

2.1 India's Biodiesel Programme: Ground Realities

The discussion in the earlier section suggests that there are not dearth of policy intentives for promoting biodiesel sector in the economy. However, even after more than a decade of the launch of the biodiesel programme, India's biodiesel sector does not seem to take off from the experimentation stage. By contrast, Brazail has been able to make significant progress in biofuel sector within a short span of time [4]. Before identifying the bottlenecks, let us make a overview of the achievement till date.

The land under jatropha cultivation rose to about 0.5 million hectares by 2010. Since 65% of these are new plantations, only after 2–3 years, one can extract feedstock them. If we assume jatropha feedstock is only used for biodiesel production, this available feedstock may only meet <0.01% of total biodiesel required for 5% blending by 2010–2011 [5]. Apart from few captive plantations managed by oil marketing companies, no significant additional wastelands have been brought under jatropha cultivation in post 2010–2011 period.

Consequently, the production of biodiesel from jatropha feedstock is still commercially insignificant. According to the Ministry of Petroleum and Natural Gas, GOI, oil marketing companies has not procured biodiesel produced from jatropha recently for blending with diesel [5].

On the other hand, biodiesel production from multiple feed-stocks (crude oil, used cooking oils, animal fats, etc.) seems to be economically viable. Most of the plants utilizing this technology have made commercial sales in last few years (see [5]). However, they were operating only at 25% of their capacity due to shortage of feedstock. It is estimated that about 110–115 million litres of biodiesel is presently

Table 11.3 State biofuel policy initiatives with focus on biodiesel

Initiatives	Andhra Pradesh	Chhattisgarh	Karnataka	Madhya Pradesh	Rajasthan	Tamilnadu	Uttarakhand	Uttar Pradesh
Formal state biodiesel policy	None	None	Yes	None	None	None	None	None
State biodiesel strategy	Yes	Yes	Yes	Yes	Yes	Yes	Yes	Yes
Existence of state coordination body	Yes for public parties	Yes for public parties	Proposal Stage	Yes	Yes	No ne	Yes for public and private parties	Yes
Allocation of government land for plantations	Forest land	Forest land	Communal as well as forest land	Forest degraded land; private land under lease	Wasteland allotted to govt. undertakings; private land under lease	Insignificant	Forest land	Wasteland
Input subsidies/ distribution of input	Yes	Yes	Yes	None	Yes	Partly subsidised	Partly subsidised	None
Governmental funding sources	Yes.	Yes	Not clear	Not clear	Not clear	Yes	Yes	Promotion of public–private, partnership

Provision of extension services (free of cost)	Yes	Yes	Yes	Facilitation cell to train farmers	Data not available	Yes	Yes	Awareness activities by department
Subsidies for/govt. provision of processing facility	None	Yes	Yes	None	Free allotment of land to a few categories	None	None	None
Minimum support price	Jatropha seeds: about Rs. 6/kg	Jatropha seeds: about Rs. 6/kg	Planning stage	Yes	Jatropha seed: about Rs. 7/kg	None	Jatropha seeds: about Rs. 3/kg,	None
Blending requirement	None	Blending requirement of 5%	None	Not clear	Not clear	None	None	None
Tax exemptions on biodiesel	Reduced VAT	None	None	Considering exemption	Yes from VAT	Yes from VAT	Yes from VAT	Considering exemption

Source: ADB [1]; Pohit et al. [3]

being produced in India using multiple feedstocks [5]. The principal buyers are small to medium scale industries who used it by blending it with conventional diesel. The average purchase price of biodiesel in India hovers around INR 45–48 per litre. Since diesel prices is about 18–20% higher than this, it is economically viable for blending it with biodiesel.

Another development is worth mentioning regarding biodiesel production sector. Industry bodies argue that that India has no commercial market for biodiesel. Moreover, producers are permitted to sell biodiesel only to oil marketing companies with regulated prices. Consequently, petroleum ministry in conjunction with state-run oil marketing companies are using their monopoly to control the market for biodiesel. Earlier, biodiesel producers such as Royal Energy Ltd., had raised complaint against the restriction. Even though sections 3, 4, and 26 of the Competition Act allow the regulator to probe including government enterprises and ministries, this has not happened in respect of this sector. In India, the price of biodiesel is regulated by the petroleum ministry, which uses the retail price of diesel for its calculation. Industry argues that this price (Rs. 30 a litre) is too low. Consequently, industry operates the plant a low capacity. Following ban of commercial sale of fuel in 2009, domestic biodiesel production industry has registered a flat growth. The industry's view point is that companies should be permitted to sell it to bulk consumers and not to restricted only to oil companies at regulated price.

Since the onset of biodiesel programme, many of the state transport bodies have been using biodiesel for running a few buses for demonstration/testing purposes. By contrast, Indian railways, one of the large consumers of diesel (2.2 billion litres a year), have been slow in adopting biodiesel for running diesel engines [6]. The Indian Railways has now planned to set up four biodiesel plants costing about Rs. 1.2 billion. Two of these plants will be built at Raipur and Chennai during the next two years, the other two units will be built later. The cost of each plant is expected to be about Rs. 300 million. The plant would have the capacity to produce up to 30 tonnes of biodiesel a day using multiple feedstocks. The engine would be run on biodiesel blended with the HSD oil. Apart from railways saving on HSD fuel expenses, the railways expect to earn Rs. 20 million a year in carbon credits by using environment friendly fuel.

In India, microalgae grow naturally on the western coast. In a recent project undertaken jointly by researchers from the Ministry of Earth Sciences (MoES) and Council of Scientific and Industrial Research (CSIR), a Chevrolet diesel Tavera was run on a 20 per cent biodiesel blend made from marine micro algae [6]. It was found that The B-20 prepared from these microalgae mats provided a mileage of 12.4 km in Tavera in comparison a 10–11 km when it is run by 100% petro-diesel. Thus, efforts are being made to think beyond jatropha as a feedstock, which is a positive development.

3 Lacunas of India's Biodiesel Programme

The earlier discussion has highlighted that availability of feedstock is a problem in the Indian setup. This calls into question whether India has enough land for cultivating jatropha.

Understandably, the demand for biodiesel is linked to growth in demand for fossil fuel and blending target rate. In the past decade, the trend rate of growth of diesel consumption was 6%. If we assume the same rate of growth in the coming years, India would need about 87.30 million tonnes of the petro-diesel by 2017–2019 [1]. This would imply a demand for 20.54 million kilo-litres of biodiesel per year if the blending proportion is 20%. This in turn imply feedstock for biodiesel need to be cultivated in 32 million hectares of wasteland assuming currend rate of oil yield per hectare [1]. It would also need marginal yield improvement over the years. By contrast, Integrated Energy Policy 2006 has argued that a plantations on 20 million hectare of wasteland would yield of 25.71 million kilo-litres (KL) of biodiesel [7].

The main issue whether sufficient land is available for planting feedstock for biodiesel. There is no consensus among the policy-makers in this regard. According to Wasteland Atlas of India, 31.26 million hectare wasteland out of available 55.37 million hectare of lands is considered to be suitable for planning biodiesel feedstock [8]. Additionally, some categories of land have been earmarked by National Biofuel Mission for biofuel production (see Table 11.4). Combining all these, it may be possible to find enough land for growing feedstock to meet blending target for biodiesel [18].

It should be noted that the mere physical availability of land does not necessarily imply that these are left unused and readily available for planting jatropha or such other plants. Given population pressure, one can expect that a significant portion of these lands have been encroached by marginal section of people or the powerful ones [10]. Also, wastelands are also used by poor people for collecting fuel wood and grazing animals [11]. It is only possible to grow feedstock on these wastelands, provided users are convinced that this would improve their economic conditions. In essence, one needs to provide tangible economic incentives. The same is true in case of marginal or unoccupied wasteland if one wants farmer's active participation. The failure of the state to convince the farmers with demonstrated benefits or

Table 11.4 Quantum of land suitable for biodisel feedstock

Type of land	Million hectares
Total wasteland suitable for biodiesel	31.26
Other categories for land for feedstock production (includes land used for (a) agricultural border fences, (b) agro forestry, (c) public land along roads, railways, canal lands under integrated watershed development program)	8.0
Total available land	39.26

Source: Biswas and Pohit [9]

provision of other incentives on a long-term basis is one of the important reasons for the very slow progress of the biodiesel production in India.

To achieve economies of scale in the production of biodiesel, one needs continuous stretch of wastelands. However, this is missing in many Indian states. In addition, smaller land holdings, ownership issues with government or community-owned wastelands, lacklustre progress by state governments and negligible commercial production of biodiesel have hampered the efforts and investments made by both private and public sector companies. Given the role of multiple institutions regarding ownership of lands in villages, the issue of availability of land and use thereof for feedstock plantation is a complex problem in India.

The other bottlenecks in case of biodiesel are limited availability of feedstock, absence of high-yielding drought tolerant jatropha seeds. There have been limited R&D efforts to develop high-yielding drought tolerant jatropha seeds. It must be mentioned that the jatropha production program was launched without any planned varietal improvement program. Understandably, this makes worse for smallholders. For instance, the difference in yield between jatropha and other dominant feedstock for biodiesel is significant. While biodiesel from jatropha produces at best 1250 litre per hectare per year, palm oil, dominant feedstock for biodiesel, yields 4594 litres per hectare per year [12].

No attempt has been made under the biodiesel mission to develop appropriate institutions for development of high yielding plant varieties. Also, sufficient attention was not paid for two-way information flow between farmers and agricultural lab in respect of yield improvement. Till now, plantations are primarily of natural selection. What is need of the hour is quicker creation of new high yielding plant varieties with and suiable for various agro-climatic zones. Simultaneously, farmers need to be educated regarding information about the crop, its risk and return, and more importantly its cultivation practices. Under the modern system of crop innovation and diffusion, agricultural extension centres including demonstration farms pass on the relevant information about method of farming, risk and return of a new crop innovated in a national research centre. This has been the practice during the process of green revolution in India in respect of crops like paddy, wheat, maize, and cotton. However, this process has not been replicated in case of jatropha. Nor could the government develop alternative system for the propagation of jatropha cultivation in a short time span. Instead, NGOs have been used for propagation of knowledge. However, the later neither has developed demonstration field from which farmers can gather the knowledge of cultivation nor are fully familiar with risk and return associated with it. NGOS usually are not well versed in respect of matching variety with soil type or the application of various inputs at different stages of plant growth. As a result, NGOs had provided wrong information to the farmers leading to crop failure in jatropha plantation Rajasthan, Uttarakhand [11, 13]. Unlike agricultural extension centre, NGOs may move out or close down after the plantation. So, the farmers would have no recourse for help if the plantation fails [11].

Earlier, we have mentioned that a significant portion of the targeted people for jatropha cultivation are not traditional educated farmers but are the marginal farmers and landless people living on the fringes of government/community wasteland. These people are illiterate and have limited or no knowledge regarding cultivating new crops, such as jatropha. These people have diverted the marginal/wasteland for planting jatropha as result of pursuance of NGOs and government officials, particularly forest officials. Consequently, their traditional source of livelihood is dislocated. In return, they have gained little due to poor yield of jatropha.

Thus, we argue that there is need to develop dedicated appropriate institutions which can fulfil the following three roles: (a) conduct field trials/experiments of the plant varieties developed at the research centres (b) set up demonstration farms at various agro-climatic regions and (c) provide a channel for information to the farmers. The importance of green fuel or biodiesel in case of India would grow in the coming years. Hence, more efforts should be made to nurture wide plant varieties, improve knowledge of cultivation practices, and strengthen intermediary organisation for effective communication with the farmers. Of course, minimum support price has to be there, and the private entrepreneurs can take care of the rest, like setting up of oil extraction and oil processing plants, as well as marketing. What the mission has overlooked is building these institutions and tried to expand jatropha plantation at one go over large areas by whatever readily available means such as NGOs. On the basis of this initial experience, it is now possible to focus more on developing the various institutions that would help on a long-term basis.

The gestation period of biodiesel crops are 3–5 years jatropha and 6–8 years for Pongamia. This implies that the payback period is long which further adds to problems for farmers. In this respect, the state support is not readily available. Since there are no specific markets for Jatropha seed supply and government sector do not play a proactive role, the middlemen play a major role in taking the seeds to the processing centres and this inflates the marketing margin [19].

Biodiesel distribution channels are virtually non-existent as most of the biofuel produced is used either by the producing companies for self-use or by certain transport companies on a trial basis. Further, the cost of biodiesel depends substantially on the cost of seeds and the economy of scale at which the processing plant is operating. The lack of assured supplies of feedstock has hampered efforts by the private sector to set up biodiesel plants in India. In the absence of seed collection and oil extraction infrastructure, it becomes difficult to persuade entrepreneurs to install trans-esterification plants.

Probably, India needs to think beyond jatropha as feedstock. The next-generation biofuel crops depend on feedstock comprising lignocellulosic biomass, which is harvested for its total biomass. These are available in abundance in India and can be harvested at a much lower cost [14] some of the potential crops that can be grown are perennial grasses (such as switch grass and miscanthus), short rotation willows. Of course, biomass is now the principal source of energy for poor in rural India and consequently depriving them of these freely available resources without providing an alternative source of energy is an ethical issue. Furthermore, their participation is must for success of growth of feedstock. Otherwise, one would be doing the same mistakes it has been done in the case jatropha cultivation. The conversion of cellulose to biodiesel is achieved through thermo-chemical process. In this process, pyrolysis/gasification technologies produce a synthesis gas ($CO + H_2$) from which a wide range of long carbon chain biofuels such as synthetic diesel or aviation fuel can be produced. An indicative yield through the process of Syngas-to-Fisher Tropsch diesel ranges from 75–200 litres per dry ton of feedstock [12]. This technology is now at a demonstration stage in many countries including India It would take some more time till they are used for commercial purpose.

4 Concluding Observations and Summary Recommendations

Farming is an integral part of rural culture; changes in land use pattern, particularly of the use of marginal, fellow, and wasteland which are often occupied by the poor people for survival, are very difficult and time-consuming. Further the incentives are not adequate for these poor people to shift from growing short duration multiple crops for self-consumption to planting Jatropha/Pongamia with long gestation lag. For them supplementary earning opportunities are to be created. Only the patches of very low productive wasteland, used for cattle grazing, are found to be planted with Jatropha but the yield is too low to recover the working capital. Threat of food shortage has already preempted diversion of farmland to jatropha or any such crops.

Although there is a reasonable improvement in the yield of Jatropha grown in the farm land, the R&D should focus on growing Jathropha/Pongamia on marginal and wasteland. India has a highly diverse agro-climatic regions and the Mission has also identified more than 400 oilseed-bearing wild plants many of which are region-climate specific, R&D activity needs to extended to many of these plants to benefit from diversity. There is also need to diversify from seed based feedstock to biomass, particularly lignocellulosic based ones that are abundant in nature, although used by the common people as fuel. This requires continued R&D efforts to improve technology. India has access to a vast reserve of marine resources, but it exploits a little. Biodiesel feedstock of the country should include, among others, marine algae.

Finally, institutional constraints need to be overcome: (1) encouraging development of the markets for feedstock at the local level; (2) multiple private operators should be allowed to operate side-by-side state agencies in procuring, process, and marketing; (3) agricultural extension workers should be trained and used along with NGOs for diffusion of knowledge/information; (4) decision making as regards choice of growing particular feedstock should be decentralised but coordinated among different agencies and stakeholders.

References

1. Asian Development Bank (2011) India: study on cross-sectoral implications of biofuel production and use. Project Number 42525, Report prepared for department of economic affairs
2. Achten WMJ, Verchot L, Franken YJ, Mathijs E, Singh VP, Aerts R, Muys B (2008) Jatropha bio-diesel production and use. Biomass Bioenergy 32(12):1063–1084
3. Pohit S, Biswas PK, Ashra S (2011) Incentive structure of India's biofuel programs: status, shortcomings and implications. Biofuels 2(4):389–404
4. Pohit S, Biswas PK, Kumar R, Jha J (2009) International experiences of ethanol as transport fuel: policy implications for India. Energy Policy 37(11):4540–4548
5. USDA Foreign Agricultural Service (2013) India: biofuel annual. GAIN Report No. IN3073
6. Vasandani, B (2012) India's bio-diesel policy and the current turmoil. Energetica India, January/February, pp 2–4
7. Government of India (2006) Integrated energy policy. Planning Commission, New Delhi
8. Government of India (2005) Wastelands atlas of India. Ministry of Rural Development
9. Biswas PK, Pohit S (2011) What ails India's biodiesel programme? Energy Policy 52 (C):789–796
10. Pohit S, Biswas PK, Bedi JS, Kumar R (2007) Biodiesel production: institutional constraints. Macrotrack IX(12):2
11. Findlater KM, Kandlikar M (2011) Land use and second-generation biofuel feedstocks: the unconsidered impacts of Jatropha biodiesel in Rajasthan, India. Energy Policy 39:3404–3413
12. Ravindranath NH et al (2011) Biofuel production and implications for land use, food production and environment in India. Energy Policy 39:5737–5745
13. Altenburg T, Dietz H, Hahl M, Nikolidakis N, Rosendahl C, Seelige K (2008) Biodiesel policies for rural development in India. D.I.E, Bonn
14. Sim R, Taylor M, Saddler J, Mabee W (2008) From 1st to 2nd generation biofuel technologies: an overview of current industry and R&D activities-extended executive summary. OECD, Paris
15. Biswas PK, Pohit S, Kumar R (2010) Biodiesel from jatropha: can India meet the 20% blending target? Energy Policy 38:1477–1484
16. Government of India (2003) Report of the committee on the development of biofuel. Planning Commission, New Delhi
17. Government of India (2009) National Policy on biofuels. Ministry of New and Renewable Energy, New Delhi
18. Gunatilake H, Pohit S, Sugiyarto G (2011) Economy-wide impacts of biodiesel production and use in India. South Asia working paper series, No. 4, May, Asian Development Bank
19. Pohit S, Biswas PK, Rajesh K, Goswami A (2010) Pricing model for biodiesel feedstock: a case study of Chhattisgarh in India. Energy Policy 38(11):7487–7496

Chapter 12
Feedstock Availability, Composition, New Potential Resources for Biohydrogen, Biomethane, and Biobutanol Production via Biotechnological Routes

Navya Thomas, Musthafa O. Mavukkandy, Eanna Farrell, Hassan A. Arafat, and Sudip Chakraborty

Abstract Booming population, higher standard of living, and growing industrialization continue to exert a demand for energy. The continuous and unlimited use of fossil fuels to meet the growing energy demand would deplete the finite oil reserves, and have detrimental environmental impacts causing global warming and climate change. Given the present circumstances and future projections, it is imperative that an alternative sustainable fuel source is identified. Biofuels, produced from biomass, are a good substitute to conventional fuels. Based on the feedstock used for biofuel production, biofuel can be classified into first, second-, or third-generation biofuels. Production of second-generation biofuels from non-food feedstock needs to deal with the challenges associated with the complexity of biomass, issues related to producing, harvesting, and transporting less-dense biomass to centralized biorefineries. The third generation is yet to dominate the society due to various environmental and technological issues. This chapter discusses the need for increased focus on biofuel production, various feedstock options for biofuel production and specifically discusses the biotechnological options for conversion of biomass to bioenergy such as biomethane, biohydrogen, and biobutanol.

Keywords Renewable energy • Biomass • Lignocellulose • Sustainable biofuel • Energy demand • Biomass conversion technology

N. Thomas • M.O. Mavukkandy • E. Farrell • H.A. Arafat
Institute Center for Water and Environment (i-Water), Masdar Institute of Science and Technology, P.O. Box 54224, Abu Dhabi, United Arab Emirates

S. Chakraborty (✉)
Institute Center for Water and Environment (i-Water), Masdar Institute of Science and Technology, P.O. Box 54224, Abu Dhabi, United Arab Emirates

Department of Informatics, Modeling, Electronics and Systems Engineering (D.I.M.E.S.), Laboratory of Transport Phenomena and Biotechnology, University of Calabria, Via P. Bucci, Cubo-42/A, 87036 Rende, Cosenza, Italy
e-mail: sudip.chakraborty@unical.it

A.K. Chandel, R.K. Sukumaran (eds.), *Sustainable Biofuels Development in India*,
DOI 10.1007/978-3-319-50219-9_12

1 Introduction

The increasing demand for alternative and cleaner sources of energy has compelled governments and companies to invest in resources and technologies that are efficient and sustainable. In comparison to 2011, the energy demand is expected to increase by 50% in 2025 [1, 2]. Due to global warming, an effect of the combustion of fossil fuels amongst many others, there is a cause for focus on a clean and renewable form of energy, to act as an alternative to fossil fuel use. The declining amount of oil reserves has resulted in the emphasis of channeling investments towards post-fossil fuel era, using available resources economically with existing or new technologies. The growing energy demand and related impact of fossil fuel emissions on global climate change emphasizes the relevance to identify sustainable and economical alternatives to fossil fuels, thus paving the way for heightened research interest in biofuels [3, 4].

Biofuel, derived from biomass (a renewable source of energy) is the sustainable alternative for the increasing global fossil fuel demand. Biofuels can be broadly categorized based on either their source or their type. As per the *Classification of Biofuel Sources* by Food and Agricultural Organization (FAO), different kinds of biomass can be obtained from either energy crops and by-products (which includes residues and waste by-products from the processing activities of biomass) or end use material (such as used wood, fibre, fruits and seed products, or municipal by-products) [5]. Based on the type of biofuel, they can be classified as solid (example—bio-char), gaseous, or liquid biofuel. Gaseous biofuels include biogas, biosyngas, and biohydrogen. Liquid biofuels are hugely attractive as they are promising alternatives for liquid transportation fuels. The use of liquid biofuels as transportation fuels would translate to a sustainable approach accompanied with regional development, creation of rural manufacturing jobs, and decrease in greenhouse gas emissions [6]. The common liquid biofuels produced are ethanol, biodiesel, vegetable oils, and biobutanol. Based on the data from Renewable Fuels Association, in the year 2015 [7], US is the world leader in fuel ethanol production (14,806 million gallons) followed by Brazil. India ranked eighth with the production of 211 million gallons in the year 2015. The top three biodiesel producing locations are the European Union, Brazil, and Argentina [5, 8]. These biofuels are used as transportation fuels by blending biofuels with liquid fuels—ethanol is blended with gasoline supply, whilst biodiesel is blended with traditional diesel. In comparison with ethanol, biobutanol is less expensive and has higher flash point making it safer to handle. Biobutanol also contains relatively more energy and can be easily mixed with gasoline in varying ratios [9].

One of the major drivers for the shift towards biofuels stems from the climate policies of different countries. Biomass considered to be carbon neutral source of energy, could assist in meeting the world energy demand without affecting food supply. Though the reported use of biofuels was less than 1% in 2012, the targeted use by the year 2020 is 10% based on the emerging climate policies [10]. The beaconing call to shift our attention to biofuels presents three major benefits—

energy security, economic stability, and environmental gains. Utilization of domestic renewable energy sources would boost energy independence and security for countries. On the other hand, increased use of biofuels would alleviate the perpetuation of monopoly by fossil fuel rich countries, provide more employment opportunities, create rural development, and reduce supply–demand gap. The use of biofuels further reduces greenhouse gas (GHG) emissions, augments waste utilization, reduces local pollution, and most importantly directs our path towards sustainable fuel source [11]. However, the complete utilization and benefits of biofuel is curtailed by certain challenges that need to be addressed and tackled such as availability of feedstock, food–fuel competition, technological challenges, problems associated with land use changes, and need for stronger biofuel policies. A major component of the overall production cost for biofuels is the cost associated with feedstock. Feedstock accounts for 45–58% of the total production cost for second-generation fuels [12], whilst for first-generation biofuels feedstock accounts for more than two-thirds of the total cost [13]. To promote the use of biofuels and to make them economically viable, it is important that the biofuels be cost competitive in comparison with fossil fuels. This in turn demands the need to device measures and strategies to enhance the effectiveness and profitability of biomass feedstock supply [14]. Cost, availability, and quality of feedstock play a major role in driving the biofuel economics. This chapter initially discusses the need for growing emphasis on biofuels and then discusses the various aspects of biofuel production specifically for biohydrogen, biomethane, and biobutanol.

2 Need for CO_2-Neutral Energy Production

There exists a dilemma of energy-economy-environment, which is due to the following: (1) consistent and cheap energy is crucial to meet the basic human necessities and to fuel the economic growth, (2) current processes used for the production and conversion of energy contribute heavily towards the environmental problems.

Most of the environmental problems such as air pollution, acid rain, and heavy metal contamination of soil and groundwater are directly or indirectly caused by the production and supply of energy. On the other hand, most of the existing energy production is limited by the concerns over environmental degradation. Hence, the relationship between energy production and environmental impact is a vicious cycle. Resolving this energy-economy-environment dilemma is one of the toughest challenges of global sustainable development [15].

In the period of 1850–2000, the world population has multiplied approximately five-fold, quadrupling the per capita energy consumption. As an avalanche effect, this has resulted in an approximately 20-fold increase in the global energy use [16]. At present, most of our energy requirements are satisfied by burning fossil fuels like coal, oil, and natural gases. Rapid consumption of these non-renewable resources due to the exploding population and the industrial development will lead

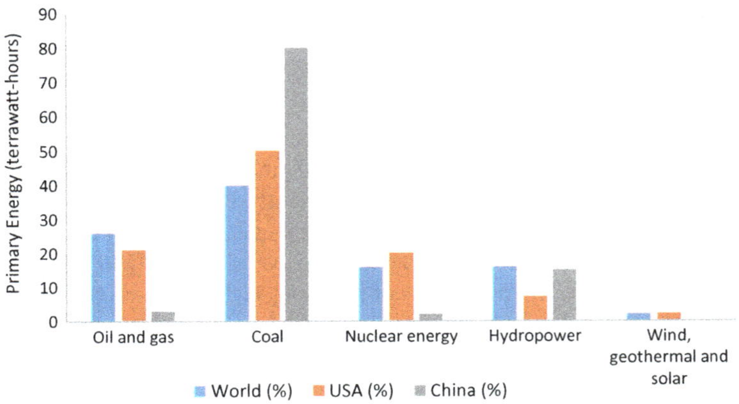

Fig. 12.1 A bar chart of global energy supply in 2005, showing the over-dependence on fossil fuels. Approximately one-third of the energy is used for the production of electricity, of which a significant portion is used for production of water [15]

to a scarcity of energy resources in the long-term. Likewise, burning of these fossil fuels result in the production and emission of various pollutants, CO_2 being of high concern due to potential role in global warming and associated climate change. Burning of fossil fuels contribute a major chunk of the world's energy supply. A bar chart showing the energy supply around the world in 2005 is presented in Fig. 12.1. It is evident that major part of the global energy supply is met by energy produced utilizing coal and oil and gas.

There exists an immense debate about "peak oil", which will imply the huge increase in the oil price along with expensive and unavoidable struggles for alternative sources of energy. Many countries (such as U.A.E and Germany) have already taken the path of developing renewable sources of energy. However, the gap between the supply and demand is so high that we need much more efforts globally.

New forms of clean energy such as wind, solar, hydro, bio, geothermal, and ocean power are required which will ensure that environmental pollution (including atmospheric emission of pollutants and contaminants, thermal pollution) is minimized. It will also boost the economy of the country by the creation of new jobs and businesses in renewable energy industry. Current global energy use is around 400 exajoules (EJ), with a predicted growth to 850–1100 EJ in 2050. The current global energy demand is just 0.01% of solar energy supply as the earth's surface gets 3.8 million EJ of solar energy per year [17, 18]. Although there is abundant supply of solar energy, its utilization is limited mainly due to techno-economic limitations. Nevertheless, it should be noted that it is neither wise nor practical to shift to the renewables rapidly because of their immaturity and exorbitant nature. Generation and transportation of renewable fuels is expensive. However, over the last decade, there is a decreasing trend in the production cost due to the advancements in the research and development and this trend is expected to continue in the coming decade as well.

3 Feedstock for Biofuel Production

The biofuel production cost depends on many factors which include production scale, process efficiency, availability, and cost of feedstock, along with the capital and operating costs. The cost of feedstock alone is a major component of the overall cost [6]. Biofuels are classified into first-generation, second-generation, and third-generation biofuels based on the type of feedstock for its production.

3.1 First-Generation Feedstock

Feedstock for first-generation biofuels includes (1) starch and sugar crops and (2) oil seeds. The growing use of food crops for biofuel production resulted in growing debate of food versus fuel and the practice was considered unsustainable with arable land for food production being transformed to a fuel source. This paved the way to evaluate alternate feedstock for biofuel generation.

3.2 Second-Generation Feedstock

Second-generation feedstock consists of lignocellulosic material such as wood chips from energy crops, agricultural and forest residues and low-valued municipal and industrial waste. Biofuel derived from lignocellulosic feedstock is termed as advanced biofuels. Energy crops, for second-generation biofuel, include perennial grasses and short rotation wood crops grown specifically for biofuel generation. Perennial grasses used for biofuel production include miscanthus, switchgrass, and reed canary. Agricultural residues include wheat straw, corn leaves, cobs, stalks and sugarcane bagasse whilst forestry residues include logging residues and wood processing mill residues. Biodegradable components of municipal and industrial waste such as residues from food and paper industry, waste from fruits and vegetable processing, meat and poultry waste can also serve as feedstock for biofuel production [19].

The significant advantage of advanced biofuel production is the easily available second-generation feedstock. Second-generation biofuels are also expected to generate better yield per unit land. Though lignocellulosic biomass is available abundantly, the variable composition and lignin content present some challenges. Conversion of lignocellulosic feedstock can be achieved via either (a) biochemical processes or (b) thermo-chemical processes. The choice of second-generation biofuel will be guided by cost involved in the conversion of lignocellulosic material to usable biomass in comparison with the low cost feed for first-generation biofuel.

Another potential feedstock for biofuel production would be municipal and industrial waste. The increasing amount of waste due to population boom, industrialization and urbanization has led governments and international organizations in taking a stronger stand to reduce waste generation. In an effort to brace this concept several developed countries have already taken stern steps to drastically reduce their waste by imposing heavy taxation on non-biodegradable packaging of goods. Moreover, they have also pushed the waste management sector to attain better efficiencies in sorting the waste to increase prospects of recycling and reusing this waste. The properties of a good waste management plan are to maximize the full potential of resources and to minimize long term impact on the society and habitat. Recycling and using waste to produce energy or value added products is a beneficial method for environment friendly waste management approach. Organic waste such as manure and industrial waste such as food processing remains are examples of low value waste products which could serve as potential feedstock for the production of biofuels.

The production of biofuels from waste has the double benefit of providing a cleaner energy as well as better and sustainable management of waste. It can produce liquid biofuels such as ethanol and biodiesel, biogas such as methane and CO_2, syngas (hydrogen and CO), or pure hydrogen [20].

3.3 Third-Generation Feedstock

Third-generation biofuel refers to biofuel derived from algal sources and aquatic biomasses. One of the biggest advantage of algal feedstock is their short doubling time (2–5 days) relative to other feedstock sources. On the flip side, the full potential of algal biomass is limited by challenges associated with algal cultivation, production process, bioreactor design for algal production and downstream processes.

4 Biotechnological Production of Biomethane, Biohydrogen, and Biobutanol

Biomass feedstock could be either dry (e.g., wood) or wet (e.g., organic fraction of municipal solid waste). Various thermal processes (like gasification and combustion) could be employed in order to convert dry biomass to bioenergy because of low water requirement. On the other hand, wet feedstock require high amount of energy for the drying process and hence thermal conversion is not recommended.

Therefore, wet biomass is preferentially converted through biological route, because an aqueous environment is favorable for microorganisms. A major cost and energy consumption is for the transport of wet feedstock to the drying and conversion processes facilities. However, these biotechnological conversion units could well be decentralized on the location of biomass availability so that the transportation cost and energy could be minimized.

4.1 Biomethane

Biomethane, also known as green gas, is a biofuel produced by the breakdown of organic material in an anaerobic environment. Anaerobic digestion process produces biogas from organic matter via three main steps (Fig. 12.2). In the first step, hydrolysis/liquefaction reactions, the insoluble complex organic molecules are broken down into soluble monomeric substances such as fatty acids, monosaccharides, and amino acids with the help of various hydrolytic enzymes excreted by acidogenic microbes. An optimum pre-treatment will result in opening the cellulose and lignocellulose structures. Hydrolysis is then followed by acidogenesis wherein the acid forming bacteria transform the hydrolysis products into simple compounds organic acids, carbon dioxide and hydrogen. The major acids produced in this phase are namely acetic acid, propionic acid, butyric acid and ethanol. Methanogenesis is the final stage wherein methane production occurs either by breaking down the acetic acid molecules to produce carbon dioxide and methane or causes reduction of carbon dioxide molecules with hydrogen [22]. The produced biogas is composed of 50–75% methane, 25–50% carbon dioxide, the rest is composed of water vapour, and traces of oxygen, nitrogen and hydrogen sulphide. Purification of biogas by removing hydrogen sulfide, ammonia, siloxane and carbon dioxide results in the

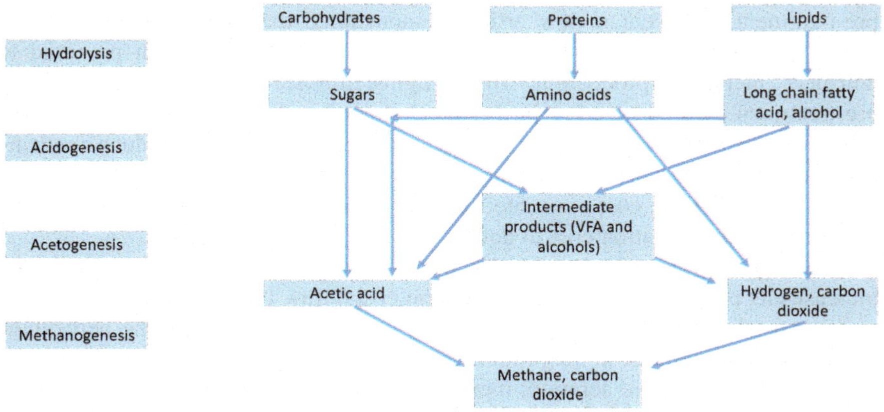

Fig. 12.2 Schematic representation of anaerobic digestion (adapted from [21])

formation of biomethane. The methane composition is generally hiked to above 96% in biomethane as a result of the purification process to match the quality standards of natural gas. Biomethane is also referred to as synthetic natural gas.

All kinds of biomass containing mostly carbohydrates, proteins, fats, cellulose, and hemicellulose can be used as biogas feedstock. The biogas components and methane yield is influenced by factors such as feedstock type, digestion system and retention time influence [23]. Lignocellulose refers to the biomass from plant material composed of cellulose, hemi-cellulose and lignin. An evaluation of a variety of energy crops by Lehtomäki et al (2006) has demonstrated that feedstock with greater lignin content had lower methane potential (0.18–0.32 m^3 methane/kg volatile solids). In addition to methane potential based on solid content, another significant factor governing the overall methane potential is the crop yield [3]. Switch grass and other perennial grasses such as miscanthus and prairie grass are high yielding crops adaptable to poor soil requiring low fertilizer applications. Genetic modification could be another potential pathway to obtain high yielding, low fertilizer and irrigation demand crops suitable for biofuel production. In addition to considering lignocellulosic crops, agricultural feed stocks and residues (such as bagasse) is a potential low cost lignocellulosic feedstock. Forest feedstock such as softwood and hardwood from wood processing plants can also be considered as a potential feedstock [24].

The predicament with lignocellulosic biomass for biofuel or biomethane production is the potential difficulty and high associated cost in converting them to fermentable sugars. The cost of conversion of lignocellulosic material to fermentable sugars is relatively higher than cost associated with production of first-generation biofuels. Pre-treatment of ligno-cellulosic material is performed to hydrolyze or liquefy the substrate to maximize the methane yield potential. Sufficient information is available on pre-treatment of lignocellulosic material for bioethanol production. The pre-treatment options for bioethanol production include (1) enzymatic liquefaction and saccharification, (2) ammonia fiber explosion, (3) ammonia recycle percolation (4) steam explosion and (5) Organo-Solv pre-treatment. A recent review article demonstrated that pre-treatment methods based on use of steam, lime, liquid hot water and ammonia high potential to enhance methane production from lignocellulosic biomass [25]. These pre-treatment methods need to further be evaluated and established in the context of energy cost of the pre-treatment versus net energy gain to produce methane [3]. The overall cost benefit analysis of using lignocellulosic biomass should also take into account the cost associated with purification of biomass to biomethane, storage and transportation cost to deliver the fuel at end point of use.

The large volume of carbon dioxide in biogas depreciates its calorific value and increases its transportation and compression costs. Raw biogas needs to be further purified to obtain biomethane by removing impurities to increase the percentage composition of methane in the product. Hydrogen sulfide removal is accomplished by chemical absorption, aeration, membrane separation, biological filters, adsorption on activated carbon and chemical oxidation. Siloxanes, compounds containing organic radicals and silicon–oxygen bond attached to the silicon atom, can be

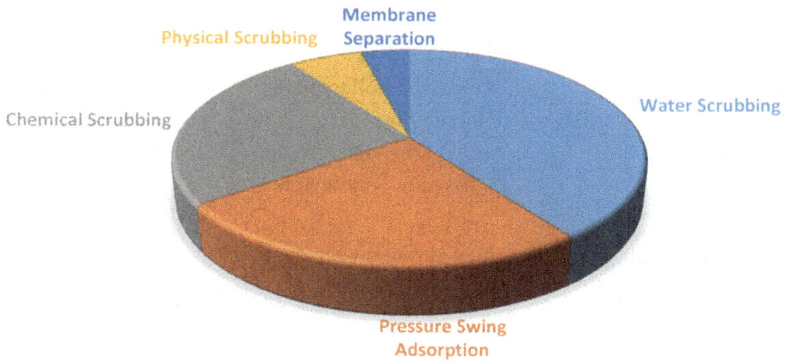

Fig. 12.3 Current European market share of biogas purification technologies [27]

removed by absorption, adsorption on activated carbon, and cryogenic separation. Ammonia can be removed from biogas by washing with dilute nitric or sulfuric acid, activated carbon filtration and also removed as part of carbon dioxide operations such as wet and chemical scrubbing [26]. The current most significant purification technologies (Fig. 12.3) for carbon dioxide removal include (1) water scrubbing, (2) pressure swing adsorption, (3) chemical scrubbing, (4) physical scrubbing, and (5) membrane separation [27]. Water scrubbing is the most common purification technology with 40% market share in the European region.

The conventional carbon dioxide removal technologies such as pressure swing adsorption, water scrubbing are energy intensive processes and thus become cost-effective only when the raw biogas production is in excess of 500 cubic metres per hour (Nm^3/h). Hence, such technologies are not suitable for decentralized purification facility or small-scale purification plants. Commercial membranes are now available for onsite purification for conditioned biomethane production. Table 12.1 presents a few commercial membranes available for biogas upgradation to biomethane. Though commercial polymeric membranes are available for biogas upgradation, these membranes are prone to damage by biogas components such as ammonia and hydrogen sulfide content. Hence, it necessitates the need for expensive pre-treatment system which adds to the overall purification cost [22].

Cost and economics play a vital factor for the large-scale implementation and adoption of any technology. For a market like India, the cost of biogas upgradation technologies needs to be well established to present the production of biomethane as an economically feasible option in comparison with the natural gas prices. A study [22] presents price comparison of natural gas and production cost of biogas in Italian market. Depending on the organic feed source, the biogas production cost shown by different research groups is 8–10 € cents per cubic metre of biogas. The upgradation cost, based on energy cost of 20 € per kWh, is 7–8 € cent per cubic metre of biomethane compressed to national grid at 30 bar. In comparison with natural gas prices of 40.09 € cent per cubic metre (based as on January 2010), the

Table 12.1 Commercial membranes for biogas upgradation to biomethane

Company	Membrane type	Product methane content
Air liquide	Polymeric hollow fibre	96.5–99%
DMT (carborex MS®)	Hollow fibre	97–99%
Evonik (sepuran®)	Polymeric hollow fibre	Up to 99%
PoroGen (PEEK-SEP™)	Polymeric hollow fibre	>95%

total cost of biomethane production at 20–22 € cent per cubic metre works out to be an economical option.

4.2 Biohydrogen

Amongst the various biofuels, hydrogen is an attractive fuel with high energy density and low pollutant generation [28]. Hydrogen is a very important feedstock to chemical industry and also used for detoxification of various water pollutants. Currently, the requirement for hydrogen for various processes is obtained by reforming fossil fuels and consequently hydrogen production has a large greenhouse gas footprint [29]. Biological hydrogen production provides an avenue to generate hydrogen via carbon neutral methods and have a sustainable hydrogen source [29, 30].

Biomass for hydrogen production include agricultural crops and their residues, lignocellulosic products, food processing waste, aquatic plants and algae, and municipal effluents. There is substantial research focused on engineering algae for hydrogen production. Genetic engineering techniques are employed to identify relevant bioenergy genes to optimize the production of targeted biofuel production [31].

Three (3) major ways of producing biohydrogen are (1) photosynthesis, (2) fermentation and (3) microbial electrolysis cells (MEC) [29]. In the photosynthetic process, hydrogenase enzyme facilitates the splitting of water molecules to obtain hydrogen and oxygen with energy derived from light source. A major challenge for photosynthetic biohydrogen production is the low oxygen tolerance of hydrogenase enzyme which facilitates the conversion of water to hydrogen and oxygen. In fermentative biohydrogen production process, protons are utilized as electron acceptors during dark fermentation of organic substrates. The biggest advantage of fermentative hydrogen production is that the process utilizes complex organic substrates and thus serves as an alternate pathway to degrade organic substrates from municipal wastes or other organic wastes. The biggest advantage of this process is that the relative hydrogen yield via this process is magnitudes greater than the other biohydrogen production process options. However, the hydrogen yield is also the greatest opportunity for improvement in fermentative hydrogen production. Though the theoretical maximum yield of substrate electron equivalent to biohydrogen is 25%, only 17% conversion is actually achieved. Possible

Table 12.2 Summary of various biohydrogen production processes [32]

Process	General reaction	Microorganisms used
Direct bio photolysis	$2H_2O + \text{light} \rightarrow 2H_2 + O_2$	Microalgae
Photo-fermentations	$CH_3COOH + 2H_2O + \text{light} \rightarrow 4H_2 + 2CO_2$	Purple bacteria, microalgae
Indirect bio photolysis	$6H_2O + 6CO_2 + \text{light} \rightarrow C_6H_{12}O_6 + 6O_2$ $C_6H_{12}O_6 + 2H_2O \rightarrow 4H_2 + 2CH_3COOH + 2CO_2$ $2CH_3COOH + 4H_2O + \text{light} \rightarrow 8H_2 + 4CO_2$ Overall reaction: $12H_2O + \text{light} \rightarrow 12H_2 + 6O_2$	Purple bacteria, microalgae microalgae, cyanobacteria
Water gas shift reaction	$CO + H_2O \rightarrow CO_2 + H_2$	Fermentative bacteria, photosynthetic bacteria
Two-phase fermentations $H_2 + CH_4$	$C_6H_{12}O_6 + 2H_2O \rightarrow 4H_2 + 2CH_3COOH + 2CO_2$ $2CH_3COOH \rightarrow 2CH_4 + 2CO_2$	Fermentative bacteria + fermentations
High-yield dark fermentations	$C_6H_{12}O_6 + 6H_2O \rightarrow 12H_2 + 6CO_2$	Fermentative bacteria

strategies to increase the percentage conversion of substrate to biohydrogen include coupling the fermentation process with other pathways such as microbial electrolysis cells. As per recent publications, the combination of dark fermentation with microbial electrolysis cell can improve the yield up to 81%. However, the microbial electrolysis cell technology is relatively new, developing technology and thus poses a limitation [29]. Table 12.2 presents different processes for production of biohydrogen.

At present, biohydrogen production is relatively more expensive than other biofuels. Though there are different process pathways available to produce the fuel of the future—biohydrogen, the low hydrogen yield is the key underlying issue that needs to be addressed to provide an economically feasible solution [33]. Other challenge for biohydrogen production includes lack of commercial scale technology demonstration. As for all biofuels, cost factors associated with storage options for hydrogen, transport systems and utilization schemes needs to be developed and organized [34].

4.3 Biobutanol

Butanol is a four-carbon alcohol having higher energy content, lesser hygroscopicity and volatility, and lower corrosivity when compared to ethanol [35]. Hence it has great potential as a transport fuel in the current fossil-fuel driven transportation sector. In the recent years, there has been a growing interest for the utilization of biobutanol (butanol/gasoline blends) in the form of as a transport fuel. It can also be noted that the existing gasoline pipeline could be used for its transportation.

It was conventionally produced via ABE fermentation which involves the anaerobic digestion of carbohydrates into a mixture of acetone, butanol, and ethanol with the help of *Clostridium* strains. However, production of butanol via this biological route faces lots challenges such as low yield, compatibility of feedstocks, product inhibition, slow rates of fermentations and high energy requirement for the separation and purification of the products [36]. Therefore, it could not, as such, compete with the synthetic production of butanol from petrochemicals. This propelled the urge for developing butanol via several novel routes other than ABE fermentation, which would help its commercialization. These includes catalytic condensation of ethanol, integration of genetically modified butanol-producing microbial strains into novel fermentation processes, application of strategies involving metabolic engineering, introduction of continuous processes with simultaneous fermentation, and recovery of the products (such as gas stripping, pervaporation, perstraction, adsorption). Biobutanol can be produced from first-generation feedstocks such as sugar cane, sugar beet, and cereal crops. It can also be produced from second-generation feedstocks such as cellulosic materials, which has unlocked a new area for the developments in cost-effective biobutanol production. Some examples of feedstocks utilized in ABE fermentation include barley straw, wheat straw, corn fiber, corn stover, switchgrass (panicumvirgatum), domestic organic waste, sago, defibrated-sweet-potato-slurry (DSPS), degermed corn, extruded corn, liquefied corn starch, cassava, and whey permeate [36].

5 Biofuels and Sustainability

Biomass is already one of the most commonly used fuel sources. The clear segregation of biofuel into primary, secondary, and tertiary classifications may detract from the historical reliance that humans have had on biomass. Presently, there are close to 3 billion people who still rely on wood, dung, and agricultural residues as a source of energy for cooking and heating. Majority of these people reside in least developed and developing countries, and as such, the importance of household energy is a key parameter in the achievement of KPI's under frameworks such as the UN Millennium Development Goals. The capacity of biomass to provide solutions in such a diverse range of settings is testament to the specialized research that has been devoted to the field of biofuels in the recent past.

One key point surrounding the sustainability of biofuel, is the socio-technical aspect of continuing use of a liquid fuel. Considering this aspect within the transport sector, biofuel does not represent an immediate behavioral change to the end user, or motorist in this case. The means of refueling is the same as it is for conventionally fuelled internal combustion engines operating on petroleum or diesel. The length of time to refuel, as well as the time between refueling, is approximately the same. There is no significant technological change for the end user other than buying or upgrading their vehicle to a specification capable of operating with the chosen biofuel. This is in sharp contrast to other current transport alternatives which can use renewable energy to power their operation, such as electric vehicles. This technology represents a step change to the end user whereby refueling method, time and interval are all different. The system architecture is also different. This is not to say that the option for renewable energy powered transport will not, or should not, be majorly derived from electric vehicles, but rather, biofuels allow the option for the end user to continue "business as usual" without addressing the point that a behavioral change, as represented by the use of electric vehicles, might well be warranted for a more sustainable transport sector. Opting for the preferred technology, rather than the ideal, does little to challenge the status-quo of this socio-technical issue, which may in turn elongate the inherent unsustainable characteristics of modern transportation.

The sensitivity of biofuel within the context of sustainability has been well documented, perhaps most notably in the "Fuel vs. Food" debate. However, an increasing international focus on water resources is starting to develop the framework within which we gauge the sustainability of biofuel. One of the key relationship dynamics that is currently being explored is known as the Food–Water–Energy nexus. IRENA has estimated that the food production will have to be increased by 60% whilst the water availability and energy generation will need to be increased by 55% and 80% respectively to meet the global demand by the year 2050 [7]. This serves to underline the dichotomy of biofuels in that they can be viewed as a potential threat to the sustainable operation of this system, but may also serve as a means of achieving these targets.

Ensuring that biofuel forms part of the solution will stem from the advent of knowledge generation and application of best practices. This resonates with each part of the nexus. To ensure that such solutions will form part of the growing momentum toward sustainability, correct and efficient policy must help forge this path. The strategy of current biofuel technology development takes place in an atmosphere of uncertainty, as decision makers for such policies are reliant on data which may not account for the necessary impacts of certain options, as important externalities are housed away from ample boundaries. Examples of turgid policy making environments include that of the California Low Carbon Fuel Standard, where up to 12 different pathways for bioethanol were incomprehensibly analyzed by its regulatory body [37].

Removing the veil of uncertainty surrounding correct biofuel strategies must begin with an unbiased account of each scenario. Suffice to say, for biofuels to be classed as a renewable energy, or indeed sustainable, the source from which we derive this biofuel energy must not become depleted. This culminates in an account of supply versus demand, as well as energy in versus energy out. While there have been several case studies which have documented a net energy consumption for the production of some biofuels [38], it further serves the importance of adopting a correct and efficient approach to ensure that this potential renewable energy source is in fact renewable. For example, a recent review of economic assessments [39], report a selling cost of biofuel that ranges between \$1.64 and over \$30 gal^{-1}. This is consistent with the large variability reported in the life cycle literature, of −75 to +534 g CO_2-eq MJ^{-1}. This variability has been identified to come from inconsistencies in system boundaries, differences in production pathway architecture and productivity assumptions [39]. To avoid scenario's where the use of biofuels is unsustainable, a universal approach needs to be taken to guarantee the validity of lifecycle and techno-economic assessment methodologies. As has been discussed previously, it is not only energy provision and reduction of GHG emissions that renewable energy is tasked with, but also to operate as a means to deliver a more sustainable future. This is represented by improving air quality, reducing stress on water resources, improving health of global ecosystems, etc. Having such methodologies will allow for better tracking of how biofuels are actually operating in the context of sustainability. Transitioning the latest technology into viable alternatives will be greatly aided by such an approach as the merits of adopting the correct solution will be more readily evaluated.

6 Conclusions

Changing global climate, depleting oil reserves, and emerging government policies are key drivers for growing attention on biofuels and renewable energy. Biofuels have not only become an alternate for fossil fuel but also the generation of biofuels results in other significant advantages such as reducing greenhouse gas emissions, rural development, creation of new job markets, utilization of waste resources (waste-to-energy scenario), and new possibilities for forest-based livelihoods. However, the biofuel development and deployment requires that we also take into account economical aspect of biofuel production. A key factor for the biofuel production cost is associated with feedstock availability, year round supply, logistics, and technologies for efficient conversion of feedstock to biofuel. Thus, one key approach for economically sustainable bioenergy project is to intelligently select the site and project size. Another aspect is to consider the supporting government policies for biofuel production and the ease of trade to other markets. Nations should work together to fabricate a coherent roadmap for bioenergy development with common goals and objectives.

References

1. Deublein D, Steinhauser A (2010) Biogas from waste and renewable resources: an introduction, 2nd edn
2. Khanal SK (ed) (2008) Anaerobic biotechnology for bioenergy production. Wiley, Oxford
3. Frigon JC, Guiot SR (2010) Biomethane production from starch and lignocellulosic crops: a comparative review. Biofuels Bioprod Biorefin 4:447–458. doi:10.1002/bbb.229
4. Yangin Gomec C, Ersahin ME, Dereli RK, Arikan O, Ozturk I (2011) Biomethane production as an alternative bioenergy source from codigesters treating municipal sludge and organic fraction of municipal solid wastes. J Biomed Biotechnol. doi:10.1155/2011/953065
5. FAO (2004) Unified bioenergy terminology (UBET). For prod econ div for dep 58

6. Demirbas A (2009) Biofuels securing the planet's future energy needs. Energy Convers Manag 50:2239–2249. doi:10.1016/j.enconman.2009.05.010
7. Renewable Energy Agency (2015) Renewable energy in the water, energy and food Nexus
8. Guo M, Song W, Buhain J (2015) Bioenergy and biofuels: history, status, and perspective. Renew Sustain Energy Rev 42:712–725. doi:10.1016/j.rser.2014.10.013
9. García V, Päkkilä J, Ojamo H, Muurinen E, Keiski RL (2010) Challenges in biobutanol production: how to improve the efficiency? Renew Sustain Energy Rev 15:964–980. doi:10.1016/j.rser.2010.11.008
10. Nonhebel S (2012) Global food supply and the impacts of increased use of biofuels. Energy 37:115–121. doi:10.1016/j.energy.2011.09.019
11. Bani-Melhem K (2003) Using concentration polarization factors to determine the mass transfer resistance in vacuum membrane distillation. Desalination 157:333. doi:10.1016/S0011-9164(03)00413-2
12. Hamelinck CN, Faaij APC (2006) Outlook for advanced biofuels. Energy Policy 34:3268–3283. doi:10.1016/j.enpol.2005.06.012
13. Carriquiry MA, Du X, Timilsina GR (2011) Second generation biofuels: economics and policies. Energy Policy 39:4222–4234. doi:10.1016/j.enpol.2011.04.036
14. IEA (2013) Economic sustainability of biomass feedstock supply
15. Holdren JP (2008) Science and technology for sustainable well-being. Science 319 (80):424–434. doi:10.1126/science.1153386
16. Holdren JP (1991) Population and the energy problem. Popul Environ 12:231–255. doi:10.1007/BF01357916
17. Charlier RH (2009) In: Johansson TB, Kelly H, Reddy AKN, Williams RH (eds) Renewable energy—sources for fuel and electricity. Burnham L Earthscan Publications Ltd, London, England, UK, and Island Press, Washington, DC. Environ Conserv 22:380. doi:10.1017/S0376892900035219
18. Ltd SI, Environment SIGB (2001) Energy needs, choices and possibilities: scenarios to 2050
19. Ho DP, Ngo HH, Guo W (2014) A mini review on renewable sources for biofuel. Bioresour Technol 169:742–749
20. Fatih Demirbas M, Balat M, Balat H (2011) Biowastes-to-biofuels. Energy Convers Manag 52:1815–1828. doi:10.1016/j.enconman.2010.10.041
21. Gujer W, Zehnder AJB (2011) Conversion processes in anaerobic digestion
22. Molino A, Nanna F, Ding Y, Bikson B, Braccio G (2013) Biomethane production by anaerobic digestion of organic waste. Fuel 103:1003–1009. doi:10.1016/j.fuel.2012.07.070
23. Ranalli P (2007) Improvement of crop plants for industrial end uses
24. Sims R, Taylor M, Jack S, Mabee W (2008) From 1st to 2nd generation bio fuel technologies: an overview of current industry and RD&D activities. IEA Bioenergy:1–124
25. Hendriks ATWM, Zeeman G (2009) Pretreatments to enhance the digestibility of lignocellulosic biomass. Bioresour Technol 100:10–18
26. Ryckebosch E, Drouillon M, Vervaeren H (2011) Techniques for transformation of biogas to biomethane. Biomass Bioenergy 35:1633–1645. doi:10.1016/j.biombioe.2011.02.033
27. Niesner J, Jecha D, Stehl KP (2013) Biogas upgrading technologies: state of art review in european region. Chem Eng Trans 35:517–522. doi:10.3303/CET1335086
28. Hallenbeck PC, Ghosh D (2009) Advances in fermentative biohydrogen production: the way forward? Trends Biotechnol 27:287–297. doi:10.1016/j.tibtech.2009.02.004
29. Lee HS, Vermaas WFJ, Rittmann BE (2010) Biological hydrogen production: prospects and challenges. Trends Biotechnol 28:262–271. doi:10.1016/j.tibtech.2010.01.007
30. Redwood MD, Paterson-Beedle M, MacAskie LE (2009) Integrating dark and light bio-hydrogen production strategies: towards the hydrogen economy. Rev Environ Sci Biotechnol 8:149–185. doi:10.1007/s11157-008-9144-9
31. Beer LL, Boyd ES, Peters JW, Posewitz MC (2009) Engineering algae for biohydrogen and biofuel production. Curr Opin Biotechnol 20:264–271

32. Benemann J (1996) Hydrogen biotechnology: progress and prospects. Nat Biotechnol 14:1101–1103. doi:10.1038/nbt0996-1101
33. Nath K, Das D (2004) Improvement of fermentative hydrogen production: various approaches. Appl Microbiol Biotechnol 65:520–529
34. Balat H, Kirtay E (2010) Hydrogen from biomass—present scenario and future prospects. Int J Hydrogen Energy 35:7416–7426
35. Wu Y-D, Xue C, Chen L-J, Bai F-W (2013) Effect of zinc supplementation on acetone–butanol–ethanol fermentation by Clostridium acetobutylicum. J Biotechnol 165:18–21. doi:10.1016/j.jbiotec.2013.02.009
36. Kumar M, Gayen K (2011) Developments in biobutanol production: new insights. Appl Energy 88:1999–2012. doi:10.1016/j.apenergy.2010.12.055
37. Sheehan JJ (2009) Biofuels and the conundrum of sustainability. Curr Opin Biotechnol 20:318–324
38. Pearce F, Aldhous P (2007) Is the biofuel dream over? New Sci 2634:6–7
39. Quinn JC, Davis R (2015) The potentials and challenges of algae based biofuels: a review of the techno-economic, life cycle, and resource assessment modeling. Bioresour Technol 184:444–452. doi:10.1016/j.biortech.2014.10.075
40. Lehtomäki A, Björnsson L (2006) Two-stage anaerobic digestion of energy crops: methane production, nitrogen mineralisation and heavy metal mobilisation. Environ Technol 27 (2):209–218

Part III
Technical Aspects, Commercial View, Capture and Storage of Biofuels

Chapter 13
Technological and Commercial Update for First- and Second-Generation Ethanol Production in India

Prasanna Borse and Amol Sheth

Abstract The development of ethanol production technology is old and ever changing since last 40 years. As we all know that centuries ago, man discovered and began employing fermentation technology to produce ethanol; today ethanol is produced from a variety of sugar and starch-bearing feedstocks for use as an industrial chemical, beverages, alcohol, and fuel ethanol. The Fuel ethanol technology has been developed through extensive research and development in more than a decade. Conventional processes have been improved while advances continue to be made in lignocellulosic biomass conversion. Advanced technologies in Fermentation, Distillation, and wastewater treatment have been developed through consistent efforts and understanding of process details as well as effective plant design. This chapter covers the various Fermentation, Distillation, and wastewater sustainable and environment friendly technologies that have been developed and commercialized in the market. Fermentation and Distillation of sugar-rich feedstocks like molasses to ethanol is a technology that currently dominates ethanol production in India. Technology developers continue to look for ways to improve and to make the process more efficient and cost-effective. This chapter focuses on recent trends and developments in current and emerging ethanol technologies in India.

Keywords Distillation • Lignocellulosic Biomass • Advanceed Ethanol technology

1 Introduction

The sugarcane molasses is most preferred feedstock for ethanol production in India. This valuable coproduct from Sugar mills gives ethanol as a value product which ensure better price realization. It also gives benefit to cane Farmers.

P. Borse, MBA, CFA, B.E. Mechanical (✉) • A. Sheth, MBA, B.E. Chemical
Praj Industries Ltd, Pune 411057, India
e-mail: prasannaborse@praj.net

A.K. Chandel, R.K. Sukumaran (eds.), *Sustainable Biofuels Development in India*,
DOI 10.1007/978-3-319-50219-9_13

The annual ethanol production is about 2700 million litres in India. It is the fourth largest ethanol producer after United States, Brazil, and China. There are three main uses of ethanol—potable industry, chemical and solvent industry and the balance is used as blending fuel in petrol [1].

1.1 Rational for Use of Ethanol as Blending in Gasoline

The rise in energy consumption cost combined with increasing volatility in the world's energy markets, calls for India, as a nation, to pursue aggressive goals to reduce its dependency on foreign sources of energy in a sustainable, reliable, and cost-effective manner.

Ethanol, when blended with petrol, improves engine efficiency due to complete combustion of fuel. Thus, it reduces the air pollution by controlling Carbon monoxide emissions and other poisonous gases.

India's petrol demand rose to 27 billion liters in 2015–2016. At 10% ethanol in petrol, the demand for ethanol blending would be 2.7 billion liters in 2015–2016. The Government of India made 5% ethanol blending in petrol mandatory in nine sugarcane growing states, effective from 1 January 2003. Today, the 5% blending has been taken nationwide with specific excise and price-related benefits [2].

Being a renewable source has the potential to contribute significantly to GHG emissions savings when used as fuel. As ethanol has the potential to replace fossil sources, most of which is imported in India, its use can also yield various other benefits, e.g., Reducing import Bill.

Bio-ethanol can be either wholly or partially substituted for petrol. Upto 10% bioethanol blending requires no engine modification. Higher proportions of ethanol can be used but engine modifications are needed.

2 Benefits of Bio-Ethanol Program

2.1 Crude Oil Price Volatility

Reduced dependence on imported oil through increased ethanol blending levels would help the nation alleviate challenges posed by Crude Oil Price Volatility in International Markets and minimize Current Account deficit.

2.2 GHG Emission Reduction Potential/Climate Change

Today ethanol blending can be increased upto 10% without any change in automotive engines. For higher blends, FFV technologies are already in place in several countries like Brazil. Hence, potentially ethanol from lignocellulosic biomass can replace significant amount of Petrol and hence Fossil Fuels thereby resulting in significant reduction in GHG emissions. Adopting a uniform 10% blending regimen for gasoline and a 5% biodiesel regimen for diesel could potentially reduce CO_2 emissions by 3–4%, creating a total abatement of 10–12 million tons of CO_2. An effective blend regime could also deliver annual economic savings of $1.2–1.5 billion in a country like India [3].

2.3 Positive Impact on the Rural Economy

1. Agriculture plays a significant role in the Indian economy. One argument for promoting ethanol production is that it supports the agricultural sector by creating demand.
2. A report compiled by Association of Biotechnology Led Enterprises (ABLE), India for the Department of Biotechnology, Government of India, estimates that if the national biofuels program becomes successful, it has the potential to create 18 million jobs and hence will significantly improve the future of the Indian bio-economy [4]. These circumstances necessitate India to strategize for a planned reduction in our dependence on fossil fuels through biofuels as well as bio-energy initiatives The biofuel 2020 India report [5].

3 Technology: Production of bio-ethanol

See Figs. 13.1 and 13.2.

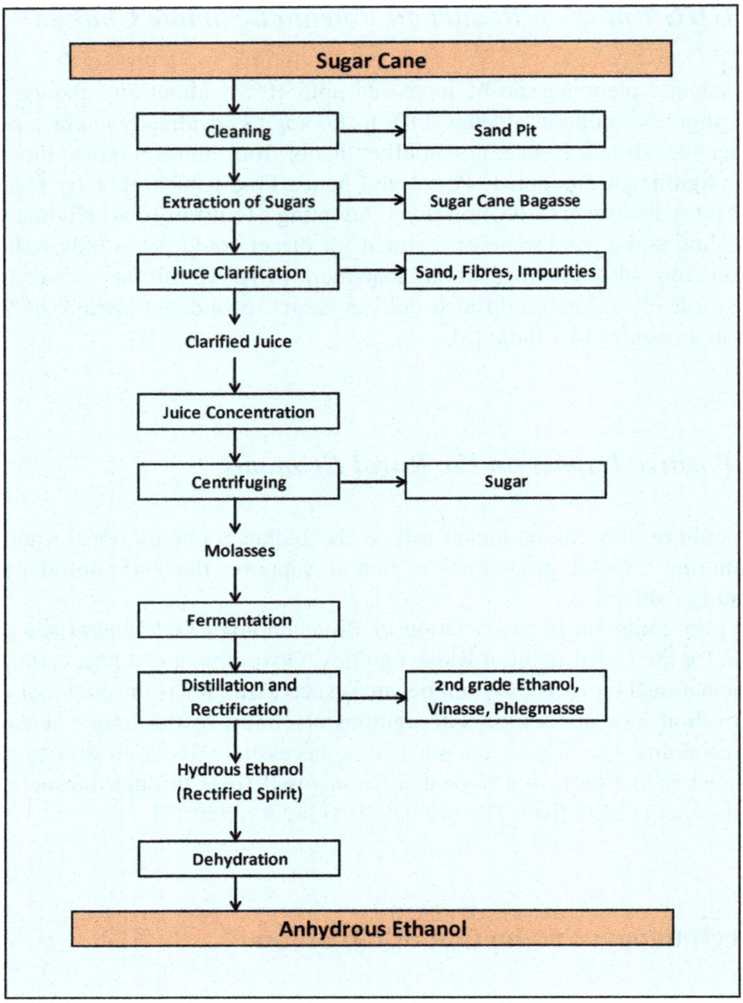

Fig. 13.1 Process flow diagram—sugarcane to ethanol production [6]

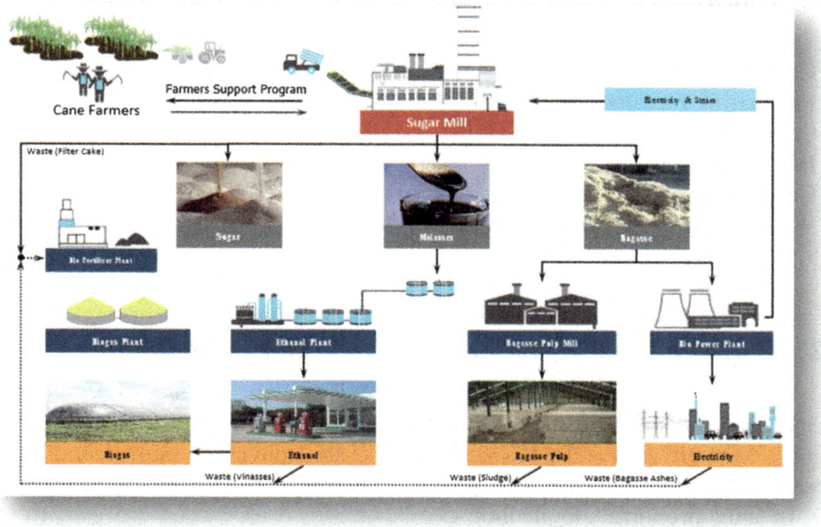

Fig. 13.2 Process flow diagram—ethanol plant

4 Fermentation Technology

Fermentation process, thus, significantly impacts the viability of distillery operations and quality of the final product.

Biochemistry of fermentation

$C_{12}H_{22}O_{11}$	H_2O		$C_6H_{12}O_6$	$C_6H_{12}O$
Sucrose	Water	**Yeast** **Invertase** →	Glucose	Fructose
342	18		180	180

$C_6H_{12}O_6$		$2C_2H_5OH$	CO_2
Glucose	→	Ethanol	Carbon dioxide
180		92	88

4.1 Challenges

- Regional and seasonal changes in the feedstock properties
- Fluctuating feedstock prices
- Contamination
- Feedstock infection
- Congener profile
- Water and energy conservation

4.2 Fermentation Technology Advancement Roadmap

Fermentation is the first step of alcohol production process, where sugar from sugary- as well as starchy-based feedstock are converted into alcohol in the presence of water and yeast. This process also produces valuable by-products like CO_2 in equal proportion to alcohol.

There are many traditional fermentation processes/methods which have evolved over a period of time to produce alcohol from sugar, e.g., batch, continuous, and fed-batch modes having various pros and cons and techno-economic viability (Fig. 13.3).

Period: 1960–1980s
In the past, most of the distilleries adopted the simpler batch Fermentation technology which gave low performance benchmarking with present advanced technologies. The major focus is on production of alcohol with minimum investment

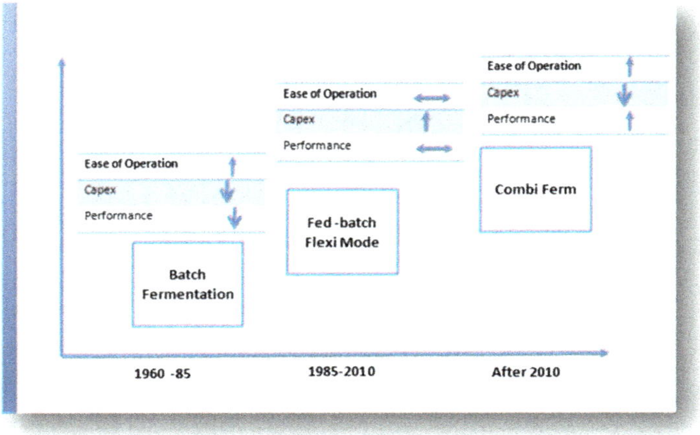

Fig. 13.3 Fermentation technology advancement roadmap

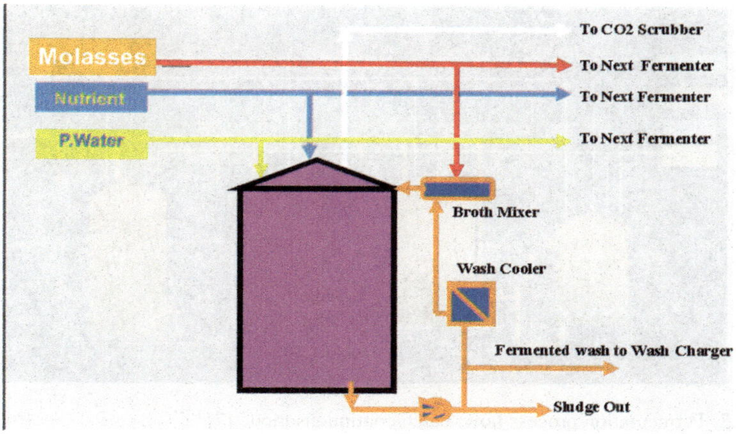

Fig. 13.4 Batch fermentation process flow [7, 8]

(Capex) and lower focus on technology. Major emphasis is on alcohol production and not on operating efficiency.

i. Batch fermentation technology

Batch fermentation is conventional method of ethanol fermentation. In Batch type, fermentation undergoes natural pattern of yeast growth cycle, stationary, and death phase (Fig. 13.4).

The technology is characterized by higher residence time for reaction to reach completion and more by-product formation. Higher by-product formation results in lower production of alcohol and increase of separation cost of by product. During the process of batch fermentation, parameters such as sugar concentration, alcohol concentration, cell count, etc. vary with respect to time. So it impacts on overall fermentation parameters like efficiency and product quality. So it was the unsteady process.

Period: 1985–2010s

After operational experience of 20–25 years in alcohol sector, producers realized the importance of operational efficiency of the plant and economics of the end product. This challenge triggers the major pull for more efficient technologies. Praj had identified the importance of alcohol production in efficient manner results indevelopment of continuous as well as Flexi mode fermentation technology which can run either on batch or continuous mode to achieve required higher efficiency (Fig. 13.5).

ii. Continuous/flexi mode fermentation technology

Continuous fermentation can be operated with single fermenter or multiple fermenters connected in series. Alcohol concentration can be increased in consecutive fermenters.

The process substrate flows into fermenter and fermented wash flows out of the fermenter continuously. As opposed to batch fermentation, parameters such as

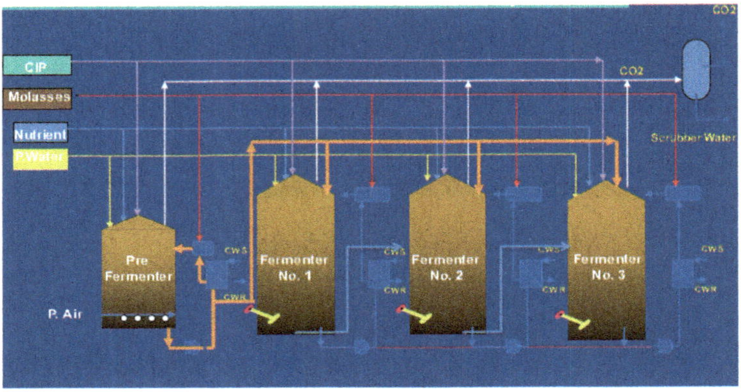

Fig. 13.5 Fermentation process flow: batch/continuous mode [7]

alcohol concentration, sugar concentration, pH, and cell count remain constant with respect to time. Manual intervention is very low in this technology compared to batch fermentation. Thus, continuous fermentation process is characterized by "Steady State process," because of its key advantages like continuity of operation, higher efficiency, consistent performance over a long period, and ease of operation. Many modern ethanol production plants have adopted this technology. However, continuous fermentation systems require good quality of feedstock and are susceptible to contamination.

Period: After 2010: More Ethanol From Same Sugar

After 2010 scenario, ethanol producers were struggling to increase their profit margin. Increase in feedstock price, unavailability of best in class feedstock composition, and appropriate fermentation technology complying to ZSD (Zero Spent Ash Discharge) norms are major challenges for the distilleries. As a need of hour, existing distilleries are exploring reliable fermentation technology and throughput enhancement opportunities by sweating their assets.

It formed the need for advanced technology solutions. To mitigate these challenges, through its R&D center, Praj has developed the advanced technology through its dedicated R&D center (Fig. 13.6). The name of technology is "CombiFerm."

CombiFerm fermentation technology is developed to get the best of both—continuous and synchronous—fermentation technologies. It is suitable for greenfield and brownfield distilleries using sugary feedstocks that can be integrated with an existing fermentation plant operating on continuous, batch, fed-batch, or synchronous mode for throughput enhancement and consistent performance. It can operate with higher recycle streams (ZSD compliance) thereby reducing freshwater footprint.

iii. Technologies involving coproducts

In addition to animal feed, other potential coproducts are produced along with ethanol like CO_2. Some research directions may alter the entire process, using different fermentation processes to produce entirely different products. Researchers are also hoping to turn much of the non-fuel product into ethanol or even something

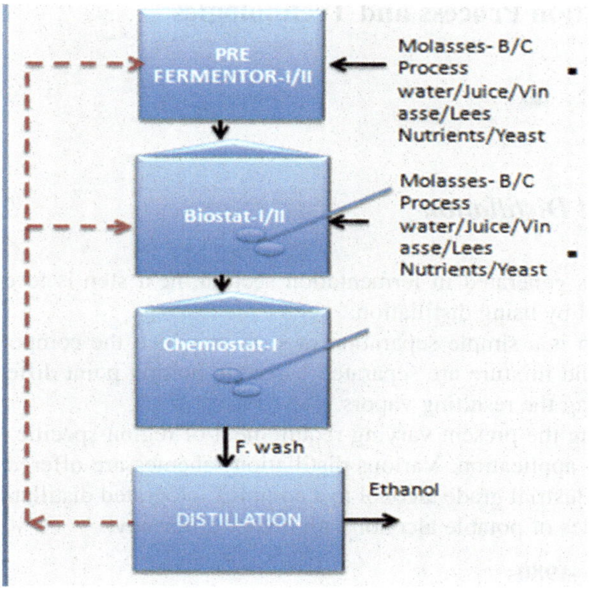

Fig. 13.6 CombiFerm process flow diagram [7]

Table 13.1 Fermentation technology advancement road map

Fermentation technology	1960–1985	1985–2010	After 2010
Fermentation process	Batch type—conventional	Batch type/continuous	Flexi mode—continuous/fed batch, combiferm
Fermentation efficiency	82–86%	87–88%	90–92%
Alcohol concentration	6–7% v/v	7–8% v/v	8–15% v/v
Recycling of spent wash	No recycling	Partially recycled	Fully recycled
Molasses composition	Detailed analysis was not utilized for optimizing process	Detailed analysis is utilized for process design	Detailed analysis is utilized for process design and optimization
Feedstock	Deisgn of Process with only molasses or grain	Design with Multi-feedstocks—Molasses. grain, etc.	Design with multi-feedstocks—molasses, grain, etc.
Wastewater generation	Not accounted—Once through	Partially recycled	Accounted; recycled and reused, ZSD
Control over contamination	Not monitored	Partially monitored	Monitored regularly for better process operation

more valuable than the fuel that is currently the primary product. Possibilities include protein and fiber that could be added to human foods to increase nutritional value (Table 13.1).

5 Distillation Process and Technologies

See Fig. 13.7.

5.1 *About Distillation*

Once Mash is generated in fermentation section, next step is to concentrate and purify ethanol by using distillation.

Distillation is a simple separation process in which the components of a substance or liquid mixture are separated based on boiling point difference/volatility and condensing the resulting vapors.

Considering the present varying requirement of region-specific product quality based on end application. Various distillation schemes are offered from a simple system for industrial grade alcohol to a complex integrated distillation schemes for different grades of potable alcohol with higher efficiency.

Period: 1960–1980s
Most of the distilleries and technology provider adopted the simpler Distillation technology which gave low performance in terms of energy consumption and product quality. The major focus is on production of alcohol with minimum investment (Capex) and lower focus on energy and heat integration.

i. Atmospheric pressure distillation technology

This process separates alcohol from fermented wash and concentrates it to 95–96% (v/v). In this distillation, separation of components is carried out close to

Fig. 13.7 Distillation process and technologies [7]

atmospheric pressure. The distillation columns consist of number of bubble cap plates where wash is boiled and alcoholic vapors are separated and concentrated on each plate stage by stage. There was no heat integration. It required low capital investment. The main disadvantages are severe fouling and scaling issues which result in higher downtime.

Period: 1985–2010s

Atmospheric Pressure distillation technology acceptance was low during this period due to high energy consumption. During this period, Multi-pressure (Multi-Pressure-Vacuum) technology was prevalent in India. This technology became popular due to heat integration and lower downtime. It was introduced much earlier in India—ahead of Brazil in energy efficiency.

ii. Multi-pressure distillation technology

MPR distillation is based on the concept of heat integration in which vapours generated in one column are used to drive other column(s). Pressure–Vacuum Distillation technology was attractive due to lower steam consumption (more than 50% reduction) with marginal increase in investment than Atmospheric distillation.

Advantages:

- Substantially Lower Operating Cost
- Consistency in Performance
- Trouble free operations
- Can be integrated with Distillation for further
- Reduction in steam cost
- Reduced spent wash generation
- Improved alcohol quality
- Reduced scaling problem in analyzer column

Period: After 2010

Market forces are creating multiple challenges for ethanol plant owners and operators in India. Lowest energy and water consumption as well as zero liquid effluent discharge are on top of the list. This is especially true in case of Greenfield and expansion of ethanol plant.

An advanced technology which enables ethanol plants to enhance their throughput with further no increase in energy consumption which results in lower capex. As a leading technology supplier globally, Praj has identified the benefit of use of energy in an efficient manner that results in development of Next-Gen technologies like Ecofine—SD (Split Distillation) and EcoSmart—ED (Evaporative Distillation) which give the product flexibility as well as lower energy and water consumption.

Advantages:

- Minimization of deposition of scale on trays
- Improved turn down
- Steam pressures can be as low as 1 bar (g) for vacuum Plants.
- Reduction in water foot print

- Lower steam consumption
- Multiproduct flexibility
- Increase in plant capacity by 30–40%
- Increase in profitability due to reduced water and energy consumption

iii. Distillation system with reboilers

To concentrate spent wash and reduce effluent generation, reboilers along with distillation column have been installed by distilleries. Use of reboilers results in indirect heating of distillation columns and restricts the mixing of steam condensate with spent wash. Reboilers coupled to multi-pressure distillation systems have resulted in direct reduction of spent wash quantity.

iv. Distillation with integrated evaporation system

To concentrate the spent wash as per the requirement of downstream bio-methanation or bio-composting systems, integrated evaporation plants are now being installed in distilleries. Integrated evaporation system uses alcohol vapors as heating media for heating the spent wash thus saving good amount of steam.

v. Mechanical vapor recompression (MVR)

Higher pressure and temperature applied to compressed vapor with the help of compressor or fan leads to drastic reduction in energy and water consumption.

5.2 Distillation Technology Advancement Road Map

See Table 13.2.

6 Fuel Ethanol Technologies

To reduce energy consumption and to ensure low level of moisture in final ethanol product; use of Molecular Sieves (zeolites) has proved to be economic and robust. Zeolites are synthetic adsorbents (Fig. 13.8).

Period: 1960–1980s

i. Azeotropic distillation technology/extractive distillation technologies

Azeotropic distillation usually refers to the specific technique of adding another component to generate a new, lower-boiling azeotrope that is heterogeneous (e.g., producing two, immiscible liquid phases), such as the example below with the addition of benzene to water and ethanol. It was old technology. The main disadvantages of azeotropic distillation are the larger diameter column required to allow for increased vapor volume due to the azeotropic agent and high energy consumption. It also carries third component along with final product and may be rejected by oil companies. Slowly, this conventional technology phased out over a period of time. And it replaces with Atmospheric Pressure Distillation technology. The

Table 13.2 Distillation technology advancement road map

Distillation process	1960–1985	1985–2010	After 2010
Distillation process	Atmospheric—"direct steam injection"; No reboiler	Atmospheric/multi-pressure-"direct steam injection"; with reboiler	Multi-pressure- "heat integration"/split distillation, ecosmart, use of reboiler
Distillation efficiency	95–97%	97–98%	98.5%–99%
Automation	Not automated	Semi-automatic	Available and practiced
Water consumption	Not monitored	Not much focus	Zero liquid discharge/ waterless
Product preference	Industrial alcohol/ rectified spirit	Multi product—RS/ ENA/AA	Extra neutral alcohol, export quality rectified spirit, absolute alcohol, fuel ethanol, etc.
Steam consumption	5.5–6 kg/Ltr of alcohol for ENA	3.0–3.8 kg/Ltr of alcohol for ENA	1.8–2 kg/Ltr of alcohol for ENA
Focus on energy conservation	No	Partially focus	Yes
TA cut	30–40% of total alcohol for ENA	5–8% of total alcohol for ENA	2–6.5% of total alcohol for ENA
Spent lees—PRC/RC lees/FOC lees	Drained off	Drained off	Accounted; treated and recycled to process
Spent wash from distillation	16–18 Ltrs/Ltr of alcohol	10–12 Ltrs/Ltr of alcohol	5–10 Ltrs/Ltr of alcohol
Fresh water requirement	30–35 Ltrs/Ltr of alcohol	20–22 Ltrs/Ltr of alcohol	1.5–2 Ltrs/Ltr of alcohol
Utilities and other waste streams	Drained off	Drained off	Treated and recycled to process

Fig. 13.8 Fuel ethanol technologies [7]

disadvantages were quality of product, entrainer in final product (Purification), higher water and energy consumption, and ease of operation. Earlier systems operated in liquid phase and used thermal swing regeneration process which did not make them very energy efficient.

Period: 1980–2010 and beyond
Further development on the adsorbent saw introduction of vapor phase operation with pressure swing regeneration system. This proved to be highly energy-efficient.

The vapor phase pressure swing regeneration system employs molecular sieve beds which act as adsorbents. These beds are made of zeolites with an effective pore size opening of about 3 Angstrom.

In order to understand the process of dehydration of ethanol, consider a column packed with freshly activated molecular sieve. As rectified spirit (hydrated ethanol) vapor first enters the bed, water is diffused and adsorbed within the pores of the adsorbent structure in a thin layer. As more ethanol enters the column, it passes through this layer to slightly lower level where another incremental amount of water is adsorbed. This continues until a point is reached where all possible water adsorption from this slug of alcohol is accomplished.

7 Technologies for Treating Wastewater Treatment

Distillery effluent is one of the most challenging effluents in the chemical processing industry. Also, for sugary feedstock, effluent volume can be as high as 12 liter per liter of ethanol production. With norms getting more stringent day by day, we have a bouquet of solutions to not only treat wastewater but to convert it into either boiler fuel, potash rich fertilizer, or animal feed—depending on the feedstock.

Effluent streams in a distillery

Liquefaction	Pump sealing, occasional PHE cleaning
Fermentation	CIP, process air blower, seals, floor washings, occasional PHE cleaning
Distillation	Spent lees, vacuum, and other pumps sealing water, equipment CIP
Evaporation	Process condensate, vacuum and other pumps sealing water, equipment CIP
Utility reject	Boiler blow down, cooling tower blow down, water treatment plant reject

Different effluent treatment technologies
See Fig. 13.9.

Period: 1960–1980s
During the 1980s, land disposal was practiced as one of the main treatment options, since it was found to enhance yield of certain crops. However, for the high strength molasses-based spent wash, the odor and unpleasant landscape due to unsystematic

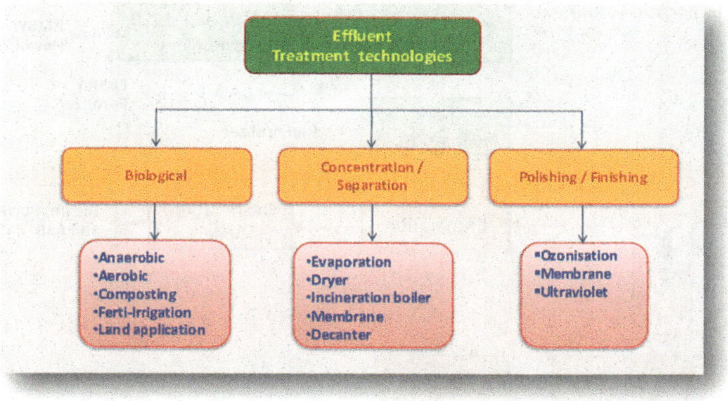

Fig. 13.9 Different Effluent treatment technologies [7]

disposal are concerns in land application in India. The Pollution control norms were not stringent. Ethanol producer preferred low cost technology like biological solutions. Lack of awareness of value-added products like biogas, bio-fertilizer, and power from Effluent treatment plant.

Period: 1980–2010

The most effective and ecological technological systems developed during the past 20 years are as a rule based on a combination of the chemical, physical, and biological solutions. Anaerobic digestion, anaerobic filters, lagoons, activated sludge, and trickling filters have all been successfully applied to the treatment of distillery wastewater.

i. Biomethanated spent wash

Stricter pollution control norms and awareness about value-added products push for new technologies. As the spent wash generated in the distilleries has high potential to produce methane, it is subjected to anaerobic digestion for methane recovery. The process of biomethanation is now well-engrained in distilleries. It was an additional revenue stream for distilleries.

Period: After 2010

ii. Zero liquid discharge for a distillery

2014–2016 saw a number of promising steps taken towards curbing pollution in the Ganga and other water bodies. Zero liquid discharge guidelines were introduced by Central Pollution Control board for distillery sector in India. It would be required to set up systems which will treat the wastewater as well as recover dissolved chemicals so that water can be recycled again in distillery operations.

The different kinds of technologies that need to work together to create a Zero liquid discharge system are Effluent treatment plant (ETP), Reverse osmosis(RO), ultra-filtration(UF), Multi-effect evaporator, Forced evaporator(FE), and

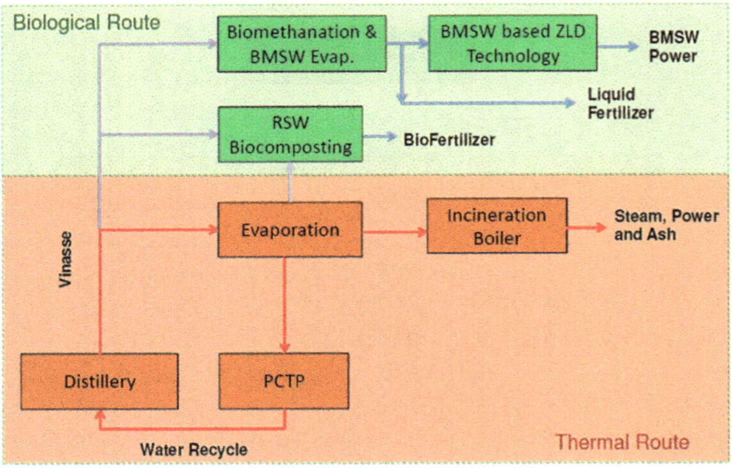

Fig. 13.10 Zero liquid process flow [7]

Incineration boiler. The selection of technologies depends upon the composition of effluent as different technologies remove different contaminants at different rates.

In recent years, membranes and membrane separation techniques have developed from a simple laboratory level to an industrial process with considerable technical and commercial impact. Today membranes are used to clean distillery effluent. Membrane processes are faster, more efficient, and more economical than conventional separation techniques (Fig. 13.10).

iii. Components involved in ZLD

- Effluent Pretreatment—Filters/Clarifiers
- Effluent COD/BOD Reduction—Biological Treatments
- Effluent Volume Reduction—Membranes
- Effluent Concentration—Evaporation
- Solids Separation/Discharge—Centrifuge/Drying
- Treatment of Recovered water by using Distillation/Membrane Separation

7.1 Wastewater Technology Advancement Road Map

See Table 13.3.

Table 13.3 Wastewater treatment technology road map

ETP plants			
Spent wash treatment options	Lagooning/ katcha pit/ferti irrigation	Bio-composting, lagooning, raw spent evaporation	Zero spent wash discharge
PRC/RC and FOC lees treatment	Not practiced	Not practiced	Treated and recycled
Utilities reject and other streams	Not practiced	Not practiced	Recycled
Thin slop treatment options	Not practiced	Not practiced	Partly recycled biomethanation followed by secondary and tertiary treatment followed by land application, concentration, and drying
Evaporation process condensate	Not practiced	Partially recycled	Fully treated and recycled
Utilities reject and other streams	Not practiced	Not practiced	Fully treated and recycled
ZLD solution	Not practiced	Not practiced	Incineration boiler

8 Second-Generation Cellulosic Ethanol Technology and Bio-refinery

It is increasingly understood that first-generation bio-ethanol produced primarily are limited in their ability to achieve targets for oil product substitution, climate change mitigation, and economic growth. The cumulative impacts of these concerns have increased the interest in developing Bio-ethanol produced from nonfood biomass.

Second-generation cellulosic technology differs from first generation in terms of feedstock which in this case is agriculture residue like corn cobs, corn stovers, bagasse, grasses, etc. The lignocellulosic content of these materials can be converted to ethanol.

Ethanol made from cellulosic biomass in the coming years it is believed that cellulosic biomass will be the largest source of ethanol. The broad category of biomass for the production of ethanol includes agricultural crops and residues and wood. Biomass resources are abundant and have multiple application potential. Among the various competing processes, bio-ethanol from lignocellulosic biomass appears to have economic potential.

Bolstered by emerging bioethanol blending mandates and improved access to low-cost feedstocks, the concept of bio-refinery is entering a period of renewed innovation and investment in India.

A bio-refinery is a facility that integrates biomass conversion processes and equipment to produce fuels, power, heat, and value-added chemicals from biomass.

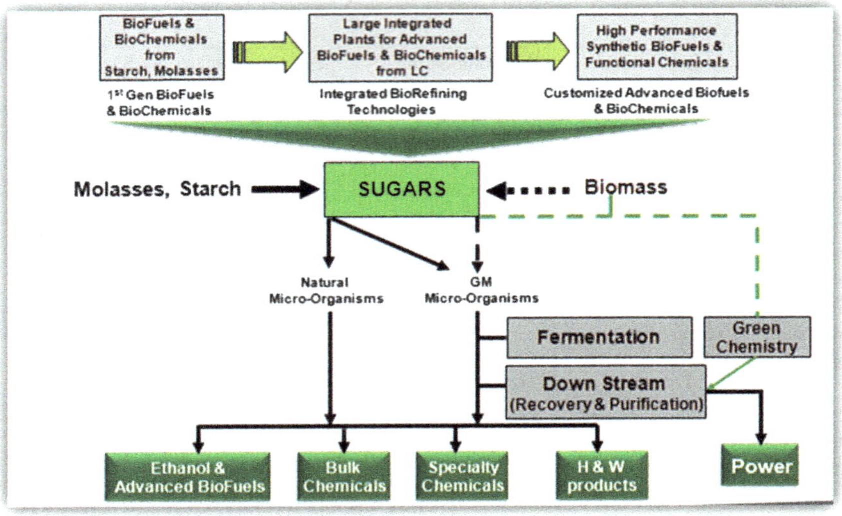

Fig. 13.11 Bio-refinery model [9]

Bio-refinery concept is analogous to today's petroleum refinery which produce multiple fuels and products from petroleum. Here, instead of hydrocarbons are processing carbohydrates.

i. Future bio-refinery models evolution

The Bio-refinery is to produce bio-ethanol and or biochemicals sustainably by using lignocellulosic agriculture crop residues like cane bagasse, cane trash, Corn cobs, Rice (paddy) straw, Cotton stalk, wheat straw, empty fruit bunches of palm, bamboo, grasses, etc. It is fully integrated "Smart Bio refinery process" of converting multiple feedstocks to multiple products and by-products Fig. 13.11.

ii. Commercialization challenges for Second-Generation Ethanol technology

Cellulosic fuels offer economic, environmental, and national security benefits, but scaling-up the technology has proven difficult and costly. Given the current investments being made to gain improvements in second-generation technology, some expectations have arisen that, in the near future, this cellulosic ethanol will reach full commercialization.

9 Conclusions

Ethanol, the dominant biofuel in world markets, is gaining importance as an alternative fuel source, as a part of renewable fuels initiatives adopted by India and globally.

The industry is still improving technologically. It is far more mature than in 1980s, and new developments appear poised to bring costs down further and to reduce the environmental impact of producing ethanol. Better process control and automation also reduce downtime and maintenance in ethanol plants.

However, recent developments in technologies are more totally environment friendly and ensure much less pollution in ethanol production. These technologies ensure minimum energy and water consumption. Other coproducts technology also developed along with ethanol technologies.

Development of bioenergy on a large scale requires the deployment of environmentally acceptable as well as sustainable technology solutions to harness them with the least environment impact.

10 About Praj Industries Ltd

Praj, a global company driven by innovation and integration capabilities, offers solutions to add significant value to bio-ethanol/alcohol plants; brewery plants; water and wastewater treatment systems; critical process equipment and systems; hipurity solutions for Pharma, Biotech, F & B, and Cosmetics industry; and bio-products. Praj has rich experience from technology development (Praj—Matrix R&D centre) at lab scale to commercialize and scale up the technology. It is currently engaged in scaling up of second-generation bio-ethanol technology based on agri-residues. Praj Matrix is built on a 5-acre property with a built-up area of about 85,000 square feet. Praj has well-equipped manufacturing facilities—one in Pune and two at Kandla (Gujarat), port of India. In the field of Ethanol Technology, PRAJ is perhaps the only company in the world having its own dedicated Research & Development Center.

References

1. AIDA (All India Distillers Association), presentation, 2013–2014, http://www.indiansugar.com/uploads, 2014–2015
2. India Stat. Database on gasoline consumption. www.indiastat.com
3. CII report on estimation of energy and carbon balance of biofuels in India, page number 3 of executive summary
4. ABLE (Association of Biotechnology Led Enterprises) report, 2012–2013
5. The biofuel 2020 India report, http://www.unep.org/transport/lowcarbon/PDFs/Biofuel_Roadmap_for_India.pdf, Page 108
6. A report on "opportunities for green chemistry initiatives: molasses based distilleries" by GOI
7. Praj proprietary content, www.praj.net (Applicable for entire document)
8. http://www.financialexpress.com/article/fe-columnist/column-a-green-blend/245833/
9. Wikipedia – internal source, secondary database

Chapter 14
Technological Advancements in Sustainable Production of Second Generation Ethanol Development: An Appraisal and Future Directions

Amit Kumar, Deepti Singh, Anuj K. Chandel, and Krishna Kant Sharma

Abstract With the dwindling oil prices around the world and the realization that the world's oil supply is limited, the quest for alternative fuels began and researchers along with policy makers initiated economical ways to produce ethanol, preferably from bioresources. The integration of agro-energy crops and advance technologies offers the potential for the development of cost-effective substrate for sustainable biofuel production. To develop renewable and sustainable biofuel from lignocellulosic polysaccharides or energy grasses, several pathway engineering approaches have been reported. These include the generation of new metabolic pathways by combining genes from distinct organisms in desired host to synthesize products. In the post-genomic era, synthetic biology seeks to model and construct biological components, functions, and organisms that do not exist in nature or to redesign existing biological systems to perform new functions. Even model plants are being investigated and modified to improve the enzymatic deconstruction potential of lignocellulosics into biofuel. Therefore, metabolic engineering approaches in synthetic biology are converging with synthetic genomics. Here we intend to discuss technological advancement and its probable role in the development of economically feasible biofuel. The techniques include protein engineering, imaging, fermentation technology, chromatography, enzyme technology, chemical engineering, genome sequencing, bioinformatics, and synthetic biology.

Keywords Biofuels • Bioethanol • Techniques • Biomass • Lignin • Cellulose

A. Kumar • D. Singh • K.K. Sharma (✉)
Laboratory of Enzymology and Recombinant DNA Technology, Department of Microbiology, Maharshi Dayanand University, Rohtak 124001, Haryana, India
e-mail: kekulsharma@gmail.com

A.K. Chandel
Centro de Tecnologia Canavieira (CTC), Fazenda Santo Antonio s/n-Caixa Postal 162, 13400-970 Piracicaba, Sao Paulo, Brazil

A.K. Chandel, R.K. Sukumaran (eds.), *Sustainable Biofuels Development in India*,
DOI 10.1007/978-3-319-50219-9_14

1 Introduction

Modern civilization is based upon energy utilization and consumption. Till now, most of the needs are met by fossil fuels, especially, petroleum and coal products, but with their limited reserves and nonrenewable, nature, an alternative substitute for petroleum and gasoline, need to be considered [1–3]. This demand is being fulfilled by expanding biofuel production from low-cost feedstocks (e.g., corn and sugarcane) and lignocellulosic biomass [4], but their uses depend on the prices dictated by the international market. Although the price for lignocellulosic biomass is much less than the price of vegetable oils, corn grains, and sugarcane juice, such biomass are generally more complex to convert into biofuel and also, its production is dependent on new technologies [5]. Considering its positive implication on environment and economy, it is impossible to disagree upon the urgency of introducing biofuel to the global community. But in developing countries, it is hardly feasible to satisfy the ever-growing population, with the available agricultural land and other resources [6], making it necessary to use lignocellulosic biomass for biofuel production.

The plant cell wall contains cellulosic biomass, complexed in a framework of lignin and other recalcitrant material, i.e., crystalline cellulose microfibrils [7–9], which needs to be separated. The deconstruction of biomass for efficient utilization of cellulose fibril can be achieved by thermochemical and biological approaches. Thermochemical approaches lead to conversion of biomass into three fractions: biochar, bio-oil, and syngas [5], whereas biological approach includes enzymes, such as cellulase for cellulose [10], xylanase and mannanase for the hemicelluloses [11], laccase and other peroxidases for lignin [12] are greatly in use for efficient conversion of biomass into sugar hydrolysate. Different microscopic techniques such as atomic force microscopy (AFM), scanning electron microscopy (SEM), transmission electron microscopy (TEM), and confocal microscopy have also paved a path towards a broader understanding of the basic molecular architecture of plant cell wall [13], observation and differentiation among untreated and pretreated biomass [14]. Chromatography tools, such as high-performance liquid chromatography (HPLC), and gas chromatography (GC) have also proved a cutting edge technique for quantitative analysis [15], and estimation [16] of sugars and ethanol, respectively. Various other advanced techniques, namely, nuclear magnetic resonance (NMR) [17], Fourier transform infrared spectroscopy (FTIR) [18], genome sequencing [19], and microarray [20], have also contributed in process upgradation.

A wide range of approaches such as metabolic engineering, genomics, and transcriptomics have been explored [21], to improve the efficiency of ethanol production and reduction in cost of production. Furthermore, synthetic biology, systems biology, and post-genomics biology have also been applied for the modification in monolignol synthesis pathway by transgenically reducing or altering lignin content and up regulating cellulose biosynthesis [21, 22]. We are still in the early stages of engineering organisms for biofuel production, as manifested by the many different organisms being explored simultaneously for their potential to serve as cell factories. *Escherichia coli* is historically the most favorable organism for genetic modifications, which has been successfully applied for the production of different metabolite by introducing exogenous genes to develop hybrid pathways [23], whereas *Saccharomyces cerevisiae* with a natural ability to produce and tolerate secreted ethanol is an attractive and versatile alternative model organism. Moreover, synthetic biology has also been successfully used in *S. cerevisiae*, to introduce an ability to ferment nontraditional sugars or to increase the organism's tolerance to sugar [24]. Other microbes like *Pichia stipitis* [25], *Pachysolen tannophilus* [26], *Klebsiella oxytoca* [27], and *Clostridium acetobutylicum* [28] have also been employed by different workers. The chapter provides comprehensive information about the technological innovation and its implication in the development of cost-competitive biofuels from lignocellulosic biomass to meet the present day global demand.

2 Biofuel Scenario of India

Ministry of Petroleum and Natural Gas, Government of India, in the year 2003 made 5% ethanol blending in petrol mandatory across nine states and five union territories. However, it is still subject to availability and market fluctuation of ethanol [29]. The National Biofuel policy [30] predicts utilization of a wide range of crops, i.e., sugarcane, sweet sorghum, cassava, maize, and oilseeds like *Jatropha* and *Pongamia* for production of biofuels. The aim of Indian biofuel policy is to focus on wastelands, avoid the use of fertile crop lands, and provide price incentives to biofuel producers as well as ban foreign companies in investing directly in plantation activities for biofuel production [31]. The national policy recommended blending targets of 5% by 2012, 10% by 2017 and 20% after 2017 [29]. The projected demand for gasoline is 2.2 billion liters by 2017, with 10% ethanol blending [32]. But domestic ethanol production of India is declining from about 2 billion liters in 2013 to 1.9 billion liters in 2014, due to fall in sugar production cycle [33]. Even converting all the molasses into ethanol, produces approximately 222 liters of ethanol per ton of molasses, which cannot meet the demand even at 5% blending. Also, the fact that India accounts for only 0.5% of oil and gas resources of world, while 70% needs of the country are fulfilled by import of crude oil and natural gas, with an average of annual 7% increase in motor gasoline demands made India's oil import expenditure to over $144 billion in 2012 [32, 33]. So the lignocellulosic biomass is the best alternative substrate for ethanol production in developing countries, like India [32].

Although the biomass itself is a cheap source, the cost of biomass conversion to simple monomers is very high. Biomass to ethanol conversion involves a series of techniques and requires considerable research and development efforts [32]. India has long been involved in the second-generation biofuel development with many labs across the country, Biochemical Engineering Research Centre, IIT Delhi, National Institute for Interdisciplinary Science and Technology campus [32], etc. (Table 14.1).

Table 14.1 Research and development programme in India on cellulosic ethanol

Substrates	Laboratory	Techniques	Ethanol yield	References
Prosopis juliflora wood	Department of Microbiology, University of Delhi South Campus, New Delhi, India	Acid pretreatment, delignification, and enzymatic hydrolysis	0.39 and 0.49 g/g	Gupta et al. [16]
Sugarcane bagasse	-do-	Acid pretreatment, anion exchange resin	0.48 g/g	Chandel et al. [34]
Lantana camara	Indian Institute of Technology, Kharagpur, India and Indian Oil Corporation Ltd., R & D Centre, Faridabad, India	Enzymatic pretreatment	9.63 g/L	Kuila et al. [35]
Sugar cane bagasse, rice straw and water hyacinth biomass	National Institute for Interdisciplinary Science and Technology, CSIR, Trivandrum, India	Enzymatic pretreatment	0.093 g/g for rice straw	Sukumaran et al. [36]
Lantana camara	Microbiology, University of Delhi South Campus, New Delhi, India	Acid pretreatment, Sequential application of overliming and activated charcoal	*Pichia stipitis* and *Saccharomyces cerevisiae* gave rise to 5.16 and 17.7 g/L of ethanol with corresponding yields of 0.32 and 0.48 g/g after 24 and 16 h	Kuhad et al. [37]
Water hyacinth	PG and Research Department of Botany, Alagappa University, Tamilnadu, India and Department of Microbiology, VHNSN College, Tamilnadu, India	Alkali pretreatment	0.411 g/g	Pothiraj et al. [38]
Lantana camara	Department of Microbiology, Osmania University Andhra Pradesh, India	Acid pretreatment, enzymatic pretreatment	0.431 ± 0.018 g/g	Pasha et al. [39]
Parthenium sp.	Division of Microbiology and Agricultural Chemicals, Indian Agricultural Research Institute, New Delhi, India	Alkali pretreatment, enzymatic pretreatment	76.6% fermentable sugar yield	Pandiyan et al. [40]

2.1 *Role of Technology: Efficiency Improvement and Cost Reduction*

With the emergence of new techniques, ethanol production has increased and become less energy intensive [41]. First-generation biofuels show a net benefit in reduced greenhouse gas (GHG) emission, but had many drawbacks, i.e., a rise in food price, limited GHG reduction benefits, accelerating deforestation and negative impact on biodiversity, whereas second-generation biofuels can help to solve the basic problems of previous technologies and can also sustain larger proportion of global fuel supply with greater environmental benefits. It has already helped to bring down both the world's dependence on oil and CO_2 production. GHG emission reduces as net CO_2 use by plants during photosynthesis is balanced by CO_2 produced during their combustion. It has good potential for cost reduction and increase in production efficiency. The co-product formation along with biofuel, offers the increase in revenue, which eventually reduces the production cost [41].

Second generation biofuel development should also be focused on the water scarcity other than substrate recalcitrance, land availability, food production, energy consumption, biodiversity, and CO_2 balance. Global annual biofuel water footprint is expected to be 970 km^3/year by 2030. The 48% of irrigation water footprint (IWF) of biofuels is expected to represent the total biofuel water footprint by 2030. The biofuel water footprint grows 5.5% of total IWF by 2030, thus causing extra load on natural water resources. So nations should consider the water factor, while investigating for biofuels to satisfy the future energy demand in the transport sector [42].

Second generation biofuel processes needs to address few technological and methodological challenges in extracting useful feedstocks from the woody or fibrous bioresource, where the useful sugars in the form of hemicellulose and cellulose are locked in by recalcitrant lignin. Improved enzymatic treatment is needed for effective hydrolysis of cellulose, hemicelluloses, and lignin. Ligninolytic enzyme treatments such as laccase-mediator treatment with an alkaline peroxide extraction lead to decrease of both aromatic and aliphatic lignin [43]. This step makes biomass free from lignin, so that it can be efficiently hydrolyzed into sugar molecules (Fig. 14.1). The conversion of biomass to liquid using thermochemical approach faces the challenge of developing a gasification process for biomass at commercial scale to produce synthetic gas. Raw bio-syngas produced from biomass contain H_2, CO, CO_2, H_2O, and hydrocarbons, which is suitable for combustion purposes but not for biofuel productions. Thus, reliable technologies should be developed to focus on the higher availability of clean and conditioned syngas [44].

The cost reduction in biofuel production and potential improvement can be achieved by using technologies like cellulosic saccharification-fermentation, thermochemical pretreatment, integrated pulp-biofuel biorefineries, and consolidated bioprocessing (CBP), as they reduce operation steps as well as chemical inhibitors [45, 46]. This concept of integrated bioprocessing (IBP) approach had been

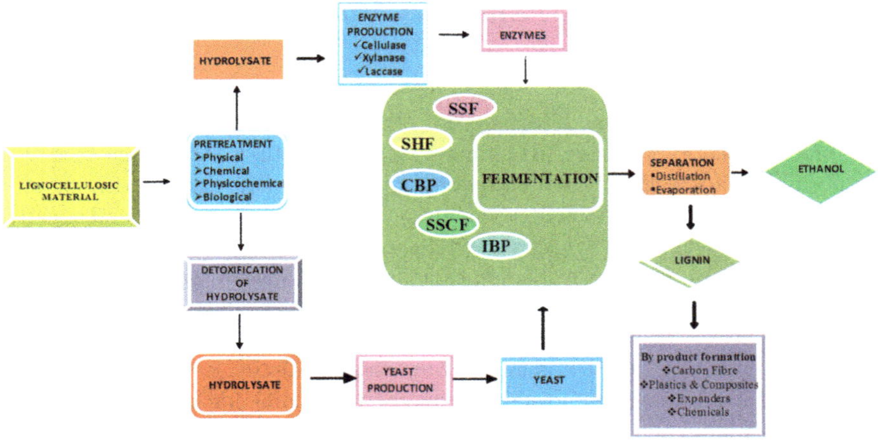

Fig. 14.1 Biomass conversion process for second generation biofuel

introduced for economical ethanol production. IBP involves the incorporation of microbial assisted delignification followed by the cellulose production and fermentation of released sugar into ethanol. This approach delineates cost-effective, economical, and green second-generation ethanol production along with co-utilization of by-products [47–49].

Biorefinery concept is to utilize lignocellulosic biomass to produce biofuels and other value added products. It combines the necessary technologies between raw material, industrial intermediate and final products. Currently, three biorefinery systems, (1) the whole crop biorefinery, (2) the green biorefinery and (3) the lignocellulosic feedstock biorefinery are favoured in research and development. A recent review provides an understanding of lignin valorization, i.e., conversion of lignin to higher value products [50]. Efforts have been made for genetically engineered lignin with abilities to form lignin based carbon fiber [50], plastics and composites [47–49] and expanders for lead acid batteries [51]. Unutilized lignin fractions in biorefinery can further be converted into fuels and specialty chemicals [50]. Recently, Alonso et al. has reported γ-Valerolactone (GVL), a reaction intermediate formed from the precursor levulinic acid with a role in chemicals formation and biomass deconstruction [52]. Further, Luterbacher and coworkers used a novel nonenzymatic method of biomass conversion into soluble carbohydrate (70–90%) hydrolysis by using GVL as a solvent along with water and H_2SO_4. CO_2 extracted carbohydrate further fed to industrial strain *S. cerevisiae* PE2, showed 87% ethanol yield compared to the theoretical yield [53]. The concept of lignin valorization [50] and biomass deconstruction via GVL [52] aid up in our understanding of by-product utilization and complete solubilization of biomass for release of complete soluble carbohydrates, which opens new dimensions for techno-economical admissible biofuel production. Conversion of these by-products to high-value co-products will decrease the cost of biofuel, which will eventually improve the economy of lignocellulose biorefinery, and minimize the waste discharge. It will also reduce the dependence on petroleum-based products and offer new economic opportunities for agricultural, chemical, and energy sectors [54].

Moreover, energy crop species require advanced conventional breeding with high-throughput transformation systems [55]. Some high-throughput technologies, i.e., genotyping, phenotyping, transcriptomics, proteomics, and metabolomics, play important role in associating genotype with phenotype. Further, whole-genome sequences and deep expressed sequence tag (EST) libraries of potential energy crops will prove a landmark to help in marker-aided selection, single nucleotide polymorphisms (SNPs) selection, for favorable alleles using molecular markers, and transformation studies in energy crops [56]. Successful transformation of white-rot fungi can further contribute in the improvement of lignin degrading enzyme production and biological detoxification [57, 58].

3 Cell Wall Architecture: Technological Understanding

The plant cell wall, composed of primary and secondary cell walls [59], is a chemically complex, dynamic, and organized, extra-cytoplasmic matrix. It consists of various polysaccharides and structural proteins, which show the diversity in their composition depending upon plant species, age, and growth conditions [60]. Lignin is found in the highest concentration in the middle lamella, but is most abundant in the secondary walls of the vascular plants along with crystalline cellulose [61, 62]. The plant cellulose consists of D-glucopyranose monomer units bound by β-1,4 glycosidic linkages [63] and packed together to form highly crystalline microfibrils each measuring 3–6 nm in diameter [64]. Intra-chain hydrogen bonding in a cellulose microfibril determines "linearity" of the chain and inter-chain introduces crystalline or amorphous state [65]. Hemicellulose is the second most common polysaccharide in nature. It is amorphous, random, and susceptible to acid hydrolysis, compared to cellulose [61]. Hemicelluloses are observed to be associated cross-linked to pectins, proteins or lignin [7]. Hemicellulose in hardwood and annual plants is mainly xylans (15–30%) [66], whereas in softwood, galactoglucomannans (15–20%) and xylans (7–10%). Hardwood xylan is composed of β-D-xylopyranosyl units, which contain 4-O-methyl-alpha-D-glucuronic acid and acetyl side groups [67]. Lignin is a cross-linked, highly functionalized hydrophobic macromolecule consists primarily of aliphatic and aromatic hydroxyl groups [2]. The hydrolysable linkages in lignin are suggested to be of two types: β-aryl ether and α-aryl ether [62]. It is an integral part of cell wall and act as filler in the spaces between cellulose and hemicelluloses [46]. It is a polymer of phenylpropene units that are guaiacyl (G) units, syringyl (S) units and p-hydroxyphenyl (H) units consisting of diverse functional groups, i.e., phenolic, hydroxyl, methoxyl, and benzyl alcohol, which affects the reactivity of lignin [67]. The lignin present in the biomass not only serves as a shield for cellulose but also interferes with the activity of cellulolytic enzymes by surface adsorption [68]. Presumably, the lignin binds cell wall polysaccharides through both covalent and non-covalent interactions to form a lignin–carbohydrate complex (LCC) [69]. The knowledge about cell wall composition, synthesis, and function can be upgraded to understand the structural complexity, with the help of new emerging technologies. Analysis of these polysaccharides considerably aids our understanding about plant and its efficient deconstruction for biofuel development [60].

3.1 Spectrometry

Earlier, Adapa and coworkers reviewed infrared spectroscopy and Raman spectroscopy for their different aspects to study the recalcitrant property of lignocellulosic biomass for their possible applications in industries [61]. Spectroscopic methods also have been reported in the analysis of biodiesel and monitoring of the transesterification reaction. Nuclear magnetic resonance (NMR) is frequently used to study the structure of lignin and its monomers along with the profiling of plant cell wall polysaccharide [70, 71]. Ultra-low field-magnetic resonance (ULF-MR) having proton Larmor frequencies overlaps with "slow" dynamic processes and biological processes, such as ligand binding and catalysis. Thus, ULF provides a probe for Biomolecular dynamics on a millisecond time scale relevant to biofuels [72]. Due to the lack of economically beneficial feedstocks, the production of biodiesel is limited. With the help of time domain (TD)-NMR analysis, various types of biowaste sources have been analyzed for their oil content and fatty acid composition. TD-NMR is non-destructive and rapid method which yields good correlations with the conventional methods [73]. Low and high field NMR with a fast field cycling setup is used in the solubilization mechanism of crystalline cellulose in H_3PO_4 [74]. 1H-NMR spectroscopy is used in the quantification of different methyl esters in diesel from different sources [17]. To monitor transesterification yield, protons of the methylene group adjacent to the ester moiety in triglycerols and the protons in the alcohol moiety of the product methyl esters were used [75]. ^{32}P-NMR analysis has been employed to quantitatively characterize the structure of hydroxyl groups in lignin and biodiesel [76], whereas 13C NMR employed to study monolignol content and interlink lignin–hemicellulose linkage distribution, in biomass recalcitrance, without its component isolation [77]. Fourier transform infrared spectroscopy (FTIR), specifically, fourier transform infrared-attenuated total reflectance (FTIR-ATR) is a valuable method for energy estimation in a number of data sets. The NaOH consumption by pretreatment and sugar productions from combined pretreatment and enzymatic hydrolysis can be predicted by FTIR-ATR spectroscopy combined with partial least squares (PLS) regression. To analyze the constituents of the cell wall [68, 78], PLS regression is applied to FTIR-ATR spectra of raw biomass [68]. With this technique, it has been concluded that the intra-specific variations in lignin and energy contents are not related to each other and thus the lignin content is not a predictor of the energy content [68]. FTIR is also used in identification of adulterations and even traces of contaminants [79] (Table 14.2), functional groups, e.g., carbonyl groups

Table 14.2 Modern techniques used in the development of biofuel

Technique	Tools	Outcome	References
Microscopy	Atomic force microscopy	Broader understanding of molecular architecture of plant cell walls	Tetard et al. [13] and Kirby [80]
	Scanning electron microscopy	Image the specimen with some natural moisture and observe the difference of untreated and pretreated biomass	Xiao et al. [14], Himmel et al. [81], Wu et al. [82] and Gomez et al. [83]
	Transmission electron microscopy	Ultrahigh, clean vaccum allows contamination-free observation and characterization of catalysts for methanol production	Himmel et al. [81] and Houghton et al. [84]
	Confocal Raman microscopy	Monitor the changes upon NaOH treatment of lignocellulosic biomass	Himmel et al. [81]
Chromatography	Gas chromatography	Quantitative analysis of fatty acid based biofuels, Ethanol estimation with FID detector	Guan et al. [15] and Gupta et al. [16]
	High performance liquid chromatography	Analysis of hydrolysates for presence of carbohydrates and ethanol.	Gupta et al. [16]
Spectroscopy	Nuclear magnetic resonance	Study of the structure of lignin, Quantification of different methyl esters in diesel, analysis of bio-waste sources	Monteiro et al. [17], Li et al. [70] and Conte et al. [74]
	Fourier transform infrared spectroscopy	Prediction of results of genetic engineering, identification of adulterations in B2 and B5 blends	Xu et al. [68]
	Microarray analysis	To study the genes of plant stress response pathway	Zuo et al. [85]
	Micro RNA expression	To study stress adaptation in Switchgrass	Sun et al. [86]
	Custom microarrays	Profiling expression of genes during endosperm development	Hayden et al. [87]
	Whole genome microarray	Study of increasing methyl ketone production	Goh et al. [88]
	Heterologous expression of efflux pumps	To study reduced biofuel toxicity export system of cell	Jin et al. [4]
Sequencing	Expressed sequence tags sequencing	cDNA library is constructed from developing seeds of *J. curcas*	Natarajan et al. [19]
	Whole genome sequencing	Enhanced upgradation of *J. curcas*	Johnson et al. [89]
	Transcriptional profiling with next generation sequencing	Improvement of cellulase in cellulose-degrading fungi	Coradetti et al. [90]
	Single molecule sequencing	Capabilities to sequence DNA up to 1000 million bp/h	Trevors and Masson [91]

[92], Si-O in ash [82], and the bands corresponding to various stretching and bending vibrations in the samples of oil and biodiesel. FTIR has also been reported to quantitatively determine and measure the degradation efficiency [93], the carbohydrate content, i.e., mannans and glucans in wheat straw and predict the cell wall constituents [68].

Mass spectrometry (MS) is a selective, sensitive, and powerful technique to obtain structural information of compounds present in complex mixtures [94]. Since it requires only small sample amount, it is an excellent tool for researchers interested in detecting changes in composition of complex carbohydrates of plants [95]. MS has been used to elucidate the molecular mass and structure of oligosaccharides derived from hemicelluloses and pectins [96]. MS technique can be used to determine the triacylglycerol profiles of oleaginous salt water microalgae, and offer significant advantages in biofuel research. Earlier, Mittelbach has used MS as a detection method besides flame ionization. For determination of free glycerol in biodiesel by GC-MS, selected ion monitoring (SIM) mode was used to track the ions. The detection limit was also improved for rapeseed methyl ester (RME), by using MS in SIM mode (10–5%) compared to the FID detector (10–4%) [75]. An ultra-high pressure liquid chromatography-mass spectrometry method has been developed for profiling lipid extracts to aid in identifying algae strains suitable for biofuels applications [97]. A simple Solid Phase Extraction (SPE) step with MALDI-TOF provides a direct MS method to characterize intact nonpolar lipid samples from microalgae. SPE removes chlorophyll, making microalgae, triacylglycerol (TAG) profiles with desirable fatty acid composition for biofuel purposes [98].

3.2 *Imaging Techniques*

Lignocellulosic biomass used for second generation biofuels, requires advance imaging techniques for better understanding of polymer interaction and lignin distribution of cell wall architecture [99]. Polymer component organization in biomass structures must be analyzed in three-dimensions by using improved and non-penetrating or noninvasive tools [100]. Sant'Anna and de Souza have extensively reviewed microscopic techniques for their use to evaluate the structural effects of pretreatments on the deconstruction of plant cell walls [101]. Imaging needs for specific research areas including advances in imaging techniques by using nuclear magnetic resonance, X-ray and electron based techniques as well as atomic force and scanning tunneling microscopy [102, 103]. AFM is a very powerful scanning probe microscopy surface analysis tool for investigating lignin degradation in waste straw by ruminal microorganisms [93], and for imaging surface structure at the sub-nanometer level [64]. Anaerobic fermentation of lignocellulosic wastes is a very complex and difficult bioprocess to investigate on the micro-scale level. The AFM images showed the removal of lignin by rumen microorganisms, and demonstrate the potential utilization of rumen microorganisms for the bioconversion of lignocellulosic wastes. AFM reveals surface morphology, while X-ray-photoelectron spectroscopy gives chemical characteristics of straw surface regions, in combination could lead to a better understanding of the performance and mechanisms of the biodegradation of lignocelluloses [93]. Lignocellulose, i.e., sugarcane bagasse, has a complex cross-linked structural cell wall [101]. AFM image of pretreated sugarcane bagasse with oxalic acid fiber explosion showed irregular shape with hydrophilic deposition because of cellulose exposure [47–49]. The ability of an AFM image for mapping hydrophilic and hydrophobic regions, due to the physical property of AFM tip, makes it additionally advantageous to study the effect of pretreatment [101]. This modern imaging tool is used for broader understanding of molecular architecture of cell walls with tapping probes, providing better images of both developing and mature plant cells and polysaccharides of algal origin [13]. AFM performed on various forms of *Populus* and switchgrass samples gives an idea of the actual lignocellulosic structure in *Populus* and switchgrass, which has immense importance in acquiring a systematic understanding of the plant cell wall architecture, crucial in the development of biomass-based alternative sources of energy [13]. It gives high-resolution images on a wide variety of materials to visualize the complete architecture and spatial arrangement of individual cellulose fibrils [80]. To investigate the multifaceted complexity of plant cells dynamically, mode synthesizing atomic force microscopy (MS-AFM) is used. This new modality of AFM allows a direct assessment of the mechanical properties of the plant cell walls, where unique features within native and treated plant cell walls could be observed. Results show the potential of MS-AFM to impact biology by unveiling the structure of complex natural systems at the subcellular level [104]. MS-AFM can map the ultrastructure of the cell wall, as the prospect of lignocellulosic biomass for biofuel production is tightly linked to extraction of the sugars necessary for fermentation into ethanol [47–49]. Moreover, scanning probe allows the study of dry or hydrated surface of plant cell wall, which provides tremendous opportunities to trace the interaction dynamics of biocatalysts and substrates as well as analyze the results of pretreatment and enzyme action on biomass surfaces [81] (Table 14.2).

Transmission electron microscopy (TEM) is a platform for high resolution structural biology and provides ultrahigh and clean vaccum for contamination-free observation [84] (Table 14.2). TEM permits determination of the internal structure and compositional analysis of biomass by monitoring energy dispersive X-ray microanalysis [81]. Biofuel produced from the pyrolysis of biomass is not suitable for direct combustion in modern diesel engines due to the presence of a large amount of oxygen components, which require upgrading of biofuel by hydrodeoxygenation. Large surface carbon supported Pd nanoparticles (Pd/C) is a good catalyst for biofuel hydrodeoxygenation [105]. Therefore, TEM and other associated techniques help to reveal the new structure of superhydrophilic mesoporous sulfonated melamine-formaldehyde resin supported Pd catalyst, which is highly active and extremely recyclable, compared with a conventional Pd/C catalyst [106]. The effects and the role of Fe^{2+} ions in the dilute acid pretreatment have been studied on filter paper, corn litter and corn stover. TEM analysis of dilute acid/Fe^{2+} ion cocatalyst-pretreated corn stover showed delamination and fibrillation of the cell wall. Further, interaction of Fe^{2+} with other plant cell wall components using prussian blue staining, analyzed by TEM and SEM showed that Fe^{2+} ions may remain in close proximity to heavily lignified regions of the biomass during pretreatment, and these regions are strongly affected morphologically by the presence of the Fe co-catalyst [107]. Immune-TEM images of thermochemical pretreated corn stover, showed direct visualization of the dramatic effect on altering the condensed ultrastructure of biomass cell walls, i.e., collapse of the cell lumen, delamination of secondary cell walls, and loss of density [108]. Further, high voltage TEM and field emission SEM in combination with back-scattered electron detector is used to analyze the lignin distribution pattern. The lignin distribution of the secondary wall layers of tracheids and beech fibres was found to be in the direction of the cellulose microfibrils, which are responsible for the lamellar structure. Thus, the polymerization of monolignols seems to be affected by the arrangement of the polysaccharides that constitute the cell wall [109].

Scanning Electron Microscopy (SEM) is one of the most versatile and widely used tools that permit the study of the surface morphology, like increased surface roughness and porosity of biological materials and its by-products after pretreatment [82]. Pretreatment is a promising technology for increasing the enzymatic digestibility of lignocellulosic biomass. Therefore, SEM has been used to study the effect of different pretreatment conditions, i.e., enzymatic, chemical, or physical [47–49]. Liquid hot water (LHW) pretreatment deconstructs the pith cells and results in the detachment of the phloem from the xylem. In addition to that, LHW pretreatment before enzyme hydrolysis, of the S-lignin-rich tissue gave a much higher glucose yield than either the wild-type or G-lignin-rich tissue. SEM is used to determine whether any anatomical changes have occurred during LHW pretreatment and enzyme hydrolysis [110, 111]. For instance, sequential acid–base treatment of sugarcane bagasse leads to pith detachment, loose matrix, pore and lignin removal [112, 113]. Furthermore, fluorescence microscopy (FM) using filipin staining and low temperature scanning electron microscopy (LT-SEM) has been used for chemical and microscopic studies of *Eucalyptus globulus* wood to study topochemistry of lignin and lipophilic extractives [114].

Confocal Raman microscopy (CRM) is another non-destructive technique, which provides insight into chemical composition at the micrometer level [115], through point by point illumination, without staining or labeling of tissues. The majority of images are generated by reflecting light off the specimen or by stimulating fluorescence from dyes (fluorophores) tagged to the sample [116]. The spatially resolved chemical and structural changes are important for development of simple pre-processing procedures that can produce complex molecular profiles by differential rates of attack on major cell wall components [81] (Table 14.2).

Modern imaging technology is needed to study the rate of formation and yields of sugars. Cellulase action on cellulose is often repressed by xylan, which acts as a barrier, so the effect of dilute acid pretreatment on xylan distribution across plasma membrane and middle lamella has been studied by confocal laser scanning microscopy (CLSM) [117]. Fluorescently labeled, *Thermobifida fusca* cellulases, Cel5A (classical endocellulase), Cel6B (classical exocellulase), and Cel9A (processive endocellulase), have been studied for their binding on pretreated biomass with the help of time-lapse CLSM. This provides the opportunity to study the different kinetics of adsorption and diffusion into the porous structures of the substrate. The study showed that the cellulases quickly bind to certain areas and slowly diffuses to less accessible area [118]. Ionic liquid pretreatment of lignocellulosic biomass cause swallowing of wood. This property of controlled expansion and contraction can be used to incorporate enzymes deep into the wood structure for improved pretreatments and accelerated cellulose hydrolysis. Nanoparticle treated biomass confocal surface-enhanced Raman images at different depths show that a significant number of nanoparticles were incorporated into the pretreated sample, and they remained on the samples after rinsing [119]. Confocal images of safranin labeled sugarcane showed lignin distribution in the sclerenchyma fibers, cell wall corners, and middle lamella. Lower level of lignin was observed in the xylem and phloem conducting tissue, whereas cellulose concentration was found to be higher in protoxylem and lower in high lignin rich area [120]. Confocal and fluorescence life time imaging microscopy (FLIM) is used to map the lignin distribution within bagasse fibers pretreated with acid and alkali. Here, fluorescence spectra decay time correlate with the delignification yield and the lignin distribution. Two-photon excitation of lignin fluorescence state is used to study well-organized nano-environment of lignin arrangement in untreated bagasse fiber which favors a very low level of interaction between the molecules [121]. Recently, Ragauskas and coworkers have beautifully accentuated the modern imaging techniques, i.e., coherent anti-Stokes Raman scattering (CARS), time-of-flight secondary ion mass spectrometry (ToF-SIMS), small-angle neutron scattering (SANS), fluorescence-tagged monolignol analogs, and molecular dynamics simulations, for lignin characterization, to better analysis of its chemical structure [50]. The action of fungal cellulase and multienzyme complex cellulosome was studied on untreated and delignified plant cell walls under controlled digestion conditions in real time with the use of CLSM. Imaging by CLSM of delignified sclerenchyma-type secondary walls depicted how fungal cellulase is more accessible to delignification than cellulosome [64].

4 Role of Modern Molecular Techniques in Biofuel Development

4.1 Genome Sequencing

Advances in DNA sequencing technology has been used for discovery of novel enzymes and metabolic engineering of organisms in biomass deconstruction and conversion to biofuels [122]. This technique is very useful in many fields such as archaeology, genetics, biotechnology, molecular biology, and microbiology

[123]. Next-generation sequencing (NGS) techniques available to researchers are 454 Genome Sequencer, Illumina (Solexa) Genome analyzer, the SOLiD system, and the Heliscope. In these techniques, base level incorporation event across a wide number of reactions is collected by real-time data collection [124]. A rapid and novel DNA and RNA sequencing technique, i.e., single molecule sequencing (SMS) are under developmental stages. Fluorescent or physical methods are used for the determination and identification of consecutive angle bases in very long read length [91]. To determine the process of improved production through mutagenesis, for identification of mutants in the genomes of hyper-producing genes and for generating *T. reesei* strains for increased biofuel production, a massively parallel sequencing method is adopted [90]. Transcriptional profiling through next-generation sequencing has provided the information that CLR-1 and CLR-2 are required for induction of cellulose genes and are conserved in the genome of fungi [90].

In the past 12 years, fungal genome-sequencing programs have been widely developed (Table 14.3). A number of online resources are available which could be used for accessing the fungal genomic data. These resources include the Broad Institute Fungal Genome Initiative Web site (http://www.broad.mit.edu/FGI/), the Joint Genome Initiative (JGI) Integrated Microbial Resource database (http://img.jgi.doe.gov/pub/main.cgi/), the TIGR fungal database (www.tigr.org.tbd/fungal), National Center for Biotechnology Information (NCBI) Entrez (http://www.ncbi.nlm.nih.gov/entrez/query), the Munich Information Center for Protein Sequences (MIPS-http://mips.gsf.de/projects/fungi/), MetaDB (http://www.neurotransmitter.net/metadb/), and the Genomes Online database (http://www.genomesonline.org/) [125].

The publication of the white rot fungus, *Phanerochaete chrysosporium* genome is a major achievement of the JGI [139]. Like FGI, a priority for the JGI is to make the genome sequence data freely available to the scientific community. TIGR has also made an important contribution to fungal genome project by sequencing some of the economically important fungi, i.e., *Aspergillus fumigatus* [140]. Whole genome sequence of *P. chrysosporium*, the model lignin degrading white-rot fungus, represents an important advance in the molecular genetics of basidiomycetes and provides a framework for future investigations on the mechanism of lignocellulose degradation [139]. The annotation of the *P. chrysosporium* genome has revealed several new isozymes for many previously identified lignocellulolytic enzymes, such as manganese peroxidase (MnP) and copper radical oxidases. It has also revealed four, new putative flavine adenine dehydrogenase (FAD)-dependent oxidases, which have been predicted to take part in lignocellulose degradation [141].

Assembled cDNA sequences of *Dunaliella tertiolecta,* a microalga and singletons annotated with Gene Ontology (GO) and Kyoto Encyclopedia of Genes and Genomes (KEGG) orthology (KO) identifiers showed lipid and starch biosynthesis and catabolism pathways in *D. tertiolecta* [142]. For isolating the functional genes from *J. curcas*, by large-scale sequencing of expressed sequence tags (ESTs), gene discovery projects have been initiated. These genes serve as an important genetic reserve for crop improvement to make it an ideal and beneficial crop for the production of biodiesel [19]. Molecular markers, transcriptional profiling, and

Table 14.3 Whole genome sequencing of some of the economically important biofuels relevant microorganisms

S. no.	Organism	Genome size (Mb)	Center/Consortium	References
1.	*Aspergillus niger CBS 513.88*	34	DSM, The Netherlands	Sharma [125]
2.	*Aspergillus oryzae RIB40*	37	NITE (National Institute of Advanced Industrial Sciences and Technology)	Machida et al. [126]
3.	*Phanerochaete chrysosporium RP-78*	30	DOE Joint Genome Institute	Sharma [125]
4.	*Pichia pastoris DSMZ 70382*	9.4	Natural Resources and Applied Life Sciences, Vienna, Austria	Steensels et al. [127]
5.	*Pichia stipitis CBS 6054*	15.4	DOE Joint Genome Institute	Jeffries et al. [128]
6.	*Schizophyllum commune*	38.5	DOE Joint Genome Institute	Ohm et al. [129]
7.	*Schizosaccharomyces pombe 972h*		Sanger Institute, Cold Spring Harbor Laboratory	Steensels et al. [127]
8.	*Saccharomyces cerevisiae JAY291*	11.6	International Effort	Argueso et al. [130]
9.	*Saccharomyces cerevisiae*	12	International Effort	Sharma [125]
10.	*Trichoderma reesei QM6a*	33	DOE Joint Genome Institute	Sharma [125]
11.	*Acetivibrio cellulolyticus CD2 DSM 1870*	6.157	DOE Joint Genome Institute	Dassa et al. [131]
12.	*Butyrivibrio fibrisolvens 16/4*	3.164	Sanger Institute	Unpublished
13.	*Clostridium acetobutylicum EA 2018*	4132 Kb	International Effort	Brown et al. [132]
14.	*Clostridium acetobutylicum DSM 1731*	4146 Kb	International Effort	Brown et al. [132]
15.	*Clostridium acetobutylicum ATCC 824*	4133 Kb	Genome Therapeutics	Brown et al. [132]
16.	*Clostridium cellulolyticum H10*	4069 Kb	DOE Joint Genome Institute	Unpublished
17.	*Clostridium cellulovorans 743B, ATCC 35296*	5262 Kb	DOE Joint Genome Institute	Unpublished
18.	*Clostridium cellulovorans 743B, ATCC 35296*	5123 Kb	International Effort	Brown et al. [132]
19.	*Clostridium sp. C7*	4401 Kb	Weizmann Institute of Science	Zepeda et al. [133]
20.	*Fibrobacter succinogenes S85, ATCC 19169*	3843 Kb	DOE Joint Genome Institute	Suen et al. [134]
21.	*Ruminococcus albus 7*	4482 Kb	DOE Joint Genome Institute	Suen et al. [135]
22.	*Ruminococcus flavefaciens FD-1*	4574 Kb	Keck Center, Univ of Illinois	Berg et al. [136]
23.	*Bacillus pumilus SAFR-032*	3704 Kb	Baylor College of Medicine	Gioia et al. [137]
24.	*Cellulomonas flavigena 134, DSM 20109*	4123 Kb	DOE Joint Genome Institute	Abt et al. [138]

whole genome sequencing provide data which can enhance the upgradation of *J. curcas* by molecular breeding [89] (Table 14.2).

Hemi-ascomycetous yeast, *P. stipitis* has been isolated from the guts of wood-inhabiting beetles and has been reported to ferment cellobiose to ethanol [143]. *P. stipitis* has been exhaustively studied for xylose fermentation [16]. The mechanism and regulation of xylose metabolism in *P. stipitis* has been characterized and genes from *P. stipitis* have been used to engineer xylose metabolism in *S. cerevisiae*. Considering its biotechnological significance, Jeffries and his coworkers sequenced and assembled the complete genome of *P. stipitis* [128]. The sequence data has revealed the abundance of genes for NADP (H) oxido-reductase reactions suggests that *P. stipitis* adopt various strategies for balancing NAD and NADP-specific cofactors. For example, FAS2 appears to be highly active when cells are growing under oxygen-limited conditions on xylose. Thus, the genome sequence provides insight into how *P. stipitis* regulates its redox balance while very efficiently fermenting xylose under micro-aerobic conditions [128]. The genome of *Aspergillus oryzae*, a fungus important for the production of hydrolytic enzymes, traditional fermented foods and beverages, has been sequenced. The ability to secrete large amounts of proteins and the development of a transformation system have facilitated the use of *A. oryzae* in modern biotechnology [144]. Machida and coworker had reported that the 37-megabase (Mb) genome of *A. oryzae* contains 12,074 genes and is expanded by 7–9 Mb in comparison with the genomes of *Aspergillus nidulans* and *Aspergillus fumigatus*. Comparison of the three aspergilli species, i.e., *A. oryzae, A. nidulans*, and *A. fumigatus,* has revealed the presence of syntenic blocks and specific blocks (lacking synteny with *A. fumigatus*) in a mosaic manner throughout the genome of *A. oryzae* [126]. The blocks of *A. oryzae*-specific sequences are enriched for genes involved in metabolism, particularly those for the synthesis of secondary metabolites. Specific expansion of genes for secretory hydrolytic enzymes, amino acid metabolism, and amino acid/sugar uptake transporters supports the idea that *A. oryzae* is an ideal microorganism for fermentation [126]. The genome of *Schizophyllum commune* (38.5 Mb) encodes at least one gene in each family of hydrolases, involved in the degradation of cellulose, hemicellulose, and pectin from woody cell wall component [129]. Genome sequencing of basidiomycetes, *S. commune* has resulted the highest number of glycoside hydrolases and polysaccharide lyases [129], enzymes responsible in the hydrolysis of woody biomass.

Further, whole genome sequencing of *Eucalyptus* [145], *J. curcas* L. [146], *S. cerevisiae* strain [130], *P. stipitis* [128], Clostridial genomes [147], and Foxtail millet (*Setaria italica*) [148] is a turning point in efforts for integrated biofuel production. Transcriptome analysis of *Camelina sativa* L. using next-generation sequencing has led to discovery of various genes along with 521 transcripts participating in lipid metabolism [149]. The whole genome (12 Mb) sequence of the industrial strain *S. cerevisiae* strain CAT-1, when compared with the reference S228c genome, showed gene polymorphisms related to bioethanol production. Two genes, *IRA1* and *IRA2* participate as inhibitors of the Ras-cAMP-PKA pathway and determine the stress resistance of yeast cells involved in ethanol fermentation [150].

4.2 Recombinant DNA Technology and Metabolic Engineering

Recombinant DNA technology (RDT) involves expression of heterologous genes into a desired host for improving a trait or desired function in the host organism (Fig. 14.2). Various ways have been suggested to increase the ethanol productivity applying RDT, like reprogramming of gene transcription, increasing the amount of "rate limiting" enzymes, deregulation of enzymes, cofactor replenishment, and increase of precursor supply [151]. The recombinant products are often toxic to microbial cells. To solve this problem, strains for ethanol tolerance or sugar tolerance have been engineered [4]. Efforts for improving the microorganism tolerance by recombinant gene manipulation have been hampered by the limited capacity to introduce multiple changes in a gene [152, 153]. Therefore, whole-genome sequence analysis has been used to map quantitative trait loci (QTL), to identify ethanol tolerance genes, i.e., *MKT1*, *SWS2*, and *APJ1* to develop an engineered strain [154].

Transcription machinery engineering and its extensions should be explored systematically [155], for identification of transcription factor mutants that can enhance the tolerance of industrial species to the final fermentation product. For simultaneous utilization of glucose and xylose and to avoid carbon catabolite

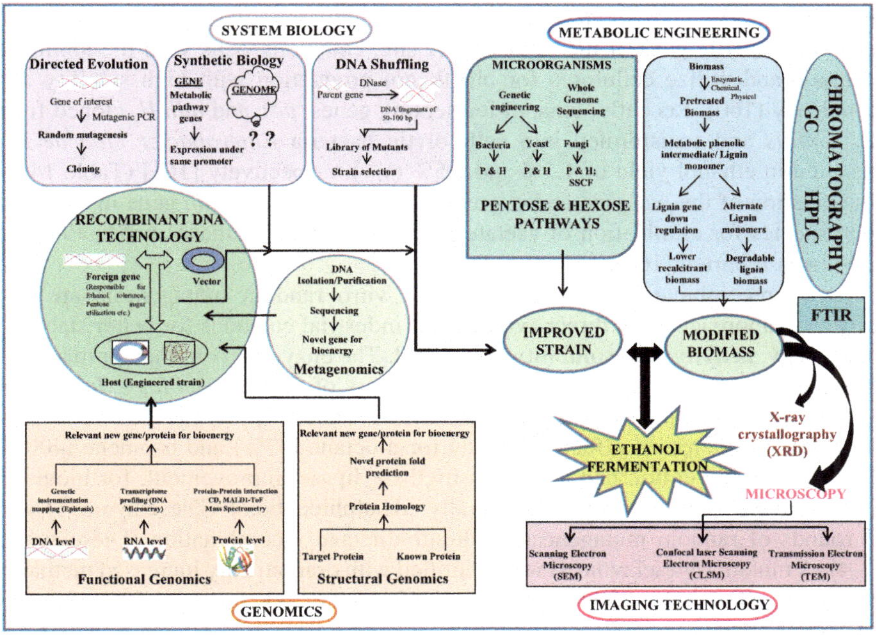

Fig. 14.2 Technology web to depict the application of modern tools used to enhance the productivity and efficiency for second-generation biofuel

repression (CCR), strains of *Escherichia coli*, *S. cerevisiae*, and *Zymomonas mobilis* have been engineered through mutagenesis. Further, Verho and coworkers reported increase in L-arabinose uptake rates in *S. cerevisiae* when genes encoding L-arabinose transporters (*LAT1* and *LAT2*) from *Ambrosiozyma monospora* were cloned in it [156]. Also, two heterologous sugar transporters from *Scheffersomyces stipitis* (AraT) and *Arabidopsis thaliana* (Stp2) support the uptake and utilization of L-arabinose in L-arabinose fermenting *S. cerevisiae* cells, especially at low L-arabinose concentrations had been reported. Gal2 from *S. cerevisiae* have a relatively low affinity but high capacity for L-arabinose, whereas AraT and Stp2 exhibited higher affinities but lower capacities. Together, with the known Gal2 low affinity L-arabinose uptake system, this set of transporters improves fermentations of lignocellulosic hydrolysates by recombinant L-arabinose fermenting *S. cerevisiae* strains [157]. Moreover, *Saccharomyces* strains were also engineered for xylose fermentation via chromosomal integration of xylose reductase (*XYL1*), xylitol dehydrogenase (*XYL2*), and xylulokinase (*XKS1*) genes for conversion of xylose to xylulose-5-P. The engineered strains showed influenced aerobic growth rate on xylose, high xylose consumption rate with ethanol yield 0.18–0.27 g/g and readily fermented the biomass-derived sugars into 13–17% more ethanol than the wild type [158]. Laboratory strains of *S. cerevisiae* were genetically engineered for improving productivity, substrate range extension and by-product elimination [159]. Furthermore, *S. cerevisiae* was genetically engineered for utilization of celluloses for bioethanol production by genetic immobilization of cellulases on the cell surface of yeasts. It was achieved by multicopy integration of a heterologous cellulolytic gene at the yeast rDNA site, endowing cells with the ability to degrade and utilize celluloses for bioethanol production with high stability and efficiency [160]. Recently, ethanol fermenting genes, *pdc* and *adh II*, cloned from *Z. mobilis* and transformed into cellulolytic bacteria *Enterobacter cloacae* JV, resulted in ethanol yield of 4.5% and 3.5% (v/v), respectively [161] (Table 14.4). Inactivation of the *pta* gene and overexpression of synthetic *adh* gene in *Clostridium* species for elimination of acetate production and for ethanol production had also been reported [168].

DNA recombination techniques and in vitro random mutagenesis are two powerful approaches for direct evolution of industrial enzymes for better stability, selectivity, activity, and solubility [169, 170]. The enzyme evolution approach has been successfully reported to increase the fitness of *Thermobacillus xylanilyticus* (Tx-Xyn) for biomass hydrolysis. By directed evolution approach, *E. coli* has been modified for increased efflux efficiency for n-octane (47%) and α-pinene (400%) [4]. Directed evolution had also been used for lipase improvement, for biodiesel production by introducing combitorially, disulphide bond (Dieselzyme 1) and 3 rounds of random mutagenesis with site-directed recombination (Dieselzymes 2–4). A mutant (Dieselzyme 4) was identified with dramatically increased methanol tolerance, showing a 50-fold longer half-inactivation time in 50% aqueous methanol [162].

Metabolic engineering provides an alternative approach in which synthetic pathways are engineered and manipulated into user-friendly hosts for efficient

Table 14.4 Genetic engineering approaches for biofuel production

Organisms	Gene/Strategy	Outcome	References
S.cerevisiae	Mapping of quantitative trait loci (QTL)	*APJ1* gene found to be first causative gene for ethanol tolerance	Swinnen et al. [154]
Ambrosiozyma monospora, *S.cerevisiae*	*LAT1*, *LAT2*	Engineered strain showed uptake of L-arabinose	Verho et al. [156]
S.cerevisiae	*AraT*, *Stp2*	Engineered strain showed improvement in fermentation of lignocellulosic hydrolysate	Subtil and Boles [157]
P. stipitis, *S.cerevisiae*	*XYL1*, *XYL2*, *XKS1*	Engineered strain showed high aerobic growth rate and 13-17% more ethanol than parent strain	Hector et al. [158]
Z.mobilis, *Enterobacter cloacae* JV	*Pdc*, *adh II*	Engineered strain showed twofold higher ethanol than wild type	Piriya et al. [161]
E.coli	Mutagenic polymerase chain reaction (PCR)	Increased efflux efficiency for n-octane and α-pinene	Jin et al. [4]
Proteus mirabilis	Random mutagenesis and site directed recombination	A mutant Dieselzyme 4 with high methanol tolerance	Korman et al. [162]
S.cerevisiae	Genome shuffling	Genetically stable strain GS3-10 having xylose and glucose fermentation capacity	Steensels et al. [127]
S.cerevisiae	Genome shuffling	Engineered strain showed high ethanol tolerance with increased ethanol yield	Steensels et al. [127]
Z. mobilis, *E.coli*	*Pdc*, *adhB*	Fermentation of hemicelluloses derived hydrolysate	Jarboe et al. [27] and Trejo et al. [163]
E.coli	Overproduction of β-ketoacyl-CoAs and thioestrase	High titer of methyl ketone	Goh et al. [88]
C. acetobutylicum	Synthetic acetone operon	High acetone (95%) formation	Mermelstein et al. [164]
C. cellulolyticum	Targeting of 2-keto acid intermediate	Isobutanol production (660 mg/l) directly from crystalline cellulose	Higashide et al. [165]
S.cerevisiae	Engineering of n-butanol biosynthetic pathway	n-butanol production improved to tenfolds	Steen et al. [166]
S.cerevisiae	*ILv2*, *ILv3* and *ILv5*, *Aro10*, *Adh2*	Highest isobutanol yield of 15 mg/g glucose	Steensels et al. [127]
E.coli	Metabolic pathway engineering	Production titer of 2 g/l (1:1 ratio of butanol and propanol)	Shen and Liao [167]

and cost-competitive biofuel production (Fig. 14.2) [171, 172]. It can also be used to evolve the microorganisms for increasing the affinity to specific substrate, tolerance to end product inhibition [173], and extreme conditions of pH, temperature [174]. This allows the use of existing infrastructure, saving enormous amount of capital required for replacing the current infrastructure to accommodate biofuels having properties different from petroleum-based fuels [28]. The development of high yielding strain of *S. cerevisiae* with xylose metabolic pathway using metabolic engineering in combination with genome shuffling had resulted into 47.08% greater yield than the wild type strain [127]. Also, the modification of *S. cerevisiae* for high tolerance to ethanol and osmolarity, resulted in increased metabolic rate, and 15.12% and 15.59% increased ethanol yield, respectively [127]. Genes that encode for pyruvate decarboxylase and alcohol dehydrogenase in *Z. mobilis* have been integrated into the chromosome of *E. coli* B [27], to allow the fermentation of hemicelluloses derived hydrolysate. Further, it is possible to change the fermentation products of an organism, by introducing genes encoding appropriate enzymes and form an alternative system for the regeneration of NAD^+ [27]. The amplification of the acetone formation pathway in *C. acetobutylicum* was the first successful demonstration of metabolic engineering process for biofuel production, which resulted in increased butanol, acetone, and ethanol by 37%, 95%, and 90%, respectively, as compared to the wild type strain [28, 164, 175]. Moreover, the production of isobutanol (660 mg/L) directly from crystalline cellulose with metabolically engineered *Clostridium cellulolyticum* strain had been achieved by diverting its 2-keto acid intermediates toward alcohol synthesis [165]. Recently two heterologous genes from the Ehrlich pathway were expressed in the *Synechocystis* sp. strain PCC 6803 to modify the strain for synthesizing isobutanol under both autotrophic and mixotrophic conditions [176].

Currently, lignocellulosic biomass is driving attention as a potential substrate for engineered n-butanol biosynthetic pathway in *S. cerevisiae* for biofuel production [21, 166]. Isobutanol yield have been reported to increase by over expressing and relocating the *ILV2*, *ILV3*, and *ILV5* genes of *S. cerevisiae* [127, 171]. *E. coli* had been engineered for the production of 1-butanol and 1-propanol from glucose. The strain first converts glucose to 2-ketobutyrate [177], thereafter, 2-ketobutyrate is converted to 1-propanol or 1-butanol depending upon the route and catalyst used [28]. The synthesis of 1-propanol and 1-butanol were systematically improved by deregulation of amino acid biosynthesis and elimination of competing pathways [167] (Table 14.4). Synthetic biology is viewed as ambitious approach to metabolic engineering, as it mostly involves de novo creation of biochemical pathway in host organism. A synthetic butanol production system had been engineered into *E. coli* by introducing *C. acetobutylicum* ATCC 824 genes, for acetyl-CoA acetyltransferase, β-hydroxybutyryl-CoA dehydrogenase, 3-hydroxybutyryl-CoA dehydratase, butyryl-CoA dehydrogenase, butyraldehyde dehydrogenase, and butanol dehydrogenase [28]. Overexpression of genes, i.e., *alsS* from *Bacillus*, native genes *ilvA* and *leuABCD* and deletion of genes *ldhA*, *frdAB*, *pta*, and *pflB* had resulted into improved isobutanol production under micro-aerobic conditions [28].

In comparison to conventional corn grain ethanol, bioethanol derived from low-input high-diversity (LIHD) mixtures of native grassland perennials can provide more usable energy, less greenhouse gases, and less agrichemical pollution, per hectare [21]. By controlling the genes that regulate the juvenile to adult phase transition in plants biomass properties of a wide range of bioenergy feedstocks can been improved. The maize Corngrass1 (*Cg1*) gene, which encodes a microRNA that promotes juvenile cell wall had been successfully transferred into bioenergy crop *Panicum virgatum* (switchgrass) [178]. Transfer of *Cg1* gene into switch grass resulted into production of up to 250% more starch and complete inhibition of flowering, resulting in higher glucose release after the saccharification of biomass with or without pretreatment. Genetic engineering of crop plants to improve the biofuel properties is a feasible way to quickly establish it as a viable bioenergy crop. The increased saccharification efficiency and ethanol yield through downregulation of single lignin biosynthetic gene had been proven successful. The overexpression of a general transcriptional repressor of the phenylpropanoid biosynthesis pathway produce a bioenergy crop with reduced cell wall recalcitrance, slightly increased polysaccharide content and reduced levels of phenolic inhibitors during fermentation [179].

Plant genetic engineering can reduce biomass conversion cost through development of modified plant varieties having less lignin content, high cellulose and crops producing cellulase and ligninase enzymes [22, 180]. Non-plant glycosyl hydrolase genes can be expressed in plant tissue to improve the biomass production and thus improving the lignocellulosic conversion of biomass to glucose and finally to fuels [181]. Genetically modified plants are being explored and tested for different aspects such as increased starch content, modified lignin content or production of cellulase enzymes to enhance production of biofuels [182]. Recently, Boerjan and coworkers have described caffeoyl shikimate esterase (CSE) as an enzyme central to the lignin biosynthetic pathway. *A. thaliana cse* mutants deposit less lignin than do wild-type plants, and the remaining lignin is enriched in p-hydroxyphenyl units. Several gene regulation techniques have already been reported to target lignin synthesis pathway for modified lignocellulosic content (Table 14.4).

4.2.1 Microarray Techniques

Many compounds used for advanced biofuel production, are toxic to microorganisms. To reduce this toxicity, export system of cell, i.e., efflux pumps are used. Bioinformatics can be used to produce a number of efflux pumps from sequenced bacterial genomes for identifying the novel biofuel pumps [4]. Whole genome microarray has led to a new finding that Fad M, which is a native thioesterase in *E.coli*, is an important catalyst for increasing methyl ketone production for biofuels [88]. Gene array analysis is a valuable tool to differentiate the expression levels for the genes involved in xylose catabolism in the parent and the engineered strain [183]. Micro-RNA expression analysis suggests that the micro-RNAs are good candidates to improve switchgrass for biofuel production [86]. It has been shown by microarray analysis that various calcium-dependent protein kinases and its closely related genes are expressed differently across different tissues and under a number of stresses [85]. Pyrosequencing followed by custom microarrays is conducted to profile expression of gene during endosperm development for identification of molecular basis of carbon partitioning between starch and oil. Further, transcriptome analysis and metabolic profiling expands the network, regulating carbon partitioning to those involved in metabolism of cofactors, suggesting a close association between cofactors and carbon partitioning, thereby providing a biotechnological aspect for the conversion of starch to oil [87] (Table 14.2). Further, oligonucleotide based microarrays investigation of global transcription responses of *S. cerevisiae* suggests that multiple metabolic processes are reprogrammed at the transcriptional level during both continuous and fed-batch fermentation processes, including glycosylation, ergosterol synthesis, reserves metabolism, and glucose metabolism [110, 111].

4.2.2 Chromatography Techniques

A wide range of chromatographic techniques have been applied for the characterization of biofuels, to trace the impurities in biodiesel and to determine the ratio of biodiesel-diesel blends [184]. GC is rapid and accurate, configured to separate and analyze volatile compounds from a mixture [185]. This technique is used in the determination of fatty acid methyl esters (FAME) of biodiesel, with methyl salicylate as internal standard using flame ionization detector (FID). FAME distribution of biodiesel can be determined by a high speed GC method. For the analysis of biodiesel, a two-dimensional GC is used [186, 187]. It is a modified version of the conventional chromatography, with a simple, inline fluidic modulator. The primary column used is 5% phenyl polymethylsiloxane which is coupled with a secondary column, polyethylene glycol [188]. The concentration of ethanol is estimated by GC with an elite wax column at oven temperature at 85 °C and FID at 200 °C [189, 190]. Analytical techniques like GC-MS not only determine the detailed chemical structures of cell wall polysaccharides but also estimate the overall polysaccharide composition of cell wall [191]. For reliable analysis of fatty acid based biofuels produced from wild type strain and genetically engineered cyanobacteria, GC-MS has proved to be beneficial [15].

Other liquid chromatography technique like HPLC is designed so that some important parameters can be easily monitored and quantitatively analyzed, such as the amount of ethanol produced, the amount of fermentable sugars in the fermentation broth, and the concentration of unwanted by-products [16]. Thus, HPLC allows the ethanol plant operator to end the fermentation process by means of knowing its carbohydrate content and ethanol or suggests adding antibiotics against bacterial contamination (Table 14.2). Earlier, Holcapek and colleagues, have used gradient elution reversed-phase HPLC for diverse compounds determination, i.e., tri-, di-, mono-acylglycerols, free fatty acids, methyl esters of oleic, linoleic, and linolenic acids, during the production of biodiesel from rapeseed oil [192]. They had also reported atmospheric pressure chemical ionization-mass spectrometry (APCI-MS) for the analysis of rapeseed oil and biodiesel. It is also used for analyzing the hydrolysates for determining the traces of carbohydrates in the sample.

5 Microbial Detoxification and Fermentation

Different pretreatment processes (dilute acid, hot water, ammonia fiber explosion (AFEX), ammonia recycle percolation, and lime) for the liberation of sugars from bioresource has been reported [193]. Chemical pretreatment of ligninocellulose produces many aromatics, aliphatic acids, inorganic compounds, and furan aldehydes [194]. There are a variety of chemicals, and liquid–liquid extraction, liquid–solid extraction, heating and evaporation, enzymatic and in situ microbial detoxification methods have been applied for removal of inhibitors [194] (Fig. 14.3). A separate procees step is required for detoxification of ligninocellulose hydrolysates. Physiochemical and chemical methods are faster, but factors like time, sugar loss, high cost, and several filtration steps and negative environment impact make their limited use on large scale [47–49]. However, enzymatic method is efficient, but time consuming process. In situ microbial detoxification is a process in which

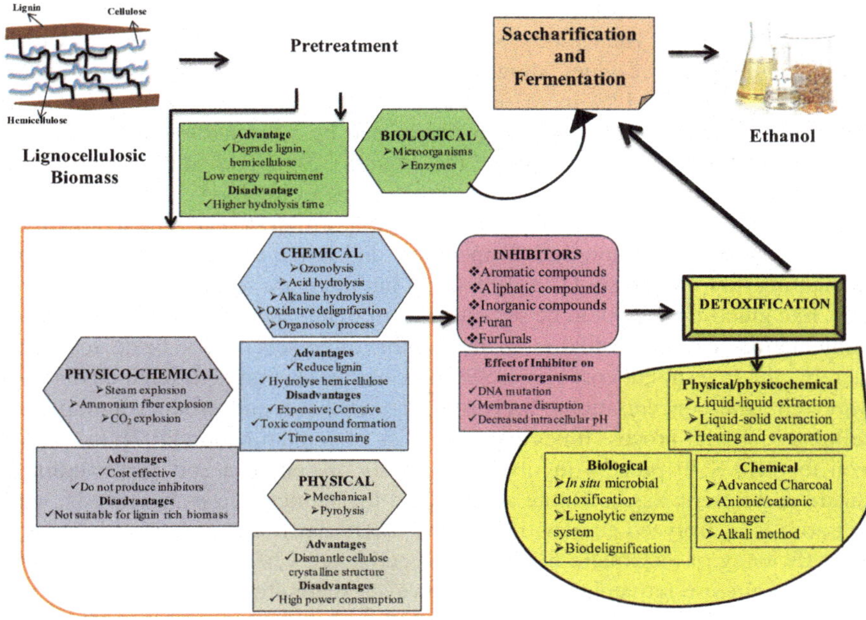

Fig. 14.3 Integrated bioprocessing of lignocellulosic biomass for biofuel production

microorganism directly involved in the detoxification of ligninocellulose hydrolysates. Microorganism like bacteria (*Methylobacterium extorquens*, *Pseudomonas* sp., *Flavobacterium indologenes*), fungus (*Coniochaeta ligniaria* C8 NRRL30616) and yeast strains were observed to deplete the toxic compounds [47–49].

Fermentation includes microbes capable of fermenting sugars, typically hexose sugars, e.g., glucose, and produce ethanol and other valuable by-products [195]. Improvement of the fermentation process is just one of several factors that need to be fully optimized. For efficient fermentation of lignocelluloses, essential traits therefore are emphasized, i.e., broad substrate utilization range, high ethanol yields and productivity, minimal by-product formation, high ethanol tolerance, increased tolerance to inhibitors, and tolerance to process hardiness [21]. Fermentative production of biofuels and chemicals requires the engineering of biocatalysts that can quickly and efficiently convert sugars to cost-competitive products comparable with existing petrochemical-based processes [196]. The enzymatic hydrolysis step is often in close collaboration with the simultaneous saccharification and fermentation (SSF), separate hydrolysis and fermentation (SHF), simultaneous saccharification and co-fermentation (SSCF), and consolidated bio-processing (CBP) fermentation step in the ethanol production. Each process has their own merits and demerits [7, 197]. Efforts have been made to develop a economically viable process for lignocellulosic biofuels production (Table 14.4).

Cellulose content from lignocellulose can be converted into fermentable sugar by using SSF, which decreases the number of steps in the saccharification and fermentation process [198, 199]. Oscillatory Baffled Reactor, fitted with equally spaced orifice plate baffles resulted in 7% more saccharification over the shake flask and 81.29% ethanol yield [200]. SSF experiments with coculture system carried out on agricultural residues showed efficient utilization of hydrolyzed sugars with 30–38% and 10–13% increase in ethanol production [201]. Hydrolysis is the rate limiting step in SSF and can be overcomed by a pre-hydrolysis at optimum temperature for enzymes to improve the process [199]. To avoid the problem of final product inhibition, equilibrium is maintained in hydrolysis reaction through glucose removal from the reaction medium [202]. Problems encountered with SSF are end product inhibition and temperature difference of the cellulases and the fermenting microorganism [173, 199, 203], but there are several other advantages also, like glucose need not to be separated from lignin and when the hydrolysis and fermentation occurs in a single step, the number of vessels need to be decreases and thereby, the investment cost [199]. SSF process has several additional advantages: lower enzyme concentrations, higher sugar yields (no end product inhibition of cellulase), shorter process times and lower risk of contamination [197]. In comparison to SSF, SHF process involves hydrolysis of cellulose and fermentation of sugars at different stages. Dilute acid is used for hemicelluloses hydrolysis, while cellulose is hydrolyzed enzymatically [197]. SHF has been left behind because of less efficiency of enzymatic hydrolysis of cellulose, when it occurs in a separate step from glucose fermentation [202]. However, there are also advantages of SHF in comparison of SSF. The optimum temperature for hydrolysis is typically higher than that of fermentation at least in which yeast is used as the fermenting organism. In SHF, the hydrolysis temperature can be adjusted independent of that of fermentation and yeast can be reused, whereas a compromise is there in case of SSF [199]. Moreover, yeast is subject to end product inhibition [7], which makes SSF an expensive process [204]. The problem of catabolic repression can be overcome using SHF because ethanol production can be increased by adding higher sugar load in cellulosic hydrolysates [16].

SHF has few limitations due to high capital cost, higher contamination risk, and longer processing time than SSCF [197]. SSCF is accomplished firstly by lignocellulosic biomass pretreatment and then neutralized and exposed to enzymes that hydrolyze cellulose and hemicelluloses to sugars and finally to ethanol. This process is performed using genetically engineered microorganisms, which reduces its production time and cost. SSCF is better than SSF in terms of cost-effectiveness, yields, and processing time [198]. This process is advantageous because only one reactor is used for ethanol production [198]. Xylose fermentation is affected during adverse conditions, but glucose fermentation is not influenced [205]. In comparison to SHF, productivity of SSCF is high, but optimal at low temperature, hence require thermotolerent and ethanol tolerant microorganism [197, 206]

In recent years, CBP has been proposed to be the most advanced process in which ethanol and all required enzymes are produced in the same reactor [202, 207]. In CBP, cellulase production, substrate hydrolysis, and fermentation are carried out in a single step [208] and in one reactor [209]. For this purpose, the organisms were developed, which are naturally cellulolytic and have been engineered to improve properties related to product or non-cellulolytic organisms, having high product yield, can be engineered to make them cellulolytic [210, 211]. Only one microbial consortium is used for both the production of cellulase and fermentation with lignocellulose as the substrate. It is an economically attractive process that involves third generation biofuel production [198]. Enzyme production from *S. cerevisiae* is difficult in comparison to enzyme production by aerobic fungi, such as *Trichoderma* or *Aspergillus* [212], because glucose suppresses respiration in yeast and thus reduces the amount of ATP available for biosynthesis of protein. To solve this problem, the hydrolysis and fermentation steps must be coordinated well inside a single cell, such that neither one limits the conversion process to proceed at maximum capability [7]. Recently, helical ribbon stirrer bioreactor used for CBP has been reported to provide high mixing rate of substrate *Jerusalem artichoke* tuber (Jat), which leads to high loading of Jat and ultimately high ethanol yield [213], whereas multispecies biofilm membrane (MBM) reactor enabled in CBP allows combined action of microbial consortium, i.e., *T. reesei*, *S. cerevisiae*, and *S. stipitis*. MBM immobilization to microbial cell, result in less cell biomass production and more resistance toward toxicity [214]. Therefore, CBP has the capability to provide cost-competitive method for biological conversion of cellulosic biomass to ethanol [171, 210, 215] (Table 14.5).

Table 14.5 Fermentation strategies used for the production of bioethanol from different substrates

Substrate	Pretreatment	Reactors	Microorganism	Fermentable sugar	Fermentation methods	Ethanol (g/L)	References
Sugarcane bagasse	Alkali	Continous stirred tank reactor (CSTR)	*S. cerevisiae, T. reesei*	Hexose	Batch, continous	31.58	Brethauer and Studer [214]
Oakchips	Steam explosion	CSTR Batch	*S. cerevisiae*	Hexose	Batch, continous	77.57	Brethauer and Studer [214]
Spruce	Steam explosion	CSTR, CSTR with cell recyclable	*S. cerevisiae*	Hexose	Continous	20.23	Brethauer and Studer [214]
Spruce	Dilute acid	CSTR	*S. cerevisiae* ATCC 96581	Hexose	Continous	17	Brethauer and Studer [214]
Spruce, Wheat straw	Dilute acid	CSTR	*S. cerevisiae* ATCC 96581	Hexose	Continous	–	Brethauer and Studer [214]
Forest Residual	Dilute acid	CSTR	*S. cerevisiae*	Hexose	Continous	–	Brethauer and Studer [214]
Sugar	Not needed	CSTR, fluidized bed bioreactor (FBBR)	*S. cerevisiae* CBS 8066	Hexose	Continous	–	Brethauer and Studer [214]
Coniferous wood, Corn steep liquor	Concentrated acid	Tower type Reactor	*S. cerevisiae* strain KF-7	Hexose	Continous	56.63	Brethauer and Studer [214]
Sugar	Not needed	Magnetically stabilized fluidized bed reactor	*S. cerevisiae*	Hexose	Continous	18.8, 9.1	Liu et al. [216]
Yeast extract Corn stover	Dilute acid	CSTR, FBBR	*P. stipitis, Thermoanaerobactor* BA1L1	Hexose and pentose	Continous	11.6	Brethauer and Studer [214]
Wheat straw	Enzymatic	FBBR	*Thermoanaerobactor* BA1L1	Hexose and pentose	Continous	17	Brethauer and Studer [214]
Spruce	Dilute acid	CSTR	*Mucor indicus*	Hexose and pentose	Continous	26.9	Brethauer and Studer [214]

Yeast extract, Synthetic media	Not needed	FBBR	*Z. mobilis*	Hexose and pentose	Continous	20	Brethauer and Studer [214]
Acid hydrolysate, Corn steep liquer	Dilute acid	CSTR	*Z. mobilis*	Hexose and pentose	Continous	21	Brethauer and Studer [214]
Hardwood floor	Dilute acid	CSTR	*S. cerevisiae*,*T. reesei*	Hexose	SSF	42	Brethauer and Studer [214]
Paper sludge	Not needed	Solid fed reactor	*S. cerevisiae*	Hexose	SSF	40	Brethauer and Studer [214]
Corn fiber	Dilute acid	CSTR	Recombinant yeast strains	Hexose and pentose	SSF	8.12	Brethauer and Studer [214]
Wheat straw	Sequential acid–base	Membrane bioreactor	*S. cerevisiae*, *T. reesei*, *P. stiptis*	Hexose and pentose	CBP	8	Brethauer and Studer [214]
Sigma cell cellulose, Cellobiose	Not needed	Oscillatory baffled reactor	*S. cerevisiae*	Hexose	SSF	128.7	Ikwebe and Harvey [200]
Jerusalem artichoke tuber	Mechanical	Helical ribbon reactor	*S. cerevisiae* DQ1	Hexose	CBP	66	Guo et al. [213]

6 Conclusion

For efficient conversion of lignocellulose to biofuel, various techniques have been used depending upon the requirement and feasibility. Technologies such as next-generation genome sequencing, designer enzymes, metabolic engineering, and even synthetic biology are no more dreams. Combination of different high-throughput methods will create a platform for technologies that are critical for overcoming obstacles in biofuel production from cellulosic biomass or algal lipids. The development of biofuels from biomass can lead naturally to diverse facilities that incorporate value-added products into biofuel production, forming adaptive systems that can be modified to meet our needs and constraints. In light of technological progress and powerful approaches available for research-driven improvements, a cost-competitive process appears to be a reality.

Acknowledgements The authors would like to acknowledge Council of Scientific Industrial Research and DST, New Delhi, India for their financial support.

References

1. Hagerdal BH, Galbe M et al (2006) Bio-ethanol-the fuel of tomorrow from the residues of today. Trends Biotechnol 24:549–556
2. Kumar P, Barrett DM et al (2009) Methods for pretreatment of lignocellulosic biomass for efficient hydrolysis and biofuel production. Ind Eng Chem Res 48:3713–3729
3. Wang H, Srinivasan R, Yu F et al (2011) Effect of acid, alkali, and steam explosion pretreatments on characteristics of bio-oil produced from pinewood. Energy Fuel 25:3758–3764
4. Jin H, Chen L, Wang J, Zhang W (2014) Engineering biofuel tolerance in non-native producing microorganisms. Biotechnol Adv 32:541–548
5. Lee RA, Lavoie JM (2013) From first-to third-generation biofuels: challenges of producing a commodity from a biomass of increasing complexity. Anim Front 3(2):6–11
6. Chu LQ, Masyuko R, Sweedler JV, Bohn PW (2010) Base-induced delignification of *Miscanthus xgiganteus* studied by three-dimensional confocal Raman imaging. Bioresour Technol 101:4919–4925
7. Brethauer S, Wyman CE (2010) Review: continuous hydrolysis and fermentation for cellulosic ethanol production. Bioresour Technol 101:4862–4874
8. Himmel ME, Ding SY, Johnson DK, Adney WS, Nimlos MR, Brady JW, Foust TD (2007) Biomass recalcitrance: engineering plants and enzymes for biofuels production. Science 315:804–807
9. Yu Z, Zhang B, Yu F, Xu G, Song A (2012) A real explosion: the requirement of steam explosion pretreatment. Bioresour Technol 121:335–341
10. Maurya DP, Singh D, Pratap D, Mourya JP (2011) Optimization of solid state fermentation conditions for the production of cellulase by *Trichoderma reesei*. J Environ Biol 33:5–8
11. Stalbrand H, Saloheimo A, Vehmaanpera J et al (1995) Cloning and expression in *Saccharomyces cerevisiae* of a *Trichoderma reesei* β-mannanase gene containing a cellulose binding domain. Appl Environ Microbiol 61:1090–1097
12. Robert V, Mekmouche Y, Pailley PR, Tron T (2011) Engineering laccases: in search for novel catalysts. Curr Genet 12:123–129

13. Tetard L, Passian A, Farahi RH et al (2010) Spectroscopy and atomic force microscopy of biomass. Ultramicroscopy 110:701–707
14. Xiao LP, Sun ZJ, Shi ZJ, Xu F, Sun RC (2011) Impact of hot compressed water treatment on the structural changes of woody biomass for bioethanol production. Bioresources 6:1576–1598
15. Guan W, Zhao H, Lu X, Wang C, Yang M, Bai F (2011) Quantitative analysis of fatty-acid-based biofuels produced by wild-type and genetically engineered cyanobacteria by gas chromatography-mass spectrometry. J Chromatogr A 1218:8289–8293
16. Gupta R, Sharma KK, Kuhad RC (2009) Separate hydrolysis and fermentation (SHF) of *Prosopis juliflora*, a woody substrate, for the production of cellulosic ethanol by *Saccharomyces cerevisae* and *Pichia stipitis*-NCIM 3498. Bioresour Technol 100:1214–1220
17. Monteiro MR, Ambrozin ARP, Liao LM, Ferreira AG (2008) Determination of biodiesel blend levels in different diesel samples by 1H NMR. Fuel 88:691–696
18. Zhou S, Ingram LO (2001) Simultaneous saccharification and fermentation of amorphous cellulose to ethanol by recombinant *Klebsiella oxytoca* SZ21 without supplemental cellulose. Biotechnol Lett 23:1455–1462
19. Natarajan P, Kanagasabapathy D, Gunadayalan G et al (2010) Gene discovery from *Jatropha curcas* by sequencing of ESTs from normalized and full length enriched cDNA library form developing seeds. BMC Genomics 11(1):606
20. Jourdren L, Duclos A, Brion C, Portnoy T, Mathis H, Margeot A, Le Crom S (2010) Teolenn: an efficient and customizable workflow to design high-quality probes for microarray experiments. Nucleic Acid Res 38(10):e117–e117
21. Chandel AK, Singh OV (2011) Weedy lignocellulosic feedstock and microbial metabolic engineering: advancing the generation of 'Biofuel'. Appl Microbiol Biotechnol 89:1289–1303
22. Fu C, Mielenz JR, Xiao X et al (2011) Genetic manipulation of lignin reduces recalcitrance and improves ethanol production from switchgrass. Proc Natl Acad Sci USA 108 (9):3803–3808
23. Martin VJ, Pitera DJ, Withers ST, Newman JD, Keasling JD (2003) Engineering a mevalonate pathway in *Escherichia coli* for production of terpenoids. Nat Biotechnol 21:796–802
24. Ha S (2011) Engineered *Saccharomyces cerevisiae* capable of simultaneous cellobiose and xylose fermentation. Proc Natl Acad Sci USA 108:504
25. Almeida JRM, Modig T, Röder A, Lidén G, Gorwa-Grauslund M (2008) *Pichia stipitis* xylose reductase helps detoxifying lignocellulosic hydrolysate by reducing 5-hydroxymethyl-furfural (HMF). Biotechnol Biofuels 1:12
26. Liu X, Mortensen UH, Workman M (2013) Expression and functional studies of genes involved in transport and metabolism of glycerol in *Pachysolen tannophilus*. Microb Cell Factories 12:27
27. Jarboe LR, Grabar TB, Yomano LP, Shanmugan KT, Ingram LO (2007) Development of ethanologenic bacteria. Adv Biochem Eng Biotechnol 108:237–261
28. Lee SY, Park JH, Jang SH, Nielsen LK, Kim J, Jung KS (2008) Fermentative butanol production by *Clostridia*. Biotechnol Bioeng 101:209–228
29. Raju SS, Parappurathu S, Chand R, Joshi PK, Kumar P, Msangi S (2012) Biofuels in India: potential, policy and emerging paradigms. Policy Paper—National Centre for Agricultural Economics and Policy Research
30. Ministry of New and Renewable Energy (2009) National Policy on biofuels. http://mnes.nic.inS
31. Ravindranath NH, Lakshmi CS, Manuvie R, Balachandra P (2011) Biofuel production and implications for land use, food production and environment in India. Energy Policy 39:737–5745
32. Sukumaran RK, Surender VJ, Sindhu R, Binod P, Janu KU, Sajna KV, Rajasree KP, Pandey A (2010) Lignocellulosic ethanol in India: prospects, challenges and feedstock availability. Bioresour Technol 101:4826–4833

33. Aradhey A (2013) Biofuels annual. GAIN Report Number: IN3073
34. Chandel AK, Kapoor RK, Singh A, Kuhad RC (2007) Detoxification of sugarcane bagasse hydrolysate improves ethanol production by Candida shehatae NCIM 3501. Bioresour Technol 98:1947–1950
35. Kuila A, Mukhopadhyay M, Tuli DK, Banerjee R (2011) Production of ethanol from lignocellulosics: an enzymatic venture. EXCLI J 10:85–96
36. Sukumaran RK, Singhania RR, Mathew GM, Pandey A (2009) Cellulase production using biomass feed stock and its application in lignocellulose saccharification for bio-ethanol production. Renew Energy 34:421–424
37. Kuhad RC, Gupta R, Khasa YP, Singh A (2010) Bioethanol production from lantanacamara (red sage): pretreatment, saccharification and fermentation. Bioresour Technol 101:8348–8354
38. Pothiraj C, Arumugam R, Gobinath M (2014) Sustaining ethanol production from lime pretreated water hyacinth biomass using mono and co-cultures of isolated fungal strains with *Pichia stipitis*. Bioresour Bioprocess 1:27
39. Pasha C, Nagavalli M, Rao LV (2006) *Lantana camara* for fuel ethanol production using thermotolerant yeast. Lett Appl Microbiol 44:666–672
40. Pandiyan K, Tiwari R, Singh S, Nain PKS, Rana S, Arora A, Singh SB, Nain L (2014) Optimization of enzymatic saccharification of alkali pretreated *Parthenium* sp. using response surface methodology. Enzyme Res. doi:10.1155/2014/764898
41. McAloon A, Taylor F, Yee W, Ibsen K, Wooley R (2000) Determining the cost of producing ethanol from corn starch and lignocellulosic feedstocks. National Renewable Energy Laboratory Report NREL/TP-580-28893
42. Gerbens-Leenes PW, Lienden van AR, Hoekstra AY, van der Meer TH (2010) Biofuel scenarios in a water perspective: the global blue and green water footprint of road transport in 2030. Glob Environ Chang 22:764–775. doi:10.1016/j.gloenvcha.2012.04.001
43. Gutiérrez A, Rencoret J, Cadena EM, Rico A, Barth D, del Río JC, Martínez AT (2012) Demonstration of laccase-based removal of lignin from wood and non-wood plant feedstocks. Bioresour Technol 119:114–122
44. Van der Drift A, Boerrigter H (2006) Synthesis gas from biomass for fuels and chemicals, ECN-C—06-001. Report following the workshop "Hydrogen and Synthesis Gas for Fuels and Chemicals" organized by IEA bioenergy Task 33 in conjunction with the SYNBIOS conference held in May 2005 in Stockholm, 30 p
45. Limayema A, Ricke SC (2012) Lignocellulosic biomass for bioethanol production: current perspectives, potential issues and future prospects. Prog Energy Combust Sci 38:449–467
46. Pu Y, Zhang D, Singh PM, Ragauskas AJ (2008) The new forestry biofuels sector. Biofuels Bioprod Biorefin 2(1):58–73
47. Chandel AK, Antunes FF, Anjos V, Bell MJ, Rodrigues LN, Singh OV, Rosa CA, Pagnocca FC, da Silva SS (2013a) Ultra-structural mapping of sugarcane bagasse after oxalic acid fiber expansion (OAFEX) and ethanol production by *Candida shehatae* and *Saccharomyces cerevisiae*. Biotechnol Biofuels 6:4
48. Chandel AK, da Silva SS, Singh OV (2013b) Detoxification of lignocellulose hydrolysates: biochemical and metabolic engineering toward white biotechnology. BioEnergy Res 6:388–401
49. Chandel AK, Gonçalves BC, Strap JL, da Silva SS (2013c) Biodelignification of lignocellulose substrates: an intrinsic and sustainable pretreatment strategy for clean energy production. Crit Rev Biotechnol 5(3):281–293. doi:10.3109/07388551.2013.841638
50. Ragauskas AJ, Beckham GT, Biddy MJ et al (2014) Lignin valorization: improving lignin processing in the biorefinery. Science 344(6185):1246843. doi:10.1126/science.1246843
51. Hirai N, Kubo S, Magara K (2009) Combined cyclic voltammetry and in situ electrochemical atomic force microscopy on lead electrode in sulfuric acid solution with or without lignosulfonate. J Power Sources 191:97–102

52. Alonso DM, Wettstein SG, Mellmer MA, Gurbuzab EI, Dumesic JA (2013) Integrated conversion of hemicellulose and cellulose from lignocellulosic biomass. Energy Environ Sci 6:76–80
53. Luterbacher JS, Rand JM, Alonso DM, Han J, Youngquist JT, Maravelias CT, Pfleger BF, Dumesic JA (2014) Nonenzymatic sugar production from biomass using biomass-derived γ-valerolactone. Science 343:277
54. Fitzpatrick M, Champagne P, Cunningham MF, Whitney RA (2010) A biorefinery processing perspective: treatment of lignocellulosic materials for the production of value-added products. Bioresour Technol 101:8915–8922
55. Busov V, Matthias F, Andrew G, Steven S (2005) Insertional mutagenesis in *Populus*: relevance and feasibility. Tree Genet Genomes 1:135–142
56. Houghton J, Weatherwax S, Ferrell J (2005) Breaking the biological barriers to cellulosic ethanol: a joint research agenda. US Department of Energy Office of Science and Office of Energy Efficiency and Renewable Energy DOE/SC-0095
57. Sharma KK, Kuhad RC (2010) Genetic transformation of lignin degrading fungi facilitated by *Agrobacterium tumefaciens*. BMC Biotechnol 10:67
58. Sharma KK, Gupta S, Kuhad RC (2006) *Agrobacterium*-mediated delivery of marker genes to *Phanerochaete chrysosporium* mycelial pellets: a model transformation system for white-rot fungi. Biotechnol Appl Biochem 43(3):181–186
59. Keegstra K (2010) Plant cell walls. Plant Physiol 154:483–486
60. Fangel JU, Ulvskov P, Knox JP et al (2012) Cell wall evolution and diversity. Front Plant Sci 3:152
61. Adapa P, Karunakaran C, Tabil L, Schoenau G (2009) Potential applications of infrared and Raman spectromicroscopy for agricultural biomass. Agric Eng Int CIGR Ejournal Manuscript 1081, XI:1–25
62. Adler E (1997) Lignin chemistry-past, present and future. Wood Sci Technol 11:169–218
63. Brown RM (2004) Cellulose structure and biosynthesis: what is in store for the 21th century? J Polym Sci A Polym Chem 42:487–495
64. Ding SY, Liu YS, Zeng Y, Himmel ME, Baker JO, Bayer EA (2012) How does plant cell wall nanoscale architecture correlate with enzymatic digestibility? Science 338(6110):1055–1060
65. Laureano-Perez L, Teymouri F, Alizadeh H, Dale BE (2005) Understanding factors that limit enzymatic hydrolysis of biomass: characterization of pretreated corn stover. Appl Biochem Biotechnol 124:1081–1099
66. Saratale GD, Oh SE (2012) Lignocellulosics to ethanol: the future of the chemical and energy industry. Afr J Biotechnol 11:1002–1013
67. Palonen H (2004) Role of lignin in the enzymatic hydrolysis of lignocelluloses. Dissertation, Aalto University
68. Xu F, Yu J, Tessob T, Dowellc F, Wang D (2013) Qualitative and quantitative analysis of lignocellulosic biomass using infrared techniques: a mini-review. Appl Energy 104:801–809
69. Obembe OO, Jacobsen E, Visser RGF, Vincken JP (2006) Cellulose-hemicellulose networks as target for in planta modification of the properties of natural fibres. Biotechnol Mol Bio Rev 1:76–86
70. Li MF, Sun SN, Xu F, Sun RC (2012) Mild acetosolv process to fractionate bamboo for the biorefinery: structural and antioxidant properties of the dissolved lignin. J Agric Food Chem 60:1703–1712
71. Mansfield SD, Kim H, Lu F, Ralph J (2012) Whole plant cell wall characterization using solution state 2-D NMR. Nat Protoc 7:1579–1589
72. Volegov P, Flynn M, Kraus R et al (2010) Magnetic resonance relaxometry at low and ultra low fields. IFMBE Proc 28:82–87
73. Willson RM, Wiesman Z, Brenner A (2010) Analyzing alternative bio-waste feedstocks for potential biodiesel production using Time domain (TD)-NMR. Waste Manag 30:1881–1888
74. Conte P, Maccotta A, De Pasquale C, Bubici S, Alonzo G (2009) Dissolution mechanism of crystalline cellulose in H_3PO_4 as assessed by high field NMR spectroscopy and fast field cycling NMR relaxometry. J Agric Food Chem 57:8748–8752

75. Meher LC, Vidyasagar D, Naik SN (2006) Technical aspects of biodiesel production by transesterification-a review. Renew Sustain Energ Rev 10:248–268
76. David K, Kosa M, Williams A et al (2010) 31P-NMR analysis of bio-oils obtained from pyrolysis of biomass. Biofuels 1:839–845
77. Foston M, Samuel R, Ragauskas AJ (2012) ^{13}C cell wall enrichment and ionic liquid NMR analysis: progress towards a high-throughput detailed chemical analysis of the whole plant cell wall. Analyst 17:3904–3909
78. Lacayo CI, Malkin AJ, Holman HYN et al (2010) Imaging cell wall architecture in single zinnia elegans tracheary elements. Plant Physiol 154:121–133
79. Sonia A, Dasan PK (2013) Chemical, morphology and thermal evaluation of cellulose microfibers obtained from *Hibiscus sabdariffa*. Carbohydr Polym 92:668–674
80. Kirby AR (2011) Atomic force microscopy of plant cell walls. Methods Mol Biol 715:169–178
81. Himmel M, Vinzant T, Bower S, Jechura J (2005) BSCL use plan: solving biomass recalcitrance. Technical Report NREL/TP-510-37902 National Renewable Energy Laboratory. http://www.nrel.gov/docs/fy05osti/37902.pdf
82. Wu DL, Wang W, Zhang JH et al (2012) Preparation of mulberry branch biomass char and its usage in wastewater treatment. Water Environ Res 84:2060–2069
83. Gomez LD, Bristow JK, Statham ER, McQueen-Masonet SJ (2008) Analysis of saccharification in *Brachypodium distachyonstems* under mild conditions of hydrolysis. Biotechnol Biofuels 1:15
84. Houghton J, Weatherwax S, Ferrell J (2006) DOE: US: Breaking the biological barriers to cellulosic ethanol: a joint research agenda. US Department of Energy Office of Science and Office of Energy Efficiency and Renewable Energy DOE/SC-0095
85. Zuo R, Hu R, Chai G et al (2012) Genome-wide identification, classification and expression analysis of CDPK and its closely related gene families in poplar (*Populus trichocarpa*). Mol Biol Rep 40:2645–2662
86. Sun G, Stewart CN, Xiao P, Zhang B (2012) Micro RNA expression analysis in the cellulosic biofuel crop switchgrass (*Panicum virgatum*) under abiotic stress. PLoS One 7:e32017
87. Hayden DM, Rolletschek H et al (2011) Co-factome analyses reveal enhanced flux of carbon into oil for potential biofuel production. Plant J l67:1018–1028
88. Goh EB, Baidoo EEK, Keasling JD, Beller HR (2012) Engineering of bacterial methyl ketone synthesis for biofuels. Appl Environ Microbiol 78:70–80
89. Johnson TS, Eswaran N, Sujatha M (2011) Molecular approaches to improvement of *Jatropha curcas* Linn. as a sustainable energy crop. Plant Cell Rep 30:1573–1591
90. Coradetti ST, Craig JP, Xiong Y, Shock T, Tian C, Glass NL (2012) Conserved and essential transcription factors for cellulase gene expression in ascomycete fungi. Proc Natl Acad Sci USA 109:7397–7402
91. Trevors JT, Masson L (2010) DNA technologies: what's next applied to microbiology research? Antonie Van Leeuwenhoek 98:249–262
92. Safar M, Bertrand D, Robert P, Devaux MF, Genot C (1994) Characterization of edible oils, butters and margarines by Fourier transform infrared spectroscopy with attenuated total reflectance. J Am Oil Chem Soc 71(3):371–377
93. Hu ZH, Liu SY, Yue ZB, Yan LF, Yang MT, Yu HQ (2008) Microscale analysis of in vitro anaerobic degradation of lignocellulosic wastes by rumen microorganisms. Environ Sci Technol 42:276–281
94. Glish GL, Vachet RW (2003) The basics of Mass Spectrometry in the twenty first century. Nat Rev Drug Discov 2:140–150
95. Bauer S (2012) Mass spectrometry for characterizing plant cell wall polysaccharides. Front Plant Sci 3:45
96. Brüll L, Huisman M, Schols H et al (1998) Rapid molecular mass and structural determination of plant cell wall derived oligosaccharides using off-line high-performance anion-exchange chromatography/mass spectrometry. J Mass Spectrom 33:713–720

97. MacDougall KM, McNichol J, McGinn PJ, O'Leary SJ, Melanson JE (2011) Triacylglycerol profiling of microalgae strains for biofuel feedstock by liquid chromatography-high-resolution mass spectrometry. Anal Bioanal Chem 401:2609–2616
98. Danielewicz MA, Anderson LA, Franz AK (2011) Triacylglycerol profiling of marine microalgae by mass spectrometery. J Lipid Res 52:2101–2108
99. McCann MC, Carpita NC (2008) Designing the deconstruction of plant cell walls. Curr Opin Plant Biol 11:314–320
100. Shaw AD, Winson MK, Woodward AM et al (2000) Rapid analysis of high-dimensional bioprocesses using multivariate spectroscopies and advanced chemometric. Adv Biochem Eng Biotechnol 66:83–113
101. Sant'Anna C, de Souza W (2012) Microscopy as a tool to follow deconstruction of lignocellulosic biomass. In: Méndez-Vilas A (ed) Current microscopy contributions to advances in science and technology. FORMATEX Microscopy Series N° 5, vol 1. Formatex Research Center, Spain, pp 639–645
102. Cross SJ, Ciruela A, Poomputsa K, Romaniec MP, Freedman RB (1996) Thermostable chaperon in from *Clostridium thermocellum*. Biochem J 316:615–622
103. Pichler BJ, Wehrl HF, Judenhofer MS (2008) Latest advances in molecular imaging instrumentation. J Nucl Med 49:5S–23S
104. Tetard L, Passian A, Farahi RH et al (2011) Nanometrology of delignified Populus using mode synthesizing atomic force microscopy. Nanotechnology 22(46):465702
105. Zhao C, Kou Y, Lemonidou AA, Li X, Lercher JA (2009) Highly selective catalytic conversion of phenolic bio-oil to alkanes. Angew Chem Int Ed 48(22):3987–3990
106. Zhongfei LV, Sun Q, Meng X, Xiao F (2013) Superhydrophilic mesoporous sulfonated melamine–formaldehyde resin supported palladium nanoparticles as an efficient catalyst for biofuel upgrade. J Mater Chem A 1(30):8630–8635
107. Wei H, Donohoe BS, Vinzant TB et al (2011) Elucidating the role of ferrous ion cocatalyst in enhancing dilute acid pretreatment of lignocellulosic biomass. Biotechnol Biofuels 4:48
108. Donohoe BS, Selig MJ, Viamajala S, Vinzant TB, Adney WS, Himmel ME (2009) Detecting cellulase penetration into corn stover cell walls by immuno-electron microscopy. Biotechnol Bioeng 103(3):480–489
109. Fromm J, Rockel B, Lautner S, Windeisen E, Wanner G (2003) Lignin distribution in wood cell walls determined by TEM and backscattered SEM techniques. J Struct Biol 143:77–84
110. Li BZ, Chenq JS, Qiao B, Yuan YJ (2010a) Genome-wide transcriptional analysis of *Saccharomyces cerevisiae* during industrial bioethanol fermentation. J Ind Microbiol Biotechnol 37:43–55
111. Li X, Ximenes E, Kim Y, Slininger M, Meilan R, Ladisch M, Chapple C (2010b) Lignin monomer composition affects *Arabidopsis* cell-wall degradability after liquid hot water pretreatment. Biotechnol Biofuels 3:27
112. Chandel AK (2014) Multi-scale structural and chemical analysis of sugarcane bagasse in the process of sequential acid–base pretreatment and ethanol production by *Scheffersomyces shehatae* and *Saccharomyces cerevisiae*. Biotechnol Biofuels 7:63
113. Rezende CA, de Lima MA, Maziero P, deAzevedo ER, Garcia W, Polikarpov I (2011) Chemical and morphological characterization of sugarcane bagasse submitted to a delignification process for enhanced enzymatic digestibility. Biotechnol Biofuels 4:54
114. Speranza M, Gutierrez A, Río JCD et al (2009) Sterols and lignin in *Eucalyptus globulus* Labill.wood: spatial distribution and fungal removal as revealed by microscopy and chemical analyses. Holzforschung 63:362–370
115. Gierlinger N, Keplinger T, Harrington M (2012) Imaging of plant cell wall by confocal resonance microscopy. Nat Protoc 7:1694–1708
116. Semwogerere D, Weeks ER (2005) Confocal microscopy. In: Wnek GE, Bowlin GL (eds) Encyclopedia of biomaterials and biomedical engineering, 2nd edn. Taylor Francis, London, pp 705–714

117. Brunecky R, Vinzant TB, Porter SE, Donohoe BS, Johnson DK, Himmel ME (2009) Redistribution of xylan in maize cell walls during dilute acid pretreatment. Biotechnol Bioeng 102:106
118. Zhu P, Moran-Mirabal JM, Luterbacher JS, Walker LP, Harold G, Craighead HG (2011) Observing *Thermobifida fusca* cellulase binding to pretreated wood particles using time-lapse confocal laser scanning microscopy. Cellulose 18:749–758
119. Lucas M, Macdonald BA, Wagner GL, Joyce SA, Rector KD (2010) Ionic liquid pretreatment of poplar wood at room temperature: swelling and incorporation of nanoparticles. ACS Appl Mater Interfaces 2:2198–2205
120. Sant'Anna C, Costa LT, Abud Y, Biancatto L, Miguens FC, De Souza W (2013) Sugarcane cell wall structure and lignin distribution investigated by confocal and electron microscopy. Microsc Res Tech 76:829–834
121. Coletta VC, Rezende CA, da Conceição FR, Polikarpov I, Guimarães FEG (2013) Mapping the lignin distribution in pretreated sugarcane bagasse by confocal and fluorescence lifetime imaging microscopy. Biotechnol Biofuels 6:43
122. Henry RJ (2012) Advances in DNA sequencing enabling more rapid development of improved biomass and biofuel conversion technologies. Biofuels 3:507–509
123. França LTC, Carrilho E, Kist TBL (2002) A review of DNA sequencing techniques. Q Rev Biophys 35:169–200
124. Richardson P (2010) Special issue: next generation DNA sequencing. Genes 1:385–387
125. Sharma KK (2015) Fungal genome sequencing: basic biology to biotechnology. Crit Rev Biotechnol 27:1–17. doi:10.3109/07388551.2015.1015959
126. Machida M, Asai K, Sano M et al (2005) Genome sequencing and analysis of *Aspergillus oryzae*. Nature 438:1157–1161
127. Steensels J, Snoek T, Meersman E et al (2014) Improving industrial yeast strains: exploiting natural and artificial diversity. FEMS Microbiol Rev. doi:10.1111/1574-6976.12073
128. Jeffries TW, Grigoriev J, Grimwood JM et al (2007) Genome sequence of the lignocellulose-bioconverting and xylose-fermenting yeast *Pichia stipitis*. Nat Biotechnol 25:319–326
129. Ohm RA, de Jong JF, Lugones LG et al (2010) Genome sequence of the model mushroom *Schizophyllum commune*. Nat Biotechnol 28:957–963
130. Argueso JL, Carazzolle MF, Mieczkowski PA et al (2009) Genome structure of a *Saccharomyces cerevisiae* strain widely used in bioethanol production. Genome Res 19:2258–2270
131. Dassa B, Borovok I, Lamed R et al (2012) Genome-wide analysis of acetivibrio cellulolyticus provides a blueprint of an elaborate cellulosome system. BMC Genomics 13:210
132. Brown SD, Nagaraju S, Utturkar S, De Tissera S, Segovia S, Mitchell W, Land ML, Dassanayake A, Köpke M (2014) Comparison of single-molecule sequencing and hybrid approaches for finishing the genome of *Clostridium autoethanogenum* and analysis of CRISPR systems in industrial relevant *Clostridia*. Biotechnol Biofuels 7:40
133. Zepeda V, Dassa B, Borovok I, Lamed R, Bayer EA, Cate JHD (2013) Draft genome sequence of the cellulolytic bacterium *Clostridium papyrosolvens* C7 (ATCC 700395). Genome Announc 5:e00698–e00713
134. Suen G, Stevenson DM, Bruce DC et al (2011a) Complete genome of the cellulolytic ruminal bacterium *Ruminococcus albus* 7. J Bacteriol 193:5574–5575
135. Suen G, Weimer PJ, Stevenson DM et al (2011b) The complete genome sequence of *Fibrobacter succinogenes* S85 reveals a cellulolytic and metabolic specialist. PLoS One 6: e18814
136. Berg ME, Antonopoulos DA, Rincon MT et al (2009) Diversity and strain specificity of plant cell wall degrading enzymes revealed by the draft genome of *Ruminococcus flavefaciens* FD-1. PLoS One 4:e6650
137. Gioia J, Yerrapragada S, Qin X et al (2007) Paradoxical DNA repair and peroxide resistance gene conservation in *Bacillus pumilus* SAFR-032. PLoS One 2:e928
138. Abt B, Foster B, Lapidus A et al (2010) Complete genome sequence of *Cellulomonas flavigena* type strain (134). Stand Genomic Sci 3:15–25

139. Martinez D, Larrondo LF, Putnam N et al (2004) Genome sequence of the lignocellulose degrading fungus *Phanerochaete chrysosporium* strain RP78. Nat Biotechnol 22:695–700
140. Ronning CM, Fedorova ND, Bowyer P et al (2005) Genomics of *Aspergillus fumigates*. Rev Iberoam Micol 22:223–228
141. Kersten P, Cullen D (2007) Extracellular oxidative systems of the lignin-degrading Basidiomycete *Phanerochaete chrysosporium*. Fungal Genet Biol 44:77–87
142. Rismani-Yazdi H, Haznedaroglu BZ, Bibby K, Peccia J (2011) Transcriptome sequencing and annotation of the microalgae *Dunaliella tertiolecta*: pathway description and gene discovery for production of next-generation biofuels. BMC Genomics 12:148
143. Parekh SR, Yu S, Wayman M (1986) Adaptation of *Candida shehatae* and *Pichia stipitis* to wood hydrolyzate for increased ethanol production. Appl Microbiol Biotechnol 25:300–304
144. Ward PP, Lo JY, Duke M, May GS, Headon DR, Conneely OM et al (1992) Production of biologically active recombinant human lactoferrin in *Aspergillus oryzae*. Biotechnology 10:784–789
145. Grattapaglia D, Vaillancourt RE, Shepherd MM et al (2012) Progress in Myrtaceae genomics; *Eucalyptus* as the pivotal genus. Tree Genet Genomes 8:463–508
146. Sato S, Hirakawa H, Isobe S et al (2011) Sequence analysis of the genome of an oil-bearing tree, *Jatropha curcas* L. DNA Res 18:65–76
147. Hemme CL, Mouttaki H, Lee YJ et al (2010) Sequencing of multiple clostridial genomes related to biomass conversion and biofuel production. J Bacteriol 192:6494–6496
148. Zhang G, Liu X, Quan Z et al (2012) Genome sequence of foxtail millet (*Setaria italica*) provides insights into grass evolution and biofuel potential. Nat Biotechnol 3:549–554
149. Mudalkar S, Golla R, Ghatty S, Reddy AR (2014) De novo transcriptome analysis of an imminent biofuel crop, *Camelina sativa* L. using Illumina GAIIX sequencing platform and identification of SSR markers. Plant Mol Biol 84:159–171
150. Babrzadeh F, Jalili R, Wang C et al (2012) Whole-ge nome sequencing of the efficient industrial fuel-ethanol fermentative *Saccharomyces cerevisiae* strain CAT-1. Mol Gen Genomics 287:485–494
151. Stephanopoulos G (2007) Challenges in engineering microbes for biofuel production. Science 315:801–804
152. Curtiss R (1976) Genetic manipulation of microorganisms: potential benefits and biohazards. Annu Rev Microbiol 30:507–533
153. Keasling JD (1999) Gene-expression tools for the metabolic engineering of bacteria. Trends Biotechnol 17:452–460
154. Swinnen S, Schaerlaekens K, Pais T et al (2012) Identification of novel causative genes determining the complex trait of high ethanol tolerance in yeast using pooled-segregant whole-genome sequence analysis. Genome Res 22:975–984
155. Esvelt KM, Wang HH (2013) Genome-scale engineering for systems and synthetic biology. Mol Syst Biol 9:641
156. Verho R, Penttilä M, Richard P (2011) Cloning of two genes (*LAT1,2*) encoding specific L-arabinose transporters of the L-arabinose fermenting yeast *Ambrosiozyma monospora*. Appl Biochem Biotechnol 164:604–611
157. Subtil T, Boles E (2011) Improving L-arabinose utilization of pentose fermenting *Saccharomyces cerevisiae* cells by heterologous expression of L-arabinose transporting sugar transporters. Biotechnol Biofuels 4:38
158. Hector RE, Dien BS, Cotta MA, Qureshi N (2011) Engineering industrial *Saccharomyces cerevisiae* strains for xylose fermentation and comparison for switchgrass conversion. J Ind Microbiol Biotechnol 38:1193–1202
159. Borgne SL (2012) Genetic engineering of industrial strains of *Saccharomyces cerevisiae*. Methods Mol Biol 824:451–465
160. Yang J, Dang H, Lu LR (2013) Improving genetic immobilization of a cellulase on yeast cell surface for bioethanol production using cellulose. J Basic Microbiol 53(4):381–389

161. Piriya PS, Vasan PT, Padma VS, Vidhyadevi U, Archana K, Vennison SJ (2012) Cellulosic ethanol production by recombinant cellulolytic bacteria harbouring *pdc* and *adh II* genes of *Zymomonas mobilis*. Biotechnol Res Int. doi:10.1155/2012/817549
162. Korman TP, Sahachartsiri B, Charbonneau DM, Huang GL, Beauregard M, Bowie JU (2013) Dieselzymes: development of a stableand methanol tolerant lipase for biodiesel production by directed evolution. Biotechnol Biofuels 7:70
163. Trejo MO, Flores N, Escalante A et al (2008) Metabolic regulation analysis of an ethanologenic *Escherichia coli* strain based on RT-PCR and enzymatic activities. Biotechnol Biofuels 1:8
164. Mermelstein LD, Papoutsakis ET, Petersent DJ, Bennett GN (1993) Metabolic engineering of *Clostridium acetobutylicum* ATCC 824 for increased solvent production by enhancement of acetone formation enzyme activities using a synthetic acetone operon. Biotechnol Bioeng 42:1053–1060
165. Higashide W, Li Y, Yang Y, Liao JC (2011) Metabolic engineering of *Clostridium cellulolyticum* for production of isobutanol from cellulose. Appl Environ Microbiol 77:2727–2733
166. Steen EJ, Chan R, Prasad N, Myers S, Petzold CJ, Redding A, Ouellet M, Keasling JD (2008) Metabolic engineering of *Saccharomyces cerevisiae* for the production of n-butanol. Microb Cell Factories 7:36
167. Shen CR, Liao JC (2008) Metabolic engineering of *Escherichia coli* for 1-butanol and 1-propanol production via the keto-acid pathways. Metab Eng 10:312–320
168. Berzin Y, Kiriukhin M, Tyurin M (2012) Elimination of acetate production to improve ethanol yield during continuous synthesis gas fermentation by engineered biocatalyst *Clostridium* sp. MTEtOH 550. Appl Biochem Biotechnol 167:338–347
169. Arnold FH, Moore JC (1997) Optimizing industrial enzymes by directed evolution. Adv Biochem Eng Biotechnol 58:1–14
170. Johannes TW, Zhao H (2006) Directed evolution of enzymes and biosynthetic pathways. Curr Opin Microbiol 9:261–267
171. He MX, Wu B, Qin H et al (2014) *Zymomonas mobilis*: a novel platform for future biorefineries. Biotechnol Biofuels 7:101
172. Laitigue C, Vashee AMA et al (2009) Creating bacterial strains from genomes that have been cloned and engineered in yeast. Science 325:1693–1696
173. Zhang Y, Zhu Y, Zhu Y, Lee Y (2009) The importance of engineering physiological functionality into microbes. Trends Biotechnol 27:664–672
174. Dalby PA, Baganz F, Lye GJ, Ward J (2009) Protein and pathway engineering in biocatalysis. Chim Oggi 27:18–22
175. Green EM (2013) Fermentative production of butanol-the industrial perspective. Curr Opin Biotechnol 22:337–343
176. Varman AM, Xiao Y, Pakrasi HB, Tang YJ (2013) Metabolic engineering of *Synechocystis* sp. strain PCC 6803 for isobutanol production. Appl Environ Microbiol 79:908–914
177. Cann AF, Liao JC (2010) Pentanol isomer synthesis in engineered microorganisms. Appl Microbiol Biotechnol 85:893–899
178. Chuck GC, Tobiasb C, Sun L et al (2011) Overexpression of the maize Corngrass1 microRNA prevents flowering, improves digestibility, and increases starch content of switchgrass. Proc Natl Acad Sci USA 108:17550–17555
179. Shen H, Poovaiah CR, Ziebell A et al (2013) Enhanced characteristics of genetically modified switchgrass (*Panicum virgatum* L.) for high biofuel production. Biotechnol Biofuels 6:71
180. Sticklen M (2006) Plant genetic engineering to improve biomass characteristics for biofuels. Curr Opin Biotechnol 17:315–319
181. Taylor LE, Dai Z, Decker SR et al (2008) Heterologous expression of glycosyl hydrolases in planta: a new departure for biofuels. Trends Biotechnol 26:413–424
182. Saha S, Ramachandran S (2012) Genetic improvement of plants for enhanced bioethanol production. Recent Pat DNA Gene Seq 7:36–44

183. Tao H, Gonzalez R, Martinez A et al (2001) Engineering a homo-ethanol pathway in *Escherichia coli*: increased glycolytic flux and levels of expression of glycolytic genes during xylose fermentation. J Bacteriol 183:2979–2988
184. Pauls RE (2011) A review of chromatographic characterization techniques for biodiesel and biodiesel blends. J Chromatogr Sci 49:384–396
185. Kriz GS, Lampman GM (2007) Introduction to organic laboratory techniques: a microscale approach. Thomson Brooks/Cole, Belmont
186. Adama F, Bertoncini F, Coupard V (2008) Using comprehensive two-dimensional gas chromatography for the analysis of oxygenates in middle distillates I. Determination of the nature of biodiesels blend in diesel fuel. J Chromatogr A 1186:236–244
187. Tiyapongpattana W, Wilairat P, Marriott PJ (2008) Characterization of biodiesel and biodiesel blends using comprehensive two dimensional gas chromatography. J Sep Sci 31:2640–2649
188. Seeley JV, Seeley SK, Libby EK, McCurry JD (2007) Analysis of biodiesel/petroleum diesel blends with comprehensive two dimensional gas chromatography. J Chromatogr Sci 45:650–656
189. Pursch M, Eckerle P, Biel J, Streck R, Cortes H, Sun K, Winniford B (2003) Comprehensive two-dimensional gas chromatography using liquid nitrogen modulation: set-up and applications. J Chromatogr A 1216:43–51
190. Ragonese C, Tranchida PQ, Sciarrone D, Mondello L (2009) Conventional and fast gas chromatography analysis of biodiesel blends using an ionic liquid stationary phase. J Chromatogr A 1216:8992–8997
191. Pettolino FA, Walsch C, Fincher GB, Bacic A (2012) Determining the polysaccharide composition of plant cell wall. Nat Protoc 7:1590–1607
192. Holcapek M, Jandera P, Fischer J, Prokes B (1999) Analytical monitoring of the production of biodiesel by high performance liquid chromatography with various detection methods. J Chromatogr A 858:13–31
193. Eggeman T, Elander RT (2005) Process and economic analysis of pretreatment technologies. Bioresour Technol 96:2019–2025
194. Jönsson LJ, Alriksson B, Nilvebrant NO (2013) Bioconversion of lignocellulose: inhibitors and detoxification. Biotechnol Biofuels 6:16
195. Badger PC (2002) Ethanol from cellulose: a general review. In: Janick J, Whipkey A (eds) Trends in new crops and new uses. ASHS Press, Alexandria, pp 17–21
196. Lezinou V, Christakopoulos P, Kekos D, Macris BJ (1994) Simultaneous saccharification and fermentation of sweet sorghum carbohydrates to ethanol in a fed-batch process. Biotechnol Lett 16:983–988
197. Taherzadeh MJ, Karimi K (2007) Enzyme based hydrolysis processes for ethanol from lignocellulosic materials: a review. BioResource 2:707–738
198. Joshi B, Bhatt MR, Sharma D, Joshi J, Malla R, Sreerama L (2011) Lignocellulosic ethanol production: current practices and recent developments. Biotechnol Mol Biol Rev 6:172–182
199. Olofsson K, Bertilsson M, Liden G (2008) A short review on SSF—an interesting process option for ethanol production from lignocellulosic feedstocks. Biotechnol Biofuels 1:14
200. Ikwebe J, Harvey AP (2011) Intensification of bioethanol production by Simultaneous Saccharification and Fermentation (SSF) in an Oscillatory Baffled Reactor (OBR). In World renewable energy congress 2011. Linköping, Sweden, pp 381–388
201. Palnitkar SS, Lachke AH (1990) Efficient simultaneous saccharification and fermentation of agricultural residues by *Saccharomyces cerevisiae* and *Candida shehatae*. Appl Biochem Biotechnol 26:151–158
202. Pereira JN, Couto MAP, LMM SA (2008) Series on biotechnology: biomass of lignocellulosic composition for fuel ethanol production within the context of biorefinery. Amigadigital Press, Rio de Janeiro, p. 47. isbn:978-85-903967-3-4
203. Aiba S, Shoda M, Nagatani M (2004) Kinetics of product inhibition in alcohol fermentation. Biotechnol Bioeng 10:845–864

204. Balat M (2011) Production of bioethanol from lignocellulosic materials via the biochemical pathway: a review. Energy Convers Manag 52:858–875
205. Ma Y, Dong H, Zou S, Hong J, Zhang M (2012) Comparison of glucose/xylose co-fermentation by recombinant *Zymomonas mobilis* under different genetic and environmental conditions. Biotechnol Lett 34:1297–1304
206. Jin M, Gunawan C, Balan V, Lau MW, Dale BE (2012) Simultaneous saccharification and co-fermentation (SSCF) of AFEXTM pretreated corn stover for ethanol production using commercial enzymes and *Saccharomyces cerevisiae* 424A(LNH-ST). Bioresour Technol 110:587–594
207. Olson DG, McBride JE, Shaw JA, Lynd LR (2012) Recent progress in consolidated bioprocessing. Curr Opin Biotechnol 23:396–405
208. Carere CR, Sparling R, Cicek N, Levin DB (2008) Third generation biofuels via direct cellulose fermentation. Int J Mol Sci 9:1342–1360
209. Park EY, Naruse K, Kato T (2012) One-pot bioethanol production from cellulose by co-culture of *Acremonium cellulolyticus* and *Saccharomyces cerevisiae*. Biotechnol Biofuels 5:64
210. Hasunuma T, Kondo A (2012) Consolidated bioprocessing and simultaneous saccharification and fermentation of lignocellulose to ethanol with thermotolerant yeast strains. Process Biochem 47:1287–1294
211. La Granqe DC, Haan RD, Van Zyl WH (2010) Engineering cellulolytic ability into bioprocessing organisms. Appl Microbiol Biotechnol 87:1195–1208
212. Xu Q, Singh A, Himmel ME (2009) Perspectives and new directions for the production of bioethanol using consolidated bioprocessing of lignocelluloses. Curr Opin Biotechnol 20:364–371
213. Guo L, Zhang J, Hu F, Dy Ryu D, Bao J (2013) Consolidated bioprocessing of highly concentrated *Jerusalem Artichoke* tubers for simultaneous saccharification and ethanol fermentation. Biotechnol Bioeng 110:2606–2615
214. Brethauer S, Studer MH (2014) Consolidated bioprocessing of lignocellulose by a microbial consortium. Energy Environ Sci 7:1446
215. Amorea A, Faraco V (2012) Potential of fungi as category I Consolidated Bio Processing organisms for cellulosic ethanol production. Renew Sustain Energ Rev 16:3286–3301
216. Liu CZ, Wang F, Ou-Yang F (2009) Ethanol fermentation in a magnetically fluidized bed reactor with immobilized *Saccharomyces cerevisiae* in magnetic particles. Bioresour Technol 100:878–882

Chapter 15
Intensive Technological Analysis for Biodiesel Production from a Variety of Feedstocks: State-of-the-Art

R. B. N. Prasad

Abstract The biodiesel industry depends on several unpredictable parameters like raw material availability, and due to this reason this industry has not spread to several countries as expected. Globally transesterification is the main technology used in the commercial production of biodiesel by reacting refined vegetable oils with methanol using homogeneous acid and alkaline catalysts depending on the quantity of the free fatty acid content present in the feedstock. During homogeneous catalyst-based process, large quantities of liquid effluents and salts result, which are responsible for environmental pollution. Due to this reason, several research groups are aiming to develop heterogeneous catalysts in place of homogeneous catalysts for developing green processes without generating effluents. These newer processes certainly help countries like India to handle the low-quality non-edible oils as feed stock. Heterogeneous catalysts provide high activity, high selectivity, and high water-tolerance properties, and these properties depend on the amount and strengths of active acid or basic sites present in these catalysts. In this direction, several heterogeneous metal-based catalysts, biomass-based carbon acid catalysts, enzyme-based catalysts are being employed by several researchers for the production of biodiesel from multi-feedstock including used cooking oils and animal fats.

Keywords Vegetable oils • Animal fat • Biodiesel • Glycerol • Homogeneous catalysts • Heterogeneous catalysts • Esterification • Transesterification

1 Introduction

Biodiesel has been focused as the most attractive alternate or complementing renewable fuel in the last two–three decades in place of petroleum fuels as the present engines need not require any modifications for biodiesel use [1]. Its

R.B.N. Prasad (✉)
Centre for Lipid Research, CSIR-Indian Institute of Chemical Technology, Hyderabad 500007, India
e-mail: rbnprasad@gmail.com

A.K. Chandel, R.K. Sukumaran (eds.), *Sustainable Biofuels Development in India*,
DOI 10.1007/978-3-319-50219-9_15

advantages over petroleum diesel cannot be overemphasized: it is safe, renewable, non-toxic, and biodegradable; it contains no sulphur; and it is a better lubricant. In addition, its use provides numerous societal benefits: rural revitalization, creation of new jobs, and reduced global warming [2]. In addition the kinetic viscosity, combustion efficiency, cetane number and biodegradability of biodiesel prepared from any type of oils and fats are almost similar compared to petrodiesel. It is interesting to note that the concept of using vegetable oil as a fuel dates back to 1895 when Rudolf Diesel developed the first diesel engine to run on vegetable oil, and he has demonstrated his engine at the World Exhibition in Paris in 1900 using peanut oil as fuel [3]. The direct use of vegetable oils and its blends as fuel in diesel engines were found to be unsatisfactory and impractical, primarily due to high viscosity and free fatty acid content of such oils, as well as gum formation due to oxidation and polymerization during storage and combustion. Carbon deposits and lubricating oil thickening are two of the more obvious problems [4]

As of now, majority of biodiesel industries are employing homogeneous catalysts for the commercial production of biodiesel from any type of feedstock. However, the use of homogeneous catalysts poses an environmental concern and also resulting with low-quality glycerol. Hence, the quest for more innovative and efficient processes employing variety of heterogeneous catalysts is being reflected in the form of several research publications on biodiesel production and some advances in this area have also been reported. This chapter reviews various newer alternative options reported for the preparation of biodiesel till date with a view to comparing commercial suitability of these methods on the basis of available feedstocks and associated challenges giving more thrust to Indian context.

2 Feedstocks Used for Biodiesel Production

Biodiesel is being produced globally from a variety of oils and fats of vegetable and animal origin, used oils, microbial oils, acid oil, fatty acid distillates [5]. However, the selection of feedstock generally depends on the availability and its commercial viability. Most of the industries in the United States of America produce biodiesel employing soybean, sunflower, and corn oils. Rapeseed/canola and sunflower oils are the major feedstocks in Europe and Canada. Few countries also employ used frying oils for the production of biodiesel. Palm oil and its fractions like palm stearin are used as the feedstock for the preparation of biodiesel in Malaysia, Indonesia, and some other Southeast Asian countries. Animal fats like tallow fats and lard and fish oil have also been used as feedstock in some of these countries for biodiesel preparation. Apart from all these traditional raw materials, some researchers have also evaluated various other vegetable oils like almond, babassu, camelina, coconut, piqui, poppy seed, sesame, tobacco, rice bran, castor cottonseed, groundnut, linseed, rubber seed, and mahua for their suitability as biodiesel feedstock [6].

The present biodiesel scenario clearly indicates that several edible oils are being used for biodiesel production and thus, its competition with food consumption has been a global concern. In addition, the range of feedstocks with different crop yields per hectare also creates lot of difference in production costs of biodiesel from country to country.

Apart from vegetable oils and animal fats, algal oil has also emerged as one of the most promising alternative feed stock. Algae can grow in municipal and industrial waste water ponds and near power-plant smokestacks by digesting the pollutants and produces triglyceride oil. It can also be grown in open ponds, exerting zero demand on arable land. Oil productivity of microalgae is several folds compared to the best producing oil crops with minimum use of land. Due to this reason, several global research organizations are investing more time and efforts into research for the development of appropriate and commercially viable processes for the production of algal oil.

2.1 Feedstock Options for India

India cannot afford to use the edible oils for the production of biodiesel as it is importing huge amounts of edible oils. In addition, during the last 5 years the imports are continuously increasing (Table 15.1) and during 2014–2015 about 14.2 million metric tons (MMT) of oils have been imported against the domestic production of 7.2 MMT [7].

Due to climatic advantages, India has a large number of oilseed-bearing trees in forests and isolated areas. However, the collection and extraction of oils from these sources are not yet carried out in a systematic manner for proper utilization. Hence there is lot of potential to exploit these cheaper minor oils as a possible feedstock for the production of biodiesel. The production of minor oils like sal, mahua, kusum, neem, and karanja is not being carried out in organized sectors and whatever the small quantities of oil produced in the country is being consumed either for edible or industrial applications. Hence, the Planning Commission of India had set an ambitious target of planting 11.2–13.4 million hectares for jatropha by the end of 11th Five Year Plan (by March 31, 2012). However, the Government

Table 15.1 Production and imports of vegetable oils in India

Year	Domestic edible oil production (in million metric tons)	Import of edible oils (in million metric tons)
2014–2015	7.20	14.42
2013–2014	8.62	11.61
2012–2013	8.08	10.38
2011–2012	8.15	9.98
2010–2011	8.52	8.37

Source: http://www.seaofindia.com/stats

of India's ambitious plan of producing sufficient biodiesel by 2012 to meet its mandate of 20% blending with diesel was unachievable mostly due to unavailability of sufficient feedstock (jatropha seeds) and lack of high-yielding drought-tolerant jatropha cultivars. In these circumstances, some state governments have also projected *Pongamia glabra* (karanja) in addition to jatropha as another major feedstock. Even though some state governments of India have initiated cultivation of both these crops, unfortunately, the reports related to these oil plantations are not really encouraging and karanja may have some potential in the coming few years compared to jatropha.

In these circumstances, Indian biodiesel industry presently employs very low-quality imported palm oil or its fractions like palm stearin, palm oil fatty acid distillates, acid oils, fish oil, used cooking oils, animal fats, available non-edible oils with higher content of free fatty acids (FFA), and small amounts of jatropha and karanja oils. According to GAIN Report of Indian Biofuels Annual [8], currently, India has 5–6 large capacity and few small-scale biodiesel plants with about 480,000 MT of total capacity. However, during 2014, these industries utilized only 28% of the installed capacity to produce about 140,000 MT of biodiesel from multiple feed-stocks. The biodiesel produced in the country is being sold to Indian Railways, small and medium enterprises, state and private transport and automobile organizations for carrying out experimental projects apart from minor sales to unorganized consumers such as cellular communication towers, brick kilns, progressive farmers, and to institutions that run diesel generators as source of power back-up. It is also interesting to note that recently Government of India has deregulated diesel price in line with petrol and following up, the Government has also allowed private biodiesel manufacturers, their authorized dealers and joint ventures of OMCs authorized by the Ministry of Petroleum and Natural Gas to sell biodiesel directly to consumers' subject to their product meeting prescribed BIS standards. These measures may encourage the biodiesel manufactures in the country in the coming years.

2.2 Selection of Appropriate Feedstock for Biodiesel

In the present scenario, it is little complex for any industry to choose the appropriate oil as they have to consider several issues like food versus fuel, economic feasibility, quality issues related to non-edible oils etc. The suitability of any uncommon feedstock oil for biodiesel production has to be critically evaluated based on their physico-chemical properties and fatty acid composition.

The appropriate physico-chemical properties and fatty acid profile are the crucial parameters for the selection of suitable feedstocks for the production of biodiesel. Comprehensive lists of the composition of various edible and non-edible oils and fats, their oil extraction techniques, technologies of biodiesel production have been compiled by several authors [9–11] and these publications will be useful as a good guide for choosing the appropriate oil.

Most of the seed oils usually comprise five major fatty acids namely palmitic (16:0), stearic (18:0), oleic (18:1), linoleic (18:2), and linolenic (18:3) with some exceptions like coconut oil [lauric (12:0) as major fatty acid], mustard [erucic (22:1) as major fatty acid], castor [ricinoleic (18:1) with hydroxy functionality at C12 position as major fatty acid] oils. Animal fat mostly contains palmitic, stearic and oleic acids as major fatty acids. Fish oil and some algal oils contain polyunsaturated rich fatty acids with more than 3 double bonds like eicosapentaenoic acid (20:5) and docosahexaenoic acids (22:6). Biodiesel produced from saturated fatty acid-rich oils like palm oil, palmolein, palm stearin and animal fat may not be suitable for use at cold places as it exists in semisolid or solid form at room temperature as its pour and cloud points are very high. A high percentage of mono-unsaturation in fatty acid composition may result biodiesel with better oxidative stability along with other required specifications. Even though there is no limit for a fatty acid containing two double bonds (linoleic acid), the content of methyl linolenate is restricted to a maximum limit of 12% by EN 14214 because of its tendency to oxidize. Similarly, the fish oil fatty acid esters are even more prone to oxidation than linolenic acid esters. In order to eliminate the use of fish oil or polyunsaturated fatty acid-rich oil as biodiesel feedstock, the FAME content with $\geq$3 double bonds have also been restricted to a maximum limit of 1% by EN 14214 specifications [12]. Hence one has to critically examine the fatty acid composition of any oil while choosing it as a feedstock for biodiesel production.

Kumar and Sharma [13] made a novel attempt to classify the 41 vegetable oils and 32 algal oils based on OSI as per ASTM D-6751 and EN 14214 biodiesel standards so that oils can be recommended for biodiesel which may or may not require fuel stabilization using additives. The proposed classification would help to select the oils, which can be suitably stabilized with/without antioxidants. However, no oil is found to have OSI exhibited more than 6 h as per EN 14214 biodiesel standard and therefore, all the oils/biodiesels would require stabilization using antioxidants.

3 Biodiesel Production Technologies

Several generally accepted technologies are well established for the production of biodiesel. The most common technology adopted for the preparation of biodiesel is transesterification of oils (triglycerides) with alcohol (methanol) which gives biodiesel (fatty acid methyl esters, FAME) as main product and glycerol as by-product. Transesterification works fine if the FFA and moisture contents of oil are in trace amounts. In case, the feed stock contains FFA, acid portion has to be removed from the oil by neutralization or esterification before transesterification. The most important operating variables affecting the transesterification process are reaction temperature, reaction time, reaction pressure, ratio of alcohol to oil, concentration and type of catalyst, mixing intensity and type of feedstock. Transesterification is an equilibrium reaction and needs larger amounts of alcohol to shift the reaction equilibrium forward to produce more methyl esters as the desired product. The transesterification reaction can be catalyzed by homogeneous catalysts like alkali and acid, variety of heterogeneous catalysts like metal oxides, mixed oxides, carbon catalysts, and enzymes. Certain pre-treatment methods like neutralization or esterification, degumming is also required for the smooth production of biodiesel. Several exhaustive reviews are available in the literature [4, 9, 14–23] describing various options of processing technologies for biodiesel production from clean oils, non-edible oils, used cooking oils, animal fats, and other minor feed stocks.

3.1 Pre-treatment Methods

The presence of free fatty acids, phospholipids, wax esters along with unsaponifiable matter in higher quantities in oils like jatropha oil, karanja oil, animal fats, fish oil, and used cooking oils normally create major processing problems during biodiesel preparation. Hence, pre-treatment protocols like degumming and also neutralization or esterification have to be performed before the transesterification reaction for the reduction of both phospholipids and fatty acids. Free fatty acid may not be present in case of clean oils like sunflower, soybean, and rapeseed, and hence a simple degumming process is sufficient before proceeding for transesterification.

The major component of gums is phospholipids, and these have to be removed during the degumming step as these have a tendency to form emulsion and cause problems in the separation of glycerol during the biodiesel preparation. Metal ions are also removed during the degumming step. Phospholipids present in any oils are classified into hydratable and non-hydratable phospholipids [24]. During normal water degumming, hydratable phospholipids are removed. Non-hydratable phospholipids contain calcium and magnesium salts of phosphatidic acid and phosphatidylethanolamine. They can be removed only with the help of citric or phosphoric acid after converting them into free form during acid degumming process as they do not hydrate in presence of normal water. Alcon process, dry degumming, acid degumming, super degumming, uni-degumming, special degumming, total degumming and enzymatic degumming are the commonly used degumming processes. Water and acid degumming processes are the most common processes employed by the industry. However, these methods are fine-tuned or changed as per the quality of the oil and technology used by the particular industry. Different pre-treatment methods employed before biodiesel preparation were reviewed by Chakrabarti and Prasad [9]. The phosphorus content must be <10 ppm in the biodiesel according to BIS and ASTM specifications and hence the degumming technique must be chosen for obtaining biodiesel with the required specification.

If the free fatty acid is $<2\%$ in the oil, it is generally removed through neutralization method by treating the oil with appropriate amounts of aqueous caustic alkali. The generated soap, which is sodium salts of fatty acids will be removed by centrifugation or settling down. In case, the oil is very dark, traditional bleaching operation is also employed for removing the pigments present in the oil. For oils having free fatty acid percentage more than 2%, esterification is generally carried out for converting the free fatty acids to methyl esters followed by transesterification reaction for converting triglycerides to biodiesel. Santoria et al. [25] in his review analyzed the industrial biodiesel production practice including the pre-treatment methods. However, industry has to take an appropriate decision, whether to employ neutralization initially or to convert the free fatty acid present in the oil to fatty acid methyl ester. The acid value of the biodiesel must be <0.5 as per the national and international specifications.

3.2 Homogenous Catalysts for Biodiesel Production

Homogenous catalytic methods involve the use of catalyst in liquid form, mainly acid and alkali catalysts. The use of homogeneous catalysts is the first conventional method applied in the biodiesel production industry. Homogeneous catalysts are less expensive and convert oils to biodiesel very fast.

Acid catalyzed esterification process is useful for the conversion of fatty acids and FFA present in oil to biodiesel. Most commonly employed acid catalysts are sulphuric, hydrochloric, sulfonic and phosphoric acids. Indirectly, this process can be considered as a pre-treatment process for converting FFA present in low-quality oils to biodiesel before alkali-catalyzed transesterification of triglycerides. Acid catalyzed esterification is slow and usually performed at high oil to alcohol molar ratios, low to moderate temperatures and high acid catalyst concentration. The liberated water during the esterification reaction has to be removed for enhancing the conversion of fatty acids to their respective esters. The separation and purification processes are tedious and generate effluents.

The homogeneous alkali-catalyzed transesterification reaction is the most common method being used at laboratory, pilot and industrial scale levels. This process is catalyzed by alkaline metal hydroxides and alkoxides as well as sodium or potassium carbonates. Sodium hydroxide or potassium hydroxide or corresponding alkoxides such as sodium methoxide or ethoxide are the most commonly used homogeneous alkali catalysts for transesterification of triglycerides to biodiesel. However, most popular alkaline catalyst is sodium hydroxide used as it is economically cheaper and easily available. Alkali is the commercially more viable catalyst compared to acid catalysts as the reactions very rapid [26]. At present, alkali-catalyzed transesterification is preferred globally due to lower catalyst costs and a very high reaction performance with higher yields in short reaction times when feedstocks with low FFA are used. Even though 1:3 stoichiometric ratio of triglyceride to methanol is sufficient for transesterification, 6–15 times alcohol molar proportions are generally employed to accelerate the equilibrium for getting yields of biodiesel at 60–65 °C temperature [27]. After the transesterification reaction, the glycerol is separated from biodiesel layer. The separated crude glycerol and biodiesel are purified as per the market specifications.

The process for the production of biodiesel from clean oils containing lower FFA is very easy as it involves only transesterification reaction. In case FFA is $<2\%$ in the crude feed stock, the FFA is removed as soap by alkali neutralization method followed by transesterification of triglycerides to biodiesel. Oil has to be free of moisture and FFA for alkali-catalyzed transesterification reaction as the resulting soap may create lot of problems including emulsion formation during the separation of glycerol. Salt formation is another challenge in case of homogenous acid catalyzed reactions employing catalysts like sulphuric acid while neutralizing the unreacted catalyst. Even though acid-assisted transesterification is a feasible process; the conversions are not very high like in the case of alkali-based reactions. Acid catalysts may not fully transesterify the triglyceride and this may result in low-quality biodiesel. Hence, two-step process is useful if the feedstocks contain more than 2% FFA as the alkaline catalysts may get consumed for the formation of soap without participating in transesterification reaction. Accordingly, the FFA will be converted to methyl ester in the first step employing acid catalyst followed by alkali-catalyst-based transesterification in the second step.

However, due to some issues like removal of the catalyst from the product, difficulties in the recovery and purification of glycerol, and generation waste water and its treatment and undesirable side reactions, industry is looking for alternative greener methods.

3.3 Heterogeneous Catalysts for Biodiesel Production

Biodiesel production technologies making use of homogeneous acid or base catalysts for both esterification and transesterification are quite matured and globally most of the industries are employing these catalysts. However, the major challenge during homogeneous catalyst-based technologies for biodiesel production is the generation of huge quantities of liquid effluents and salts. Biodiesel has to be washed by water after neutralization of the alkali followed by removal of moisture under reduced pressure. The industry using homogenous catalysts always face several environmental problems like corrosion, removal of catalyst residues and salts. The industry has also to take lot of care for the removal of residual catalysts from the by-product glycerol solution during the purification of glycerol. Hence, it is really necessary to introduce heterogeneous catalysts in place of homogenous catalysts to safeguard the environment in long run. During the last decade, several research groups are working in the area of heterogeneous catalysts for developing green processes for the preparation of biodiesel. The major advantage during the usage of heterogeneous catalyst is the recovery of glycerol as it does not contain water. The heterogeneous catalysts will have excellent economic advantage as these can be reused several times without losing their activity.

Heterogeneous acid catalysts will play a crucial role in the esterification of free fatty acids as low-cost oil feedstock contains high concentrations of FFA and if suitable, alkaline catalyst-based transesterification may be adopted after esterification of FFA. Several heterogeneous catalysts such as ion exchange resins, sulfated oxides, transition metal oxide and derivatives, boron group base heterogeneous catalysts, alkaline earth metal oxides and derivatives, mixed metal oxides and derivatives, alkali metal oxides and derivatives, hydrotalcites, zeolites, mesoporus silicas, heteropolyacids, waste material-based heterogeneous catalysts, carbon-based heterogeneous catalysts, biocatalysts like lipases reported for esterification, transesterification as well as simultaneous esterification and transesterification reactions and several exhaustive reviews are available in the literature describing the efforts of researchers on the development of heterogeneous catalysts for the preparation of biodiesel [16, 28–32]. It is clearly evident that heterogeneous catalysts have the ability to esterify and transesterify both fatty acids and triglycerides and hence these catalysts are more appropriate for the low-quality feedstocks which are being used for the biodiesel production in India. As of now, there are no serious reports about the commercial utilization of these heterogeneous catalysts and however, this is an exciting research area for making biodiesel industry eco-friendly. Some of the interesting aspects of heterogeneous catalysts are described here.

3.3.1 Metal-Based Heterogeneous Catalysts

Several metal oxides like barium, strontium, calcium, and magnesium oxides (BaO, SrO, CaO, MgO) were projected to have more potential for the preparation of biodiesel in quantitative yields due to their higher basicity and lower solubility in alcohols [33]. In spite of its better efficiency, Bao has not preferred over CaO due to its toxic nature and higher cost. Hence CaO has been widely used by several researchers for the transesterification of triglycerides. Some researchers also projected SrO as a potential transesterification catalyst for biodiesel preparation. All these oxides were extensively employed by several researchers for methanolysis of soybean, rapeseed, sunflower, palm, canola oils in addition to waste frying oils at 60–65 °C with 80–98% yields of biodiesel. Nanopowder calcium oxide (Nano CaO) was found to be more efficient compared to CaO for the biodiesel production from soybean oil with over 96% yields under microwave conditions [34]

Few researchers reported the use of oxides in combination of other materials for the production of biodiesel. KOH loaded on MgO, CaO supported on silicas, calcium oxide, solid super base prepared from CaO by dipping in ammonium carbonate solution, CaO impregnated with 10% KNO_3, zirconia (ZrO_2) loaded with potassium bitartrate ($C_4H_4O_6HK$), Li, Na and K ion impregnated CaO, ZnO loaded with strontium nitrate followed by calcination were some of the modified catalysts for biodiesel production from soybean, canola, sunflower, castor, jatropha, cottonseed and rape seed oils. Eventhough the yields are very high in some cases, several issues like higher catalyst load, higher molar ratio of methanol, longer reaction times, leaching of catalysts, and problems in recycling of catalysts are some of the concerns for commercialization of these catalysts [35]

Several mixed oxides like Ca and Zn, CaO supported on zinc oxide, MgAl and MgCa oxides, $CaZrO_3$ and CaO–CeO_2, CaO and ZrO_2 mixed oxides, Nanocomposite mixed oxides of strontium-titanium (Sr-Ti), Aluminium oxide modified Mg–Zn, $CaTiO_3$, $CaMnO_3$, $CaZrO_3$, Cao–CeO_2, were employed by several researchers for the preparation of biodiesel from palm kernel oil, palmolein, soybean and sunflower oils, jatropha oil, waste cooking oils using higher equivalents of methanol to oil at varying temperatures to obtain 75–98% of yields [35]

Several base catalysts like potassium, sodium, barium and calcium, monolithic potassium, potassium hydroxide, sodium hydroxide, calcium oxide, magnesium oxide, barium oxide, potassium carbonate, Potassium nitrate (KNO_3), potassium fluoride, Eu_2O_3 were loaded on alumina and transesterified soybean, jatropha, cotton seed, rapeseed, canola, and sunflower oils and reported 63–99% yields by varying oil to methanol ratio (1:12–32) and temperature (50–70 °C). Eu_2O_3/Al_2O_3 yielded lowest yields (63%) and KF/Al_2O_3 and K/Al_2O_3 yielded 99% of biodiesel. However, most of these catalysts leached their active species into methanol and could not be recycled due to this reason [35].

Several types of hydrotalcites like Mg-Al hydrotalcites, cerium modified Mg-Al hydrotalcites, potassium loaded–calcined Mg-Al hydrotalcite, Mg-Al-CO_3 hydrotalcite, hydrotalcite-like compounds containing Mg^{2+}, Fe^{3+} and Al3+ (MgAlFe), Mg-Co-Al-La layered double hydroxide, calcinated Li-Al, Mg-Al and Mg-Fe layered double hydroxides, KF/Ca-Al hydrotalcite, KF/hydrotalcite, poly (vinyl alcohol) membranes loaded with hydrotalcite, lipase (*Saccharomyces cerevisiae*) immobilized on Mg-Al hydrotalcite were also employed for the preparation of biodiesel from soybean, rape seed, canola, cottonseed, and palm oils with 60–97% yields by varying oil to methanol ratio (1: 6–30) and temperature (60–100 °C) according to a review published by Kaki and Prasad [35]. Zeng et al. [36] employed Mg-Al hydrotalcite as a base for immobilization of *Saccharomyces cerevisiae* lipase and employed this successfully as a transesterification catalyst for the preparation of biodiesel from rapeseed oil with 96% yields.

Heteropolyacids (HPAs) have been the subject of recent attention as a result of their excellent water tolerance, superacidity, and porous architecture. In their native form, heteropolyacids are unsuitable as heterogeneous catalysts for biodiesel applications due to their high solubility in polar media [37]. Ion-exchanging larger cations into Keggin type phospho- and silicotungstic acids could able to increase their chemical stability. For example, Cs salts of phosphotungstic acid $Cs_xH_{(3-x)}$ $PW_{12}O_{40}$ and $Cs_yH_{(4-y)}$ $SiW_{12}O_{40}$ are virtually insoluble in water, with proton substitution accompanied by a dramatic increase in surface area of the resulting crystallites [38, 39]. As a consequence of these enhanced structural properties, albeit at the expense of losing acidic protons, both $Cs_xH_{(3-x)}$ $PW_{12}O_{40}$ and $Cs_yH_{(y\_x)}$ $SiW_{12}O_{40}$ are active for palmitic acid esterification to methyl palmitate and tributyrin transesterification. Most of the studies carried out using HPAs as catalysts are dedicated to the esterification of fatty acids or transesterification of short chain triglycerides.

Zeolites are flexible and versatile catalytic materials, whose acidity and/or basicity can be modulated by appropriate doping. As generally recognized, the versatile catalysis ability of zeolites results from their chemical composition, pore size distribution, and ion-exchange abilities. A microporus inorganic lithium containing zeolite has been shown to be a new generation solid base catalyst for transesterification [16]. Most of these catalysts contain the basic sites (cation) generated by thermal decomposition of the supported salt. Among zeolite family Zeolite-X, titanosilicates and mesoporus zeolites have attracted the most research attention for biodiesel preparation [40]. The acidic properties of zeolites are usually improved by protonation, that is, by exchange of the cations contained in the positively charged aluminosilicate cage with protons. It is also possible to induce some hydrophobicity of zeolites by the elimination of water of hydration [21]. Several studies have been reported for the preparation of biodiesel from variety of feedstocks employing different types of zeolites.

3.3.2 Carbon-Based Heterogeneous Catalysts

Heterogeneous catalysts were also prepared from unusual natural materials like waste egg shell, snail shell, shrimp shell, and cocoa pod husks derived Na-X zeolite by varying methods like calcination, partial carbonization and employed for production of biodiesel from soya, sunflower and rapeseed oils [35] with 83–99% yields by varying oil to methanol ratio (1:6–12) and temperature (60–80 °C).

Several carbon-based sulfonated catalysts (CBSC) have been reported in recent years with extraordinary stability and several strong protonic acid sites from renewable sources such as biomass. These catalysts are being developed from number of inexpensive biomasses like sugars, glycerol, lignin, corncobs, palm shell, and oilseed cakes and a detailed review on latest developments in biodiesel production using carbon-based catalysts by Konwar et al. [41] has been published recently. Carbon-based materials are considered as ideal catalysts due to desirable features such as low material cost, high surface area and thermal stability. They are easily prepared by functionalizing carbon surface with acids or bases and in other cases carbon material was reported to be used as support. In addition, due to structural resemblance of these activated carbons with graphite or grapheme sheets, it is possible to attach different acidic or basic functional groups to the carbon material. This paves a way for creating a diverse class of new catalysts from these carbon materials with unique structural features and catalytic properties depending upon the nature of attached groups for use in biodiesel preparation. Tada's group was the first research group to report the carbon catalyst containing polycyclic aromatic carbon rings from sucrose and D-glucose by partial carbonization followed by creating sulphonite (–SO_3H) groups by reacting the incomplete carbonized material with sulphuric acid. This acidic catalyst is very attractive for the conversion of fatty acids into fatty acid methyl esters [42]. The analysis confirmed that this catalyst consists of sheets of amorphous carbon bearing hydroxyl and carboxyl (–OH and –COOH) groups and high densities of –SO_3H groups.

Carbon catalyst prepared from crude glycerol was another innovation by reacting with sulphuric acid by in situ partial carbonization and sulfonation

[43]. The researchers claimed that this is a single pot reaction and any type of crude glycerol obtained either from biodiesel process or glycerol pitch resulted from fat splitting process can be used for the preparation of this catalyst. This novel invention was aimed to develop alternate applications to crude glycerol and also make use of the catalyst for esterification of fatty acid in place of homogenous catalysts like sulphuric acid for the preparation of biodiesel [43]. The researchers further converted this glycerol-based carbon acid catalyst into base catalyst with transesterification activity by reacting with sodium hydroxide in specific reaction conditions [44]. Several vegetable oils like palm, sunflower, soya, rice bran, high oleic sunflower and castor were converted to biodiesel with quantitative yields employing this base catalyst. These glycerol-based carbon acid and base catalysts exhibited very good catalytic activity with excellent reusability along with moisture tolerant property with the capability to replace the traditional homogenous alkali and base catalysts. Rao et al. [45] also reported a similar type of carbon catalyst with acidic sulfonated groups from deoiled canola meal by partial carbonization followed by steam activation and treatment with sulphuric acid and this catalyst also exhibited esterification activity. The initial reports on carbon catalysts are very attractive from different angles and however, it may take some time to commercialize these catalysts.

3.3.3 Biocatalysts

The biocatalyst-mediated processes for biodiesel production have not yet reached a stage to claim as commercially feasible. However, the search for a truly environmentally friendly approach for biodiesel production has intensified the research into the use of enzymes as catalysts and a number of studies have been reported in the literature. The reported literature on biocatalysts reveals that lipases certainly compete alternative to the existing chemical catalysts. Significant data related to lipases and immobilized lipases used and their ratios to feedstocks, molar ratios of substrates, effect of solvent on the reaction, reaction conditions like time and temperature, and variation in acyl donors has been reported in the literature. Lipases are generally more effective over chemical catalyst as they exhibit substrate specificity, functional group specificity, and stereo specificity and hence industrial applications of the lipases in the oleochemical industry are becoming more attractive. The enzymatic preparation of biodiesel from different feed stocks employing lipases as alternative catalysts solves several disadvantages caused by chemical catalysts. There are several advantages for lipases as they are stable, do not require co-enzymes and tolerate organic solvents even though these catalysts cannot be commercially competitive as of now in comparison to existing base catalysts for transesterification. A large number of research communications are available in the literature employing lipases either in free from or in immobilized state for this purpose; however, the reaction times are longer compared to the alkali-catalyzed processes. Immobilized *Thermomyces lanuginosus*, *Candida antarctica*, *C. rugosa*, *Pseudomonas fluorescens*, *Rhizopus oryzae*, *P. cepacia* were some of the lipases extensively used at laboratory scale for the preparation of biodiesel from different oils, fats, and fatty acids [46].

The main advantage of lipases is that it can simultaneously esterify FFA and transesterify triacylglycerols and can be reused several times and hence these can be employed for the production of biodiesel from high FFA oils, acid oils, used oils, and low-quality oils extracted from spent bleaching earth. However, it is mandatory to remove the phospholipids, pigments, and other impurities from the oils and fats employing necessary pre-treatment methods like degumming and bleaching for safeguarding the activity of lipases and enhancing the biodiesel yields. In spite of all these problems, lipase-based reactions yield high-quality glycerol with minimum impurities like water and salts. A detailed overview on the use of biocatalysts for the production of biodiesel and the methodologies for enhancing the yields are reported in the literature [47].

Several studies revealed that glycerol reduces the activity of lipase by forming a coating over the lipase and blocks the active sites. Hence, glycerol has to be continuously removed from the reaction system. The lipase-catalyzed alcoholysis of oil can be achieved with and without the solvent medium. There are several initial studies reported in the literature for the transesterification of a variety of oils like sunflower, soybean, rapeseed, olive, jatropha, castor with methanol employing lipases like *Pseudomonas fluorescens, Mucor miehei, C. antarctica, Chromobacterium viscosum, C. rugosa,* and porcine pancreas for the preparation of biodiesel. The reactions were carried out directly or in the presence of solvents like hexane, and the reactions carried out in solvent medium resulted with better yields. In some cases, the biodiesel yields were improved using higher dosage of the enzyme, i.e., up to 30% based on the oil. The enzymes are reported to be unstable in the short chain alcohols like methanol in general, and the lower yields of the methanolysis could be due to the irreversible inactivation of lipase caused by the contact between the lipase and the insoluble methanol that existed as drop in the oil. These findings motivated the researchers to switch over to step-wise incremental addition of methanol to safeguard the lipase. Several findings in this direction resulted with higher yields more than 98% employing soybean, rapeseed, sunflower oils and enhancing the stability of the lipase for recycling several times immobilized *P. fluorescens and Rhizomucor miehei, C. antarctica, C. rugosa, P. cepacia* [48–50].

Several attempts were made by researchers for stabilizing the lipase by treating with glutaraldehyde or immobilizing employing various types of materials like macro porous acrylic resin, phyllosilicate, sol-gel polymer matrix and biomass support particles etc., Low-quality acid oil of the corn, rape and sunflower oils and the residual oil present in the spent bleaching earth obtained from the soya, rapeseed and palm oil contain large amounts of free fatty acids and these oils were converted to biodiesel employing *C. antarctica* and *R. oryzae* lipase-mediated esterification and transesterification and the resultant biodiesel could not be obtained in high quality. However, the lipases could able to recycle several times without loosing its activity [45, 51, 52].

Due to the high cost of lipases, it is always advantageous to commercialize lipase-based reactions employing fixed bed reactor employing immobilized lipases.

There were some isolated attempts in this direction for the preparation of biodiesel employing immobilized *C. antarctica* and *Burkholderia cepacia* lipases for the preparation of biodiesel from soybean, rapeseed and yellow grease in 90–100% yields and it was found that the activity of lipase continued for five cycles. As indicated earlier, the short-chain alcohols such as the methanol and ethanol are commonly used as acyl acceptors for the biodiesel production and however, the use of the excess alcohols and also by-product glycerol would lead to inactivation of the lipase. Novel acyl acceptors such as methyl acetate, ethyl acetate and propan-2-ol were reported recently for the successful interesterification of various oils into the biodiesel. Soybean oil was converted to biodiesel in presence of methyl acetate employing Novozyme *435* and the by-product triacetin obtained during the interesterification did not deactivate the enzyme and the enzyme could be reused continuously for 100 batches with about 92% yields [53, 54]. Ethyl acetate was also used as the acyl acceptor for the production of the biodiesel from the crude oils of jatropha, karanja, and sunflower using Novozyme 435 and the yields of ethyl esters were 91.3, 90.0, and 92.7%, respectively [55]. In a similar way, propan-2-ol was also used as an acyl acceptor for the production of biodiesel from crude jatropha, karanja, and sunflower oils with good yields employing Novozyme 435 [56].

In spite of significant number of research communications, the higher price of lipase compared to inorganic catalysts may be the real challenge to make the biodiesel production commercially viable employing biocatalysts. However, due to several advantages over the traditional catalysts, there is lot of scope for developing commercially viable biocatalysts in the area of biodiesel in the coming years.

3.4 Supercritical Fluid Methods for Biodiesel Production

The application of supercritical technology to biodiesel production is still in its infancy. Efforts have to be made through research and pilot plant experimentation for establishing this technology, and initial studies published to date do show that it is likely to be a worthy competitor to the current production techniques [57]. Under supercritical conditions, the mixture of oil and methanol becomes homogeneous where both the esterification of free fatty acids and the transesterification of triglycerides occur without the need for a catalyst. This method is suitable for any type of raw material including animal fats and high free fatty acid containing oils and also tolerant to moisture content in the feedstock [58]. The typical range of operating conditions tried over for the preparation of biodiesel have been temperatures of 280–400 °C and pressures of 10–30 MPa. However, the main challenges for this method include the high temperature and pressures in addition to the high methanol-to-triglycerides molar ratios of over 42:1 [59]. The major advantage of supercritical technology is that both the oil extraction and the preparation of biodiesel can be carried out in a single stage in the same vessel [17].

3.5 Ultrasonic Methods for Biodiesel Production

Ultrasonication provides the mechanical energy for mixing and the required activation energy for initiating the transesterification reaction. In transesterification, it causes cavitation of bubbles near the phase boundary between the alcohol and oil phases leading to intensive mixing of the system. The cavitation leads to a localized increase in temperature, and due to the formation of micro jets, neither agitation nor heating are required to produce biodiesel by ultrasound application [60]. Ultrasonication increases the chemical reaction speed and yield of the transesterification of vegetable oils/animal fats. A number of oils like soybean, palm, jatropha, sunflower, canola, tung, rapeseed, fish and waster cooking oils, beef tallow were subjected to ultrasonic-assisted transesterification reactions under different process conditions of varying temperatures (25–65C), oil to alcohol ratio (1:3–12), time (5–180 min), and Ultrasonic frequency (20–40 kHz) in presence to homogeneous and heterogeneous catalysts and obtained yields upto 99% [10, 61, 62]. However, due to problems in downstream processing, a lot of efforts are required for making this process commercially viable [17].

3.6 Process Intensification Technologies in Biodiesel Production

The majority of commercial plants are producing biodiesel employing transesterification or esterification followed by transesterification reactions in stirred tank reactors in the presence of acid/base catalyst. The industry is facing several practical problems due to longer reaction times, high molar ratio of alcohol to oil and catalyst concentration, immiscible nature of oils and alcohol, reversible transesterification reaction, high operation cost etc. Hence, some researchers developed process intensification technologies and applied to improve mixing and mass/heat transfer between the two liquid phases of oil and alcohol to improve production efficiency and reduce operating cost of the process. Qiu et al. [15] summarized the process intensification technologies which enhance physical processes including heat, mass, and momentum transfer in the context of biodiesel system. The review article provided a rudimentary assessment of reaction time, energy efficiency, operating/capital cost, the difficulty of temperature control, and current status in a very detailed manner on several novel reactors like static mixers, micro-channel reactors, oscillatory flow reactors, cavitational reactors, rotating/spinning tube reactors, microwave reactors. Data was also analyzed related to the reaction/separation coupled technologies employing membrane reactors, reactive distillation and centrifugal contactors and the studies revealed that these reactors require less downstream processing steps. Some of these process intensification technologies offer the flexibility to process a variety of feedstocks. The data revealed that most of these reactors or process technologies have potential to increase the rate of reaction by intensifying transport process and mixing between alcohol and oil. The common characteristic of process intensification technologies is the small "footprint" required compared to conventional equipment as small size of reactors and less processing processes reduce the cost of construction and maintenance. Some of these technologies have also been scaled-up and commercialized successfully.

4 Conclusions

At present about 30 billion liters per year of biodiesel is being produced globally. Some of the countries namely USA, Brazil, Germany and other European countries, Indonesia, Argentina, Canada, and Malaysia are producing biodiesel from edible oils like soybean, sunflower, corn, rapeseed/canola, palm oils along with animal fat, and used cooking oils. India cannot afford to use any edible oil for the production of biodiesel, as the country is importing huge quantities of edible oil for human consumption. Hence, Indian biodiesel industry is presently employing very low-quality imported palm oil or its fractions like palm stearin, palm oil fatty acid distillates, acids oils, fish oil, used cooking oils, animal fats, available non-edible oils with higher content of free fatty acids and small amounts of jatropha and karanja oils accounting to not more than 140,000 MT of biodiesel per year. Several organizations like ASTM, European Union, and BIS have formulated specifications for B100 for ensuring that no operational problems will be encountered for the engines when the biodiesel is blended with petrodiesel. The biodiesel business worldwide is a complex phenomenon and depends on many issues that cannot be controlled so easily. Globally most of the countries are using clean edible oils for the production of biodiesel employing homogeneous catalysts like sodium hydroxide, potassium hydroxide or their corresponding alkoxides for transesterification reaction. Esterification step is also necessary in case of FFA containing oils as the fatty acids have to be initially converted into fatty acid methyl esters employing acid catalyst like sulphuric acid followed by transesterification using homogeneous alkaline catalyst. During homogeneous catalyst-based processes, large quantities of wash water and salts are formed which are responsible for environmental issues. Due to these reasons, efforts are being made to substitute homogenous catalysts with heterogeneous catalysts to counter the pollution problems. The heterogeneous process will be economically superior on account of lower raw material costs (catalyst and fresh methanol), better by-product values, and lower variable costs. Several research groups are working in this direction to develop commercially feasible heterogeneous catalysts for reducing the effluent loads from biodiesel industry which are using different types of low-quality vegetable oils. Even though significant number of research studies have been reported employing metal-based and other type of heterogeneous catalysts, several efforts are being made in recent years to utilize low-cost biomass like carbohydrates, glycerol, and oilseed cakes for the preparation of carbon-based catalysts. Biocatalysts have the capability to convert both FFA and triglycerides to biodiesel with good yields and, however, the expensive lipases and other issues related to inactivation of lipase during the reaction in presence of methanol and glycerol are still the major bottleneck to commercialize the enzyme-catalyzed technology for the biodiesel production. Some researchers developed process intensification technologies and applied to improve mixing and mass/heat transfer between the two liquid phases of oil and alcohol to improve production efficiency and reduce operating cost of the process. Hence, there are lot of challenges and opportunities for the development of newer approaches for the preparation of biodiesel for the countries like India, where low-quality oils are being used as feedstocks.

References

1. Jayed MH, Masjuki HH, Saidur R et al (2009) Environmental aspects and challenges of oilseed produced biodiesel in Southeast Asia. Renew Sustain Energy Rev 13:2452–2462
2. Kiss AA, Dimian AC, Rothenberg G (2008) Biodiesel by catalytic reactive distillation powered by metal oxides. Energy Fuel 22(1):598–604
3. Krawczyk T (1996) Biodiesel-alternative fuel makes inroads but hurdles remain. INFORM 7:801–829
4. Fukuda H, Kondo A, Noda H (2001) Biodiesel fuel production by transesterification of oils. J Biosci Bioeng 92(5):405–416
5. Knothe G (2005) The history of vegetable oil based diesel fuels. In: Knothe G, Gerpen JV, Krahl J (eds) The biodiesel handbook. AOCS Press, Champaign, IL, pp 12–25
6. Pinto AC, Lilian LLN, Guarieiro LLN et al (2005) Biodiesel: an overview. J Braz Chem Soc 16:1313–1330
7. The Solvent Extractors' Association of India (2016) http://www.seaofindia.com/stats
8. Indian Biofuels Annual (2015) GAIN Report No. IN5079 dated 7th January, 2015, Global Agricultural Information Network, USDA Foreign Agricultural Service
9. Chakrabarti PP, Prasad RBN (2012) Biodiesel production from *Jatropha curcas* oil. In: Carels N, Sujatha M, ahadur B B (eds) Jatropha: challenges for a new energy crop, Farming, economics and biofuel, vol 1. Springer, New York, NY, pp 463–490
10. Borugadda VB, Goud VV (2012) Biodiesel production from renewable feedstocks: status and opportunities. Renew Sustain Energy Rev 16:4763–4784
11. Atabani AE, Silitonga AS, Ong HC et al (2013) Non-edible vegetable oils: a critical evaluation of oil extraction, fatty acid compositions, biodiesel production, characteristics, engine performance and emissions production. Renew Sustain Energy Rev 18:211–245
12. Knothe GH (2006) Analyzing biodiesel: standards and other methods. J Am Oil Chem Soc 83:823–833
13. Kumar M, Sharma MP (2015) Assessment of potential of oils for biodiesel production. Renew Sustain Energy Rev 44:814–823
14. Meher LC, Vidyasagar D, Naik SN (2006) Technical aspects of biodiesel production by transesterification: a review. Renew Sustain Energy Rev 10:248–268
15. Qiu Z, Zhao L, Weatherley L (2010) Process intensification technologies in continuous biodiesel production. Chem Eng Process 49:323–330
16. Saifuddin N, Samiuddin KP (2015) A review on processing technology for biodiesel production. Trends Appl Sci Res 10:1–37
17. Aransiola EF, Ojumu TV, Oyekola OO et al (2014) A review of current technology for biodiesel production: state of the art. Biomass Bioeng 61:276–297
18. Vasudevan PT, Fu B (2010) Eenvironmentally sustainable biofuels: advances in biodiesel research. Waste Biomass Valoriz 1:47–63

19. Alami CS, Saravanan CG (2015) Biodiesel production techniques: a review. Int J Res Appl Sci Eng Technol 3(6):41–45
20. Abbaszaadeh A, Ghobadian B, Omidkhah MR et al (2012) Current biodiesel production technologies: a comparative review. Energy Conver Manage 63:138–148
21. Sivasamy A, Cheah KY, Fornasiero P et al (2009) Catalytic applications in the production of biodiesel from vegetable oils. ChemSusChem 2:278–300
22. Yunus Khan TM, Atabani AE, Badrunddin IA et al (2014) Recent scenario and technologies to utilize non-edible oils for biodiesel production. Renew Sustain Energy Rev 37:840–851
23. Koh MY, Ghazi TIM (2011) A review of biodiesel production from *Jatropha curcas* L. oil. Renew Sustain Energy Rev 15:2240–2251
24. Hvolby A (1971) Removal of non-hydratable phospholipids from soybean oil. J Am Oil Chem Soc 48:503–509
25. Santoria G, Nicolab GD, Moglieb M et al (2012) A review analyzing the industrial biodiesel production practice starting from vegetable oil refining. Appl Energy 92:109–132
26. Freedman B, Pryde EH, Mounts TL (1984) Variables affecting the yield of fatty esters from transesterified vegetable oils. J Am Oil Chem Soc 61:1638–1643
27. Fangrui M, Hanna MA (1999) Biodiesel production: a review. Biores Technol 70:1–15
28. Chopade SG, Kulkarni KS, Kulkarni A et al (2012) Solid heterogeneous catalysts for production of biodiesel from transesterification of triglycerides with methanol: a review. Acta Chim Pharm Indica 2(1):8–14
29. Kawashima A, Matsubara K, Honda K (2008) Development of heterogeneous base catalysts for biodiesel production. Biores Technol 99:3439–3443
30. Lee HV, Juan JC, Abdulla NFB (2014a) Heterogeneous base catalysts for edible palm and non-edible jatropha-based biodiesel production. Chem Central J 8:30
31. Lee AF, Bennett JA, Manayil JC et al (2014b) Heterogeneous catalysis for sustainable biodiesel production via esterification and transesterification. Chem Soc Rev 43:7887–7916
32. Chouhan APS, Sarma AK (2011) Modern heterogeneous catalysts for biodiesel production: a comprehensive review. Renew Sustain Energy Rev 15:4378–4399
33. Kawashima A, Matsubara K, Honda K (2009) Acceleration of catalytic activity of calcium oxide for biodiesel production. Biores Technol 100:696–700
34. Hsiao MC, Lin CC, Chang YH (2011) Microwave irradiation-assisted transesterification of soybean oil to biodiesel catalyzed by nanopowder calcium oxide. Fuel 90:1963–1967
35. Kaki SS, Prasad RBN (2015) Base catalyzed transesterification for biodiesel: novel catalyst options. In: Raghavan KV, Reddy BM (eds) Industrial catalysis and separations: innovations for process intensification. Apple Academic Press, Oakville, pp 437–470
36. Zeng HY, Liao KB, Deng X et al (2009) Characterization of the lipase immobilized on Mg–Al hydrotalcite for biodiesel. Process Biochem 44:791–798
37. Kozhevnikov IV (1998) Catalysis by heteropoly acids and multicomponent polyoxometalates in liquid-phase reactions. Chem Rev 98:171–198
38. Narasimharao K, Brown DR, Lee AF et al (2007) Structure–activity relations in Cs-doped heteropolyacid catalysts for biodiesel production. J Catal 248:226–234
39. Pesaresi L, Brown DR, Lee AF et al (2009) Cs-doped H4SiW12O40 catalysts for biodiesel applications. Appl Catal A 360:50–58
40. Puna JF, Gomes JF, Correia JMN et al (2010) Advances on the development of novel heterogeneous catalysts for transesterification of triglycerides in biodiesel. Fuel 89:3602–3606
41. Konwar LJ, Boro J, Deka K (2014) Review on latest developments in biodiesel production using carbon-based catalysts. Renew Sustain Energy Rev 29:546–564
42. Toda M, Takagaki A, Hara M et al (2005) Biodiesel made with sugar catalyst. Nature 435:178–178
43. Devi BLAP, Gangadhar KN, Prasad RBN et al (2009) A glycerol-based carbon catalyst for the preparation of biodiesel. ChemSusChem 2:617–620

44. Devi BLAP, Vijayalakshmi K, Prasad RBN et al (2011) Novel glycerol-based heterogeneous carbon base catalyst useful for the transesterification of vegetable oils to biodiesel, a process and use thereof. Indian patent. Patent Appl No. 2291/DEL/2011 A
45. Rao BVSK, Chandramouli K, Prasad RBN et al (2011) Carbon-based solid acid catalyst from de-oiled canola meal for biodiesel production. Catal Commun 14:20–26
46. Noureddini H, Gao X, Philkana RS (2006) Immobilized *Pseudomonas cepacia* lipase for biodiesel fuel production from soybean oil. Biores Technol 96:769–777
47. Zhao X, Qi F, Yuan C et al (2015) Lipase-catalyzed process for biodiesel production: enzyme immobilization, process simulation and optimization. Renew Sustain Energy Rev 44:182–197
48. Watanabe Y, Shimada Y, Sugihara A et al (2000) Continuous production of biodiesel fuel from vegetable oil using immobilized *Candida antarctica* lipase. J Am Oil Chem Soc 77:355–360
49. Soumanou MM, Bornscheuer UT (2003) Improvement in lipase-catalyzed synthesis of fatty acid methyl esters from sunflower oil. Enzyme Microb Technol 33:97–103
50. Kaieda M, Samukawa T, Kondo A, Fukuda H (2001) Effect of methanol and water contents on production of biodiesel fuel from plant oil catalyzed by various lipases in a solvent-free system. J Biosci Bioeng 91:12–15
51. Tütter M, Aksoy HA, Gilbaz EE et al (2004) Synthesis of fatty acid esters from acid oils using lipase B from *Candida antarctica*. Eur J Lipid Sci Technol 106:513–517
52. Pizarro AVL, Park EY (2003) Lipase-catalyzed production of biodiesel fuel from vegetable oils contained in waste activated bleaching earth. Process Biochem 38:1077–1082
53. Du W, Xu Y, Liu D et al (2004) Comparative study on lipase-catalyzed transformation of soybean oil for biodiesel production with different acyl acceptors. J Mol Catal B Enzym 30:125–129
54. Xu Y, Du W, Liu D et al (2003) A novel enzymatic route for biodiesel production from renewable oils in a solvent-free medium. Biotechnol Lett 25:1239–1241
55. Modi MK, Reddy JRC, Prasad RBN et al (2007) Lipase-mediated conversion of vegetable oils into biodiesel using ethyl acetate as acyl acceptor. Biores Technol 98:1260–1264
56. Modi MK, Reddy JRC, Prasad RBN et al (2006) Lipase-mediated transformation of vegetable oils into biodiesel using propan-2-ol as acyl acceptor. Biotechnol Lett 28:637–640
57. Coniglio L, Coutinho JAP, Claver J-Y et al (2014) Biodiesel via supercritical ethanolysis within a gloval analysis "feedstocks-conversion-engine" for a sustainable fuel alternative. Prog Energy Combust Sci 43:1–35
58. Al-Zuhair S, Hussein A, Al-Marzouqi AH et al (2012) Continuous production of biodiesel from fat extracted from lamb meat in supercritical CO_2 media. Biochem Eng J 60:106–110
59. Marulanda VF, Anitescu G, Tavlarides LL (2010) Investigations on supercritical transesterification of chicken fat for biodiesel production from low-cost lipid feedstocks. J Supercrit Fluids 54:53–60
60. Santos FFP, Rodrigues S, Fernandes FAN (2009) Optimization of the production of biodiesel from soybean oil by ultrasound assisted methanolysis. Fuel Process Technol 90(2):312–316
61. Thanh LT, Okitsu K, Sadanaga Y et al (2010) A two-step continuous ultrasound assisted production of biodiesel fuel from waste cooking oils: a practical and economical approach to produce high quality biodiesel fuel. Biores Technol 101:5394–5401
62. Veljkovi VB, Avramovi JM, Stamenkovi OS (2012) Biodiesel production by ultrasound-assisted transesterification: state of the art and the perspectives. Renew Sustain Energy Rev 16:1193–1209

Chapter 16
Biotechnological Platform for Biohydrogen Production: Present Status and Future Challenges

Shantonu Roy, G. Balachandar, and Debabrata Das

Abstract Hydrogen is a clean, renewable, and efficient source of energy. Being considered as a potential alternative energy source, its economical production has gained attention in recent times. In the present study, a comprehensive insight into dark fermentative biohydrogen production process has been exemplified. This process is less energy intensive and environmentally benign and waste materials can be used as substrate. A plethora of H_2-producing microorganisms have been identified in literature. Delineation of detailed microbial characteristics of these microbes shows the potentiality of thermophiles for high rate of H_2 production. Process parameters such as pH, temperature, substrate concentration, HRT, and alkalinity play a vital role directly or indirectly on the metabolic activity of these organisms. Mathematical modeling and simulation of hydrogen production process help in better understanding of the process. Various logistic equations and unstructured models have been proposed for better understanding of H_2 production process. Moreover, it might help in scaling up of the process. To make bioH_2 production a renewable process, an improvisation in the use of feedstock is decidedly required. Organic wastes could be a potential feedstock for sustainable H_2 production. For commercial production, the overall yield of H_2 production process is still inferior. Major bottleneck for the commercialization of these processes is lower net utilizable energy product (NUEP) as hydrogen. Establishment of integrated two-stage processes might help in overall energy recovery. Widely studied second-stage processes were photofermentation, biomethanation, microbial fuel cell, microbial electrolysis cell, etc. Integration of bioH_2 with biomethanation process can maximize the NUEP from the organic wastes. Future aspect of bioH_2 relies on performance of scaled up reactors and their integration with fuel cells. This might help in decentralized energy generation in terms of electricity.

Keywords Biohydrogen • Dark fermentation • Organic wastes • Two-stage process • Scale up

S. Roy • G. Balachandar • D. Das (✉)
Department of Biotechnology, Indian Institute of Technology, Kharagpur 721 302, West Bengal, India
e-mail: ddas.iitkgp@gmail.com

A.K. Chandel, R.K. Sukumaran (eds.), *Sustainable Biofuels Development in India*,
DOI 10.1007/978-3-319-50219-9_16

1 Introduction

Most of the countries rely on fossil fuels to fulfill the growing energy demands. However, the supply of these fossil fuel reserves is inherently finite. The present rate of consumption of hydrocarbon-based fossil fuels such as petroleum and coal would lead to their exhaustion of reserve [1]. This has given impetus towards search for renewable sources of energy. A conscientious global effort would be required for the development of green technologies to fulfill the growing energy demands [2].

Hydrogen has gained importance as a clean, carbon-neutral, and renewable energy source. It has the highest energy density (143 GJ t^{-1}) and on combustion it produces only water as by-product. In the present global scenario, 368 trillion cubic meters of hydrogen is produced for various commercial purposes [3]. The major processes involved in hydrogen production are steam methane reforming, oil/naphtha reforming of refinery/chemical industrial off-gases, coal gasification, and water electrolysis [4]. It was quite evident that present hydrogen production processes are dependent on utilization of non-renewable energy sources. Moreover, the conventional process of hydrogen production consumes lot of energy and also has high carbon footprint. For hydrogen to be considered as future renewable energy source, it should be produced from renewable feedstock. Biohydrogen may be produced from the organic wastes at ambient temperature and atmospheric pressure [5, 6]. So, a sustainable process could be developed from organic wastes and subsequently helping in waste stabilization. Biological hydrogen is mainly produced by the following processes: photolysis of water (direct and indirect biopholysis) by blue green algae and microalgae and oxidation of organic acids by photofermentation and dark fermentation (using mesophilic or thermophilic bacteria). However, the above-mentioned processes have their own advantages and disadvantages. The processes such as biophotolysis of water and photofermentation are marred with very low rate of hydrogen production and require light as additional energy source. It also faces difficulty on scaling up of the process. Among the various other processes, dark fermentation appears to be more promising. It is independent of light energy, requires moderate process conditions, and is less energy consuming. Dark fermentation has relatively higher rate and yield as compared to the other biological hydrogen production processes [2, 7]. Biohydrogen Group of IIT Kharagpur has been involved in the biohydrogen research work for more than a decade. This paper deals with the review of the present status of the biohydrogen research.

2 Biochemistry of Dark Fermentative H_2 Production

In dark fermentative H_2 production, glucose is considered as principal substrate. The complex polymeric organic substrates are hydrolyzed to simple sugars like glucose. Glucose is the simplest sugar which is preferred by most of the microbes. It is further metabolized via glycolytic pathway to produce pyruvate. Subsequently under anaerobic conditions, the fate of pyruvate is to get converted into acetic acid and butyric acid. On doing so, H_2 is produced as by-product.

The obligate anaerobes such as *Clostridia* and thermophilic bacteria [8] follow an unique mechanism for H_2 production. The pyruvate-ferredoxin oxidoreductase (PFOR) enzyme oxidizes pyruvate to acetyl coenzyme A (acetyl-CoA). This pyruvate oxidation step requires ferredoxin (Fd) reduction which in its reduced form is oxidized by [FeFe] hydrogenase and catalyzes the formation of H_2. The overall reaction is shown in Eqs. (16.1) and (16.2):

$$\text{Pyruvate} + \text{CoA} + 2\text{Fd (ox)} \rightarrow \text{Acetyl-CoA} + 2\text{Fd(red)} + CO_2 \quad (16.1)$$

$$2H^+ + \text{Fd(red)} \rightarrow H_2 + \text{Fd(ox)} \quad (16.2)$$

Stoichiometry shows that 4 mol H_2 mol^{-1} of glucose can be produced if acetate is the sole end product of pyruvate oxidation. However, if butyrate is the sole end product then only two moles of H_2 could be produced per mole of glucose. The overall biochemical reaction with acetic acid and butyric acid as the metabolic end products is shown in Eqs. (16.3) and (16.4), respectively:

$$C_6H_{12}O_6 + 2H_2O \rightarrow 2CH_3COOH + 2CO_2 + 4H_2 \quad (16.3)$$

$$C_6H_{12}O_6 \rightarrow CH_3CH_2CH_2COOH + 2CO_2 + 2H_2 \quad (16.4)$$

The facultative anaerobic bacteria (such as *E. coli*, *Enterobacter* sp.) follow a different pathway for H_2 production. It involves formation of acetyl-CoA and formate from oxidation of pyruvate. This reaction is catalyzed by pyruvate formate lyase (PFL) [9] (Eq. 16.5):

$$\text{Pyruvate} + \text{CoA} \rightarrow \text{acetyl-CoA} + \text{formate} \quad (16.5)$$

The formate is then further cleaved to produce carbon dioxide and hydrogen. This reaction is catalyzed by formate hydrogen lyase (FHL) enzyme (Eq. 16.6) [10]:

$$HCOOH \rightarrow CO_2 + H_2 \quad (16.6)$$

2.1 Genetic Modification of Metabolic Pathway for Improvement of H_2 Production

Metabolic engineering and molecular biology advancement have opened a new direction towards biofuel research. Techniques such as gene silencing, gene knock-out, homologous overexpression, and synthetic biology have been used to modify metabolic pathway of microorganisms for improvement of H_2 production. Metabolic bottlenecks in biohydrogen production pathways can be overcome by implementation of above techniques to divert or redirect the flow of electrons towards hydrogen-evolving pathways. Moreover, improvement in substrate utilization efficiency, and using protein engineering to develop efficient and/or oxygen-tolerant hydrogenase, could also prove promising for the development of a potent strain.

Obligate anaerobes encode hydA gene that encodes the [FeFe] hydrogenase. This enzyme helps in transfer of electron from ferredoxin to proton. Homologous or heterologous overexpression of hydA gene was one of the preliminary researches towards improvement of H_2 production. Homologous overexpression of hydA gene of *Clostridium paraputrificum* M-21 showed 1.7-fold improvements in hydrogen production as compared to wild strain. In this study, the lactic acid pathway was suppressed and acetic acid production was increased [11]. During dark fermentative hydrogen production, one of the major bottlenecks encountered is accumulation of end metabolites that inhibit hydrogen production. Side metabolic pathways operating during fermentative hydrogen production also compete for NADH reducing the pool of NADH towards the hydrogen production via NFOR pathway resulting in decreased hydrogen production. Another approach is to block the existing side pathways that divert the pool of NADH away from hydrogenase enzymes.

Site-directed mutagenesis approach was used to develop mutants of *Clostridium* sp. Homologous recombination plays a crucial role in such mutagenesis. The ClosTron system is based on bacterial group II intron which helps in development of stable mutant in *Clostridium* species. The mobile group II intron from the ltrB gene of *Lactococcus lactis* (Ll. ltrB) disseminates towards specific target site via RNA-mediated "retrohoming" mechanism. During retrohoming, the target site is recognized on the basis of complementary base pairing between the excised intron lariat RNA and target-site DNA [12]. Intron specificity towards target DNA can be manipulated by altering the DNA sequence encoding the appropriate part of the intron. Such ClosTron-mediated mutagenesis was used to develop mutants of *Clostridium perfringens* strain W11 which had defective lactate dehydrogenase activity. The *ldh* deleted mutant strain increased 51% H_2 production [13]. Knocking out the genes whose products are responsible for competition with hydrogen production pathway has also gained importance in recent years. The knockout of *pfl* gene which encodes pyruvate formate lyase diverts the pyruvate pool towards formate formation. Similarly, knocking out adhE gene would lead to suppression of ethanol production from acetyl-CoA. Gene deletion technique by a double deletion in hydrogenases 1 and 2 or by disruption of factors involved in maturation of uptake hydrogenases has given a new approach towards improvement in H_2 production.

The H_2 production was improved in *E. coli* MC4100 by implementing deletion of gene responsible for the twin-arginine translocation (tat) system. The prevention of correct assembly of uptake hydrogenases and formate dehydrogenases (Fdh-O and Fdh-N) could be done by deletion of Tat system [14].

The strategy of random mutagenesis by using chemical mutagens was also exploited for the H_2 production. Development of random mutants having defective alcohol production pathway produced less amount of volatile fatty acid using proton suicide methods. The production of various metabolites like ethanol, butanediol, lactic acid, and butyric acid requires NADH pool. In these mutants, the metabolic pathways competing for NADH pool are suppressed by chance mutation. By following random chemical mutagenesis, improvement in H_2 production was observed in *Enterobacter aerogenes* [15]. Overall, 42% improvement in H_2 yield was observed in batch fermentation. Our group also exploited the random mutagenesis approach for improvement in H_2 production in *K. pneumoniae* IIT-BT 08 (previously known as *Enterobacter cloacae* IIT-BT 08). The metabolic redirection was done by blocking alcohol formation pathways and some organic acids, considering the hypothesis that increases in NADH concentration would further lead to enhancement H_2 production. The proton suicide conditions such as 7 mM allyl alcohol and the combined effect of NaBr and $NaBrO_3$ (40 mM each at pH 5.5) were lethal towards wild type as compared to mutants having defective ethanol pathway and less end-acid concentrations. Few screened mutants had higher H_2 yields than the wild-type strain. A 1.5 times improvement in H_2 yield was observed through mutagenesis of *K. pneumoniae* IIT-BT 08 [16, 17].

The *K. pneumoniae* IIT-BT 08 harbors many plasmids. Maintenance of plasmid requires expenditure of energy. A strategy was developed where the plasmids of *K. pneumoniae* IIT-BT 08 were successfully cured. These cured strains showed a distorted pattern of antibiotic and metal resistance. Moreover, uptake hydrogenase genes were not encoded by plasmids. It was observed that μ_{max} and K_s were higher for the cured strain than wild type. The cured strain showed lower value of maintenance coefficient as compared to wild type. The H_2 production showed very little improvement in plasmid-cured strains [18].

3 Microbial Insight on Dark Fermentative H_2-Producing Microorganisms

First review on biohydrogen production was published in Nature Biotechnology as "Biohydrogen production deserves serious funding" [19]. Followed by this, impetus on biohydrogen gained momentum in early twenty-first century.

3.1 Mesophilic Dark Fermentative H_2 Production

In nature, hydrogen is produced by a wide variety of microorganisms under anaerobic conditions. Several microbes were discovered in different parts of the world, each having unique hydrogen production ability. The different domains of microbes such as anaerobes, facultative anaerobes, methylotrophs, and photosynthetic bacteria have been reported for H_2 production [20].

Promising H_2-producing microorganisms mainly belong to strict and facultative anaerobic chemoheterotrophs belonging to *Clostridia* and *Enterobacteriaceae*, respectively. The reason for microorganisms producing H_2 is to dissipate the excess electrons within the cells which are generated during substrate oxidation.

3.1.1 Hydrogen Production by *Clostridium* sp.

These are fermentative, spore-forming obligate anaerobes. They have low G+C content and are Gram-positive, rod-shaped group of bacteria. It has lower doubling time and can withstand unfavorable conditions (heat shock, physical stress, etc.) as compared to other anaerobic bacteria. The above features could be considered industrially important. The World War I has brought fermentative application of *Clostridia* strain in consideration for solvents and alcohol production [21]. In such solventogenic process, H_2 is one of the by-products. One of the highest H_2 yields has been reported by organisms belonging to this group [22]. Few newly isolated *Clostridium* sp. such as *C. butyricum*, *C. welchii*, *C. beijerinckii*, and *C. pasteurianum* were used individually or as synthetic mixed consortia for H_2 production. The highest H_2 yield of 1.8–2.0 mol mol^{-1} glucose was reported on using *C. beijerinckii* AM21B isolated from termite gut [23]. This strain is capable of utilizing a wide range of other carbohydrates, such as xylose, arabinose, galactose, cellobiose, sucrose, and fructose. The *Clostridium* sp. was also known to use cellulose and hemicellulose as substrate for the H_2 production which are predominant in lignocellulosic biomass.

3.1.2 Hydrogen Production by *Enterobacter* sp.

These bacteria are generally Gram-negative, rod-shaped, motile (peritrichous flagellated) or nonmotile, facultative anaerobes. They have higher growth rates as compared to obligate anaerobes. They can utilize a wide range of carbon sources and are resistant to lower traces of dissolved oxygen. The H_2 production was not inhibited by high partial pressure of H_2 [24]. However, *Enterobacter* sp. have lower H_2 yield when compared to *Clostridium* sp. using glucose as substrate. Yield and rate of H_2 production of 1.0 mol mol^{-1} glucose and 21 mmol L^{-1} h^{-1}, respectively, were observed in batch fermentation [25].

3.1.3 Hydrogen Production by *Escherichia coli*

They are generally Gram-negative, rod-shaped, motile bacteria with low G+C content. Primarily, it produces H_2 from formate. Formate is converted to H_2 and CO_2 in the absence of oxygen catalyzed by enzyme complex formate lyase (FHL) [26]. FHL is a membrane-bound multi-enzyme complex having two subunits, viz. a formate dehydrogenase and a hydrogenase. The H_2 yield using *E. coli* was 0.9–1.5 mol mol^{-1} glucose [27].

3.1.4 Hydrogen Production by *Citrobacter* sp.

They also belong to *Enterobacteriaceae* family and are facultative anaerobes, Gram-negative bacilli with low G+C content. *Citrobacter* are known to produce H_2 both chemolithotrophically and organotrophically. CO and H_2O are the energy source under chemolithotrophic conditions. *Citrobacter* sp.Y19 produced 15 mmol $L^{-1}h^{-1}$ of H_2 from CO and H_2O by the water–gas shift reaction under anaerobic conditions [28]. Chemoorganotrophic conditions with glucose as substrate gave H_2 yield of 1.1 mol mol^{-1} of glucose [29].

3.1.5 Hydrogen Production by *Bacillus* sp.

Many potential H_2-producing microorganisms under genus *Bacillus* have been identified. They are generally Gram-positive, facultative mesophilic bacterium. *Bacillus* sp. grows mostly at 30 °C. However, they can survive at much higher temperatures. Many useful enzymes were secreted at 37 °C. Under unfavorable conditions, it can form spores. The *Bacillus licheniformis* isolated from cattle dung was reported to produce H_2 with lower H_2 yield of 0.5 mol mol^{-1} glucose as compared to *Clostridium* sp. [30]. It generally follows lactic acid pathway. The *Bacillus coagulans*, another species of this genus that was isolated from sewage sludge, was also reported as potential H_2-producing organism. It showed higher H_2 yield (2.2 mol mol^{-1} glucose) as compared to *Bacillus licheniformis* [31]. Use of these strains for H_2 production has certain advantages over strict anaerobes like *Clostridium* and methanogens that are industrially important such as the ease of their handling and nontoxicity towards dissolved oxygen.

Our group isolated many potential H_2-producing mesophilic microorganisms. The efficient H_2-producing strains were *Bacillus coagulans* IIT-BT S1, *Klebsiella pneumoniae* IIT-BT 08, and *Citrobacter freundii* IIT-BT L139 isolated from different sources [31, 32]. Among the above isolates highest H_2 of 2.2 mol H_2 mol^{-1} glucose was observed with *K. pneumoniae* IIT-BT 08-optimized process parameters and 10 g L^{-1} glucose as substrate, respectively [32].

3.2 Thermophilic Dark Fermentative H_2 Production

The kinetics and stoichiometry of H_2 production were more favorable at thermophilic temperatures as compared to mesophilic system. Fermentation at high temperatures also helps in reducing the risk of methanogenic and pathogenic contaminations. The methanogenic and pathogenic contaminations are inherited either from the feedstock or from the inoculum and might lead to decrease in H_2 productivity. Under thermophilic regimes, the H_2 production process is also less affected by the partial pressure of hydrogen (pH_2) in the liquid phase. High-temperature effluent rich in organic content is discharged from many industries. Industrial effluents that are expelled at high temperature are distillery industry effluents, sugar industries' wastewater, food processing, etc. These effluents on discharging in water bodies might cause a serious threat to environment. Cooling process is not cost effective and might eventually lead to losing its biological activity [33]. Such high-temperature effluents could be suitably used by thermophilic microorganisms for H_2 production. Thermophilic bacteria can be classified on the basis of optimum growth temperature:

- Moderate thermophiles growing at temperature range of 45–55 °C
- True thermophiles growing at temperature range of 55–75 °C
- Extremophiles growing at temperature above 75 °C

Many thermophilic H_2-producing species have been identified such as *Clostridia*, *Thermotoga*, *Thermoanaerobacterium*, *Thermoanaerobacter*, and *Caldicellulosiruptor* sp. [34].

3.2.1 Hydrogen Production by *Thermoanaerobacterium* sp.

This genus has interrelationships with *Clostridium* species. It was isolated from Frying Pan Springs in Yellowstone National Park in 1993. It had the ability to degrade xylan and can produce H_2 [35]. These are low G+C-containing bacteria having Gram-negative straight rod shape and motile peritrichous flagella. Under nutritionally deprived conditions, it form spores. They have diverse metabolic end products such as ethanol, acetate, CO_2, H_2, and lactate.

3.2.2 Hydrogen Production by *Thermoanaerobacter* sp.

This genus was listed under the irregular, non-spore-forming, Gram-positive rods in Bergey's Manual of Systematic Bacteriology [36]. The genera *Thermoanaerobacter* and *Thermoanaerobium* included the first thermophilic anaerobic bacteria that produce hydrogen along with ethanol and lactate as sugar fermentation products. These are obligate anaerobes, non-spore-forming (exception *Thermoanaerobacter finnii*) bacteria. These microorganisms can utilize a variety of sugars but cannot degrade cellulose. The H_2, ethanol, lactate, acetate, and CO_2 are the major products. No butyrate production was reported in these species. Some species have been reported to produce up to 4 mol H_2/mol glucose which is the theoretical maximum H_2 yield.

3.2.3 Hydrogen Production by Thermophilic *Clostridium* sp.

This genus has gained much importance in biofuel research in recent years. The genus *Clostridium* belongs to the phylum Firmicutes. They are rod-shaped, Gram-positive, motile, often spore-forming, and obligate anaerobic organisms. They can degrade cellulose using cellulase and can ferment the lignocellulosic biomass to H_2. The highest H_2 yield of 1.6 mol mol^{-1} hexose was observed using this microbe and cellulose as carbon source [37]. These types of microbes are particularly important in converting lignocellulosic biomass to biohydrogen.

3.2.4 Hydrogen Production by *Caldicellulosiruptor* sp.

The *Caldicellulosiruptor saccharolyticus* is an extremophile that grows at 70 °C and produces H_2. On the basis of its physiological characteristics and phylogenetic position, this microorganism is characterized within the *Bacillus*/*Clostridium* sub-phylum [38]. These species are obligatory anaerobic, and non-spore-forming Gram-positive bacteria. These microbes are isolated from the natural habitats like hot springs and lake sediments. These microbes can utilize a wide range of substrates like cellulose, cellobiose, xylan, and xylose with the help of vast repertoire of hydrolytic enzymes. For such properties, these species have the potential to use lignocellulosic wastes for hydrogen production. The predominant metabolite formed by this organism is acetate and lactate. Surprisingly, butyrate is not produced as metabolic end product which is in contrast to other thermophilic microbes such as *Thermoanaerobacterium* sp. Maximal volumetric hydrogen production rate of 5–6 mmol L^{-1} h^{-1} was observed using paper pulp [39].

3.2.5 Hydrogen Production by *Thermotoga* sp.

These microbes grow at the highest reported temperature (90 °C) for H_2 production. It was first isolated from geothermal heated seafloors in Italy and Azores. The name of their genus was derived from the presence of a characteristic outer sheetlike structure called toga. They are rod-shaped, Gram-positive obligate anaerobes. Their natural habitats have high temperature, pressure, and sulfur-containing environments. They can use elemental sulfur or thiosulfate or both as their electron source. Their metabolic end products mainly have acetate, hydrogen, and CO_2, with ethanol in very trace amounts. *Thermotoga maritima* and *Thermotoga neoplanita* are the two microbes that were reported for having hydrogen-producing ability [40].

IIT Kharagpur has also explored the suitability of using thermophiles for H_2 production. For this, a systematic enrichment process was followed where the raw sludge was subjected to various physical and chemical pretreatments. These pretreatments resulted in creation of a selection pressure that suppressed the activity of methanogens and promoted acidogenic H_2 formers. The microbial profile of this enriched mixed culture was further analyzed through denaturation gradient gel electrophoresis (DGGE). The predominance of species closely affiliated to *Thermoanaerobacterium* sp. For improvement of H_2 production using this mixed culture, a statistical multiparameter methodology was used. The optimized physicochemical parameters such as substrate concentration, pH, and temperature were 1% (w/v) glucose, 6.5, and 60 °C, respectively, were found suitable for hydrogen production. The maximum H_2 yield of 2.7 mol H_2 mol^{-1} glucose was observed using the thermophilic mixed culture under batch fermentation [41]. A thermophilic microorganism *Thermoanaerobacterium thermosaccharolyticum* IIT-BT ST1 was isolated from the anaerobic digester sludge. Highest H_2 yield of 2.61 mol H_2 mol^{-1} glucose was observed on using 1% (w/v) of glucose as substrate in batch fermentation at 60 °C [42]. The results thus indicate the advantages of using thermophilic H_2 production system. Scaling up of such systems might help in commercializing bioH_2 process. These results were comparable with other reported values in the literature. The potentiality of different isolates belonging to facultative and obligate anaerobes is represented in Table 16.1.

4 Process Parameters Affecting Dark Fermentative H_2 Production Process

4.1 Role of pH on Dark Fermentation

One of the most important chemical parameters associated with any biochemical process is pH. It not only governs the efficiency of enzymatic machinery of the microorganisms but also plays a crucial role in oxidation-reduction potential of the cells. The metabolic pathway responsible for H_2 production involves action of many enzymes. The glycolytic enzymes and supporting enzymes (Fe–Fe H_2ase, formate lyase, etc.) are the key players in H_2 production. The glucose is metabolized to form pyruvate. The fate of pyruvate under anaerobic conditions determines the H_2 yield [60]. Since all the enzymes have an optimum pH for their maximum activity, it becomes imperative to study the role of pH in H_2 production [61]. The pH profile changes as dark fermentation takes place. Accumulation of metabolite-like volatile fatty acids leads to drop in pH. This drop in pH might affect the functioning of enzymatic machinery involved in H_2 production. Thus very low pH (3.8–4.2) leads to ceasing of H_2 production. Accumulated volatile fatty acids could destroy the cell membrane's integrity and eventually lead to disruption of maintenance of internal pH [62]. Moreover, at low pH, a metabolic shift towards

Table 16.1 Comparative study of dark and photofermentative hydrogen production processes

Microorganism	Substrate	Substrate concentration	Hydrogen yield	References
Studies on mesophilic dark fermentative H_2 production at IIT Kharagpur				
E. cloacae IIT-BT 08	Glucose	10 g/L	2.2 mol H_2/ mol glucose	[32]
E. cloacae IIT-BT 08	Sucrose	10 g/L	6 mol H_2/mol sucrose	[32]
E. cloacae IIT-BT 08	Cellobiose	10 g/L	5.4 mol H_2/ mol cellobiose	[32]
E. cloacae IIT-BT 08	Maltose	10 g/L	1.4 mol H_2/ mol maltose	[32]
E. cloacae DM11	Glucose	10 g/L	3.3 mol H_2/ mol glucose	[43]
E. cloacae BL-21	Glucose	10 g/L	3.1 mol H_2/ mol glucose	[44]
B. coagulans IIT-BT S1	Glucose	20 g/L	2.2 mol H_2/ mol glucose	[31]
E. cloacae IIT-BT 08	Cane molasses	10 g COD/L	8.2 mol H_2/kg $COD_{removed}$	[45]
E. cloacae IIT-BT 08	Distillery effluent	40 g COD/L	7.4 mol/H_2 kg $COD_{removed}$	[46]
Studies on thermophilic fermentative H_2 production at IIT Kharagpur				
Mixed culture	Glucose	10 g/L	2.7 mol H_2/ mol glucose	[41]
Mixed culture	Distillery effluent	60 g COD/L	464 mL H_2/g reducing sugar	[41]
Thermoanaerobacterium thermosaccharolyticum IIT BTST1	Glucose	12 g/L	2.6 mol H_2/ mol glucose	[42]
Studies on photofermentative H_2 production at IIT Kharagpur				
Rhodobacter sphaeroides O.U. 001	Spent media of dark fermentation	–	1.7 mol H_2/ mol acetic acid	[47]
Rhodobacter capsulatus	Malic acid	–	80 mL H_2/L h	[48]
Comparative studies with other reported literature				
C. cellobilparum	Glucose	–	2.73 mol H_2/ mol hexose	[49]
Clostridium CGS2	Starch	–	2.03 mol H_2/ mol hexose	[50]
E. aerogenes	Starch	10 g/L	1.09 mol H_2/ mol glucose	[51]
Escherichia coli	Glucose	–	2.0 mol H_2/ mol hexose	[52]
Klebsielle oxytoca HP1	Sucrose	50 mM	1.5 mol H_2/ mol sucrose	[53]

(continued)

Table 16.1 (continued)

Microorganism	Substrate	Substrate concentration	Hydrogen yield	References
Klebsiella oxytoca HP1	Glucose	50 mM	1 mol H_2/mol glucose	[53]
Clostridium sp.	Xylose	–	2.6 mol H_2/ mol hexose	[54]
Thermoanaerobacterium	Glucose	–	2.67 mol H_2/mol hexose	[55]
Mixed microflora	Sucrose	–	1.8 mol H_2/ mol hexose	[56]
Mixed microflora	Starch	–	1.5 mol H_2/ mol hexose	[57]
Mixed culture	Food waste	–	57 mL H_2/g VS	[58]
C. freundii 01, *E. aerogenes* E10 and *R. palustris* P2	Distillery effluent	–	2.76 mol H_2/ mol hexose	[59]

solventogenesis from acidogenesis was also reported [63]. H_2 production process involves enriched mixed consortia or anaerobic sludge as inoculum, and lower pH suppresses methanogenesis and other H_2-consuming microbes [64]. Based on the importance of pH, H_2 production under controlled pH was explored by many researchers. H_2 production and substrate conversion improved under controlled pH experiments. Some reports were available on H_2 production using pentose sugars. The controlled pH ranges of 5.5–6.5 showed significant improvement in H_2 production with highest H_2 yield of 1.72 mol mol^{-1} of xylose at a pH of 6.5 [65]. *K. pneumoniae* IIT-BT 08 showed improvement in H_2 yield under controlled pH of 6.5. The H_2 yield was improved by 31% and that of substrate conversion efficiency by 10%. In case of packed bed reactors having whole-cell immobilized system, maintaining pH inside the reactor is difficult [66]. Moreover, a nonlinear variation in pH was observed along the length of the packed bed reactor. To prevent drastic pH drop inside the PBR, infusion of feed of higher pH during continuous operation was suggested [67]. Similar result was observed in thermophilic packed bed reactor, where infusion of feed of higher pH stabilized H_2 production inside the reactor. Moreover, 15% increase of rate of H_2 production was observed under such conditions [68].

4.2 *Role of Temperature on H_2 Production*

Most of the literature reports on biohydrogen production are based on mesophilic dark fermentation [69]. The temperature also plays a vital role in H_2 production. Similar to the influence of pH, temperature also governs metabolism via mediating the enzymatic reactions. Every enzyme has an optimum temperature range at which maximum productivity is observed. Extreme temperature would lead to denaturation of metabolic and life-supporting enzymes. It was observed that on increasing temperature range of 15–35 °C, H_2 yield of 1.7 mol mol^{-1} glucose was observed [70]. The shift in the metabolic pathways and microbial community was significantly influenced by operation temperature. Therefore, several studies were reported on understanding the effect of temperature variance on microbial community distribution. Very limited information were available on the role of temperature on the population dynamics of microorganisms [71]. Thermophilic H_2-producing microbes showed higher H_2 yield as compared to the mesophilic counterparts [72]. Higher temperatures make the H_2 production process thermodynamically favorable. It increases the entropy of the system; thus overall free energy increases, making the process spontaneous.

4.3 Role of Partial Pressure on H_2 Production

Hydrogen production pathways are very sensitive to hydrogen partial pressure. The partial pressure of hydrogen inside the reactor increases as it starts getting accumulated in the headspace. As H_2 is the product of dark fermentation, its accumulation would inhibit the product formation which is in accordance with Le Chatelier's principle. This increase in partial pressure also contributes towards metabolic shift during fermentation. It leads to formation of reduced end products such as ethanol, propionate, lactate, butanol, and acetone [73]. Many strategies were used for the removal of H_2 from the fermentation system. Decreasing partial pressure by intermittent N_2 sparging helped in improving the H_2 yield by 68% [74]. Similarly, sparging of methane also improved H_2 yield by removing accumulated H_2 from the reactor. Membrane made up of silane or polyvinyltrimethylsilane could selectively absorb H_2 from the system. Such properties were used to decrease partial pressure of H_2 for improvement of H_2 yield [75].

4.4 Role of Hydraulic Retention Time on H_2 Production

Hydraulic retention time (HRT) is the time that the cells and soluble nutrients are retained inside the reactor. Unlike methanogenesis, H_2 production occurs at lower HRTs. Hydraulic retention time is governed by the volume of the reactor and flow rate of feed (HRT = volume of reactor/feed flow rate). For operation of H_2 production in continuous mode, the HRT should be optimized. At suitable HRT, highest rates of H_2 production and substrate conversion were observed. Very low HRT might also lead to washout state where all the active cells escape out of the reactors. Thus optimization of HRT has close relation with specific growth rate of the organism. On working with mixed consortia, which contain methanogens and acidogenic H_2 producers, manipulating the HRT might lead to shift in microbial profile inside the reactor. Lower HRT would lead to enrichment of acetogenic H_2 producers inside the reactor and the methanogens would get washed out [76]. Thus acidic pH (6–6.5) and low HRT could completely suppress methanogens from the mixed consortia. The HRT also plays an influential role in the formation of end metabolites. This property is concomitantly associated with the changes in microbial profile in response to changes in HRT. On lowering HRT from 10 to 6 h, the H_2 production was increased using *Clostridium* sp. and followed by decrease in propionate production [77]. On using anaerobic sludge, HRT of 1 day showed higher H_2 production and increase in B/A ratio [78]. No methane was detected in the biogas. Study of HRT helps in designing the experiment and reactor for treatment of industry wastewater, where implementation of low HRT by infusing transient loading improved H_2 production and also helped in COD removal [79].

5 Mathematical Modeling and Simulation of Hydrogen Production Process

The efficiency of biological H_2 production can be analyzed by defined mathematical models. These models also increase the understanding of the effect of substrate concentration, feedback inhibition, and effect of different substrate on bioH_2 production [80]. The kinetic parameters determined from unstructured mathematical models could help in designing and scaling up of bioreactors. During H_2 production, kinetics of biomass formation and products (hydrogen, VFA, solvents, etc.) are needed to be studied. Moreover, few models were also developed for studying the effects of different physicochemical parameters (pH, temperature, dilution rate, organic loading rate, etc.) on hydrogen production.

5.1 Mathematical Models Representing Microbial Growth and Substrate Conversion Efficiency

Biohydrogen production in batch fermentation is dependent on availability of limiting substrate. Conventional Monod model is an unstructured mathematical model which describes correlation of the limiting substrate concentration and specific growth rate of the cell in batch fermentation [81]. This may be written as follows:

$$\mu = \frac{\mu_{\max} S}{K_s + S} \tag{16.7}$$

where μ (h^{-1}) represents specific growth rate, $\mu_{\max}$ (h^{-1}) represents maximum specific growth rate, K_s represents saturation constant (g L^{-1}), and S is the limiting substrate concentration (g L^{-1}). The Monod model can be linearized by Lineweaver–Burk equation. The values of μ_{max} and K_s were estimated by regression analysis. Similarly, the above kinetic constants were estimated where complex sugar such as starch has been used as limiting substrate for bioH_2 production. A Monod-type model was developed to demonstrate the influence of limiting substrate (starch) on H_2 production (Eq. 16.8) [82]:

$$\nu_{H_2} = \frac{\nu_{max, H_2} C_{starch}}{K_s + C_{starch}} \tag{16.8}$$

where ν_{max} represents the maximum rate of H_2 production rate (mL L^{-1} h^{-1}), K_s represents half-saturation constant (g COD L^{-1}), and C_{starch} represents initial starch concentration (g COD L^{-1}). There were many bottlenecks of using classical Monod model. This model does not augur the effect of substrate inhibition during bioH_2 production. On using high substrate concentration, the growth parameters could not be satisfactorily simulated with classical Monod model. Moreover, role of other factors such as pH, cell density, substrate diffusion, and presence of metal could not be modeled solely by Monod model. Various mathematical models were proposed to simulate the growth and substrate utilization pattern in which the above-mentioned parameters were also included. Various empirical models were reported to study substrate inhibition on using pure or mixed culture for H_2 production [83].

The Andrew's model was proposed to study the effect of substrate inhibition on H_2 production and was found satisfactory when compared with classical Monod model [84]. A nonlinear relationship was observed between the specific growth rate and substrate concentration and was represented by Eq. (16.9):

$$\mu = \frac{\mu_{max} S}{K_s + S + \frac{S^2}{K_i}} \tag{16.9}$$

On using mixed culture, the effect of different initial glucose concentration on bioH_2 production was modeled using Andrew's model (Eq. 16.10) [85]. The estimated kinetic constants showed significant regression coefficient (R^2) of 0.902:

$$r = \frac{67.1S}{47.7 + S + \frac{S^2}{13.5}} \tag{16.10}$$

In spite of accessing the effect of substrate inhibition, Andrew's model fails to describe the nature of inhibition shown by the substrate. Considering substrate stimulation at low concentration and substrate inhibition at high concentration, an improved Monod-type model (Eq. 16.11) was developed [86]:

$$r = k\left(1 - \frac{S}{S_{\max}}\right)^n \times \frac{S}{S + K_s\left(1 - \frac{S}{S_{\max}}\right)^m} \tag{16.11}$$

where r represents the rate of hydrogen production (mL h^{-1}); k represents the hydrogen production rate constant (mL h^{-1}); S represents the substrate concentration (g L^{-1}); and $S_{\max}$ represents the maximum substrate concentration (g L^{-1}). At $S_{\max}$ there is a ceasing of hydrogen production; K_s represents the saturation constant (g L^{-1}). Based on the nature of inhibition (noncompetitive, competitive, uncompetitive, and mixed inhibition) the values of "m" and "n" can be determined. The nature of substrate inhibition was found to be uncompetitive and the values of "m" and "n" were positive.

5.2 Mathematical Models Representing Product Formation Kinetics

There are two types of product formed during dark fermentative H_2 production. The gaseous products are primarily H_2 and CO_2 whereas metabolites formed during fermentation such as solvents and volatile fatty acids are the liquid products present in the fermentation broth.

The commonly used mathematical model to simulate the product formation kinetics in bioH_2 production is modified Gompertz equation (Eq. 16.12):

$$H(t) = P\exp\left\{-\exp\left[\frac{R_m e}{P}(\lambda - t) + 1\right]\right\} \tag{16.12}$$

where $H(t)$ represents cumulative volume of hydrogen production (mL), P the gas production potential (mL), Rm (mL h^{-1}) the maximum production rate, λ (h) the lag time, t incubation time (h), and e the exp(1) 2.718.

The nonlinear profile of cumulative H_2 production can be modeled using the above equation. In batch fermentation, the value of H changes as a function of time. Slope of the linearly increasing phase of H_2 production could be represented as the maximum rate of H_2 production (R_m) [87].

The Gompertz equation was also applied for determining the kinetic constants for acetate and butyrate production [88]. The Gompertz equation for simulation of acetate and butyrate formation has been described in Eq. (16.13) given below:

$$P(t) = P_{\max,i}\exp\left\{-\exp\left[\frac{R_{\max,i}e}{P_{\max,i}}(\lambda_i - t) + 1\right]\right\} \tag{16.13}$$

where "*i*" represents acetate or butyrate; P_i is the product "*i*" formed per liter of working volume of the reactor at time *t*; and the potential maximum product per liter of working volume of the reactor is represented by $P_{\max,i}$.

Hydrogen production in dark fermentation is considered as growth-associated product. This relation could be mathematically deciphered by using Luedeking–Piret model [89]:

$$\frac{\mathrm{d}P}{\mathrm{d}t} = \alpha\frac{\mathrm{d}x}{\mathrm{d}t} + \beta x$$

where α and β are the growth-associated and non-growth-associated coefficients, respectively. Another model based on the modification of Luedeking–Piret model was proposed where biomass production and concomitant H_2 production are modeled on the basis of growth-associated nature (Eq. 16.14) [90]:

$$\frac{1}{x}\frac{\mathrm{d}C_{H_2}}{\mathrm{d}t} = \alpha\mu = \alpha\left[\frac{1}{x}\frac{\mathrm{d}x}{\mathrm{d}t}\right] \quad (16.14)$$

where C_{H2} represents H_2 concentration (mol), x represents cell concentration (g VSS L^{-1}), and μ represents specific growth rate (h^{-1}). On plotting rate of C_{H2} versus biomass concentration, the slope represents the "α" value. On using the above model, the "α" value was calculated for fermentation done by *C. butyricum* (strains CGS2 and CGS5) and *C. pasteurianum* (strains CH1, CH4, CH5, and CH7) and the values were 0.041 and 0.039 mol g^{-1} volatile suspended solids for xylose and sucrose, respectively.

6 Suitability of Different Carbon Sources on Hydrogen Production

The most commonly used pure substrates are glucose, sucrose, and lactose as they can be easily channelized to glycolytic pathway. However, pure carbohydrate sources are not suitable as feedstock for hydrogen production due to their high costs. Many microbial species show preference towards substrates for H_2 production. One such example was *Clostridium saccharoperbutylacetonicum* ATCC 27021 which grows only on disaccharides (lactose, sucrose, and maltose). It produced 2.81 mol H_2 mol^{-1} sugar. H_2 production by *K. pneumoniae* IIT-BT 08 was reported using different carbon sources such as hexose sugars (glucose, fructose, etc.), disaccharides (sucrose, maltose, cellobiose, etc.), pentose sugars (L-arabinose, D-xylose, etc.), and complex carbohydrates (potato starch, cellulose, etc.). Highest H_2 yield of 6.0 mol H_2 mol^{-1} was observed with sucrose [32]. The maximum rates of H_2 production and yield from different C-sources are shown in Fig. 16.1. Thermophilic hydrogen production using different types of carbon sources (glucose, sucrose, L-arabinose, D-xylose, D-cellobiose, and starch) is

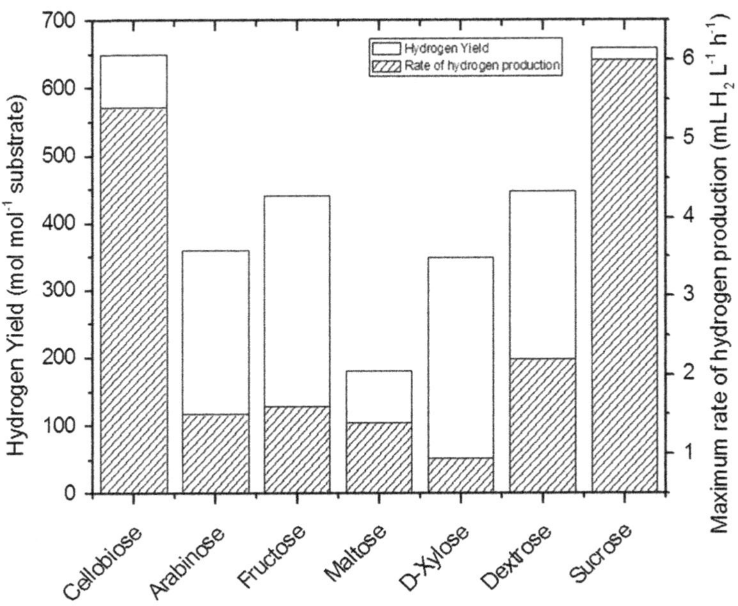

Fig. 16.1 Effect of different carbon sources on mesophilic dark fermentative hydrogen production

shown in Fig. 16.2. The broad spectrum of enzymatic machinery present in thermophilic microorganisms helps in utilizing complex carbohydrates. So, application of thermophilic dark fermentation holds promise in biohydrogen research [41].

6.1 Organic Wastes as Feedstock for Biohydrogen Production

For commercial production of biohydrogen, cheap feedstock/raw material should be used. Most of the studies on bioH_2 are based on utilization of simple sugars such as glucose, sucrose, maltose, and lactose. These simple sugars are expensive and usages of such raw material are not economically viable. To address this issue, production of bioH_2 using different organic wastes as substrate is a cheap and promising approach. There is a relatively high abundance of complex sugars (polysaccharides) in nature. Most of these polymeric sugars (cellulose, hemicellulose, etc.) are inaccessible to microorganisms. In order to tap the energy bound in these polymeric sugars, a detailed research is required targeting the pretreatment and saccharification techniques. BioH_2 could be considered as renewable and cheap when its production is based on low-value renewable resources.

Many high COD-containing wastes have been explored for bioH_2 production. This includes municipal solid wastes, food wastes, distillery wastes, cheese whey,

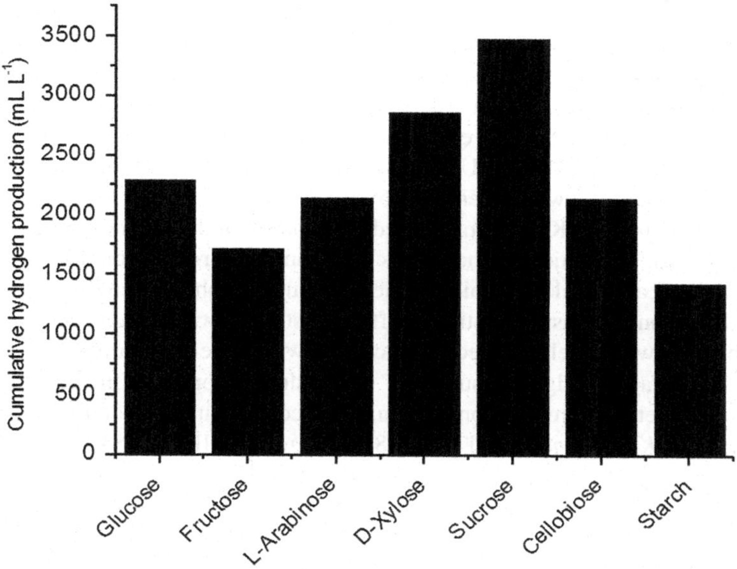

Fig. 16.2 Effect of different carbon sources on hydrogen production in a batch system using thermophilic mixed culture

etc. Organic fractions of municipal solid wastes are widely available renewable resource rich in polysaccharides and proteins [91]. Recently, usage of municipal solid waste for H_2 production showed promising results. However, the yield with raw sewage sludge was still considerably low, i.e., 0.16 mg of H_2 g^{-1} of dried solids (DS). Various pretreatment methods such as ultrasonic treatment, acidification, sterilization, and freezing and thawing were employed for improvement of H_2 yield. Boiled sludge (heat treatment) leads to solubilization of nutrients present in raw sludge. Usage of boiled sewage sludge gave 15.64 mL of H_2 g^{-1} DS. Pretreatment techniques such as sterilization and freezing and thawing gave H_2 yield of 47 mL of H_2 g^{-1} of DS [91].

On the other hand, food wastes also have a great environmental threat. It contains about 90% volatile suspended solids. High organic content makes them suitable feedstock for microbial fermentation. The institutional food wastes used for thermophilic hydrogen production gave 81 mL H_2 g^{-1} VSS as compared to 63 mL H_2 g^{-1} VSS by mesophilic dark fermentation [92]. Dairy industry effluents have high biological oxygen demand (BOD) and chemical oxygen demand (COD) which makes them hazardous for environment if discharged untreated [93]. On using dairy wastewater, the maximum hydrogen production of 5.2 mL H_2 g^{-1} COD was observed [94]. Distillery or alcoholic beverage industrial wastewaters are rich in biodegradable organic material, such as sugars, hemicelluloses, dextrin, resins, and organic acids. These wastewaters have high COD (80–160 g L^{-1}). Many reports were available on biohydrogen production using distillery wastewater. In anaerobic sequencing batch biofilm reactor, maximum hydrogen production of 156.7 L H_2 kg^{-1} COD was observed [95]. Few effluents are discharged at very high temperature, e.g., palm oil mill effluent (POME). These effluents have high organic content and could be a potential substrate for thermophilic dark fermentative H_2 production. On an average 0.9–1.5 m^3 POME was generated from 1 t of palm oil being produced. Using UASB reactor, hydrogen production rate of 4.4 L g^{-1} POME d^{-1} was observed [96].

In view of this, IIT Kharagpur studied potentiality of different organic wastes for H_2 production. The major emphasis was given on their organic content, availability, sustainability, and biodegradability. The continuous biohydrogen production using wastes like cane molasses, distillery effluent, starchy wastewater, and whey with immobilized whole-cell packed bed system was widely reported [97–99]. The usage of sewage sludge as substrate for hydrogen production was performed using a synthetic microbial consortium. This consortium was composed of three bacteria, viz. *K. pneumoniae* IIT-BT 08, *B. coagulans* IIT-BT S1, and *C. freundii* IIT-BT L139. *B. coagulans* was found to be the dominant species when sewage sludge was used as substrate [100]. This synthetic mixed microbial consortium showed promising H_2 production with conventional MYG medium. The H_2 yield of 16.1 mmol H_2 g^{-1} $COD_{removed}$ was observed with MYG medium. When compared with pure substrates, the H_2 yields with sewage sludge were low.

6.2 Suitability of Lignocelluloses as Solid Matrix and Performance of Continuous Hydrogen Production Process

Continuous hydrogen production was studied mostly by using suspended cells. These systems are prone to cell washout at higher dilution rates. Implementation of whole-cell immobilized system has several advantages over suspended cell systems [101]. The problem of cell washout could possibly be overcome by using immobilized whole cells in packed bed reactors. Immobilized whole-cell reactor showed higher substrate conversion efficiency and mean cell residence time also increases. The phenomenon such as adsorption or entrapment was used for whole-cell immobilization. The entrapment of cells in gels has certain disadvantages associated with its operation in continuous mode. This includes degradation of the gel matrix and mass transfer hindrance limitations of nutrient and metabolite. In contrast, the phenomenon of natural adsorption of cells on matrix is cost effective and easy to perform. Adsorbed whole cells showed least internal mass transfer resistance [102]. Various immobilized whole-cell systems were studied for hydrogen production, viz. in granular reactors, packed bed reactors, fluidized bed reactors, and up-flow anaerobic sludge blanket reactors (UASB).

Continuous hydrogen production process using immobilized *K. pneumoniae* IIT-BT 08 on agro-based neutral matrix was found effective. Various cheap agro-based residues (bagasse, rice straw, and coconut coir) were used as immobilization matrix. Coconut coir was found to be most suitable matrix on the basis of packing density, cell loading, chemical neutrality, and hydrogen production rate [97, 103]. One of the major problems associated with operation of continuous H_2 production in tubular packed bed reactor was gas hold-up. Gas hold-up leads to decrement in working volume. This might lead to improper distribution of substrate flux inside the reactor, thus leading to decrease in H_2 production. Different reactor configurations such as tapered and rhomboid bioreactors could address the problem of gas hold-up and thus could improve the H_2 production rate. The rhomboid bioreactor configuration helped in reduction of gas hold-up problem by 67% as compared to tubular bioreactor. Continuous H_2 production along with recycling of effluent in rhomboidal bioreactor showed improved H_2 production rate. At a dilution rate of 0.93 h^{-1} and recycle ratio of 6.4, the maximum rate of H_2 production of 75.6 mmol L^{-1} h^{-1} was observed. Overall substrate conversion efficiency was increased by 15% as compared to non-recycling systems. The similar study was conducted with cane molasses as substrate for continuous H_2 production in the tubular bioreactor. The maximum rates of H_2 production and yield obtained were 1250 mL L^{-1} h^{-1} and 11.6 mol H_2 kg^{-1} $COD_{removed}$, respectively, with COD reduction of 47% at a dilution rate of 0.6 h^{-1} [45]. The above observation thus emphasizes the importance of whole-cell immobilization on eco-friendly natural polymers over synthetic matrix which has disposal problem associated with it (Table 16.2).

7 Integration of Two-Stage Process for Improvement of H_2 Production and Maximum Energy Recovery

7.1 Integration of Dark and Photofermentation

Integration of dark and photofermentation might improve gaseous energy recovery in terms of H_2. Substrate conversion efficiency improves on using such two-stage process. Stoichiometrically one mole of glucose could yield 12 moles of H_2 through integration of dark and photofermentation. The H_2 production was carried out using glucose as substrate by *K. pneumoniae* DM11 in the first stage. In the second-stage photofermentation, *Rhodobacter sphaeroides* O.U. 001 utilized volatile fatty acid-rich spent media by a photobioreactor to produce H_2. Combined H_2 yield of 5.14 mol H_2 mol^{-1} glucose was observed using two-stage systems [47]. Figure 16.3 represents the integration of two-stage process for maximum energy recovery from dark fermentation.

Table 16.2 Comparative studies on the rate of hydrogen production in different reactors with immobilized whole cells

Organism	Substrate	Reactor type	Type of immobilized matrix	Mode of operation	Rate of H_2 production (mmol/L h)	References
Enterobacter cloacae DM 11	Glucose	Packed bed reactor	Lignocellulosic material (coconut coir)	Continuous	75.6	[98]
Enterobacter aerogenes NCIMB 10102	Glucose	Packed column reactor	Synthetic sponge	Continuous	10.2	[102]
Enterobacter aerogenes HY-2	Glucose	Column reactor	Self-flocculated cells	Continuous	58.0	[103]
Enterobacter aerogenes E.82005	Cane molasses	Column reactor	Polyurethane	Continuous	13	[104]
Enterobacter aerogenes HO-39	Glucose	Column reactor	Agar gel and porous glass beads	Continuous	37.9	[105]
Enterobacter sp. BY-29	Glucose	Packed bed reactor	Porous glass beads	Batch	49.1	[106]
Bacillus licheniformis	Glucose	Packed bed reactor	Brick dust	Continuous	1.1 mol/H_2 mol glucose	[107]
Clostridium butyricum	Glucose	Column reactor	Porous glass beads	Continuous	51.4	[108]
Clostridium butylicum and *Enterobacter aerogenes*	Starch	Stirred reactor	Porous glass beads	Continuous	58.0	[109]
Clostridium acetobutylicum ATCC 824	Glucose	Trickle bed filter	Porous glass beads	Continuous	1270 mL/g glucose L of reactor	[110]
Acclimated sewage sludge	Sucrose	Packed bed reactor	Activated carbon	Continuous	330.4	[111]
Microflora (sewage sludge)	Sucrose	Packed bed reactor	Polyethylene–octane elastomer	Continuous	59	[112]
Rhodobacter spheroid	–	Packed bed reactor	Agar gel	Continuous	2.1	[113]
Rhodospirillum rubrum	–	Nozzle loop reactor	Ca-alginate	Continuous	2.7	[114]

AGSBR Agitated granular sludge bed reactor, *DTFBR* Draft tube fluidized bed reactor, *FBR* Fluidized bed reactor, *CSTR* Continuous stirred tank reactor

Fig. 16.3 Continuous biohydrogen production using cane molasses as substrate in 20 L packed bed reactor

7.2 Integration of Dark Fermentation and Biomethanation

Gaseous energy recovery in the form of H_2 alone would not be commercially lucrative. During H_2 production from organic waste, various metabolites are produced such as volatile fatty acids and small-chain alcohols. These metabolites could be a suitable feedstock for biomethanation process. Thus CH_4 production could be considered for treatment of spent media generated from fermentative H_2 production leading to maximum net gaseous utilizable energy product (NUEP). The integration of acidogenic hydrogen production with methanogenesis has gained greater attention under the eponym of HYMET©. An anaerobic sequencing batch reactor (AnSBR) with suspended growth configuration was used for biomethanation. The pH of the spent media rich in volatile fatty acids was adjusted to 7.2. This spent media was then fed into the second-stage reactor which contained acetoclastic methanogen-enriched cow dung slurry. The reactor was operated at different organic loading rates (OLRs). Constant methane production was considered as indicator for stabilized performance of the bioreactor and subsequently the reactor was shifted to higher OLR. At OLR of 14 g L^{-1} day^{-1}, highest methane yield of 82.6 L g^{-1} COD_{added} and 66% COD removal were observed. The total gaseous energy recovery for two-stage process was found to be 53.6%. From single-stage H_2 production, gaseous energy recovery was only 28%. Thus two-stage systems not only help in improving gaseous energy recovery but also can make HYMET© (H_2 production followed by CH_4 production) concept commercially feasible [115] (Fig. 16.4).

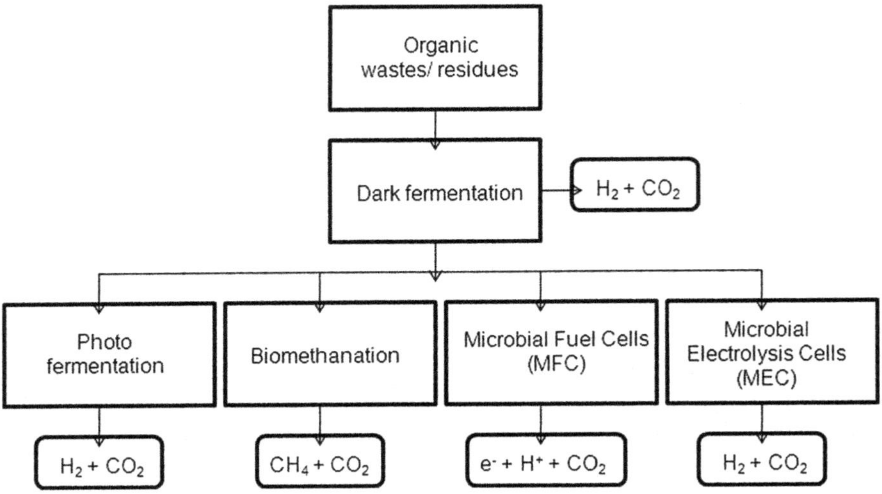

Fig. 16.4 Integration of two-stage process for maximum energy recovery

7.3 Integration of Dark Fermentation and MFC

Use of MFC as a second-stage process enables the utilization of the residual substrate energy besides generation of electricity resulting in yielding higher energy recovery rates. Easily degradable carbon sources which remain unutilized after dark fermentation can become potential substrates for the bioelectricity generation. In microbial fuel cell, the electrogenic microbes can utilize various organic materials present in spent media which eventually leads to COD removal. Reports are available on usage of vegetable wastes for bioH_2 production followed by MFC for electricity generation. The hydrogen production was performed in an acidogenic sequential batch biofilm reactor (AcSBBR). The volatile fatty acid-rich spent media was then used in a single-chambered MFC. Highest power output of 111.76 mW m^{-2} was observed [116]. About 80% COD removal, 79% volatile fatty acids, 78% carbohydrate conversion, and 65.38% turbidity removal were observed. In another study, the effluent from hydrogen-producing biofermentor (HPB) was fed to a single-chamber MFC (SCMFC). The highest power density and coulombic efficiency (CE) of 4200 mW m^{-3} and 5.3% were observed, respectively. This resulted in improvement in energy conversion efficiency by 29% [117]. IIT Kharagpur explored such two-stage integration for maximization of energy recovery. The H_2 production was carried out using cane molasses as a substrate in both batch and continuous process. The maximum H_2 yields achieved in batch and continuous process were 8.23 mol H_2 kg^{-1} and 11.6 mol H_2 kg^{-1} $COD_{removed}$, respectively, using 0.5 L double-jacketed reactor. The spent fermentation media from dark fermentation was then considered as a substrate in an MFC for electricity generation. Reductions in COD and total carbohydrates were about 85% and 88%, respectively. A power output of 3020 mW m^{-3} was obtained with an anolyte pH of 7.5 using alkali-pretreated spent media. The results showed that integrating an MFC with dark fermentation is a promising way to utilize the substrate energy. This integration was found quite beneficial in improving the energy recovery [45].

7.4 Integration of Dark Fermentation and MEC

Integrating dark fermentation with a microbial electrolysis cell (MEC) is another method which can improve H_2 yield. By implementing MEC, complete oxidation of substrate to carbon dioxide is possible. Moreover, this system can also use the residual sugar present in the spent media of dark fermentation. This process requires a low energy input (0.2 V) in comparison to the electrochemical H_2 production by water electrolysis. Extensive research is going on in order to develop membrane-less MECs for the improvement of the overall energy recovery. In an integrated system using a single-chambered MEC and an applied voltage of 0.6 V, the overall hydrogen recovery was 96%, with a production rate of 2.11 m^3 m^{-3} day when the MEC was combined with the fermentation system [118]. Some reports even showed very high VFA utilization (49.8%) besides generating a maximum cumulative hydrogen production (CHP) of 3.6 mmol [119].

8 Future of Biological Hydrogen Production

After two decades of research on biological H_2 production, the technological development, implementation of bioH_2 policy, and the road map of such energy generation methods are still naïve. In terms of large-scale or pre-commercial stage of bioH_2 production, a very few reports are available (Table 16.3). The biggest scale of bioreactor reported for H_2 production is of 100 m^3 [120]. It showed the feasibility of using distillery effluent in dark fermentative H_2 production in large scale. The maximum H_2 yield of 2.76 mol H_2 mol^{-1} glucose was observed on operating this reactor which is still less than the theoretical maximum. Other reports with bench scale 20 L bioreactor also showed promising results where molasses was used as carbon source. In CSTR configuration, maximum H_2 yield of 20.13 mol kg^{-1} $COD_{removed}$ was observed [121]. The major bottleneck of biohydrogen production is low substrate conversion efficiency, inhibition of growth due to accumulation of end metabolites, drop in pH, and inefficiency of microorganisms towards achieving theoretical maximum.

Moreover, one of the reasons of underperformance of bioH_2 production was also the absence of technologies related to direct application of it. With the advent of

Table 16.3 Hydrogen production in large-scale fermentors

Reactor type	Volume (m^3)	Microorganism	Substrate	Hydrogen yield	Hydrogen rate (L/L. d)	References
CSTR	100	*Citrobacter freundii* 01 and *Enterobacter aerogenes* E10	Distillery effluent	2.76 mol H_2/ mol glucose	–	[120]
CSTR	2.0	Mixed microbial consortia	Molasses	26.13 mol/ kg $COD_{removed}$	5.57	[121]
AGSBR	0.4	Mixed microbial consortia	Molasses	1.04 mol H_2/ mol sucrose	15.59	[122]
CSTR	0.03	*Clostridium* sp.	Waste sugar	2.93 mol H_2/ mol glucose	–	[123]
Granule-based CSTR	0.006	Mixed microbial consortia	Glucose	1.84 mol H_2/ mol glucose	78.24	[124]
DTBFR	0.008	Thermophilic mixed culture	Glucose	–	4.80	[125]
FBR	0.005	Thermophilic mixed culture	Glucose	–	6.00	[125]
DTFBR	0.008	*Clostridium* sp.	Sucrose	2.5 mol H_2/ mol glucose	54.48	[126]

technologies such as fuel cell, the hydrogen produced by dark fermentation could directly convert to electricity. This has infused new life in implementation of hydrogen-based economy. A fuel cell is a device that is similar to a continuously recharging battery which generates electricity by the low-temperature electrochemical reaction of hydrogen and oxygen. The contrasting difference in how batteries store energy is that fuel cell can produce electricity continuously as long as hydrogen and oxygen are supplied to it. Hydrogen-powered fuel cells produce water as by-product and produce virtually no pollutant. Fuel cells operate at temperatures much below the internal combustion engine. Fuel cells are not bound by the limitations of the Carnot cycle; thus it can efficiently convert fuel into electricity as compared to IC engines. The operating temperature, the type of fuel, and a range of applications of fuel cells are dependent on the electrolyte they use. The electrolyte can be acid, base, salt, or a solid ceramic or polymeric membrane that conducts ions. However, at present, fuel cells cannot compete with conventional energy conversion technologies in terms of cost and reliability. High-temperature solid oxide fuel cells (SOFCs) and molten carbonate fuel cells (MCFCs) are ideal for distributed energy supply operating today with natural gas, which enables development and use of this technology independently from the establishment of a hydrogen infrastructure. Indeed, they offer an interesting transition to the hydrogen economy. It has given a fresh breath to biofuel research, where gaseous energy or ethanol can be converted into electricity directly. Types of fuel cell technology available are as follows:

Low-Temperature Proton-Exchange Membrane Fuel Cells
The highest power density is provided by proton-exchange membrane fuel cells (PEMFCs) and alkaline fuel cells (AFCs). The requirement of costly platinum catalyst and need for highly pure hydrogen are the major bottlenecks of using such fuel cells. PEMFCs are most favored for mass-market automotive and small-scale CHP applications, and there is a massive global effort to develop commercial systems.

Phosphoric Acid Fuel Cells (PAFCs)
These types of fuel cells are more tolerant towards impurities that remain as contaminant with hydrogen. PAFCs could be potentially used for stationary power generation and large vehicles. They are commercially available today, but have relatively high cost. Direct methanol fuel cells (DMFCs) are powered by methanol and are considered for a number of applications, particularly those based around replacing batteries in consumer applications such as mobile phones and laptop computers. Thus hydrogen economy has a great potential for creating employment. Moreover, its environmental benefits also give it an upper hand in considering it as future fuel.

9 Conclusion

Research on biohydrogen has given lots of promise in terms of energy generation and waste management. Two vital aspects of biohydrogen production process are availability of cheap renewable feedstock and potential microbial strains. Different potential microbes have been isolated and found suitable for H_2 production. Mesophilic microbes such as *K. pneumoniae* IIT-BT 08, *Bacillus coagulans* IIT-BT S1, and *Citrobacter freundii* IIT-BT L139 are found suitable as H_2-producing microorganisms. Thermophilic microorganisms showed many advantages as compared to mesophiles, viz. having a vast array of hydrolytic enzymes, pathogen destruction, and higher yield and rate of H_2 production. Modification of metabolic pathway has led to the improvement of H_2 production. Techniques like chemical mutagenesis, gene knockout, and homologous and heterologous gene expression have helped in achieving higher H_2 yields. For making bioH_2 production process renewable and economically feasible, greater emphases have been given on the use of cheap and locally available agro-based feedstocks. Distillery effluent, whey, starchy wastewater, and cane molasses show promise as feedstock for H_2 production. Integration of conventional dark fermentative H_2 production with newly emerging energy production systems (MFCs, MEC, biomethanation, etc.) could improve overall energy recovery. Application of bioH_2 for electricity generation via fuel cell has given a new hope towards decentralized energy production. For realizing the goal of clean environment and clean energy, biohydrogen could prove to be the stepping stone for the coming future.

Acknowledgements Author is thankful to Council for Scientific and Industrial Research (CSIR); Department of Biotechnology (DBT); Ministry of New and Renewable Energy (MNRE); Defence Research and Development Organisation (DRDO); and Department of Science and Technology (DST), Government of India, for the financial assistance.

References

1. Soetaert W, Vandamme EJ (2009) Biofuels. Wiley, Great Britan, pp. 1–8
2. Das D, Verziroglu TN (2001) Hydrogen production by biological processes: a survey of literature. Int J Hydrog Energy 6:13–28
3. Pandu K, Joseph S (2012) Comparisons and limitations of biohydrogen production processes: a review. Int J Adv Eng Technol 2:342–356
4. Baghchehsaree B, Nakhla G, Karamanev D, Argyrios M (2010) Fermentative hydrogen production by diverse Microflora. Int J Hydrog Energy 35:5021–5027
5. Benemann JR (1997) Feasibility analysis of photobiological hydrogen production. Int J Hydrog Energy 22:979–987
6. Greenbaum E (1990) Hydrogen production by photosynthetic water splitting. In: Veziroglu TN, Takashashi PK (ed) Hydrogen energy progress VIII. Proceedings 8th WHEC, Hawaii, pp 743–54
7. Das D, Khanna N, Veziroglu TN (2008) Recent developments in biological hydrogen production processes. Chem Ind Chem Eng Q 14:57–67
8. Zeikus JG (1977) The biology of methanogenic bacteria. Bacteriol Rev 41:514–541
9. Knappe J, Sawers G (1990) A radical-chemical route to acetyl-CoA: the anaerobically induced pyruvate formate-lyase system of *Escherichia coli*. FEMS Microbiol Lett 75:383–398
10. Stephenson M, Stickland LH (1932) Hydrogenlyases. J Biochem 26:712–724
11. Morimoto K, Kimura T, Sakka K, Ohmiya K (2005) Overexpression of a hydrogenase gene in *Clostridium paraputrificum* to enhance hydrogen gas production. FEMS Microbiol Lett 246:229–234
12. Mohr G, Smith D, Belfort M, Lambowitz AM (2000) Rules for DNA target-site recognition by a lactococcal group II intron enable retargeting of the intron to specific DNA sequences. Genes Dev 14:559–573
13. Wang R, Zong W, Changli Q, Wei Y, Yu R, Zhou Z (2011) Isolation of *Clostridium pefrigens* strain W11 and optimization of its biohydrogen production by genetic modification. Int J Hydrog Energy 36:12159–12167
14. Sargent F, Macaskie LE (2006) Inactivation of the *Escherichia coli* K-12 twin-arginine translocation system promotes increased hydrogen production. FEMS Microbiol. Lett. 262:135–137
15. Ma K, Zhao H, Zhang C, Lu Y, Xing XH (2012) Impairment of NADH dehydrogenase for increased hydrogen production and its effect on metabolic flux redistribution in wild strain and mutants of Enterobacter aerogenes. Int J Hydrog Energy 37:15875–15885
16. Das D (2009) Advances in biohydrogen production processes: an approach towards commercialization. Int J Hydrog Energy 34:7349–7357
17. Kumar N, Ghosh A, Das D (2001) Redirection of biochemical pathways for the enhancement of H_2 production by *Enterobacter cloacae*. Biotechnol Lett 23:537–541

18. Khanna N, Kumar K, Todi S, Das D (2012) Characteristics of cured and wild strains of *Enterobacter cloacae* IIT-BT 08 for the improvement of biohydrogen production. Int J Hydrog Energy 37:11666–11676
19. Benemann JR (1996) Hydrogen biotechnology: progress and prospects. Nat Biotechnol 14:1101–1103
20. Nandi R, Sengupta S (1937) Microbial Production of Hydrogen: An Overview 1998, 24, 61-84. Weizmann, C., Rosenfeld, B., The activation of the butanol-acetone fermentation of carbohydrates by *Clostridium acetobutylicum* (Weizmann). Biochem J 31:619–639
21. Weizmann C, Rosenfeld B (1937) The activation of the butanol-acetone fermentation of carbohydrates by *Clostridium acetobutylicum* (Weizmann). Biochem J 31:619–639
22. Kamalaskar LB, Dhakephalkar PK, Meher KK, Ranade DR (2010) High biohydrogen yielding *Clostridium* sp. DMHC-10 isolated from sludge of distillery waste treatment plant. Int J Hydrog Energy 35:10639–10644
23. Taguchi F, Mizukami N, Hasegawa K, Saito-Taki T, Morimoto M (1994) Effect of amylase accumulation on hydrogen production by *Clostridium beijerinckii*, strain AM21B. J Ferment Bioeng 77:565–567
24. Tanisho S, Suzuki Y, Wakao N (1987) Fermentative hydrogen evolution by *Enterobacter aerogenes* strain E.82005. Int J Hydrog Energy 12:623–627
25. Tanisho S (1998) Effect of CO_2 removal on hydrogen production by fermentation. Int J Hydrog Energy 23:559–563
26. Stickland LH (1929) The bacterial decomposition of formic acid. Biochem J 23:1187–1198
27. Blackwood AC, Ledingham GA, Neish AC (1956) Dissimilation of glucose at controlled pH values by pigmented and non-pigmented strains of *Escherichia coli*. J Bacteriol 72:497–499
28. Jung G (2002) Hydrogen production by a new chemoheterotrophic bacterium *Citrobacter* sp. Y19. Int J Hydrog Energy 27:601–610
29. Vatsala T (1992) Hydrogen production from (cane-molasses) stillage by *Citrobacter freundii* and its use in improving methanogenesis. Int J Hydrog Energy 17:923–927
30. Kalia VC, Purohit HJ (2008) Microbial diversity and genomics in aid of bioenergy. J Ind Microbiol Biotechnol 1992(35):403–419
31. Kotay SM, Das D (2007) Microbial hydrogen production with *Bacillus coagulans* IIT-BT S1 isolated from anaerobic sewage sludge. Bioresour Technol 98:1183–1190
32. Kumar N, Das D (2000) Enhancement of hydrogen production by *Enterobacter cloacae* IIT-BT 08. Process Biochem 35:589–593
33. Jo JH, Lee DS, Park JM (2008) The effects of pH on carbon material and energy balances in hydrogen-producing *Clostridium tyrobutyricum* JM1. Bioresour Technol 99:8485–8491
34. Zeidan AA, van Niel EWJ (2010) A quantitative analysis of hydrogen production efficiency of the extreme thermophile *Caldicellulosiruptor owensensis* OLT. Int J Hydrog Energy 35:1128–1137
35. Lee YE, Jain MK, Lee C, Zeikus JG (1993) Taxonomic distinction of saccharolytic thermophilic anaerobes: description of *Thermoanaerobacterium xylanolyticum* gen. nov., sp. nov., and *Thermoanaerobacterium saccharolyticum* gen. nov., sp. nov.; reclassification of *Thermoanaerobium brockii, Clostridium*. Int J Syst Bacteriol 43:41–51
36. Wiegel J, Ljungdahl LG (1981) *Thermoanaerobacter ethanolicus* gen. nov., spec. nov., a new, extreme thermophilic, anaerobic bacterium. Arch Microbiol 128:343–348
37. Levin D, Islam R, Cicek N, Sparling R (2006) Hydrogen production by *Clostridium thermocellum* 27405 from cellulosic biomass substrates. Int J Hydrog Energy 31:1496–1503
38. Rainey FA, Donnison AM, Janssen PH, Saul D et al (1994) Description of *Caldicellulosiruptor saccharolyticus* gen. nov., sp. nov: An obligate anaerobic, extremely thermophilic, cellulolytic bacterium. FEMS Microbiol Lett 120:263–266
39. Kadar Z, de Vrije T, van Noorden GE, Budde MAW et al (2004) Yields from glucose, xylose, and paper sludge hydrolysate during hydrogen production by the extreme thermophile *Caldicellulosiruptor saccharolyticus*. Appl Biochem Biotechnol 114:497–508

40. Finkelstein M, McMillan JD, Davison BH (eds) (2002) Biotechnology for fuels and chemicals. Humana Press, Totowa
41. Roy S, Ghosh S, Das D (2012) Improvement of hydrogen production with thermophilic mixed culture from rice spent wash of distillery industry. Int J Hydrog Energy. doi:10.1016/2012.08.016
42. Roy S, Vishnuvardhan M, Das D (2014a) Improvement of hydrogen production by thermophilic isolate *Thermoanaerobacterium thermosaccharolyticum* IIT-BT ST1. Int J Hydrog Energy 39:7541–7552
43. Nath K, Kumar A, Das D (2006) Effect of some environmental parameters on fermentative hydrogen production by *Enterobacter cloacae* DM11. Can J Microbiol 52:525–532
44. Chittibabu G, Nath K, Das D (2006) Feasibility studies on the fermentative hydrogen production by recombinant *Escherichia coli* BL-21. Process Biochem 41:682–688
45. Pandit S, Balachandar G, Das D (2014) Improvement of energy recovery from cane molasses by dark fermentation followed by microbial fuel cells. Front Chem Sci Eng 8:43–54
46. Mishra P, Das D (2014) Biohydrogen production from *Enterobacter cloacae* IIT-BT 08 using distillery effluent. Int J Hydrog Energy 39:7496–7507
47. Nath K, Das D (2005) Hydrogen production by *Rhodobacter sphaeroides* strain O.U. 001 using spent media of *Enterobacter cloacae* strain DM11. Appl Microbiol Biotechnol 68:533–534
48. Basak N, Das D (2007) The prospect of purple non-sulfur (PNS) photosynthetic bacteria for hydrogen production: the present state of the art. World J Microbiol Biotechnol 23:31–42
49. Chung KT (1976) Inhibitory effects of H_2 on growth of *Clostridium cellobioparum*. Appl Environ Microbiol 31:342–348
50. Chen SD, Lee KS, Lo YC, Chen WM et al (2008) Batch and continuous biohydrogen production from starch hydrolysate by *Clostridium* species. Int J Hydrog Energy 33:1803–1812
51. Fabiano B, Perego P (2002) Thermodynamic study and optimization of hydrogen production by *Enterobacter aerogenes*. Int J Hydrog Energy 27:149–156
52. Turcot J, Bisaillon A, Hallenbeck PC (2008) Hydrogen production by continuous cultures of Escherichia coli under different nutrient regimes. Int J Hydrog Energy 33:1465–1470
53. Minnan L, Jinli H, Xiaobin W, Huijuan X et al (2005) Isolation and characterization of a high H_2-producing strain *Klebsialle oxytoca* HP1 from a hot spring. Res Microbiol 156:76–81
54. Taguchi F, Mizukai N, Saito-Taki T, Hasegawa K (2005) Hydrogen production from continuous fermentation of xylose during growth of *Clostridium* sp. Strain No. 2. Can J Microbiol 41:536–540
55. Thong SO, Prasertsan P, Karakashev D, Angelidaki I (2008) Thermophilic fermentative hydrogen production by the newly isolated *Thermoanerobacterium thermosaccharolyticum* PSU-2. Int J Hydrog Energy 33:1204–1214
56. Lin CN, Wu SY, Chang JS (2006a) Fermentative hydrogen production with a draft tube fluidized bed reactor containing silicone-gel-immobilized anaerobic sludge. Int J Hydrog Energy 31:2200–2210
57. Lin CY, Chang CC (2008) Hung., Fermentative hydrogen production from starch using natural mixed culture. Int J Hydrog Energy 33:2445–2453
58. Pan J, Zhang R, El-Mashad HM, Sun H (2008) Effect of food to microorganism ratio on biohydrogen production from food waste via anaerobic fermentation. Int J Hydrog Energy 33:6968–6975
59. Vatsala TM, Mohan RS, Manimaran A (2008) A pilot-scale study of biohydrogen production from distillery effluent using defined bacterial co-culture. Int J Hydrog Energy 33:5404–5415
60. Antonopoulou G, Stamatelatou K, Venetsaneas N, Kornaros M et al (2008) Biohydrogen and methane production from cheese whey in a two-stage anaerobic process. Ind Eng Chem Res 47:5227–5233
61. Lay JJ, Li YY, Noike T (1997) Influences of pH and moisture content on the methane production in high-solids sludge digestion. Water Res 31:1518–1524

62. Bowles LK, Ellefson WL (1985) Effects of Butanol on *Clostridium acetobutylicum*. Appl Environ Microbiol 50:1165–1170
63. Khanal SK, Chen WH, Sung LLS (2004) Biological hydrogen production: effects of pH and intermediate products. Int J Hydrog Energy 29:1123–1131
64. Ginkel SV, Sung S (2001) Biohydrogen as a function of pH and substrate concentration. Environ Sci Technol 35:4726–4730
65. Calli B (2008) Dark fermentative H_2 production from xylose and lactose—Effects of on-line pH control. Int J Hydrog Energy 33:522–530
66. Khanna N, Kotay SM, Gilbert JJ, Das D (2011) Improvement of biohydrogen production by *Enterobacter cloacae* IIT-BT 08 under regulated pH. Int J Hydrog Energy 152:9–15
67. Keskin T, Giusti L, Azbar N (2012) Continuous biohydrogen production in immobilized biofilm system versus suspended cell culture. Int J Hydrog Energy 37:1418–1424
68. Roy S, Vishnuvardhan M, Das D (2014b) Continuous hydrogen production in packed bed reactor. Appl Energy 136:51–58
69. Li CL, Fang HHP (2001) Fermentative hydrogen production from wastewater and solid wastes by mixed cultures. Crit Rev Environ Sci Technol 37:1–39
70. Lee KS, Lin PJ, Chang JS (2006) Temperature effects on biohydrogen production in a granular sludge bed induced by activated carbon carriers. Int J Hydrog Energy 31:465–472
71. Singh P, Pandey A (2001) An evaluative report and challenges for fermentative biohydrogen production. Int J Hydrog Energy 34:13435–13444
72. Van Groenestijn JW, Hazewinkel JHO, Nienoord M, Bussmann PJT (2002) Energy aspects of biological hydrogen production in high rate bioreactors operated in the thermophilic temperature range. Int J Hydrog Energy 27:1141–1147
73. Hawkes FR, Hussy I, Kyazze G, Dinsdale R, Hawkes DL (2007) Continuous dark fermentative hydrogen production by mesophilic microflora: principles and progress. Int J Hydrog Energy 32:172–184
74. Mizuno O, Dinsdale R, Hawkes FR, Hawkes DL, Noike T (2000) Enhancement of hydrogen production from glucose by nitrogen gas sparging. Bioresour Technol 73:59–65
75. Teplyakov VV, Ievlev AL, Durgaryan SG (1985) The gas permeability of copolymers. Polym Sci USSR 27:919–926
76. Lo YC, Su YC, Chen CY, Chen WM, Chang JS et al (2009) Biohydrogen production from cellulosic hydrolysate produced via temperature-shift-enhanced bacterial cellulose hydrolysis. Bioresour Technol 100:5802–5807
77. Zhang H, Bruns MA, Logan BE (2006a) Biological hydrogen production by *Clostridium acetobutylicum* in an unsaturated flow reactor. Water Res 40:728–734
78. Mariakakis I, Meyer C, Steinmetz H (2012) Fermentative hydrogen production by molasses; effect of hydraulic retention time, organic loading rate and microbial dynamics. In: Manic D (ed) Hydrogen energy – challenges and perspectives. In Tech. doi:10.5772/47750
79. Şentürk E, İnce M, Engin GO (2013) The effect of transient loading on the performance of a mesophilic anaerobic contact reactor at constant feed strength. J Biotechnol 164:232–237
80. Okpokwasili GC, Nweke CO (2005) Microbial growth and substrate utilization kinetics. Afr J Biotechnol 5:305–317
81. Shuler LM, Kargi F (2002) Bioprocess engineering—basic concepts, 2nd edn. New Delhi, Prentice-Hall of India, pp. 161–179
82. Lee K-S, Hsu Y-F, Lo Y-C, Lin P-J, Lin C-Y, Chang J-S (2008) Exploring optimal environmental factors for fermentative hydrogen production from starch using mixed anaerobic microflora. Int J Hydrog Energy 33:1565–1572
83. Ntaikou I, Gavala HN, Lyberatos G (2009) Modeling of fermentative hydrogen production from the bacterium *Ruminococcus albus*: definition of metabolism and kinetics during growth on glucose. Int J Hydrog Energy 34:3697–3709
84. Nath K, Muthukumar M, Kumar A, Das D (2008) Kinetics of two-stage fermentation process for the production of hydrogen. Int J Hydrog Energy 33:1195–1203

85. Jian Long W, Wei W (2008) The effect of substrate concentration on biohydrogen production by using kinetic models. Sci China Ser B: Chem 51:1110–1117
86. Han K, Levenspiel O (1988) Extended monod kinetics for substrate, product, and cell inhibition. Biotechnol Bioeng 32(4):430–437
87. Wang J, Wan W (2009) Optimization of fermentative hydrogen production process using genetic algorithm based on neural network and response surface methodology. Int J Hydrog Energy 34:255–261
88. Mu Y, Wang G, Yu H-Q (2006) Kinetic modeling of batch hydrogen production process by mixed anaerobic cultures. Bioresour Technol 97:1302–1307
89. Obeid J, Magnin JP, Flaus JM, Adrot O, Willison JC, Zlatev R (2009) Modeling of hydrogen production in batch cultures of the photosynthetic bacterium *Rhodobacter capsulatus*. Int J Hydrog Energy 34:180–185
90. Lo YC, Chen WM, Hung CH, Chen SD, Chang JS (2008) Dark H2 fermentation from sucrose and xylose using H_2-producing indigenous bacteria: feasibility and kinetic studies. Water Res 42:827–842
91. Noike T, Mizuno O (2000) Hydrogen fermentation of organic municipal wastes. Water Sci Technol 42:155–162
92. Wang CC, Chang C, Chu CP, Lee DJ et al (2003) Producing hydrogen from wastewater sludge by *Clostridium bifermentans*. J Biotechnol 102:83–92
93. Chen CC, Chuang YS, Lin CY, Lay et al (2012) Thermophilic dark fermentation of untreated rice straw using mixed cultures for hydrogen production. Int J Hydrog Energy 37:15540–15546
94. Orhon D, Gorgun E, Germirli F, Artan N (1993) Biological treatability of dairy wastewaters. Water Res 27:625–633
95. Mohan VS, Babu LV, Sarma PN (2008) Effect of various pretreatment methods on anaerobic mixed microflora to enhance biohydrogen production utilizing dairy wastewater as substrate. Bioresour Technol 99:59–67
96. O-Thong S, Mamimin C, Prasertsan P (2009) Effect of temperature and initial pH on biohydrogen production from palm oil mill effluent: long-term evaluation and microbial community analysis. Electron J Biotechnol 14
97. Kumar N, Das D (2001) Continuous hydrogen production by immobilized *Enterobacter cloacae* IIT-BT 08 using lignocellulosic materials as solid matrices. Enzym Microb Technol 29:280–287
98. Nath K, Roy N, Mukherjee L, Das D (2004a) Biohydrogenation using *Enterobacter cloacae* DM 11 and a comparative energy analysis with biomethanation. In: Energy and Environment (ed) A world of challenges and opportunities. Science Press, New York Ltd, pp 273–280
99. Saha S, Roy N, Nath K, Das D (2004) Biohydrogenation of industrial effluents by *Enterobacter cloacae* DM11. In: Marini M, Spazzafumo G (ed) Proceedings of hypothesis V. Porte Conte, pp 361–368, ISBN: 88-86281-90-0
100. Kotay SM (2008) Microbial hydrogen production from sewage sludge. Ph.D. thesis, IIT Kharagpur
101. Azbar N, Çetinkaya Dokgoz FT, Keskin T, Korkmaz KS (2009) Continuous fermentative hydrogen production from cheese whey wastewater under thermophilic anaerobic conditions. Int J Hydrog Energy 34:7441–7447
102. Palazzi E, Fabino B, Perego P (2000) Process development of continuous hydrogen production by *Enterobacter aerogenes* in a packed column reactor. Bioprocess Eng 22:205–213
103. Rachman MA, Nkashimada Y, Kakizono T, Nishio N (1998) Hydrogen production with high yield, and high evolution rate by self-flocculated cells of *Enterobacter aerogenes* in a packed-bed reactor. Appl Microbiol Biotechnol 49:450–454
104. Tanisho S, Ishiwata Y (1995) Continuous hydrogen production from molasses by fermentation using urethane foam as a support. Int J Hydrog Energy 23:559–564

105. Yokoi H, Tokushige T, Hirose J, Hayashi S, Takasaki Y (1997a) Hydrogen production by immobilized cells of aciduric *Enterobacter aerogenes* strain HO-3. J Ferment Bioeng 83:481–484
106. Yokoi H, Aratake T, Hirose J, Hayashi S, Takasaki Y (2001) Simultaneous production of H_2 and bioflocculant by *Enterobacter sp.* BY-29. World J Microbiol Biotechnol 17:609–613
107. Kumar JSR, Sharma CB, Joshi AP, Kalia VC (1995) Increased hydrogen production by immobilized microorganisms. World J Microbiol Biotechnol 11:156–159
108. Yokoi H, Maeda Y, Hirose J, Hayashi S, Takasaki Y (1997b) H_2 production by immobilized cells of *Clostridium butyricum* on porous glass beads. Biotechnol Technol 11:431
109. Yokoi H, Tokushige T, Hirose J, Hayashi S, Takasaki Y (1998) H_2 production from starch by a mixed culture of *Clostridium butyricum* and *Enterobacter aerogenes*. Biotechnol Lett 20:143–147
110. Zhang H, Bruns MA, Logan EB (2006b) Biological hydrogen production by *Clostridium acetobutylicum* in an unsaturated flow reactor. Water Res 40:728–734
111. Lee KS, Lo YS, Lo YC, Lin PJ et al (2003) H_2 production with anaerobic sludge using activated-carbon supported packed bed bioreactors. Biotechnol Lett 25:133–138
112. Wua KJ, Changa CF, Chang JS (2007) Simultaneous production of biohydrogen and bioethanol with fluidized-bed and packed bed bioreactors containing immobilized anaerobic sludge. Process Biochem 42:1165–1171
113. Zhu H, Suzuki T, Tsygankov AA, Asada et al (1999) Hydrogen production from tofu wastewater by *Rhodobacter sphaeroides* immobilized in agar gels. Int J Hydrog Energy 24:305–310
114. Seon YH, Lee CG, Park DH, Hwang KY et al (1983) Hydrogen production by immobilized cells in nozzle loop bioreactor. Biotechnol Lett 15:1275–1280
115. Nath K, Roy N, Mukherjee L, Das D (2004b) Biohydrogenation using *Enterobacter cloacae* DM 11 and a comparative energy analysis with biomethanation. In: Energy and Environment (ed) A world of challenges and opportunities. Science Press, New York, pp 273–280
116. Mohanakrishna G, Venkata Mohan S, Sarma PN (2010) Utilizing acid-rich effluents of fermentative hydrogen production process as substrate for harnessing bioelectricity: an integrative approach. Int J Hydrog Energy 35:3440–3449
117. Yogesh S, Baikun L (2007) Optimizing energy harvest in wastewater treatment by combining anaerobic hydrogen producing biofermentor (HPB) and microbial fuel cell (MFC). Int J Hydrog Energy 35:3789–3797
118. Lu L, Ren N, Xing D, Logan BE (2009) Hydrogen production with effluent from an ethanol-H_2-coproducing fermentation reactor using a single-chamber microbial electrolysis cell. Biosens Bioelectron 24:3055–3060
119. Lalaurette L, Thammanagowda S, Mohagheghi A, Maness P et al (2009) Hydrogen production from cellulose in a two-stage process combining fermentation and electrohydrogenesis. Int J Hydrog Energy 34:6201–6210
120. Vatsala TM, Mohan RS, Manimaran A (2008) A pilot-scale study of biohydrogen production from distillery effluent using defined bacterial co-culture. Int J Hydrog Energy 33:5404–5415
121. Ren NQ, Li JZ, Li BK, Wang Y, Liu SR (2006) Biohydrogen production from molasses by anaerobic fermentation with a pilot scale bioreactor system. Int J Hydrog Energy 31:2147–2157
122. Lin CY, Wu SY, Lin PJ, Chang JS et al (2010) A pilot-scale high-rate biohydrogen production system with mixed microflora. Int J Hydrog Energy 36:8758–8764
123. Krupp M, Widminn R (2009) Biohydrogen production by dark fermentation: Experiences of continuous operation in large lab scale. Int J Hydrog Energy 34:4509–4516
124. Show KY, Zhang ZP, Tay JH, Liang TD, Lee DJ, Jiang WJ (2007) Production of hydrogen in a granular sludge-based anaerobic continuous stirred tank reactor. Int J Hydrog Energy 32 (18):4744–4753

125. Peintner C, Zeidan AA, Schnitzhofer W (2010) Bioreactor systems for thermophilic fermentative hydrogen production: evaluation and comparison of appropriate systems. J Clean Prod 18:15–22
126. Lin CN, Wu SY, Chang JS (2006b) Fermentative hydrogen production with a draft tube fluidized bed reactor containing silicone-gel-immobilized anaerobic sludge. Int J Hydrog Energy 31:2200–2210

Chapter 17
Technological Development for Capturing Regeneration, Standardization, and Storage of Solar Energy: Current Status and Future Direction

Hasan Baig, Honey Brahma, Tapas K. Mallick, and Nabin Sarmah

Abstract The Indian Economy has seen a steady growth in the past few years and is expected to lead others in the coming decade. With growing population, increased energy demand, rising oil prices, and climate change, the choice of adopting renewable energy seems to be of utmost importance. The Ministry of Energy set up a separate department for Non-Conventional Energy Resources in 1982 which now operates under the name if Ministry of New and Renewable Energy. Efforts are being initiated throughout the country at different levels for tapping different renewable sources of energy. Solar energy is one of the most promising renewable energy resources available in India whose full potential still remains untapped. India has about 300 sunny days with most parts of the country receiving an average solar radiation in the range of 4–7 kWh/day, making it an ideal location for the installation of solar technologies.

Keywords Solar energy • Energy storage • Indian energy scenario

1 Introduction

The Indian economy has seen a steady growth in the past few years and is expected to lead others in the coming decade. With growing population, increased energy demand, rising oil prices, and climate change, the choice of adopting renewable energy seems to be of utmost importance. The Ministry of Energy set up a separate department for Non-Conventional Energy Resources in 1982 which now operates under the name of Ministry of New and Renewable Energy. Efforts are being

H. Baig • T.K. Mallick
Environmental and Sustainability Institute, University of Exeter Penryn Campus, Penryn TR10 9FE, UK

H. Brahma • N. Sarmah (✉)
Department of Energy, Tezpur University, Napaam, Tezpur, Assam 784028, India
e-mail: nabin@tezu.ernet.in

A.K. Chandel, R.K. Sukumaran (eds.), *Sustainable Biofuels Development in India*,
DOI 10.1007/978-3-319-50219-9_17

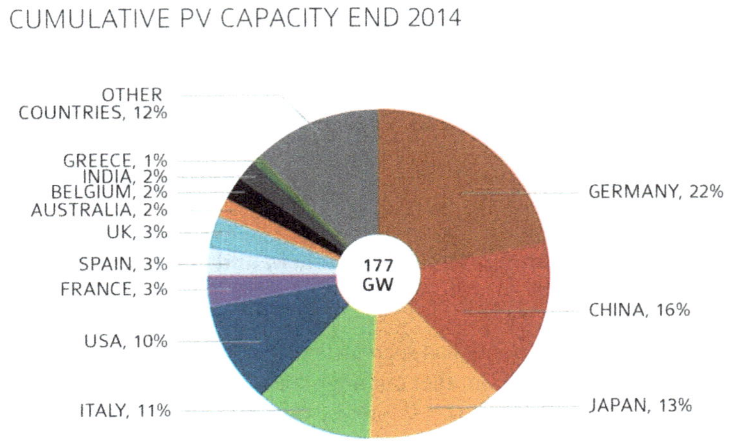

Fig. 17.1 Cumulative photovoltaic installed in the World. Source: IEA PVPS

initiated throughout the country at different levels for tapping different renewable sources of energy. Solar energy is one of the most promising renewable energy resources available in India whose full potential still remains untapped. India has about 300 sunny days with most parts of the country receiving an average solar radiation in the range of 4–7 kWh/day, making it an ideal location of installation of solar technologies.

Over the last few decades, several solar energy-based systems and devices have been developed and deployed in India as energy solutions for lighting, cooking, water and air heating, and electricity generation. The Jawaharlal Nehru National Solar Mission (JNNSM) was launched in 2010 to install 100 GW grid-connected photovoltaic systems by 2022. Currently, the global installed solar power is currently 227 GW, and India currently accounts for about 2% of installed capacity with 3.3 GW [1] (Fig. 17.1).

With more than half of its population currently living in rural areas and previously relying on dirty and polluting kerosene lamps, sustainable solutions like solar lanterns are changing the face of modern India and bringing solar energy deployment to the forefront. While the use of solar lanterns is becoming common among villagers for lighting purposes, people in the cities and towns are installing solar hot water systems and photovoltaic panels for meeting their energy demands. The stakeholders and industry are coming forward for making investments in huge power projects and creating new solar industries, overall giving us a picturesque reflection of how significant the implementation of solar energy technologies would play in not only meeting our growing energy demands but also its role in all sectors of our economy in the coming years.

Research and development activities in solar energy technologies across the nation via several industrial, research, and educational institutions both independently and collaborative agreements with international organizations help in creating the know-how about making the best use of these technologies. Projects ranging from developing small-scale solar applications like solar lanterns for villages to huge power plants are under way. This chapter aims to highlight the basic technologies using solar energy available in India today, current research going on in this area, the role of government in its implementation, and collaborative programs under way with international partners.

2 Solar Energy and Its Potential in India

The sun is a potential source of energy, a natural fusion reactor continuously radiating energy which can be classified as beam and diffuse radiations. The solar radiation received by the earth with and without scattering effects of atmospheric gases is known as diffuse and beam radiation. The sum of both these radiations is known as the total or global solar radiation. The Earth's atmosphere receives about 1366 W/m^2. This value varies with respect to the time of the year and the location on earth with highest incidence close to equatorial regions.

India is geographically very well placed on the earth's solar belt (40°S–40°N) with annual number of sunny days ranging from 250 to 300 and most of the regions receiving an average annual global horizontal radiation amounting to 1800–2000 kWh/m^2 as shown in Fig. 17.2, which is equivalent to about 6000 million GWh per year [2]. The opportunity for solar energy to be a leader in energy supply lies in the inherent climatic conditions in India and the rising energy demands.

India's electricity shortage has been increasing drastically with time and can by no means be met by using conventional power, giving the solar industry a perfect opportunity for meeting this demand.

The map shows that the highest annual global radiation is received in the North West regions of India including states like Gujarat, Rajasthan, and some parts of states like Maharashtra, Andhra Pradesh, and Madhya Pradesh. In a recent study [3], solar hotspots or regions characterized by an exceptional solar power potential suitable for decentralized commercial exploitation of energy with the favorable techno-economic prospects and organizational infrastructure were identified. It was found that more than 58% of the country has these hotspots receiving global radiation on an average greater than 5 kWh/m^2/day, making solar energy promising. Solar radiation varies throughout the day with the sun's position, declination angle, hour angle, and meteorological conditions. Irradiance, air mass, aerosol density, water vapor, and turbidity are some of the atmospheric parameters that influence the incident solar spectrum falling on the earth's surface. Advanced Measurement Stations (AMS) are installed in four locations: Chennai, Kolkata, New Delhi, and Gandhinagar, to assess the aerosol depth, turbidity, ozone label, water vapor, and NO_x in the atmosphere [4].

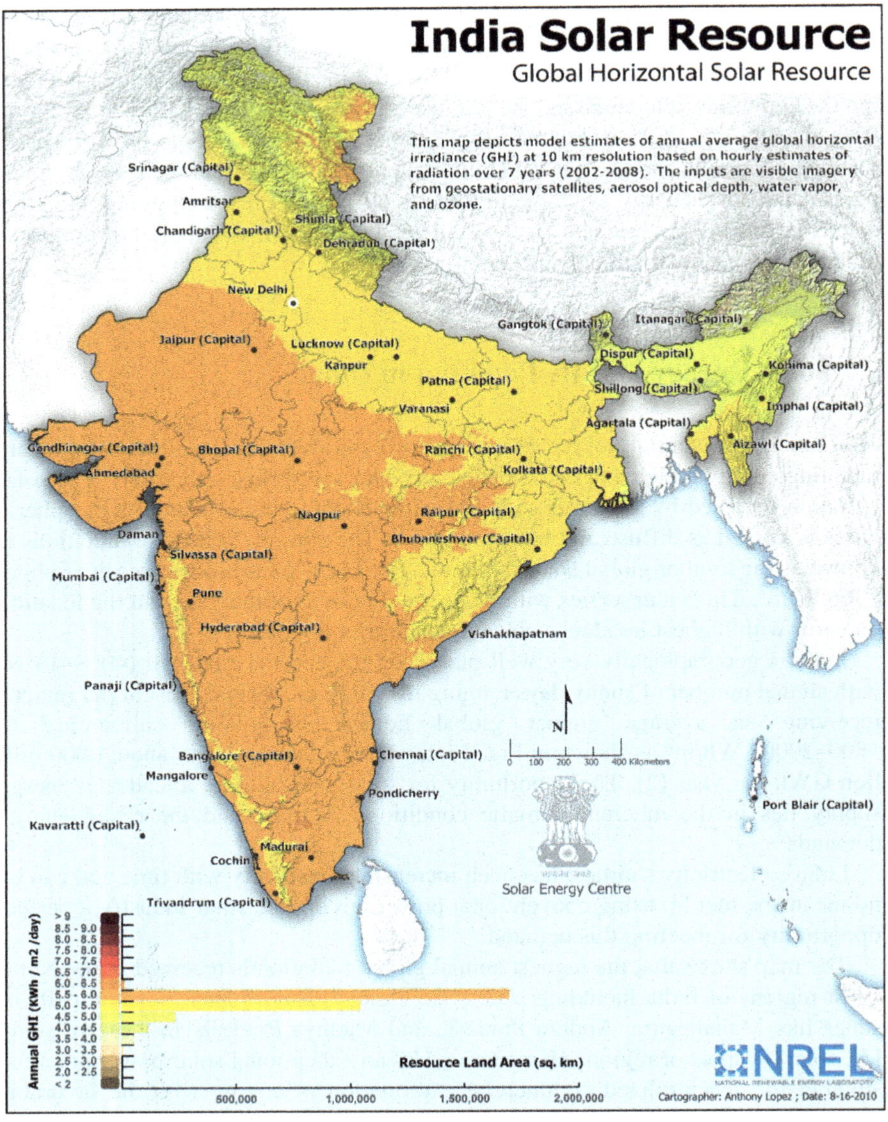

Fig. 17.2 Global horizontal irradiation map of India (Image Courtesy: NREL)

The total capacity for solar power generation in India is estimated to be in the range 510,000 kWh–800,000 kWh per acre of land. The highest solar energy generation is 750,000 kWh–800,000 kWh per acre that covers the eastern parts of both Jammu and Kashmir and Uttarakhand, western dry region, and Gujarat. The major region of India's landmass receives the solar energy potential range of 680,000 kWh–730,000 kWh per acre including the eastern part of Rajasthan, Karnataka, Tamil Nadu, Maharashtra, Madhya Pradesh, Haryana, Punjab, and

Chhattisgarh. The solar energy potential of northern part of Jammu and Kashmir, Himachal Pradesh, Sikkim, West Bengal, Bihar, the western part of Assam, Meghalaya, and Manipur ranges from 620,000 kWh–650,000 kWh per acre. The eastern Himalayan states including the eastern part of Arunachal Pradesh, of Assam, and Nagaland receives least global solar insolation below 4 kWh/m^2/day [5, 6]. Figure 17.3 shows the annual average direct normal irradiance (DNI)

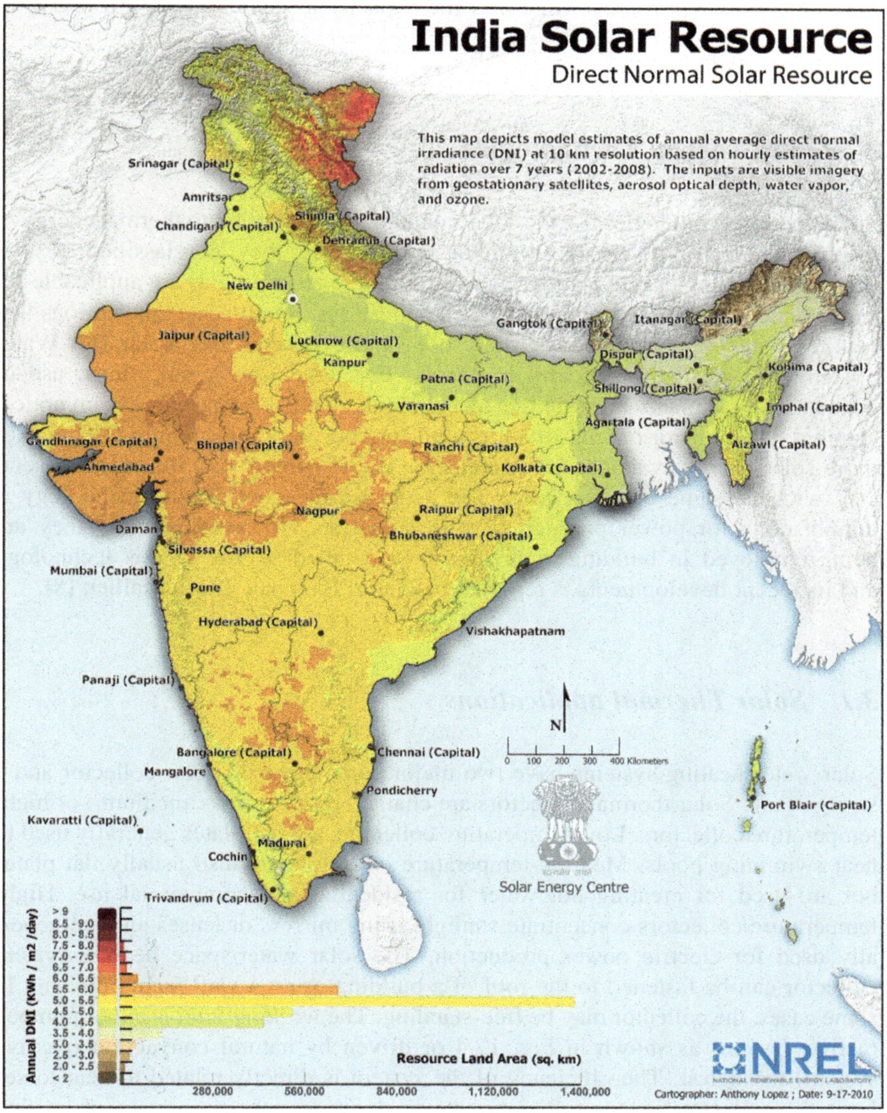

Fig. 17.3 Direct normal irradiance map for India (Image Courtesy: NREL)

(kWh/m^2/day) in India from which it can be seen that large part of the country has the potential for solar thermal projects. The major region includes the states Gujarat, Rajasthan, Maharashtra, Karnataka, Andhra Pradesh, Tamil Nadu, Jammu and Kashmir, Uttarakhand, and Himachal Pradesh. The North-eastern states receive the least direct normal irradiance. The daily sunshine hour is maximum in states like Gujarat, Rajasthan, Maharashtra, and Andhra Pradesh and lowest in the North-eastern states. The Thar Desert of Rajasthan receives the maximum annual DNI of 2000 kWh/m^2/day which has the capacity to generate 700–2100 GW of energy [7].

3 Solar Energy Technologies

Solar energy technologies use the solar radiation to convert it into thermal energy or to produce electricity. Based on this basic difference, they are classified as solar thermal technologies and photovoltaics. Both these technologies are applicable for small-scale applications and large-scale applications. Small-scale applications like heating water or providing electricity use simple systems like Solar Hot Water System (SHWS) and photovoltaic panels. For large-scale applications, use of concentrator is made in both the technologies. The concentrator is an optical element like a mirror or lens which is used for concentrating sunlight. This makes more solar energy available to the collector, thereby improving the output associated with the same collector area. The main use of concentrating technology is mainly done for power production or desalination. However, recently they are being employed in building applications. A detailed review of this technology and its recent developments is reported by Hasan Baig and Tapas Mallick [8].

3.1 Solar Thermal applications

Solar water heating systems have two major components: a solar collector and a storage tank. Solar thermal collectors are characterized as low-, medium-, or high-temperature collectors. Low-temperature collectors are flat plates generally used to heat swimming pools. Medium-temperature collectors are also usually flat plates but are used for creating hot water for residential and commercial use. High-temperature collectors concentrate sunlight using mirrors or lenses and are generally used for electric power production. The solar water/space heating system collector can be fastened to the roof of a building or on a wall facing the sun. In some cases, the collector may be free-standing. The working fluid is either pumped (active system) as shown in Fig. 17.4 or driven by natural convection (passive system) through it. The efficiency of the system is directly related to heat losses from the collector surface (efficiency being defined as the proportion of heating energy that can be usefully obtained from insulation). Heat losses are

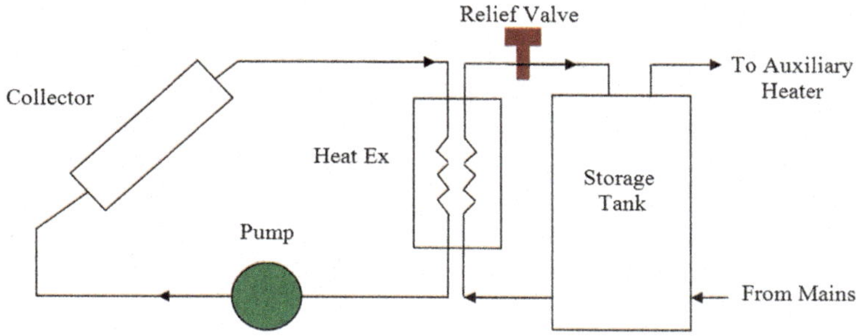

Fig. 17.4 Active solar water heating system

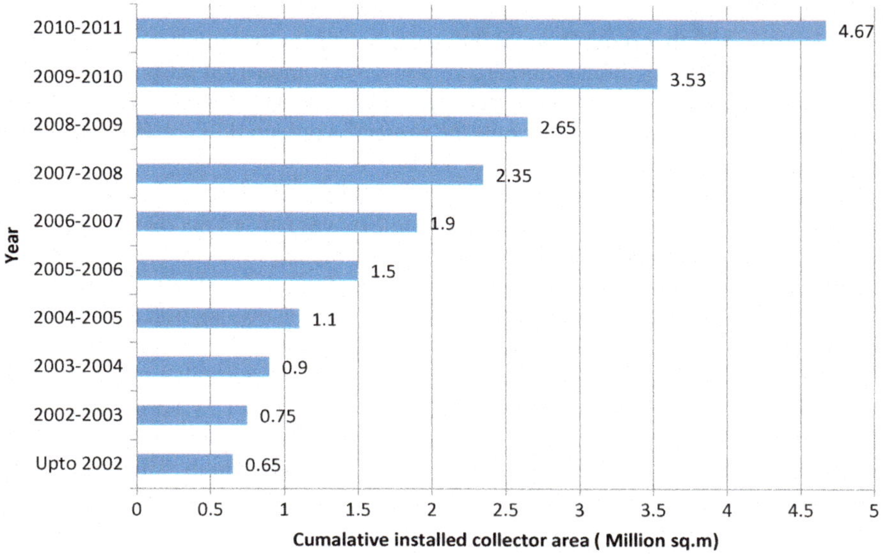

Fig. 17.5 Trends in cumulative installed SWHS in India until August 2011

predominantly governed by the thermal gradient between the temperature of the collector surface and the ambient temperature. A flat-plate collector forms the basic building block for any solar water heating system. The main function of the collector is to absorb the incident solar radiation and convert it into useful thermal energy.

Several types of such collectors are available in the market today; however, they may be classified essentially into three types, namely, the evacuated tube collector which uses a heat pipe inside a vacuum-sealed tube, compound parabolic collectors which make use of concentrator technology, and the simple flat-plate collectors which consist of finned tubes; the fins coated with black absorb solar radiation and transmit to the fluid.

The cumulative growth of SHWS in terms of collector areas installed has been multiplying as can be seen in Fig. 17.5 with only 0.119 million square meter in

Table 17.1 Projected potential of SWHS under realistic scenario for India (cumulative millions of square meter)

Sector	2013	2017	2022
Residential	4.25	7.68	15.74
Commercial/Institutional			
• Hotels	0.35	0.61	0.97
• Hospitals	0.17	0.27	0.43
• Others	0.27	0.39	0.52
Industry	0.33	0.57	1.05
Total	5.37	9.52	18.7

1989; the figure rose to 0.68 million square meter by the end of 2002, and by the end of year 2004, a total of 0.9 million square meter of collectors have been installed [9] growing up to 3.53 million square meter in 2010. According to MNRE [10] as of August 2011, around 4.67 million square meters of systems have been installed.

It is estimated that more than 20,000 solar hot water systems are installed each year in India for residential, commercial, and several industrial purposes including the production of textiles, chemicals, paper, plastics rubber and dairy goods [11]. A recent study by Greentech [12] shows the projected potential of SWHS in different sectors for the coming 20 years (Table 17.1).

In order to estimate the number the households that can afford to invest in buying the SWHS systems, a techno-economic feasibility was carried out by Chandrasekar and Kandpal [13] in 2004. The main factors affecting the investment were found to be the capital cost and the interest rate. According to a recent study by Veeraboina and Ratnam [14], the demand for hot water in India shows significant variations across regions primarily because it not only depends on climatic factors but also depends on human behavior. Some regions which do not fall under cold and moderate climate regions but exhibit high use of hot water systems are parts of Maharashtra, Kerala, and Tamil Nadu. Other factors that were found to affect the usage of SWHS include the type of house, electricity supply situation in the locality, awareness about SWHS technologies, household income, and policy program of the local municipality.

A number of manufacturers are now present across every state of India; a comprehensive list giving details can be found at MNRE [15]. According to the Ministry of New and Renewable Energy (MNRE), India, there are 60 BIS (Bureau of Indian standards)-approved manufacturers for solar flat plate collectors and 44 MNRE-approved evacuated tube collectors. The system costs Rs. 15000–20,000 for a 100 liter capacity and Rs. 110–150 per installed liter for higher capacity systems with an estimated payback period of 3–4 years.

3.1.1 Concentrating Solar Thermal Power

The general idea in this technology is to collect sunlight from many reflectors spread over a large area at one central point (solar power tower plants) or to concentrate sunlight using parabolic trough collectors (Parabolic Trough Plants) and to achieve the required high temperatures. A solar thermal power plant in

Fig. 17.6 World's largest solar cooking system in Shirdi which can cook for 100,000 people per day

principle works no differently than a conventional steam power plant and can be easily integrated with existing technologies. The important difference is that no harm is done to the environment by burning coal, oil, and natural gas or by splitting uranium to produce steam as the energy is solely produced from the sun. Parabolic systems use trough-shaped mirrors to focus sunlight onto an absorber tube (receiver) placed in the trough's focal line. The troughs are designed to track the sun along one axis, predominantly north-south. Parabolic dish systems consist of a parabolic-shaped point focus concentrator in the form of a dish that reflects solar radiation onto a receiver mounted at the focal point. These concentrators are mounted on a structure with a two-axis tracking system to follow the sun. A typical dish concentrator can be spotted in Shirdi, and is the World's largest solar cooking system in Shirdi consisting of 73 solar dishes which can generate steam of about 650 °C which is used to cook food. The solar power is used to produce steam which is then used to cook food for more than 100,000 people daily (Fig. 17.6).

Concentrated solar power (CSP) is beginning to gain impetus in India for power production; several announcements of big projects were made recently in a map by Iyican et al. [16]). The list of CSP projects ongoing and under construction/planned in India is presented in Table 17.2.

3.2 *Photovoltaics*

The conversion of the solar energy directly to electricity is referred to as photovoltaics (PV). The process primarily occurs due to the intrinsic properties of the semiconductor materials such as silicon and germanium. When doped properly, these materials can be made into n- and p-type semiconductors and combined to

Table 17.2 Utility-scale CSP plants that are currently operating, under construction/planned in India

	Project	Size MW	Location	Technology	Completion date
Planned/under construction (National Solar Mission)					
1	Rajasthan Sun Technique Energy Pvt. Ltd. Reliance Power	100	Rajasthan	TBC	March 2013
2	LancoInfratech Ltd. Project	100		Parabolic Trough	March 2013
3	KVK Energy Ventures Pvt Ltd. Project	100		Parabolic Trough	March 2013
4	Godavari Power & Ispat Ltd. Project	50		Parabolic Trough	March 2013
5	Corporate Ispat Alloys Ltd. Project	50		Parabolic Trough	March 2013
6	Megha Engineering & Infra-structures Ltd. Project	50	Andhra Pradesh	Parabolic Trough	March 2013
7	Aurum Renewable Energy 20 MW	20	Gujarat	CLFR	March 2013
	Total Planned	470			
Operational/under development					
8	Rajasthan Solar One (5 MW Operational)	10	Rajasthan	Parabolic Trough	
9	Bap Solar Power Plant	10		Parabolic Dish Sterling	
10	NTPC Pilot project	15		Parabolic Trough	
11	ACME Bikaner Project (2.5 MW Operational)	10		Power Tower	
12	MNRE R&D Project (Fully Operational)	1	Haryana	Parabolic Trough	
13	Andhra Pradesh Project	50	Andhra Pradesh	Parabolic Trough	
14	Thermal Power Project, Kutch	25	Gujarat	Parabolic Trough with Thermal Storage	
	Total operational/under development (MW)	121			

form a p–n junction. When sunlight is incident on these materials, electrons are excited by the incident photon energy and current is generated due to the movement of electrons and holes. This process is termed as the photovoltaic effect and was discovered in 1839 by Becquerel. The first silicon solar cell was manufactured by Chapmin in 1954. The PV technology found its first applications in the space missions and remote electricity stations. Although photovoltaics have been around since the 1950s, the terrestrial applications were heightened during the oil embargo period of the 1970s. The solar industry has never looked back since then and has been widely been developed across the globe.

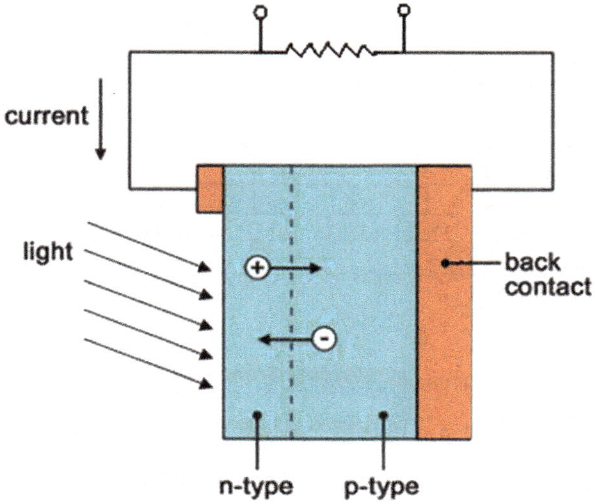

Fig. 17.7 Schematic diagram of solar cell circuit

The basic working principle of the solar cell is shown in Fig. 17.7. The solar cell is essentially a p–n junction which is formed by coupling the n- and p-type materials. The n-type materials have excessive electrons, and the p-type materials have excessive holes. At the junction of n-type and p-type semiconductors, few surplus electrons near the junction in the n-type will diffuse into the p-type leaving behind positive charge in the n-type and while holes travel from p-type to n-type leaving behind negative charge in the p-type [17]. This creates a strong electric field near the junction. The sunlight when incident on the solar cell excites an electron on the n-type material and moves it across the junction toward the p-type material, hence creating an electron–hole pair movement along the junction which leads to the production of the current [18–20].

The Photovoltaic technology on module scale can be grouped into four basic categories namely:

(a) Silicon solar cells (mono- and multicrystalline)
(b) Thin-film solar cells (CdTe, Copper indium gallium selenide (CIGS) and amorphous-Si)
(c) Concentrating PV (single-crystalline Si and III–V multijunction cells)
(d) New materials

The different types of PV technologies and subcategories of these technologies are given in Fig. 17.8.

Modules are a group of solar cells arranged together in series to produce a desired amount of current and voltage. Manufacturing of PV modules in India can be dated back in the mid-1970s during the solar photovoltaics program of the Government of India. Major companies in India like the Central Electronics Ltd., BHEL, REIL, and the other manufacturers of SPV modules were solely assembling the modules while importing the cells from other countries. Domestic solar cell manufacturing did not occur because of the high investment cost involved and

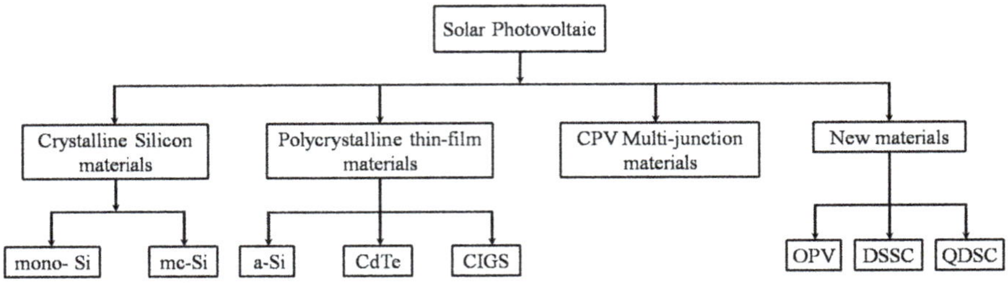

Fig. 17.8 Different types of PV technologies

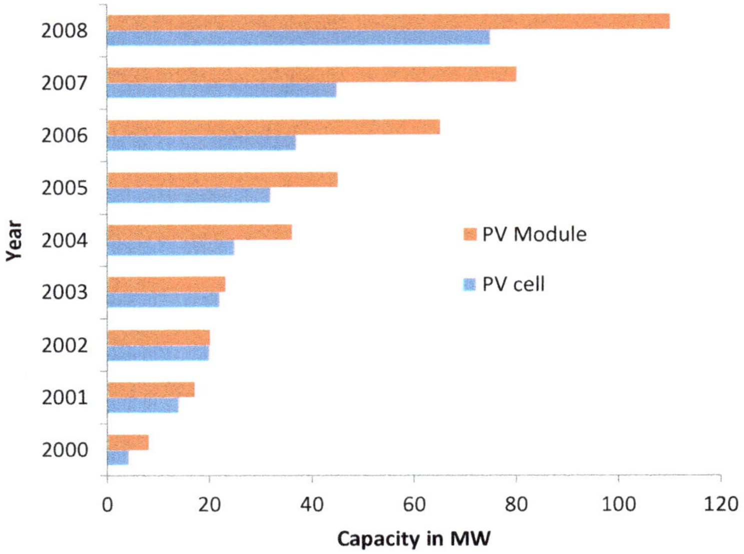

Fig. 17.9 PV cell and module manufacturing

insufficient technology. Until very recently, huge investments were being made by the government and industry to kick start solar cell manufacturing. Figure 17.9 shows the trend in PV cell and PV modules capacity over the last decade; the excess in module capacity is primarily due to the import of solar cells by module assemblers.

The climatic conditions in India and the inherent demand for electricity make India as one of the most destinations for PV installations across the world. Although India has negligible installed capacity, a promising figure could be highly probable in the near future. Comparing the global picture of solar energy potential and drivers in Fig. 17.10, it can be clearly seen that India has one of the highest amounts of solar radiation and the lowest feed in tariff schemes making the market favorable.

	Fundamental		Driver		
	Irradiation (kWh/m2/ per day)	Long Term Energy Deficit	FiT (per kWh)	RPO	Capital Subsidy
Mature markets					
Germany	2.7	No	~₹12 ($0.30)	No	Yes
Italy	3.3	No	~₹11 ($0.27)	Yes	Yes
Spain	3.7	No	~₹11 ($0.27)	Yes	Yes
New Markets					
India	5.5	Yes	₹12 ($0.30)	Yes	Yes
South Africa	5.2	Yes	₹22($0.55)	Yes	Yes
Australia	5.5	No		Yes	Yes
Ontario (Canada)	3.1	No	₹17($0.425)*	No	Yes

*For projects less than 10MW

Fig. 17.10 Drivers for PV market in India. Source: REN 21 Global Status Report, BRIDGE TO INDIA ANALYSIS

Until the year 2008, all the raw material required for making solar cells was imported from other countries. But now India has all the capability to produce ingots and wafers used to manufacture solar cells. According to BRIDGE TO INDIA – India Solar Handbook [21], the module production growth is expected to become threefold by the year 2020 as shown in Fig. 17.11. The Indian manufacturing industry currently has module production capacities of about 1300 MW, more than 90% of which is for crystalline silicon modules. A number of companies about 10–12 have entered the market for solar cell manufacturing; however, one company by the name of Maharishi Solar is the only Indian company with manufacturing facilities for ingots and wafers; it has a total capacity of 15 MW.

Installations of PV modules are progressing at a rapid rate across the country. The subsidies introduced by the government play an important role for this growth. A cumulative of 46.16 MW has been achieved till August 2011, with about 8.5 MW in the last year alone. A number of new housing complexes using solar energy to power utilities are being built in the nation. A typical housing colony using PV panels and solar heaters can be seen in Fig. 17.12.

3.2.1 Silicon Materials

The performance of wafer-based crystalline silicon solar cells over the last few decades has greatly improved in terms of stability and reliability [22]. Crystalline silicon solar cells dominate the photovoltaic (PV) market covering 90% of the total PV cell manufactured worldwide. The laboratory-based efficiency of a crystalline silicon solar cell is reported to be 25%; however, the standard industrial efficiency

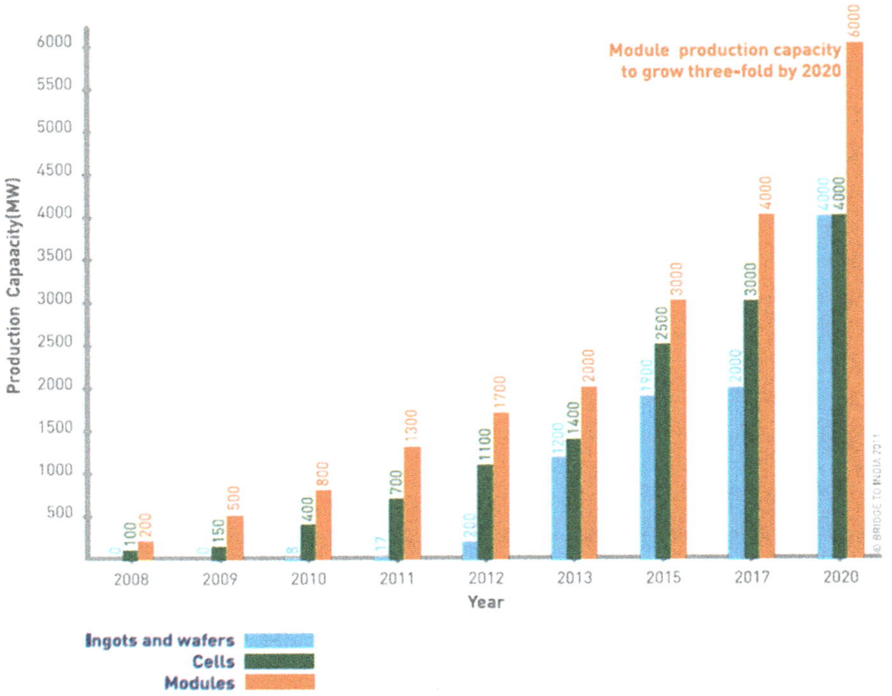

Fig. 17.11 Forecast of PV production capacities of ingots, cells, and modules by 2020

limits to 15–18% with some exceptions of high quality cells with efficiencies above 20% [23]. Figure 17.13 shows the schematic of crystalline silicon solar cells with two different metal contact approaches (finger and bus bar). There are typical two types of crystalline silicon solar cell: monocrystalline and polycrystalline silicon solar cell. The module production and balance of system equally contribute to the photovoltaic cost of production.

(a) Monocrystalline silicon solar cell: The reported terrestrial crystalline silicon cell efficiency under ASTM G-173-03 global is 25.6 ± 0.5% as tested by AIST [25]. Monocrystalline silicon solar cells with the Back Contact-Back Junction (BC-BJ) cells and heterojunction with intrinsic thin-layer (HIT) are considered to have high efficiencies for mass industrial production. However, this complex cell structure requires time consumption and technical specialization which makes it incompatible on commercial ground. The light-induced degradation (LID) effect is one of the negative effects of the monocrystalline cell, so LID effect free wafers, namely, boron-doped magnetic field Czochralski (CZ) wafers, gallium-doped CZ wafers, phosphorous-doped n-type CZ wafers, and n-type CZ-silicon wafers, are employed by the cell producers [23].

Fig. 17.12 Solar Housing Complex in Kolkata having solar panels on the rooftop

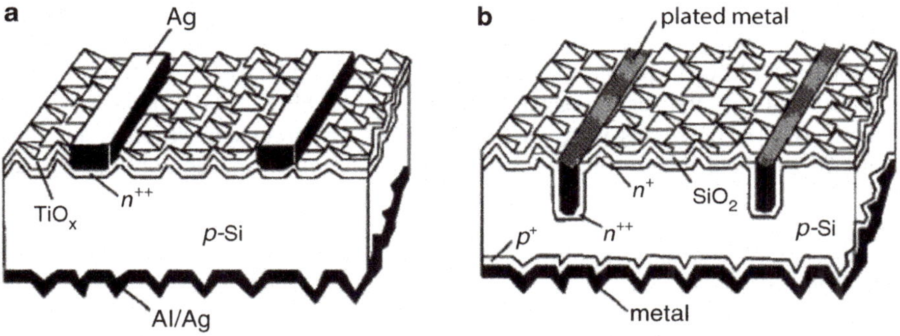

Fig. 17.13 Schematic representation of (a) screen-printed crystalline silicon solar cell and (b) laser-grooved buried contact crystalline silicon solar cell [24]

(b) Multicrystalline silicon (mc-Si) solar cell: The efficiency of a multicrystalline silicon device for cell area of 244 cm^2 as measured by Fraunhofer Institute for Solar Energy Systems (FhG-ISE) is reported to be 20.8 $\pm$ 0.6% [25]. The growth of the solar cells based on multicrystalline silicon wafers can be noticed by an increase in their market share from 30 to 48% of the world's photovoltaics market from 1998 to 2010 [26, 27]. Typically, the silicon wafer is fabricated with a technique which allows silicon ingots with large columnar grains to grow from the bottom when solidifying molten silicon [26]. Several different techniques for the manufacture and texturing of the solar cells have been reported to improve the efficiency of the multicrystalline solar cells [28–31]. The efficiency of the multicrystalline silicon solar cell is limited because of the minority carrier recombination effects. The recombination takes place due to the intragrain defects and dislocations formed during the manufacturing process.

3.2.2 Polycrystalline Thin-Film Materials

The state-of-the-art polycrystalline thin-film solar cells are cadmium telluride (CdTe) and copper indium gallium selenide (CIGS).

Cadmium telluride/cadmium sulfide (CdTe/CdS): CdTe is a quite encouraging polycrystalline thin-film solar cell and is in focus due to its direct optical band gap of 1.5 eV, high absorption coefficient (greater 5×10^5 cm^{-1}) and matching electron affinity to CdS which acts as a window material [32, 33]. A schematic of CdTe thin-film solar cell is shown in Fig. 17.14. In order to enhance the performance of the conventional SnO_2/CdS/CdTe, interdiffusion techniques are employed. The interdiffusion technique on CdS and zinc tin oxide (ZTO) layers improves the quantum efficiency and adhesion between the transparent conductive oxide (TCO) and CdS after $CdCl_2$ treatment [33]. The confirmed world-record efficiency of a CdTe terrestrial cell under AM1.5 G spectrum at 250 °C (IEC 60904-3: 2008, ASTM G-173-03 global) is 21 ± 0.4% with fill factor 79.4% [25]. Projects for control of the as-deposited defects by refining CdTe deposition and postdeposition processes were carried out [34]. Different deposition techniques, namely, physical vapor deposition, sputtering, and close-spaced sublimation, are

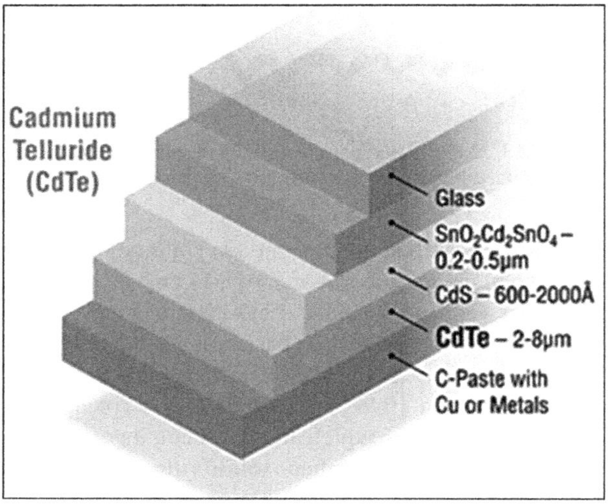

Fig. 17.14 Schematic diagram of CdTe thin-film PV device (NREL)

Fig. 17.15 Schematic diagram of CIGS thin-film PV device (NREL)

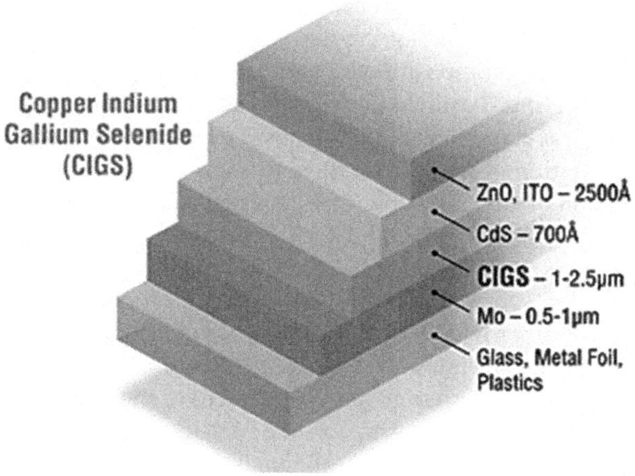

adopted for the deposition of CdTe on the material substrate [35]. CdTe/CdS based on SnO_2-coated borosilicate substrate was reported to have conversion efficiency of 15% in 1994. Techniques such as chemical bath deposition (CBD) techniques and closed spaced sublimation (CSS) are used for CdS and CdTe, respectively [36, 37].

Cadmium indium gallium selenide (CIGS): $Cu(In,Ga)Se_2$ (CIGS) is one class of polycrystalline thin-film solar cell to have emerged in last few decades; however, due to its technical complexity and the economic struggle to a develop quality device, the technology is still evolving [38]. The scarcity of indium prevents the scale-up of CIGS to terawatt development [39]. CIGS cells reported to have a conversion efficiency of $20.6 \pm 0.6\%$ under the global AM1.5 spectrum (1000 W/m^2) at 25 °C (IEC 60904-3: 2008, ASTM G-173-03 global) [25]. Figure 17.15 shows the schematic diagram of a CIGS thin-film solar cell.

Soda lime glass/Mo/CIGS/CdS/i-ZnO/ZnO:Al/MgF_2 cells have attained a conversion efficiency of 20% with MgF_2 acting as the anti-reflection coating. The improved baseline process is employed to reach a higher conversion efficiency [40]. Different buffer layers are employed to CIGS devices to achieve greater conversion efficiencies. The use of chemical bath deposition (CBD)-ZnS improves the quantum efficiency at shorter wavelengths compared to CBD-CdS with 18.1% efficiency [41]. Indium sulfide (In_2S_3) buffer layer deposition using atomic layer chemical vapor deposition (ALCVD) has achieved an efficiency of 16.4% [42].

The technological improvement in solar cell as briefly outlined above has resulted in a new world record efficiency for thin-film solar cells of 21.7% (with ARC). The area of this solar cell (line 1 in Table 17.1) is (0.4972 ± 0.0031) cm^2 as determined by an optical method. These values have been measured and certified independently by Fraunhofer ISE in Freiburg, Germany.

Vacuum processing offers a safe and controlled process and has been employed to develop CIGS devices by thermal processing of indium (In), gallium (Ga), Selenium (Se), and planar magnetron sputtering of copper (Cu). This process is termed as a hybrid process [38]. However, it requires high initial investment. The techniques including the fast non-vacuum deposition techniques and the prefixing of the film composition on a molecular level in a precursor layer provide a low-cost processing. Paste coating, selenization, and electrodeposition methods encourage non-vacuum deposition processes to develop improved quality CIGS absorber formation [43]. A three-stage process which is a co-evaporation process for fabrication of CIGS absorber has acquired a conversion efficiency of 19.3% for absorber band gap 1.15 eV and 18.4% for absorber band gap 1.12 eV [44]. The materials used as flexible substrates for the cells are metals or polymers. Metal foils make a flexible substrate due to their potential properties of high mechanical and thermal stability, provided quality diffusion barriers are introduced to suppress the entrance of diffusion impurities into the absorber [45]. CIGS on stainless chromium (Cr) steel substrate is not cost-effective; the reason is the expensive Cr, but it has an advantage over corrosion and high temperature processing compared to a Cr-free substrate. A conversion efficiency of 12.8% is recorded for even unalloyed steel (Cr-free) through controlling the diffusion of iron (Fe) by introducing SiO_x diffusion barriers [46].

Substrate characteristics include thermal expansion properties and stability, surface roughness, and surface roughness which extremely influence the development and properties of the substrate layer. A quality substrate requires the following properties: vacuum and thermal stability, good thermal expansion, chemical inertness, sufficient humidity barrier, surface smoothness, cost effectiveness, less energy consumption, an abundance of material, and a light weight [45].

Amorphous silicon (a-Si): Thin-film amorphous silicon solar cells were reported to have the potential to achieve a maximum efficiency of 14–15% in 1976. Thin-film solar cells of ~1 μm have been developed by considering the a-Si deposited using a glow discharge in silane (SiH_4). The device was fabricated by depositing a-Si on the substrates of indium-tin-oxide (ITO)-coated glass at 250–400 °C [47]. The recent recorded efficiency of a-Si terrestrial cell is 10.3% under ASTM G-173-03 global [25].

Multijunction solar cell were explored when the silicon cell needed a substitution in concentrated photovoltaic (CPV) used in space missions [48]. The first two Mars rovers, Spirit and Opportunity, were powered by multijunction solar cells as reported by Boeing Spectrolab in 2010 [49]. The popular multijunction solar cell on the market is lattice-matched InGaP/InGaAs/Ge triple junction solar cell with a recorded efficiency of 44.4% (NREL). The schematic illustration of a lattice-matched InGaP/InGaAs/Ge triple junction solar cell is shown in Fig. 17.16. In 2014, Soitec and CEA-Leti, France, and Fraunhofer Institute of Solar Energy systems ISE, Germany, together in collaboration confirmed the new record of III–V multijunction solar cell (four-junction) efficiency; this solar cell converts 46% of its incident solar light to electricity at a concentration of 508 suns as confirmed by Japanese AIST [51]. The 5-junction concentrator cell has super-efficiency potential of about 50% [52].

A multijunction solar cell consists of different subcells stacked together in series connection which have different forbidden band gaps. The band gaps of the cell decrease from top cell to bottom cell. Thus, the top cell absorbs the higher energy

Fig. 17.16 A triple-junction solar cell [50]

portion of the solar spectrum and the lower energy range of solar spectrum passes through to the lower cell. Therefore, better utilization of the available incident solar spectrum results in a higher conversion of incident sunlight into electricity. Multijunction solar cells with rough surface texture provides higher efficiencies and are cost-effective [53].

Multijunction solar cells are made by combining layers of different materials capable of absorbing different ranges of the light spectrum. By combining different layers, we are able to maximize the efficiency of the solar cell [50]. These solar cells find lots of applications in the space industry but have also been employed in terrestrial power applications. These solar cells can work under higher solar concentrations and are typically used in MCPV and almost all HCPV and Ultra-HCPV systems.

3.3 New Materials for Solar Photovoltaics

These are the third-generation solar cells developed from new materials other than silicon such as organic materials, solar ink, solar dyes, and conductive polymers.

3.3.1 Organic Photovoltaic

Organic photovoltaics (OPV) are another upcoming photovoltaic technology with a the reported efficiency of organic photovoltaic to date of 11% [25]. The schematic diagram of an organic photovoltaic solar cell is presented in Fig. 17.17. Organic photovoltaic technology has gained attention in the last three decades due to its potential for its low-cost substrate fabrication by standard coating and printing

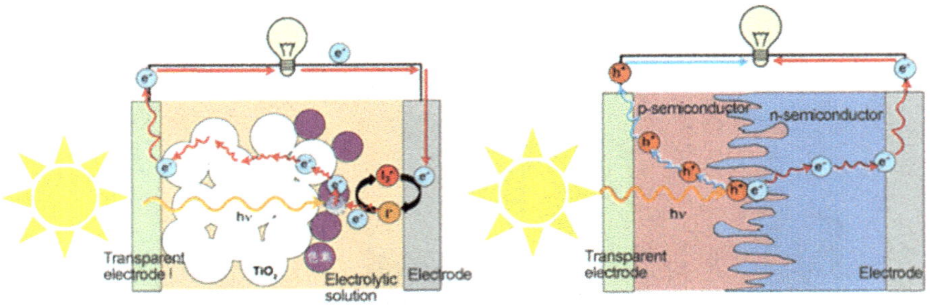

Fig. 17.17 Dye-sensitized solar cell (*left*) and organic film solar cell (*right*) (Source: www.yomiuri.co.jp/adv/chuo/dy/research/20130124.html)

techniques, flexibility, light weight, transparency, and large area coating. However, it has limitations over conversion efficiency and lifetime [54–56].

Polyethylene terephthalate (PET) substrates can be utilized to construct ultrathin, flexible, and compliant OPV solar cells. They provide a total device thickness thinner than a thread of spider silk [57]. Polyethylene naphthalate (PEN) [58] or polyethersulfone (PES) [59] is also being used for the introduction of flexible organic solar cells.

The photoconductivity of the first organic compound was observed in anthracene by Pochettino in 1903 and Volmer in 1913 [60]. Organic solar cells comprise of materials made of two organic layers or a homogeneous mixtures of two organic materials. Out of two organic materials, one is used as the electron donor and the other as the electron acceptor. Indium tin oxide (ITO)-coated substrates are mostly employed as the transparent anode [61]. However, indium being a rare metal adds to intensive energy usage, poor performance on the plastic substrate, brittleness, and a high economic cost [61–64].

A promising alternative to ITO is transparent conductor, namely, PEDOT:PSS (highly conductive), a silver grid embedded in PEDOT:PSS, silver nanowires, single-walled carbon nanotubes with effective potential for reduced module cost and a reduced energy payback time [62].

With atomic layer deposition (ALD) technique, ZnO has been grown on an inverted organic solar cell (OSC) as an electron selective layer due to its relatively higher electron mobility properties [58]. Other techniques include a sol-gel process [65], spin coated [66], and solution processing [67]. Vacuum evaporation and techniques based on solution processing are the preferred techniques in the preparation of the thin film for organic solar cell production [68].

OSC based on aluminum-doped zinc oxide (ZAO) as a replacement to indium tin oxide (ITO) shows a similar cell behavior [69].

Organic solar cell (OSC) is based on copper phthalocyanine (CuPc) and fullerene (C60) as the donor layer and acceptor respectively. The anode indium tin oxide (ITO) is replaced by aluminum doped zinc oxide (ZnO:Al), and introducing an ultrathin gold film among the anode, the transparent conductor oxide and the donor layer provides improved power conversion efficiency of the cell [61]. Flexible transparent conducting electrodes for polymer-fullerene bulk-heterojunction solar cells are fabricated using printing films of single-walled carbon nanotube (SWNT) network on plastic [70]. Optically transparent cellulose nanocrystal (CNC) is introduced as substrates for organic solar cells with suitable rectification in the shade and reaches a power conversion efficiency of 2.7%. The bottom and top electrodes used are Ag/Polymer surface and MoO_3/Ag, respectively, with employing any aqueous solution [71].

3.3.2 Dye-Sensitized Solar Cell

Dye-sensitized solar cells (DSSC) are also known as Gräetzel cells named after Professor Michael Graetzel who discovered dye-sensitized solar cell together with Dr. Brian O'Regan at École Polytechnique Fédérale de Lausanne (EPFL), Switzerland, in 1991 (http://gcell.com/dye-sensitized-solar-cells). The reported efficiency of DSSCs has reached 13% by adopting molecular engineering of porphyrin sensitizers [72]. Figure 17.17 shows the structure of a dye-sensitized solar cell. Dye-sensitized solar cells are popular because of their low cost of production, optically tunable properties (color, transparency), can work even in low light illumination that is under a cloudy atmosphere and diffused radiation, and ease of fabrication [22]. However, their long-term durability and most of the dye unresponsive beyond the visible range resulting, resulting in the wastage of the infrared spectrum that accounts around 49% of total solar radiation are of major concern. Arylamine organic dyes have received attention due to their potential property of favorable molar absorption coefficient, being cost-effective, and varied composition [73].

3.3.3 Quantum Dots Solar Cell (QDSC)

Quantum dots are nanostructured semiconductors discovered in 1980 with diameters ranging from 2 to 10 nm, i.e., 10–50 atoms. The certified efficiency of quantum dot solar cells has reached 8.55% [74]. These solar cells are nano-scale, have high absorption capacity, and reduce the need for absorber materials [75].

Quantum dots have the ability to emit light of different wavelengths from the same material by simply controlling the size of the dots (nanocrystals) as shown in Fig. 17.18. Since a decrease in the size of the nanocrystal allows a gap between the energy level of the highest valence band and the lowest of conduction band, thus greater amount of energy is required to excite the nanocrystal and these this results in high energy release on dots the dot returning to its stable state. Thus, the emitted light shows color shifts from red to blue.

Significant advances in the performance of solar cells in terms of efficiency has been reported for recently developed cell technologies. Some mature solar cell technologies have chosen various options for slow progress in efficiency improvement; while focusing more on cost reduction. As shown in Fig. 17.19, multijunction

Fig. 17.18 Alloyed quantum dot emits light of different wavelengths by tuning the composition [84]

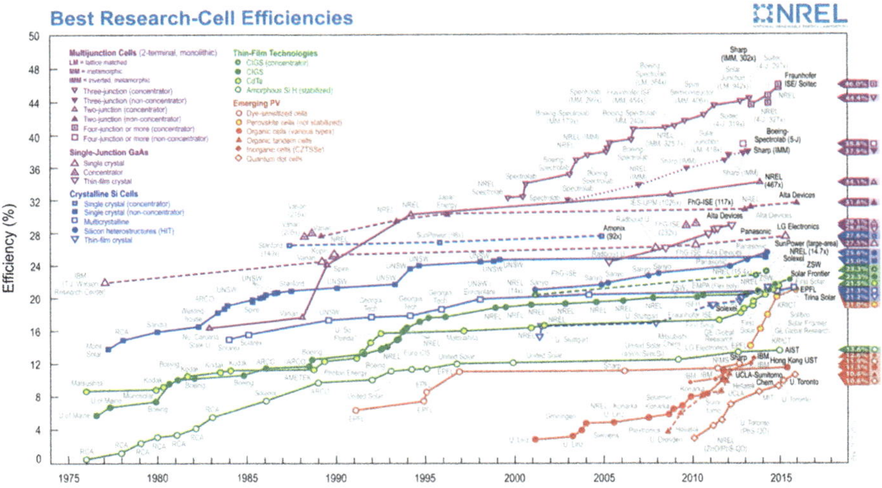

Fig. 17.19 Efficiencies for different PV solar cell technology (Solar Cell efficiencies [85])

solar cells have taken a giant leap on efficiency improvement and reduction in cost, while crystalline silicon and thin-film solar cells have had more prominence in technological enhancements in large-scale production to reduce costs rather than rapid increase in efficiency. Figure 17.20 shows the images of some of the solar cell technologies which can be differentiated from their design.

The performance level of active layers, coatings, and transparent electrodes made via low-cost processes defines the solar energy potential that a photovoltaic device can utilize. Fabrication technologies are processed to develop a high-quality textured surface so that it captures optimal optimized light. Development of high-activity supercapacitor electrodes with applicability in batteries and solar cells is being undertaken using porous metal and conducting polymers on metal organic framework.

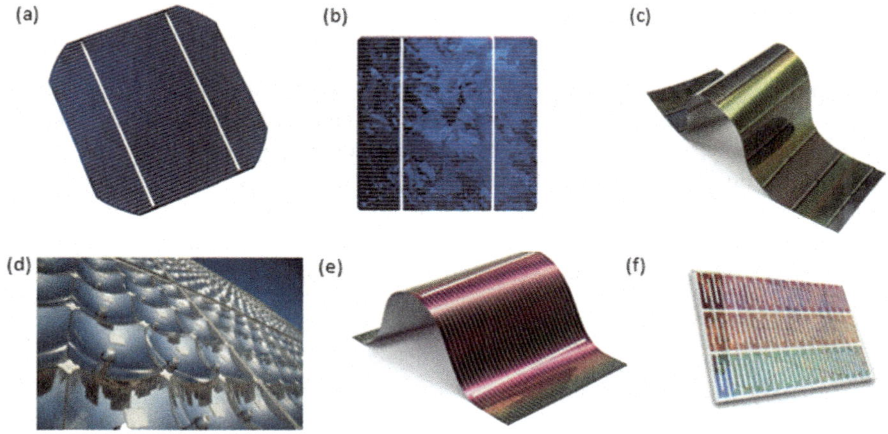

Fig. 17.20 Different PV technologies: (**a**) monocrystalline Si solar cell, (**b**) polycrystalline Si solar cell, (**c**) thin-film solar cell, (**d**) concentrated PV module, (**e**) organic solar cell, (**f**) dye-sensitized solar cell

4 Rural Development

Solar energy technology has been implemented in several rural development programs; some of the notable programs held in the last decade include the Urjagram program and Integrated Rural Energy Programme. This was launched in 2005, to make villages self-reliant in energy through an optimal mix of various renewable energy technologies (e.g., solar, picohydro, biomass). The Remote Village Electrification Programme was initiated by the MNRE in 2006 for electrification through renewable energy sources where grid connectivity is either not feasible or not cost-effective. Around 3330 unelectrified census villages and 830 unelectrified hamlets of electrified villages have so far been provided with basic electricity facilities [21]. A majority of remote census villages taken up for electrification under the programme are provided with SOLAR PV home lighting systems (about 95%). A typical application gaining lots of attention among the village community is the solar lantern. A striking example is the Green Oscar award-winning solar lantern Price-to-Performance Winner—AISHWARYA® [76], a simple lantern powered by solar energy. Before these lamps were introduced, kerosene lamps were used which not only caused pollution but also affected the vision. Costing less than Rs. 1500/ each, the lanterns are self-funding in one to two years depending on the kerosene saved. By 2009, more than 100,000 homes in the states of Andhra Pradesh and Maharashtra were benefiting from this new technology, over half a million people used solar lanterns [23]. This industry has had a very high social impact as it was primarily being manufactured by villagers themselves who were given a few days' training and worked in small-scale industries to make this product. Despite the government not having any subsidy on this product it is still one of the most popular one among the village population.

4.1 Other Applications

A number of applications are being developed on an industrial scale across the world using solar energy technologies [77]. The electricity produced by using PV panels is widely being used by the telecommunication, agricultural, water desalination, and building industries to operate lighting, pumping, heating, and cooling equipment. Special applications suitable to local requirements like paddy parboiling [78], and demonstration projects for air heating applications like food dehydration, and space heating [79] were also developed previously in India.

5 Government Initiatives

The Indian government has been playing a key role for the development and progress of solar energy technology. A number of such initiatives are highlighted in Table 17.3.

5.1 MNRE Solar Energy Initiatives

The Ministry of New and Renewable energy (earlier known as Ministry of Non-Conventional Energy Sources) is a government organization which aims to develop and deploy new and renewable energy for supplementing the energy requirements of India. It also facilitates the research, design, development, and manufacturing of these technologies. With over 30 years of presence, the organization has played a pivotal role in the development of renewable energy in India. The ministry in particular aims to cut down the use of conventional fuels and make India independent for its energy demands. Throughout its course, the ministry has taken new initiatives to promote and develop the renewable energy sector. Currently, its biggest initiative undertaken by it is the Jawaharlal Nehru National Solar Mission. Several other initiatives have also been taken under the research and development of renewable energy technologies. On a grid scale, 180 MW of MW Solar Power Plants have been commissioned in India by January 2012. Recently, "Teach 1000 Teachers" program was started by the MNRE which aims to educate teachers on solar photovoltaics through video conferencing at IIT Bombay. A Special Area Demonstration Project Scheme has also been started by the MNRE and a number of sites identified with an objective of demonstrating application of various Renewable Energy systems. These sites are places of national and international importance including world heritage sites, heritage monuments, religious locations, and places of public interest to create greater awareness of renewable and to supplement the energy requirement at such locations. The Ministry has also initiated a major

Table 17.3 Initiatives taken by the Indian government

Year	Initiative	Description
2003	Electricity Act	The act enables suitable measures for the cogeneration and generation of electricity using renewable sources. It was promoted by the State Electricity Regulatory Commissions (SERCs) providing suitable procedures for connectivity with the grid and sale of electricity to any person. Additionally, it has also provided norms for purchase of electricity from such sources and directs the percentage of the total consumption of electricity in the area of a distribution licensee
2005	National Electricity Policy	This policy mainly focused on increasing the share of energy from nonconventional energy sources. It also introduced competitive bidding for the purchase of these technologies
2006	National Tariff Policy	This policy fixed a minimum percentage for purchase of energy from renewable sources taking into account availability of such resources in the region and its impact on retail tariffs
	National Rural Electrification Policies (NREP)	This policy primarily provided electricity access for villages where grid connectivity would not be feasible or not cost-effective. It offered off-grid solutions based on stand-alone renewable energy-based systems and provided alternative solutions like solar photovoltaics for use of isolated lighting
2007	Semiconductor Policy	The Government of India would provide special incentive packages for setting up semiconductor fabrication facilities. The incentives will be applicable for manufacturing of all semiconductor materials, displays including Liquid Crystal Displays, Organic Light Emitting diodes, Plasma display panels, and any other emerging displays, storage devices, solar cells, photovoltaic, and other advanced micro- and nanotechnology products
2008	Solar PV generation based incentive	MNRE offered to provide, through IREDA, a generation-based incentive of a maximum of Rs. 12/kWh to eligible projects, which are commissioned by December 31 2009, after taking into account the power purchase rate (per kWh) provided by the State Electricity Regulatory Commission or utility for that project
2011	Generation-Based Incentive Scheme	Government-Based Incentive (GBI) is provided to support small grid solar power projects connected to the distribution grid (below 33 KV) to the state utilities. The Indian Renewable Energy Development Agency (IREDA) has selected 78 projects with a total capacity of about 98 MW for which the Ministry will provide GBI of Rs. 12.41 kWh^{-1} to the State utilities when they directly purchase solar power from the project developers. The amount of GBI paid to the utilities is kept fixed, as a difference of the CERC tariff for 2010–2011 (Rs. 17.91 kWh^{-1}) and a reference tariff of Rs. 5.5 kWh^{-1}

project on Solar Radiation Resource Assessment (SRRA) across the country to assess and quantify the solar radiation availability including weather parameters with a view to develop Solar Atlas.

5.2 National Solar Mission

The main objective of the Jawaharlal Nehru National Solar Mission (JNNSM) under the brand "Solar India" is to establish India as a global leader in solar energy, by creating the policy conditions for its dispersal across the country. The National Solar Mission was launched in January 2010 giving the much needed advancement to the solar energy utilization. The Mission has set a target of 20,000 MW and stipulates implementation and achievement of the target in three phases (Phase I up to 2012–2013, Phase II from 2013 to 2017, and Phase III from 2017 to 2022) for various components, including grid-connected solar power. One of the important objectives of the National Solar Mission is to promote domestic manufacturing. In order to achieve this, the developers are expected to procure their project components from domestic manufacturers, as far as possible. In the case of Solar PV Projects selected in the First Batch during FY 2010–2011, it was mandatory for Projects based on crystalline silicon technology to use the modules manufactured in India. For Solar PV projects to be selected in the second batch during FY 2011–2012, it has been made mandatory for all the projects to use cells and modules manufactured in India. Exceptions are allowed but only for PV modules made from thin-film technologies or concentrator PV cells that may be sourced from any country, provided the technical qualification criterion is fully met.

Fig. 17.21 shows the initial targets set by the Solar India Mission in three phases. During Phase 1, it is expected that around 7 million square meters of solar collectors would be installed which would power about 200 MW for off-grid applications and another 1500 MW for utility uses.

During Phase 2, around 15 million square meters of solar collectors are expected to be installed which would power about 1000 MW for off-grid applications and another 7000 MW for utility grid applications. In its final phase, 20 million square meters of solar collectors would be installed which would power about 2000 MW for off-grid applications and another 20000 MW for utility grid applications.

India's current solar installed capacity is 6998.85 MW in March 2016 as reported by MNRE. Resource assessment revealed that around 1.89 million square kilometers that is 58% of the geographical area in India has the solar hotspot potential [6]. Gujarat and Rajasthan have the leading solar power projects in the country with an installed capacity of 1119.17 MW and 1285.95 MW, respectively (Fig. 17.22). Solar PV rooftop projects have started in almost all the metropolitan cities like Hyderabad, Gurgaon, Jaipur, Noida, Bhubaneswar, and Raipur and are appering rapidly.

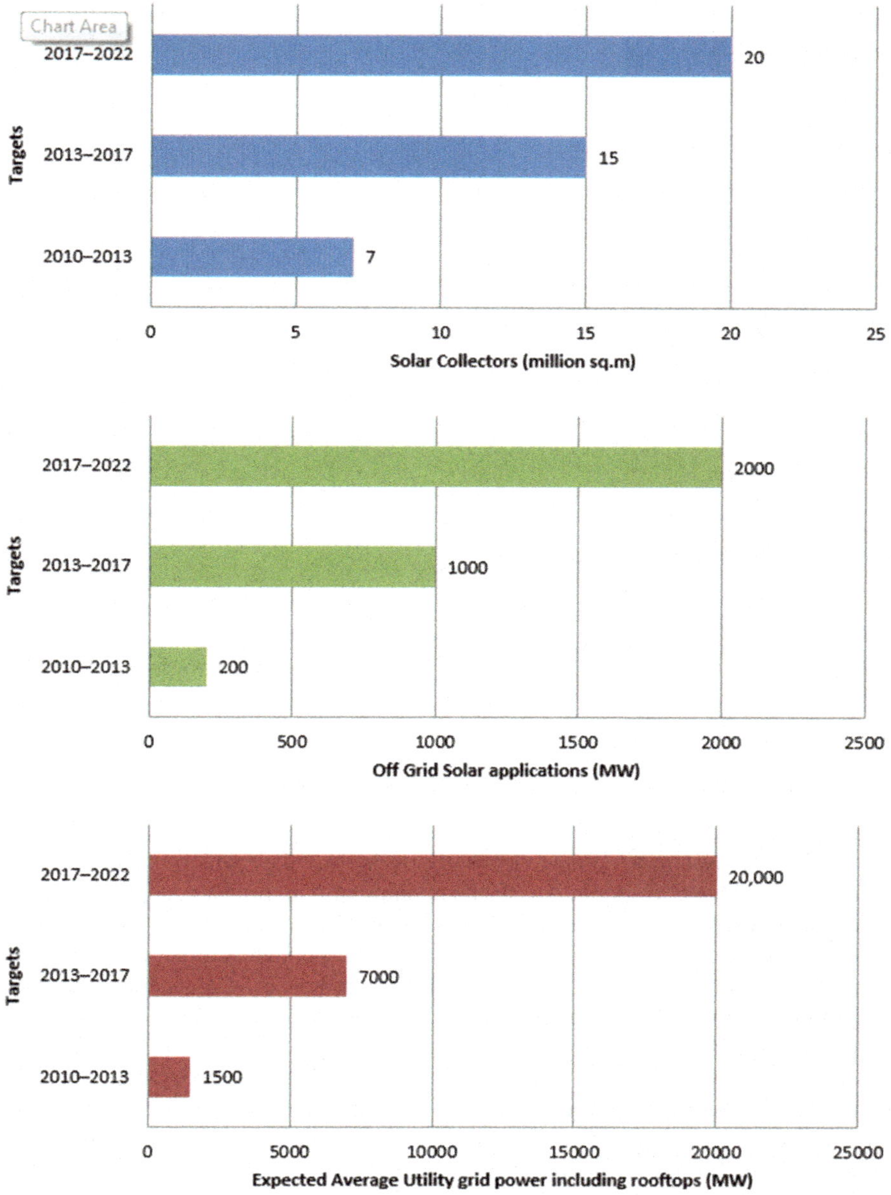

Fig. 17.21 Targets Set by JNNSM

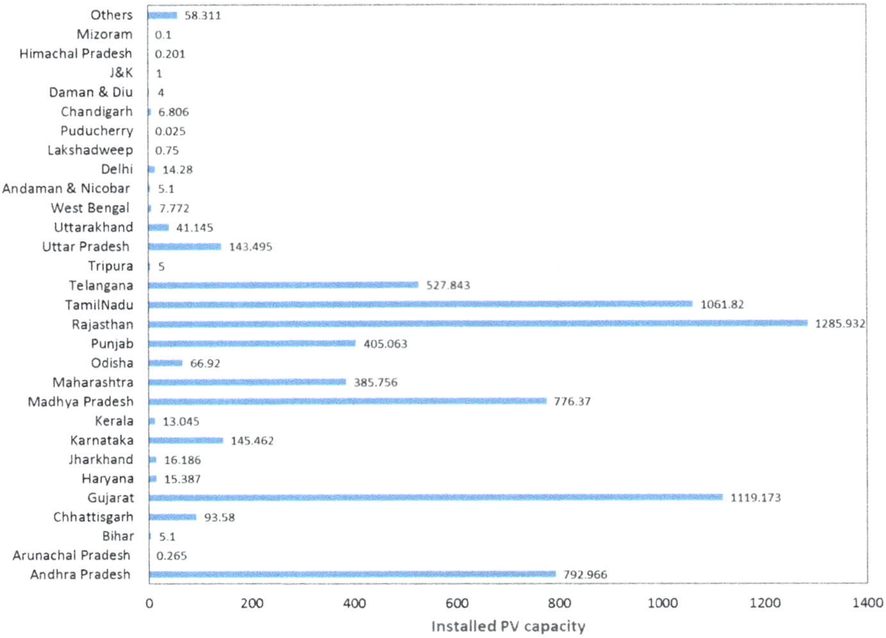

Fig. 17.22 State-wise distribution of the Solar Power Project (MW) in India (Source: MNRE 2016)

6 Electrical Storage for Solar Energy Applications

A typical PV system generates electricity during the day, which is stored by the batteries and made available during the evening and night. This helps in bridging the gap between the peak solar production and peak load demands. The power from the batteries is then passed through an inverter which converts direct electricity into alternating currents which can used by home appliances. Considering the rapid development of PV power generation, one of the greatest challenges to the integration of this solar resource into the electricity grid is its intermittent nature and, therefore, the variability of the power produced [80]. Energy storage for solar photovoltaic applications can be distributed into two main groups: small-scale home-storage systems, which are decentralized and located in the distribution grid, and large-scale stationary systems, which are connected to the distribution grid at one central point. Batteries currently form an expensive component of grid-scale electrical energy–storage technologies, especially when the costs are amortized over the >30-year lifetime of an installed solar electricity system [81]. Figure 17.23 showcases the most important storage technologies for the power industry.

These include electrochemical storage (batteries), superconducting magnetic energy storage (SMES), kinetic energy (using flywheels), potential energy (from

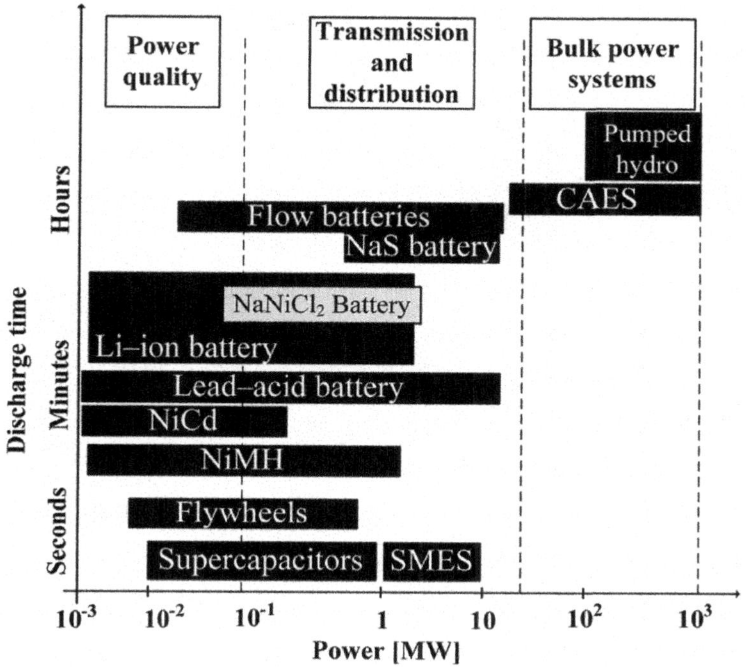

Fig. 17.23 Energy storage technologies used for the power plants

hydroelectric power plants), or compressed air [essentially from compressed air energy storage (CAES)]. Sizing of the battery storage is very important when designing the PV system. Overestimating the battery charging losses can result in larger PV arrays than required and underestimating those results in the loss of load and possible damage to batteries.

6.1 Lead-Acid Battery

Batteries charge and discharge based on a reversible chemical reaction initiated by a voltage applied to the battery terminals. One of the most important batteries considered for higher capacities of energy storage is the lead-acid battery. Typically, they consist of both a positive electrode and a negative electrode, submersed into an electrolyte solution as shown in Fig. 17.24.

The charging of the battery occurs when the electrons migrate from the positive electrode (anode) to the negative electrode (cathode). In its charged state, the negative plate is made of elemental lead, while the positive plate is prepared using lead dioxide. The electrolyte in this case is sulfuric acid (H_2SO_4). Because the redox-active species are solids, there is no need to separate the electrodes. The discharging process occurs due to the migration of the electrons from the

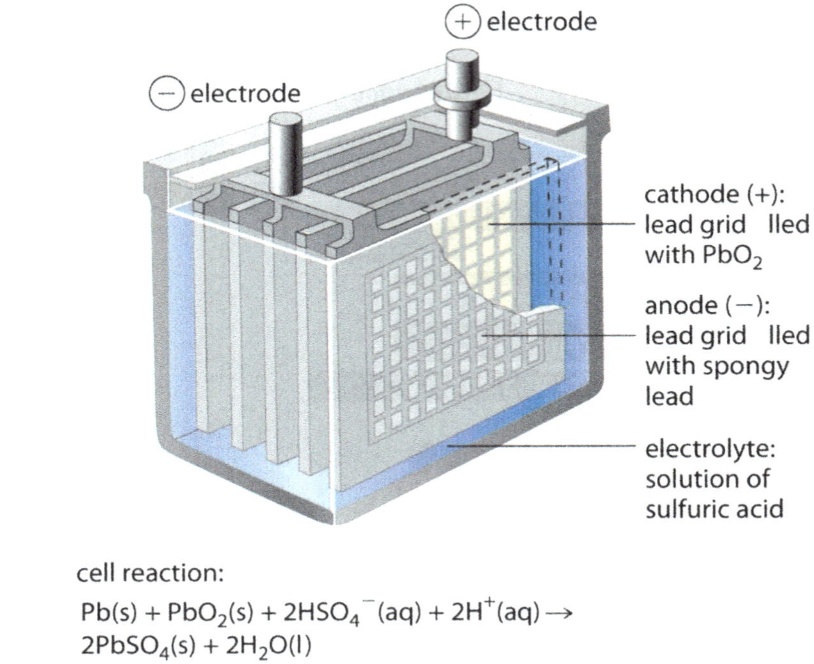

Fig. 17.24 Lead-acid battery (Chemwiki [82])

negative electrode to the positive electrode. In this instance, the grains of lead sulfate are generated by both positive and negative electrodes while the electrolyte begins to lose its concentration, thus forming water. When charging (by application of a voltage higher than E_{cell}), the reverse reaction occurs, dissolving these grain and $PbSO_4$ is converted back to metallic lead and PbO_2.

The following sets of equations describe the electrode reactions during the discharge process.

- Reduction at the cathode with E_{cathode}: $PbO_{2(s)} + HSO^-_{4(aq)} + 3H^+_{(aq)} + 2e^- \rightarrow PbSO_{4(s)} + 2H_2O_{(l)}$
- Oxidation at the anode with E_{anode}: $Pb_{(s)} + HSO^-_{4(aq)} \rightarrow PbSO_{4(s)} + H^+_{(aq)} + 2e^-$
- Overall reaction: $Pb_{(s)} + PbO_{2(s)} + 2HSO^-_{4(aq)} + 2H^+_{(aq)} \rightarrow PbSO_{4(s)} + 2H_2O_{(l)} + E_{\text{cell}}$

6.2 *Nickel Cadmium*

Ni–Cd batteries have been the second most commonly used rechargeable batteries which were invented by the Swedish engineer Waldemar Jungner in 1899. These batteries use a cadmium negative electrode, a nickel (III) oxide–hydroxide positive electrode, a separator, and an electrolyte (usually potassium hydroxide). These types of batteries are used for low energy storage solutions typically in the range of 1 kW–0.5 MW and present some environmental hazards which require complex recycling procedures.

6.3 *Lithium-Ion*

The need for storage solutions is growing and lithium-ion (Li-ion) batteries have emerged over the last few years as one of the preferred methods. However, they are more often regarded as the power sources for mobile phones and laptops. Lithium-ion battery storage technologies have steadily expanded their commercial scope and efficiencies for solar PV applications. Different Li-ion options exist in the market right now, in the form of lithium manganese oxide, lithium nickel manganese cobalt, and other forms. The key advantage these batteries have is that the energy is stored and released more efficiently with one charge-discharge efficiency above 90 percent.

Figure 17.25 shows the basic principle on which the lithium-ion batteries work. During charging, the lithium ions flow from the anode to the cathode through the

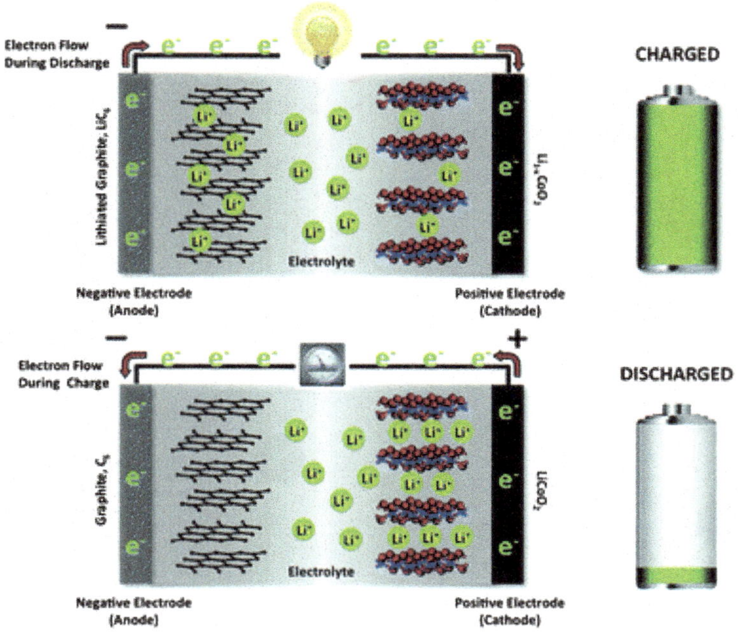

Fig. 17.25 Principle on which Lithium-Ion batteries work

electrolyte. Electrons flow in the opposite direction around the outer circuit. When no more ions are flowing, the battery is fully charged and ready to use. During discharging, the ions flow back from the cathode to the anode and electrons travel the opposite way through the circuit.

6.4 DC- and AC-Coupled Design

There are two main ways of linking a battery storage system into such a system as shown in Fig. 17.26.

(a) DC Coupled: the batteries are installed on the same side of the solar inverter as the solar panels; they charge from the panels, and their DC is only converted to AC when it's used ("DC coupled").

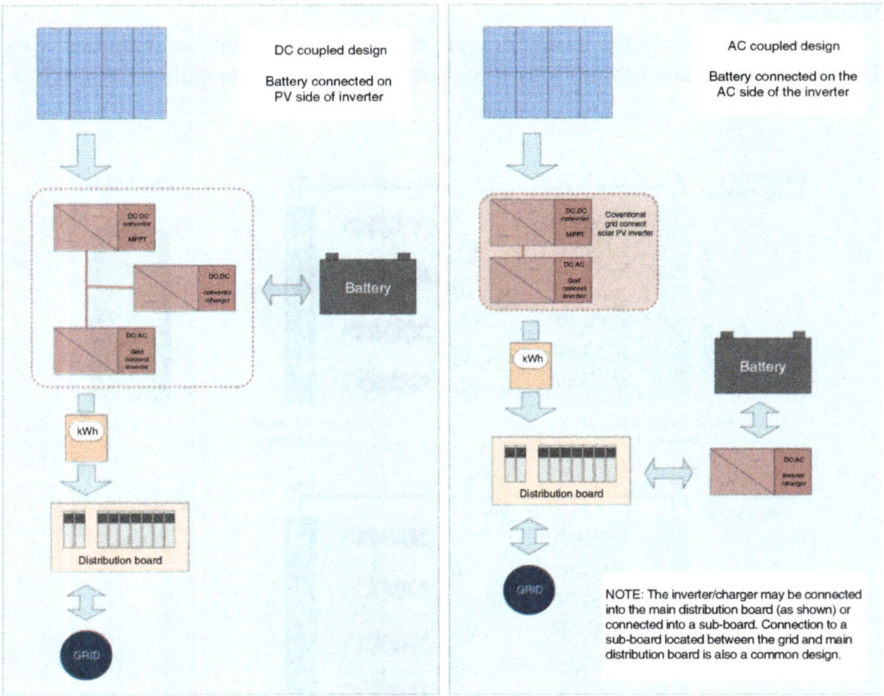

Fig. 17.26 DC- and AC-coupled design for battery

(b) AC Coupled: the batteries are installed on the grid side, where the solar PV's DC has already been converted to AC ("AC coupled"). A separate inverter converts the AC back to DC for storing in the battery. When the battery discharges, the same separate inverter converts the DC back to AC.

7 Standardization of Solar Energy Harvesting

The Bureau of Indian standards (BIS) is the responsible government agency to set up standards for the PV industry. It develops quality standards which are in line with the guidelines issued by the International Electrotechnical Commission (IEC) as shown in Table 17.4.

Table 17.4 List of standards issued for PV system components

Component	Standard description	Standard number
Solar PV systems (more than 100 Wp and up to 20 KWp capacity only): charge controller/MPPT unit Power conditioners/inverters including MPPT and protections	Environmental testing	IEC 60068-2 (1, 2, 14, 30)
		IEC 60068-2 (1, 2, 14, 30)/Equivalent BIS Std.
	Efficiency measurements	IEC 61683/IS 61683
Storage batteries	General requirements and methods of testing tubular lead acid/VRLA/GEL capacity test charge/discharge efficiency self-discharge	As per relevant BIS Std.
Cables	General test and measuring method PVC insulated cables for working voltage up to and including 1100 V and UV resistant for outdoor installation	IEC 60227/IS 694 IEC 60502/IS 1554 (Pt. I and II)
Switches/circuit breakers/connectors	General requirements connectors—safety AC/DC	IEC 60947 Part I, II, III/IS 60947 Part I, II, III EN 50521
Junction boxes/enclosures for inverters/charge controllers/luminaries	General requirements	IP 54 (for outdoor)/IP 21(for indoor) as per IEC 529

7.1 Accredited Test Centers/Laboratories for Solar Energy Applications

Several labs have been accredited by the MNRE for the certification of the different components used in solar energy harvesting systems. A list of these laboratories is shown in Table 17.5.

PV modules must qualify (enclose test reports/certificate from a IEC/NABL accredited laboratory) as per the relevant IEC standard. Additionally, the performance of PV modules at Standard Test Condition (STC) condition must be tested and approved by one of the IEC/NABL Accredited Testing Laboratories including the Solar Energy Centre.

Table 17.5 List of accredited laboratories for certifying the solar energy harvesting systems

Laboratory name	Website and components approved
UL	http://india.ul.com/ Solar Pumps
CPRI—Central Power Research Institute	www.cpri.in/ Lighting Systems, Battery, Inverter, Charge Controller, Solar Pumps
NISE—National Institute of Solar Energy	http://nise.res.in/ PV Module, Lighting Systems, Battery, Inverter, Charge Controller, Solar Pumps
ERTL—Electronics Regional Test Laboratory	http://electronicstds.gov.in/CREITG/ PV Module, Lighting Systems, Battery, Inverter, Charge Controller
TUV Rheinland India Pvt. Ltd.	http://www.tuv.com/en/uk/home.jsp Solar Pumps, Lighting, Inverter, PV modules, Charge controller
ETDC—Electronics Test and Development Centre	http://www.stqc.gov.in/testing-and-calibration-lab-main-page/344 PV Module, Lighting Systems, Battery, Inverter, Charge Controller
Intertek	http://www.intertek.com/energy/photovoltaic/modules/ PV Module, Lighting Systems, Battery, Inverter, Charge Controller
EQDC—Electronics and Quality Development Centre	http://eqdc.in/ Solar Photovoltaic Systems and Batteries, Solar Panel, CFL and LED Based Home Lighting System, Street Lighting System, Solar Lantern
TERI—Tata Energy Research Institute	http://labl.teriin.org/ Solar Lanterns
SITRAC—Scientific and Industrial Testing Research	http://www.sitarc.com/ Solar Pumps

8 Future Trend: Techno-economic Aspects

India aims to have a cumulative renewable power generation of 175,000 MW, of which solar power generation is expected to be about 99,533 MW. The region-wise distribution is shown in Fig. 17.27. The northern region of India is expected to develop 31,120 MW, with Uttar Pradesh leading among the Indian states with an expected cumulative installation of 10,697 MW.

A number of policies and programs are under preparation for implementing the targets around the country. A recent initiative announced recently is the accelerated depreciation scheme, which allows up to 80% of the assets to be depreciated in the first year, leading to a massive benefit for the investors and solar power generation companies.

8.1 Accelerated Depreciation (AD)

For the growth of the solar power industry, the government has recently capped [83] the accelerated depreciation at 40% down from the original 80%. The accelerated depreciation of the solar projects helps reduce the cost during its start-up years, lessening the burden of taxation for the project investors. For example, for a project with a total cost of Rs. 1000 millions, 80% would mean Rs. 800 million; this value can be expensed in the first year itself under the accelerated depreciation, consequently leading to tax savings as explained in the table. It can be noticed that the AD benefits the investor substantially to the amount of Rs. 339.4 million. The effective

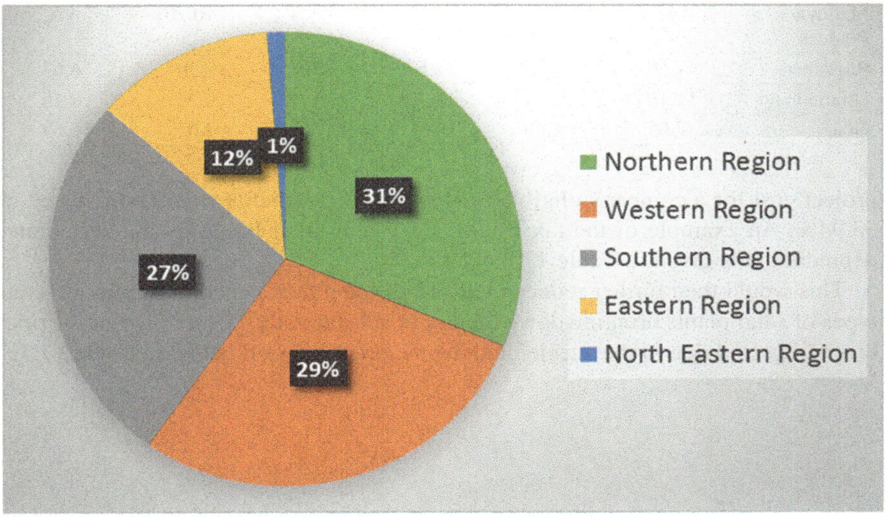

Fig. 17.27 Region-wise break-up of solar power generation by 2020

Table 17.6 Example showing the tax benefit for the PV installers based on the accelerated depreciation

Years	1	2	3	4
Opening	100%	20%	4%	0.8%
	100 Lakhs	20 Lakhs	4 Lakhs	0.8 Lakhs
Depreciation allowed (80% of opening balance)	80.00%	16.00%	3.20%	0.64%
Closing (in rupees)	80 Lakhs	16 Lakhs	3.2 Lakhs	0.64 Lakhs
%	20%	4.00%	0.80%	0.16%
	20 Lakhs	4 Lakhs	0.8 Lakhs	0.16 Lakhs
Accelerated depreciation	80 Lakhs	16 Lakhs	3.2 Lakhs	0.64 Lakhs
Tax benefit	27.192 Lakhs	5.4384 Lakhs	1.08768 Lakhs	0.217536 Lakhs
Total tax benefit due to AD	33.94 Lakhs			

Table 17.7 Tariff costs from different solar power generation plants in different states

	Tariff in rupees					
	Large scale solar (PV)		Solar thermal		Solar (PV) rooftop	
State	Without AD	With AD	Without AD	With AD	Without AD	With AD
Andhra Pradesh	17.91	14.95	15.31	12.85	NA	NA
Tamil Nadu	7.01	6.28	11.03	9.88	NA	NA
Karnataka	8.40	NA	10.92	NA	9.56	
Kerala	17.91	14.95	15.31	12.85	17.91	14.95
Gujarat	8.97	8.03	12.91	11.55	NA	NA
Maharashtra	7.95	6.79	NA	NA	8.45	7.29
Madhya Pradesh	10.44	NA	12.65	NA	10.70	NA
Rajasthan	7.50	6.63	11.67	10.27	7.50	6.63
Uttarakhand	11.10	10.15	13.30	12.15	9.20	8.15
Goa	7.05	6.69	NA	NA	7.05	6.69

project cost for a company eligible to claim the AD benefits would be reduced by 33.94%. An example of the tax benefit for PV installer based on the accelerated depreciation is given in Table 17.6.

This would then further reduce the costs of the power generation from different types of solar plants bringing down the prices substantially. A summary of the price drops expected due to the accelerated depreciation is listed below in Table 17.7.

8.2 Viability Gap Funding

Recently, the government has announced a Viability Gap Funding support stream to help in the setting up of Grid-Connected Solar PV power projects with a minimum 2000 MW capacity by the solar power developers on the Build-Own-Operate basis. The power generated from these projects will be purchased by the Solar Energy Corporation of India (SECI) at a fixed cost of Rs. 5.43 kWh^{-1} for 25 years and sold to willing state utility companies at a price of Rs. 5.50 kWh^{-1} for 25 years.

Typically, the term Performance ratio (PR) is used everywhere in the world for estimating the performance of the PV plant. Where

$$\text{PR of a plant for a period of time} = \text{Energy measured (kWh)}/(\text{Irradiance } \left(\text{kWh/m}^2\right) \text{ on the panel} \times \text{Active area of PV module } \left(\text{m}^2\right) \times \text{PV module efficiency})$$

However, in India the performance is assessed using a term called as the Cumulative Utilization Factor (CUF) which is defined as

$$\text{Capacity Utilization Factor (CUF)} = \text{Energy measured (kWh)}/(365 \times 24 \times \text{installed capacity of the plant})$$

Any excess generation over and above 10% of declared Cumulative Utilization Factor will be purchased by SECI at a tariff of Rs. 3 kWh^{-1}, provided SECI is able to get any buyer for sale of such excess generation. CUF can be understood as the real output of the plant compared to the theoretical maximum output of the plant.

8.3 Grid-Connected Solar Power Projects by 2021–2022

The government of India has now passed a resolution to increase the installed solar power target to reach 100,000 MW by the year 2021–2022. This goal is to be achieved through both Grid connected rooftop solar PV and medium and large scale grid-connected solar power. The year-wise scale-up targets is shown in Fig. 17.28.

The rooftop solar PV is aimed to reach up to 40% of the target in the next 6 years. The key sectors to contribute in this would include institutions, industrial and commercial organizations, and the residential buildings. The remaining 60% is targeted to be achieved through solar parks, projects connected to spare substations, farmers, and projects on land outside solar parks.

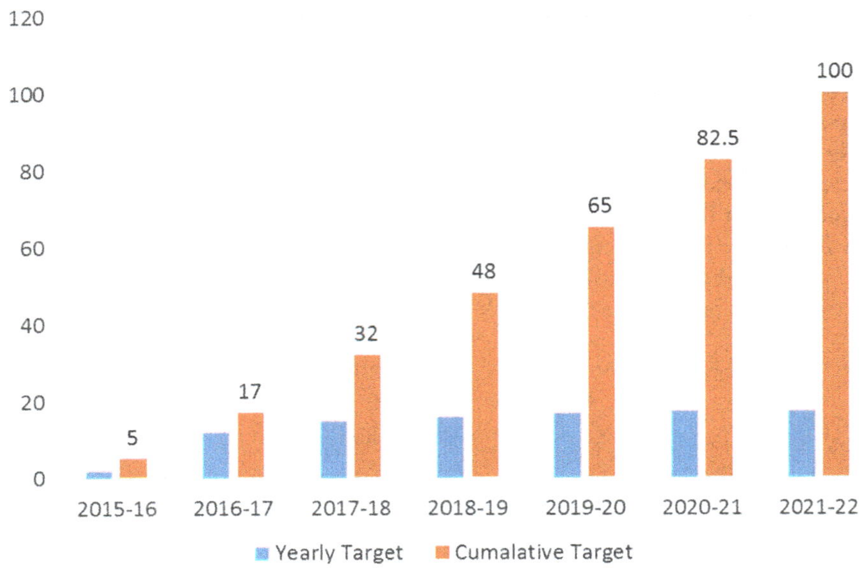

Fig. 17.28 Yearly solar PV installation target in GW

References

1. Trends in Photovoltaic Applications (2015), vol 20. IEA PVPS
2. Sharma NK, Tiwari PK, Sood YR (2012) Solar energy in India: strategies, policies, perspectives and future potential. Renew Sust Energ Rev 16(1):933–941
3. Ramachandra TV, Rishabh J, Gautham K (2011b) Hotspots of solar potential in India. Renew Sust Energ Rev 15:3178–3186
4. Solar Radiation Resource Assessment Project in India (June 1, 2014)
5. Harinarayana T, Kashyap KJ (2014) Solar energy generation potential estimation in India and Gujarat, Andhra, Telangana States. Smart Grid Renew Energy 05(11):275–289. doi:10.4236/sgre.2014.511025
6. Ramachandra TV, Jain R, Krishnadas G (2011a) Hotspots of solar potential in India. Renew Sust Energ Rev 15(6):3178–3186. doi:10.1016/j.rser.2011.04.007
7. Chandra Bhushan AKaNG (2015) The state of concentrated solar power in India: a roadmap to developing solar thermal technologies in India. Centre for Science and Environment, New Delhi
8. Baig H, Mallick TK (2010) Challenges and opportunities in concentrating photovoltaic research. Mod Energy Rev 3(2):20–28
9. Morapakala S (2011) Domestic solar hot water systems: developments, evaluations and essentials for "viability" with a special reference to India. Renew Sust Energ Rev 15 (8):3850–3861. doi:10.1016/j.rser.2011.07.006
10. Broman L (1984) Non-imaging solar concentrators with flat mirrors, pp 102–109
11. Soteris K (2003) The potential of solar industrial process heat applications. Appl Energy 76 (4):337–361
12. Greentech (2010) Solar water heaters in India: market assessment studies and surveys for different sectors and demand segments. Ministry of New and Renewable Energy

13. Chandrasekar B, Kandpal TC (2004) Techno-economic evaluation of domestic solar water heating systems in India. Renew Energy 29(3):319–332. doi:10.1016/s0960-1481(03)00198-8
14. Veeraboina P, Ratnam GY (2012) Analysis of the opportunities and challenges of solar water heating system (SWHS) in India: estimates from the energy audit surveys and review. Renew Sust Energ Rev 16(1):668–676. doi:10.1016/j.rser.2011.08.032
15. MNRE (2011) List of Manufacturers empanneled under refinancing scheme implemented through NABARD Ministry of New and Renewable Energy
16. Iyican L, Witte LC, Bayazitoğlu Y (1980) An experimental study of natural convection in trapezoidal enclosures. J Heat Transf 102(4):648–653. doi:10.1115/1.3244366
17. Chapin DM, Fuller CS, Pearson GL (1954) A new silicon p-n junction photocell for converting solar radiation into electrical power. J Appl Phys 25(5):676–677
18. Fonash S (2010) Solar cell device physics. Elsevier Science, Oxford
19. Green MA (1982) Solar cells: operating principles, technology, and system applications. Prentice-Hall, Englewood Cliffs
20. Luque A, Hegedus S (2003) Handbook of photovoltaic science and engineering. Wiley. doi:citeulike-article-id:4296078
21. BRIDGE TO INDIA – India Solar Handbook (2011)
22. Shah A, Torres P, Tscharner R, Wyrsch N, Keppner H (1999) Photovoltaic technology: the case for thin-film solar cells. Science 285(5428):692–698. doi:10.1126/science.285.5428.692
23. Saga T (2010) Advances in crystalline silicon solar cell technology for industrial mass production. NPG Asia Mater 2(3):96–102. doi:10.1038/asiamat.2010.82
24. Aberla AG (2000) Surface passivation of crystalline silicon solar cells: a review. Prog Photovoltaics Res Appl 8(5):473–487
25. Green MA, Emery K, Hishikawa Y, Warta W, Dunlop ED (2015) Solar cell efficiency tables (Version 45). Prog Photovolt Res Appl 23(1):1–9. doi:10.1002/pip.2573
26. Archer MD, Hill R (2001) Clean electricity from photovoltaics. Imperial College Press, London
27. Ardani K, Margolis R (2010) Solar technologies market report. NREL report. doi:citeulike-article-id:10690569
28. Lipiński M, Panek P, Świątek Z, Bełtowska E, Ciach R (2002) Double porous silicon layer on multi-crystalline Si for photovoltaic application. Sol Energy Mater Sol Cells 72(1–4):271–276. doi:10.1016/s0927-0248(01)00174-x
29. Shah A, Vallat-Sauvain E, Torres P, Meier J, Kroll U, Hof C, Droz C, Goerlitzer M, Wyrsch N, Vanecek M (2000) Intrinsic microcrystalline silicon (μc-Si:H) deposited by VHF-GD (very high frequency-glow discharge): a new material for photovoltaics and optoelectronics. Mater Sci Eng B 69–70(0):219–226. doi:10.1016/s0921-5107(99)00299-8
30. Shirasawa K, Yamashita H, Fukui K, Takayama M, Okada K, Masuri K, Watanabe H (1990) Large area high efficiency multicrystalline silicon solar cell. In: Photovoltaic specialists conference, 1990, Conference record of the twenty first IEEE, 21–25 May 1990, pp 668–673, vol 661. doi:10.1109/pvsc.1990.111704
31. Willeke G, Nussbaumer H, Bender H, Bucher E (1992) A simple and effective light trapping technique for polycrystalline silicon solar cells. Sol Energy Mater Sol Cells 26(4):345–356. doi:10.1016/0927-0248(92)90054-s
32. Dobson KD, Visoly-Fisher I, Hodes G, Cahen D (2000) Stability of CdTe/CdS thin-film solar cells. Sol Energy Mater Sol Cells 62(3):295–325. doi:10.1016/S0927-0248(00)00014-3
33. Wu X (2004) High-efficiency polycrystalline CdTe thin-film solar cells. Sol Energy 77:803–814. doi:10.1016/j.solener.2004.06.006
34. Gessert TA, Wei SH, Ma J, Albin DS, Dhere RG, Duenow JN, Kuciauskas D, Kanevce A, Barnes TM, Burst JM, Rance WL, Reese MO, Moutinho HR (2013) Research strategies toward improving thin-film CdTe photovoltaic devices beyond 20% conversion efficiency. Sol Energy Mater Sol Cells 119:149–155. doi:10.1016/j.solmat.2013.05.055

35. Moutinho HR, Hasoon FS, Abulfotuh F, Kazmerski LL (1995) Investigation of polycrystalline CdTe thin films deposited by physical vapor deposition, close-spaced sublimation, and sputtering. J Vac Sci Technol A 13(6):2877–2883. doi:10.1116/1.579607
36. Erra S, Shivakumar C, Zhao H, Barri K, Morel DL, Ferekides CS (2007) An effective method of Cu incorporation in CdTe solar cells for improved stability. Thin Solid Films 515 (15):5833–5836. doi:10.1016/j.tsf.2006.12.064
37. Ferekides C, Britt J (1994) CdTe solar cells with efficiencies over 15%. Sol Energy Mater Sol Cells 35:255–262. doi:10.1016/0927-0248(94)90148-1
38. Delahoy AE, Chen L, Akhtar M, Sang B, Guo S (2004) New technologies for CIGS photovoltaics. Sol Energy 77(6):785–793. doi:10.1016/j.solener.2004.08.012
39. Feltrin A, Freundlich A (2008) Material considerations for terawatt level deployment of photovoltaics. Renew Energy 33(2):180–185. doi:10.1016/j.renene.2007.05.024
40. Jackson P, Würz R, Rau U, Mattheis J, Kurth M, Schlötzer T, Bilger G, Werner JH (2007) High quality baseline for high efficiency, Cu(In1−x,Gax)Se2 solar cells. Prog Photovolt Res Appl 15 (6):507–519. doi:10.1002/pip.757
41. Tokio N, Masayuki M (2002) 18% efficiency Cd-free Cu(In, Ga)Se 2 thin-film solar cells fabricated using chemical bath deposition (CBD)-ZnS buffer layers. Jpn J Appl Phys 41(2B): L165
42. Naghavi N, Spiering S, Powalla M, Cavana B, Lincot D (2003) High-efficiency copper indium gallium diselenide (CIGS) solar cells with indium sulfide buffer layers deposited by atomic layer chemical vapor deposition (ALCVD). Prog Photovolt Res Appl 11(7):437–443. doi:10.1002/pip.508
43. Kaelin M, Rudmann D, Tiwari AN (2004) Low cost processing of CIGS thin film solar cells. Sol Energy 77(6):749–756. doi:10.1016/j.solener.2004.08.015
44. Ramanathan K, Teeter G, Keane JC, Noufi R (2005) Properties of high-efficiency CuInGaSe2 thin film solar cells. Thin Solid Films 480–481:499–502. doi:10.1016/j.tsf.2004.11.050
45. Kessler F, Rudmann D (2004) Technological aspects of flexible CIGS solar cells and modules. Sol Energy 77(6):685–695. doi:10.1016/j.solener.2004.04.010
46. Wuerz R, Eicke A, Frankenfeld M, Kessler F, Powalla M, Rogin P, Yazdani-Assl O (2009) CIGS thin-film solar cells on steel substrates. Thin Solid Films 517(7):2415–2418. doi:10.1016/j.tsf.2008.11.016
47. Carlson DE, Wronski CR (1976) Amorphous silicon solar cell. Appl Phys Lett 28 (11):671–673. doi:10.1063/1.88617
48. Jones RK, Ermer JH, Fetzer CM, King RR (2012) Evolution of multijunction solar cell technology for concentrating photovoltaics. Jpn J Appl Phys 51:10ND01. doi:10.1143/jjap.51.10nd01
49. Baig H, Sarmah N, Chemisana D, Rosell J, Mallick TK (2014) Enhancing performance of a linear dielectric based concentrating photovoltaic system using a reflective film along the edge. Energy 73(14):177–191
50. Baig H, Heasman KC, Mallick TK (2012) Non-uniform illumination in concentrating solar cells. Renew Sust Energ Rev 16(8):5890–5909
51. New world record for solar cell efficiency at 46% French-German cooperation confirms competitive advantage of European photovoltaic industry (1 December 2014)
52. Yamaguchi M (2015) Fundamentals and R&D status of III-V compound solar cells and materials. Phys Status Solidi (c) 12(6):489–499. doi:10.1002/pssc.201400216
53. Masahiko Muramoto KJ, Takashi Hayakawa K-CJ (2002) Multi-Junction Solar Cell, 1 Oct 2002
54. Brabec CJ (2004) Organic photovoltaics: technology and market. Sol Energy Mater Sol Cells 83(2–3):273–292. doi:10.1016/j.solmat.2004.02.030
55. Li G, Shrotriya V, Huang J, Yao Y, Moriarty T, Emery K, Yang Y (2005) High-efficiency solution processable polymer photovoltaic cells by self-organization of polymer blends. Nat Mater 4(11):864–868

56. Lungenschmied C, Dennler G, Neugebauer H, Sariciftci SN, Glatthaar M, Meyer T, Meyer A (2007) Flexible, long-lived, large-area, organic solar cells. Sol Energy Mater Sol Cells 91 (5):379–384. doi:10.1016/j.solmat.2006.10.013
57. Kaltenbrunner M, White MS, Głowacki ED, Sekitani T, Someya T, Sariciftci NS, Bauer S (2012) Ultrathin and lightweight organic solar cells with high flexibility. Nat Commun 3:770. doi:10.1038/ncomms1772
58. Wang J-C, Weng W-T, Tsai M-Y, Lee M-K, Horng S-F, Perng T-P, Kei C-C, Yu C-C, Meng H-F (2010) Highly efficient flexible inverted organic solar cells using atomic layer deposited ZnO as electron selective layer. J Mater Chem 20(5):862–866. doi:10.1039/b921396a
59. Zhou Y, Fuentes-Hernandez C, Shim J, Meyer J, Giordano AJ, Li H, Winget P, Papadopoulos T, Cheun H, Kim J, Fenoll M, Dindar A, Haske W, Najafabadi E, Khan TM, Sojoudi H, Barlow S, Graham S, Brédas J-L, Marder SR, Kahn A, Kippelen B (2012) A universal method to produce low–work function electrodes for organic electronics. Science 336(6079):327–332. doi:10.1126/science.1218829
60. Spanggaard H, Krebs FC (2004) A brief history of the development of organic and polymeric photovoltaics. Sol Energy Mater Sol Cells 83(2–3):125–146. doi:10.1016/j.solmat.2004.02.021
61. Bernède JC, Berredjem Y, Cattin L, Morsli M (2008) Improvement of organic solar cell performances using a zinc oxide anode coated by an ultrathin metallic layer. Appl Phys Lett 92 (8):083304. doi:10.1063/1.2888176
62. Emmott CJM, Urbina A, Nelson J (2012) Environmental and economic assessment of ITO-free electrodes for organic solar cells. Sol Energy Mater Sol Cells 97:14–21. doi:10.1016/j.solmat.2011.09.024
63. Galagan Y, JEJM R, Andriessen R, Fan C-C, PWM B, Veenstra SC, Kroon JM (2011) ITO-free flexible organic solar cells with printed current collecting grids. Sol Energy Mater Sol Cells 95(5):1339–1343. doi:10.1016/j.solmat.2010.08.011
64. Yim JH, S-y J, Pang C, Lee KM, Jeong H, Park J-Y, Ahn YH, de Mello JC, Lee S (2014) Fully solution-processed semitransparent organic solar cells with a silver nanowire cathode and a conducting polymer anode. ACS Nano 8(3):2857–2863. doi:10.1021/nn406672n
65. Kyaw A, Sun X, Jiang C, Lo G, Zhao D, Kwong D (2008) An inverted organic solar cell employing a sol-gel derived ZnO electron selective layer and thermal evaporated MoO3 hole selective layer. Appl Phys Lett 93(22):221107
66. Hau SK, Yip H-L, Baek NS, Zou J, O'Malley K, Jen AK-Y (2008) Air-stable inverted flexible polymer solar cells using zinc oxide nanoparticles as an electron selective layer. Appl Phys Lett 92(25):253301
67. Krebs FC (2008) Air stable polymer photovoltaics based on a process free from vacuum steps and fullerenes. Sol Energy Mater Sol Cells 92(7):715–726. doi:10.1016/j.solmat.2008.01.013
68. Günes S, Neugebauer H, Sariciftci NS (2007) Conjugated polymer-based organic solar cells. Chem Rev 107(4):1324–1338. doi:10.1021/cr050149z
69. Schulze K, Maennig B, Leo K, Tomita Y, May C, Hüpkes J, Brier E, Reinold E, Bäuerle P (2007) Organic solar cells on indium tin oxide and aluminum doped zinc oxide anodes. Appl Phys Lett 91(7):073521. doi:10.1063/1.2771050
70. Rowell MW, Topinka MA, McGehee MD, Prall H-J, Dennler G, Sariciftci NS, Hu L, Gruner G (2006) Organic solar cells with carbon nanotube network electrodes. Appl Phys Lett 88 (23):233506. doi:10.1063/1.2209887
71. Zhou Y, Fuentes-Hernandez C, Khan TM, Liu JC, Hsu J, Shim JW, Dindar A, Youngblood JP, Moon RJ, Kippelen B (2013) Recyclable organic solar cells on cellulose nanocrystal substrates. Sci Rep 3:1536. doi:10.1038/srep01536
72. Mathew S, Yella A, Gao P, Humphry-Baker R, CurchodBasile FE, Ashari-Astani N, Tavernelli I, Rothlisberger U, NazeeruddinMd K, Grätzel M (2014) Dye-sensitized solar cells with 13% efficiency achieved through the molecular engineering of porphyrin sensitizers. Nat Chem 6(3):242–247. doi:10.1038/nchem.1861. http://www.nature.com/nchem/journal/v6/n3/abs/nchem.1861.html – supplementary-information

73. Liang M, Chen J (2013) Arylamine organic dyes for dye-sensitized solar cells. Chem Soc Rev 42(8):3453–3488. doi:10.1039/c3cs35372a
74. Chia-Hao M, Chuang PRB, Bulović V, Bawendi MG (2014) Improved performance and stability in quantum dot solar cells through band alignment engineering. Nat Mater 13 (8):796–801
75. Ahmed R (2013) Quantum dots: the pros and cons in PV: pv magazine. Photovoltaic Market and Technology
76. Anderson TN (2013) Natural convection heat transfer in V-trough solar concentrators. Sol Energy 95:224–228. doi:10.1016/j.solener.2013.06.022
77. Mekhilef S, Saidur R, Safari A A (2011) Review on solar energy use in industries. Renew Sust Energ Rev 15(4):1777–1790
78. Solar energy for paddy parboiling in India: Financial feasibility analysis (1996) Fuel and Energy Abstracts 37(6):441
79. Bansal NK (1999) Solar air heater applications in India. Renew Energy 16(1–4):618–623
80. de la Parra I, Marcos J, García M, Marroyo L (2015) Storage requirements for PV power ramp-rate control in a PV fleet. Sol Energy 118:426–440. doi:10.1016/j.solener.2015.05.046
81. Boicea VA (2014) Energy storage technologies: the past and the present. Proc IEEE 102 (11):1777–1794. doi:10.1109/jproc.2014.2359545
82. Chemwiki Commercial Galvanic Cells (2016). http://chemwiki.ucdavis.edu/Core/Analytical_Chemistry/Electrochemistry/Case_Studies/Commercial_Galvanic_Cells. Accessed 20 May 2016
83. Kenning T (2016) Modest advances for solar and clean energy in India budget http://www.pv-tech.org/news/coal-cess-doubling-to-benefit-clean-energy-in-india-budget
84. Sigma Aldrich (2016). http://www.sigmaaldrich.com/materials-science/nanomaterials/quantum-dots.html
85. Solar Cell efficiencies (2016). http://www.nrel.gov/ncpv/images/efficiency_chart.jpg. Accessed 20 May 2016

Part IV
Proximate Techno-Economic Analysis, Biofuel Policies and Environmental Scenario

Chapter 18
Proximate Technical and Economic Aspects and Life Cycle Analysis of Biodiesel Production in India: An Overview

S. Prasad, K. R. Sheetal, Anuj K. Chandel, and M. S. Dhanya

Abstract One of the key challenges confronting the developing world is how to meet its growing energy demands and sustain economic growth without contributing to climate change and pollution. The Government of India has undertaken several policy measures to augment production and use of biodiesel during the past one decade at the national level. The government expects biodiesel will be eco-friendlier than petro-based diesel by reducing negative effects and can lead to sustainable development. Increased use of biodiesel is a significant part of the global strategy for climate change mitigation and air quality improvement. Since biodiesel is prepared entirely from biomass, it does not contain any sulfur, and having oxygen content in it improves the combustion efficiency of ignition engines and lowers the emissions. The new alternative feedstock and improved process technologies may provide a solution for the existing challenges of biodiesel production with sustainable impact in next decade in Indian perspective. The life cycle studies have also shown positive energy balance and GHG emissions for biodiesel compared to fossil diesel. The development of a domestic biodiesel, marketing, and its use is also expected to improve lives of the common people by generating more rural employment opportunities and reduce the reliance on petroleum fuels for transportation in a developing country like India.

Keywords Biodiesel • Policies • Pollution • Climate change • Mitigation • India

---

S. Prasad (✉) • K.R. Sheetal
Centre for Environment Science and Climate Resilient Agriculture, Indian Agricultural Research Institute, New Delhi 110012, India
e-mail: shiv_drprasad@yahoo.co.in

A.K. Chandel
Centro de Tecnologia Canavieira (CTC), Fazenda Santo Antonio s/n-Caixa Postal 162 CEP, 13400-970 Piracicaba, Sao Paulo, Brazil

M.S. Dhanya
Centre for Environmental Science and Technology, Central University of Punjab, Bathinda 151001, India

A.K. Chandel, R.K. Sukumaran (eds.), *Sustainable Biofuels Development in India*,
DOI 10.1007/978-3-319-50219-9_18

1 Introduction

India was the fourth largest energy consumer in the world after China, the USA, and Russia in 2011 [1]. With the demand for transport fuel growing at the rate of 6.8% a year, it is predicted that by 2020, India will become the third-largest consumer in the globe, after the USA and China [2]. Increased use of fossil fuels also causes environmental problems both locally and globally [3, 4]. Over the past few decades, there has been a tremendous increase in efforts to reduce the reliance on fossil fuels, along with the reduction of pollution and climate change mitigation in many parts of the world, including India [5, 6].

In general terms, the term "biodiesel" refers to an oxygenated fuel made from biologically derived sources that have properties closer to those of petroleum-based diesel fuels [7]. The American Society for Testing and Materials (ASTM) defines biodiesel fuel as monoalkyl esters of long chain fatty acids obtained from a renewable lipid feedstock, such as vegetable oil or animal fat which is the most suitable substitute for diesel [8, 9].

Vegetable oil fatty acid methyl esters (FAME) are the most used type of biodiesel fuel, produced by transesterification of high-quality vegetable oil with methanol [8, 10, 11]. There are technical and economic challenges that need to be addressed to make biodiesel profitable. The cost of the source of triglycerides plays a primary role in process profitability. To make biodiesel competitive with petroleum diesel, low cost feedstocks such as nonedible oils from *Jatropha curcas* and *Pongamia pinnata* and other tree borne oilseeds, waste oils, and triglyceride containing feedstocks could be used [12, 13] for biodiesel production.

The techno-economic analysis is performed to determine the economic viability of biodiesel production and process. This analysis can be useful in determining the highest potential of biodiesel for near, mid, and long-term success. The techno-economic analysis is also helpful in directing research toward biodiesel production, advancement, and commercialization.

2 Biodiesel as Automotive Fuels

Biodiesel and ethanol have gained much consideration as alternatives, because of its renewability and feasibility in production from a variety of feedstocks using proven technology [8, 12]. Petro-diesel fuel is made up of hundreds of different hydrocarbon chains (roughly in the range of 14–18 carbons in length) and contains aromatic hydrocarbons (benzene, toluene, xylenes, etc.), sulfur, and crude oil residue contaminants. But biodiesel hydrocarbon chains are generally 16–20 carbons in length and contain oxygen at one end and with no sulfur content [14]. It is known as an "oxygenate" fuel because it holds 10–12% oxygen by weight, which contributes to lesser exhaust emissions and fuel use, by enhancing fuel combustion.

As an alternative fuel, biodiesel can be used in neat form or blended with petroleum-based diesel. A blend of 20% biodiesel with 80% diesel, by volume, termed "B20" is the most commonly used blend in transport fuel. Similarly, a blend of 5% biodiesel with 95% diesel is labeled as "B5" and so on. The blending of biodiesel improves combustion efficiency and emission profile of petroleum-based diesel [3, 15]. Biodiesel fuel blends decrease emissions of particulate matter (PM), hydrocarbons (HC), carbon monoxide (CO) and sulfur oxides (SOx) [16, 17]. Biodiesel is nontoxic, biodegradable, and does not contribute to global warming [6, 18]. Hence it is a clean and renewable alternative energy source [19, 20].

3 The Biodiesel Policy in India

The Government of India has initiated various policy measures to expand production and use of biofuels during the last decade; with several Ministries associated with policy making, regulation, promotion and development of the biodiesel sector at the national level [21, 22]. India's pro-biofuels policy program emphasizes an optimistic outlook of biodiesel, showing that these renewable alternatives will make the country more energy secure and less dependent on other countries for transport fuels [21, 22]. The development of a domestic biodiesel market is anticipated to improve lives of the poor by creating more rural employment opportunities [2, 23].

3.1 National Mission on Biodiesel in India

In April 2003, the Government of India launched a National Mission on Biodiesel that recognized *Jatropha curcas* as the most relevant tree-borne oilseed (TBO) for producing biodiesel and focused on promoting its plantations on "wastelands" [24]. In October 2005, the Ministry of Petroleum and Natural Gas (GoI) announced a "biodiesel purchase policy" in which oil companies would buy biodiesel and blend it with high-speed diesel at a 5% mixing ratio [25], to be carried out at 20 procurement centers located across major producing areas in India, from January 2006 [26].

The biodiesel was to be procured at a pre-determined price (reviewed every 6 months by the Ministry). The government does not grant any straight financial support for the biodiesel production or investment in biodiesel plants and essential infrastructure. Although exempted from the central excise tax by central and some state governments, most state governments do not provide any sales tax exemptions for biodiesel or biodiesel blended diesel [27, 28]. Nevertheless, the central government and several state governments offer monetary incentives for supporting the planting of such nonedible oilseed crops [29].

3.2 National Policy on Biofuels

The Indian biofuel sector policies emerged over a decade ago to improve energy security by becoming independent from foreign oil imports. The National Policy on Biofuels passed on December 24, 2009, by Government of India [29], recommends an objective of 20% blending of both biodiesel and bio-ethanol by 2017 [30]. India's biofuel plan continues to focus on the use of non-food sources, including molasses for the ethanol production and non-edible oils for the biodiesel production [31]. To monitor production potential and set blending targets, the policy proposes establishing a 'National Registry' on feedstock availability and also a periodical review of biodiesel blending targets.

National Biofuel Policy drafted by the MNRE focuses on non-food sources; avoiding competition with food security, and protecting fertile farmlands from being diverted for bioenergy plantations [27]. The biofuel policy deals with some of the critical issues like minimum support prices (MSPs) for biofuel crops. This policy also deals with subsidies for farmers, marketing, grants and fiscal concessions for the biofuel industry, research and development, mandatory blending of auto-fuel with biofuel, quality norms, testing and certification of biofuels [32].

Concurrently, some states have evolved their own biodiesel policies which are also not in direct conflict with the National Biofuel Policy. The state governments are steered to act within the broad contours and provisions of the National Biofuel Policy [33]. The state government of Chhattisgarh has started a program on *Jatropha* production in wastelands and forest lands.

Chhattisgarh government has also formed the CBDA (Chhattisgarh Biofuel Development Authority), for monitoring various biodiesel-related activities in the state. Similarly, biofuel policies have been framed in Rajasthan, Odisha, Uttarakhand, Tamil Nadu, Andhra Pradesh, and Karnataka for encouraging cultivation of biofuel crops; along with several tax concessions and subsidy schemes to cultivators and biofuel processing industries in these states [33].

The Government is considering the creation of a National Biofuel Fund for providing financial incentives like subsidies and grants for novel and second generation feedstock; advanced technologies and conversion processes; and production units based on new and second generation feedstock. Besides, the biofuel technologies and projects would be allowed 100% foreign equity through automated approval routes to bring Foreign Direct Investment (FDI), provided such biofuels produced are put only to domestic use [21, 22].

4 Biodiesel Production and Current Status in the World

Biodiesel has experienced a significant surge worldwide [27]. A rapid jump in production volume is being seen not only in developed nations such as Germany, Italy, the USA, and France, but also in developing countries like Brazil, Argentina,

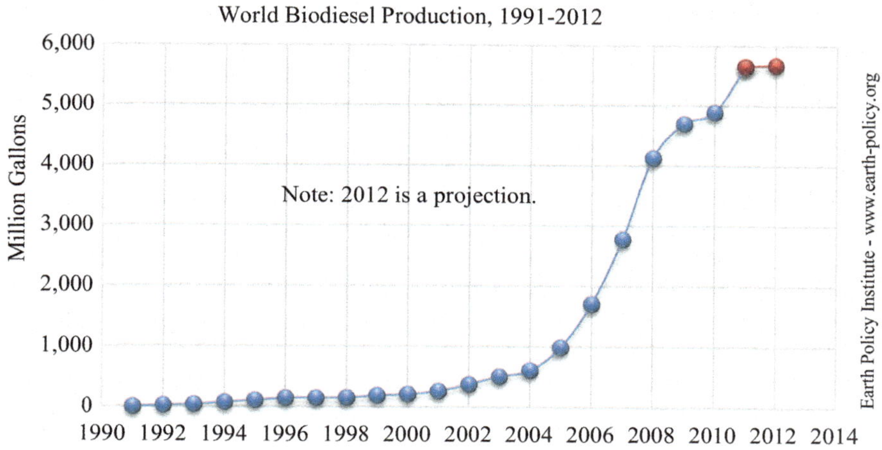

Fig. 18.1 World biodiesel production, 1991–2012. *Source*. Licht [35], World ethanol and biofuels report

Indonesia, and Malaysia [34]. The number of biodiesel producing and exporting countries increased from 20 to more than 200 from 2007 to 2010, indicating a transition in global biofuel production. In 2009, biodiesel accounted for 75% of biofuels produced in Europe. The per day biodiesel production in 2009 totaled 308.2 thousand barrels globally and the European Union with 172.6 thousand barrels. The world production of biodiesel between 1991 and 2012 is presented in Fig. 18.1.

4.1 India's Energy Demand and Biodiesel Production: Current Status

In India, primary energy consumption doubled between 1990 and 2012, reaching an estimated 32 quadrillion Btu, with the second-largest population of the globe (more than 120 crores in 2012), growing at a rate of 1.3% each year since 2008. According to World Bank, India's per capita energy expenditure is one-third of the global average and as per, International Energy Agency (IEA) potentially higher energy requirement in the long term is required as the nation continues its path of economic development [1].

In India, the consumption of crude oil was 168.13 MT for the year 2010–2011. But 80% of this consumption was met by import [36]. In India, the energy demand is increasing at a rate of 6.5% per annum. Currently, India is the fourth largest consumer and also the importer of oil and petroleum products [1] globally. Due to the rapid increase in the demand for diesel and other petroleum products, imports are expected to rise to 92% by 2030 [37]. Domestic production of crude oil can only fulfill 25–30% of national consumption [38]. Thus the energy security has become an important issue for the nation as a whole, necessitating look for alternative fuels, which can be produced from feedstock available within the country [3, 27].

Demand projections suggest that nearly 4.61 million tons (MT) of biodiesel would be required for 5% blending by the year 2016–2017. To realize this, assuming *Jatropha* as the primary feedstock for biodiesel with an average seed yield and biodiesel recovery rate of 2.5 t/ha and 30%, respectively, the area required for the crop has been worked out to be 4.91 Mha. An estimated area of 26.25 Mha would be needed under *Jatropha* cultivation to meet 20% blending target by the year 2020–2021, assuming no new and better feedstock are introduced with more yields or oil content [39].

4.2 Biodiesel Production and Utilization in India

Various developmental activities have been carried out with respect to the production and use of biodiesel through transesterified nonedible oil [23, 25]. Aatmiya Biofuels Pvt. Ltd., Por-Vadodara, Gujarat had set up a biodiesel plant with a commercial production capacity of 1000 L/day from *Jatropha* [23]. One of the first commercial biodiesel production plants is being established in Andhra Pradesh by Southern Online Biotechnologies (P) Ltd. The project is proposed with an initial capacity of 30 tons of biodiesel per day, which is expandable to 100 tons per day. The technologies for the unit is provided by Lurgi Life Science Engineering, Germany, along with their local partner, Chemical Constructions International Private Limited, New Delhi [25].

In India, the Indian Railways has been experimenting with the new eco-friendly biodiesel fuel to run passenger trains and was successful in its trial run by using 5% biodiesel as fuel in the Delhi–Amritsar Shatabdi Express on 31 December 2002. It has plans to use the vast tracts of land all along the railway tracks for growing *Jatropha* and other nonedible tree borne oilseeds. RSDO (Railways' Research Design and Standards Organization) located in Varanasi (Uttar Pradesh, India) are conducting initial trials on test beds. The GOI's Planning Commission set an enthusiastic target of 11.2–13.4 Mha of wasteland to be planted with *Jatropha* by the year 2012, to provide sufficient biodiesel to blend at 20% with petrodiesel [29].

As per the recent Wastelands Atlas of India [40], about 63.85 Mha of the area is classified as wasteland (including 14.06 Mha of degraded forest lands). Also, unless the growers are offered assured returns or financial viability, it will not be possible to promote plantation of TBOs on private wastelands, unlike on government land or community area. Therefore, there is a need to exercise caution before one makes any assumption about the potential use of existing wastelands for raising biodiesel plantations [33].

5 Raw Materials for Biodiesel Production

All over the world, edible oils are widely used for the production of biodiesel. Oil crops such as rapeseed, mustard, sunflower, olive, soybean, canola, cottonseed, palm, coconut, peanut, and jojoba can be used for biodiesel production. The choice of raw materials depends primarily on its availability and cost [19, 41–43]. Since more than 95% of the biodiesel is synthesized from the edible oil, there are several claims that many difficulties may arise [44]. It is believed that large-scale production of biodiesel from edible oils may bring global imbalance to the food supply and demand market [45].

Countries such as the USA and those belonging to European community are self-dependent in the production of edible oils and even have the surplus quantity to export. Hence, edible oils especially soybean and rapeseed are used in the USA and European Nations, respectively [46] for biodiesel production. Similarly, countries with the coastal area such as Malaysia and Indonesia have surplus coconut oil, and is utilized for the production of biodiesel [47].

5.1 Production and Consumption of Edible Oil in India

India accounts for 9.3% of world's total oilseed production and contributes as the fourth largest edible oil producing country. Even then, about 46% of edible oil is imported for catering to the domestic needs [21, 22]. Table 18.1 gives the production and consumption of edible oil in India. The possibilities of production of biodiesel from edible oil resources are almost impossible [48], as the primary need is to meet the demand for feeding the population, that at present imported from other countries. So only the tree borne oilseeds and nonedible oil resources like *Jatropha* and *Pongamia* seem to be the possibility for biodiesel production in India [3].

Table 18.1 Production and consumption of edible oil (Lakh tonne)

Oil year	Production of oil seeds	Net availability of oil from domestic sources	Import of edible oil	Consumption of edible oil
1998–1999	247.48	69.60	26.22	95.82
1999–2000	207.15	60.15	41.96	102.11
2000–2001	184.40	54.90	41.86	96.76
2001–2002	206.63	61.46	43.22	104.68
2002–2003	150.58	47.28	43.65	90.93
2003–2004	251.42	71.09	52.95	124.04
2004–2005	248.42	73.10	44.0	117.10
2005–2006	254.56	78.30	57.9	136.2
2006–2007	258.48	80.5	64.2	144.7
2007–2008	276.22	85.4	72.48	157.88

Source. GOI [21, 22], Production of oilseeds: Ministry of Agriculture, India

5.2 Biodiesel Production from Major TBO(s) and Nonedible Oil Sources

As a sustainable solution, biodiesel has to be produced from the different nonedible tree borne oilseeds [49–51]. In order to overcome this devastating food to fuel phenomenon in a country like India, systematic research, and advances, has been made to produce biodiesel by using alternative resources like nonedible oils [47]. In India, the different types of nonedible tree borne oilseeds already have been identified [23], and can be grown in wasteland unsuitable for other crop cultivation. In India, over 100 Mha are presently classified as degraded land, and this can be utilized to grow such plants for biodiesel production.

India, even though endowed with a vast land area covering the coastal zone, does not produce sufficient edible oils and has to import them to satisfy the food requirements. Hence, in the Indian context, the raw materials used for biodiesel production has to be the unutilized and underutilized materials [52]. The raw materials used at present in India are *Jatropha curcas* and *Pongamia pinnata,* with seeds possess 40% and 33% oil, respectively. Oils from both the plants contain toxins and hence are nonedible. *Jatropha curcas* contains phorbol esters and curcin as toxins [53], whereas *Pongamia pinnata* contains furanoflavonols, chromenoflavones, flavones, and other furanodiketones as toxins which make the oil nonedible [43, 54].

It is estimated that the potential availability of nonedible oil in India is about 1 million ton per year. India has more than 300 different species of trees, which produce oil-bearing seeds [17]. Of these, around 75 plant species have 30% or more fixed oil in their seeds/kernel. The other promising nonedible oil sources are *Azadirachta indica*, *Shorea robusta*, *Madhuca indica*, *Schleichera oleosa*, *Sapindus mukorossi*, *Garcinia indica*, *Diploknema butyracea*, and *Aleurites fordii* [55]. The oil content from the seeds of the plant varies. The oil content in the castor bean, hemp, and neem seed is around 50%, 35%, and 30%, respectively in comparison to the oil content of soybean (20%) and rapeseed (40%).

Table 18.2 gives the potential quantity (tons/year), current utilization (tons/ year), and percent of the use of these most important nonedible oil resources in India. Additional research is also needed concerning the agroclimatic and soil

Table 18.2 Nonedible oil sources in India

Oil source	Botanical name	Potential quantity (TY^{-1})	Current utilization (TY^{-1})	% of utilization
Rice-bran	*Oryza sativa*	474,000	101,000	21
Sal	*Shorea robusta*	720,000	23,000	3
Neem	*Azadirachta indica*	400,000	20,000	5
Karanja	*Pongamia pinnata*	135,000	81,000	6
Kusum	*Schleichera oleosa*	25,000	–	–
Mahua	*Madhuca indica*	180,000	–	–

Source. Subramanian et al. [17]

requirements of TBOs, as well as inputs and maintenance activities that are necessary to make its cultivation profitable. Particular emphasis should be given to breeding drought-resistant varieties of different oil-bearing tree species that provide acceptable yields. Such varieties would help to put unproductive dry-lands to productive use and further reduce the competition between biodiesel and food production.

5.3 Waste Cooking Oil (WCO)

The usage of waste cooking oil and waste materials can be an economical way of biodiesel and reduce environmental problems [56, 57]. Waste cooking oil is widely produced, inedible, and could serve as an almost ready-to-use substitute for fossil origin fuel. The crude vegetable oil and waste cooking oil has a high viscosity. This viscosity problem is usually bypassed by blending WCO with petrol diesel or by using transesterification to produce biodiesel [56, 58]. Pugazhvadivu and Jeyachandran [59] proposed solving the injection and filter clogging problems by preheating the waste cooking oil to 135 °C improved the overall performance of the engine.

5.4 Biodiesel from Microalgae

The comprehensive research initiatives have proved that microalgae biomass appear to be one of the promising sources of renewable biodiesel which is capable of meeting the global demand for transport fuels [60]. The eukaryotic green microalgae possess chlorophyll and accessory pigments to capture the energy from sunlight and use photosynthetic systems (PSII and PSI) to carry out plant like oxygenic photosynthesis. The excess reduced carbon is stored inside the cells as carbohydrates (CH_2O) and/or lipids [61, 62]. Currently, microalgae are being promoted as an ideal third generation biofuel feedstock because of their rapid growth, CO_2 fixation ability, and high production capacity of lipids [10, 16, 63].

Many algal strains are exceedingly rich in lipids or oil contents (Table 18.3). The average oil in microalgae is about 20–50% (dry weight), oil content itself can be estimated to be 64.4% of the total lipid component. The oil content of some microalgae exceeds even 70% of the dry weight of algae biomass [10, 64]. Microalgae are capable of producing manifold oil per unit area of land, compared to conventional terrestrial oilseed crops (Table 18.4). Using microalgae to produce biodiesel will not undermine the production of food, fodder, and other products derived from crops [10], and can avoid freshwater competition as well [65]. In fact, microalgae have the highest oil yield compared to various plant oils. It can produce up to 100,000 L oil ha^{-1} $year^{-1}$, whereas palm, coconut, castor, and sunflower produce up to 5950, 2689, 1413, and 952 L ha^{-1} $year^{-1}$, respectively (Table 18.4).

Table 18.3 Oil content of microalgae

Microalgal species	Oil content (% dry wt.)
Botryococcus braunii	25–75
Chlorella sp.	28–32
Crypthecodinium cohnii	20
Cylindrotheca sp.	16–37
Dunaliella primolecta	23
Isochrysis sp.	25–33
Nannochloris sp.	20–35
Nannochloropsis sp.	31–68
Neochloris oleoabundans	35–54
Nitzschia sp.	45–47
Phaeodactylum tricornutum	20–30
Schizochytrium sp.	50–77

Source. Chisti [10]

Table 18.4 Microalgae oil potential compared to other terrestrial oil seed crops

Crop	Oil content (%)	Oil yield (L ha^{-1} $year^{-1}$)
Corn	3.1–5.7	172
Soybean	18–22	446
Canola/Mustard	40–45	1190
Jatropha	20–40	1892
Coconut	17	2689
Oil palm	23	5950
Microalgae	30 (Min. scenario)	58,700
Microalgae	70 (Max. scenario)	1,36,900

Source. Chisti [10]

5.5 *Major Components of Plant Oil and Lipids*

The major components of plant oil and lipids are triglycerides. Triglycerides are esters of glycerol with long-chain acids, commonly called fatty acids. For example soybean oil has five fatty acids: approximately equal amounts of palmitic acid, oleic acid, and linolenic acid (about 13% each), linoleic acid (almost 55%), and stearic acid (approximately 4%). The quality of feed vegetable oil mainly free fatty acid (FFA) content plays a major role in identifying the suitable technology. The FFA has a significant effect on the transesterification of glycerides with alcohol using the catalyst [66].

The free fatty acid content must be reduced to below 0.1 wt.% by neutralizing oil with an alkali [67]. The high free fatty acid content (>1% w/w) provoke soap formation and the separation of products would be extremely difficult, and as a result low yield of biodiesel product would be obtained [64, 68]. According to

Anggraini and Wiederwertung [69] also if the FFA content is more than 3%, then the conversion efficiency decreases gradually.

6 Biodiesel Production Technologies

The selection of suitable technology for production of biodiesel calls for the careful selection of processing steps, catalyst, and downstream process integration [23, 70]. There are several processes used for biodiesel synthesis and processing of oil. One process of biodiesel production is pyrolysis technique; it refers to a chemical change caused by application of thermal energy in the absence of air or nitrogen. The process is known as cracking of vegetable oils, which helps to reduce viscosity and improve cetane number [71]. Limitations of this process include high equipment cost and the need for separate distillation equipment for separation of various fractions [23]. The process of production also leads to removal of oxygen that negates some of the environmental benefits usually associated with biodiesel.

Micro emulsification is another alternative technique that has been stated to produce biodiesel. The components of a biodiesel micro-emulsion include vegetable oil, alcohol and surfactant and cetane improver in suitable proportions. However, fuel produced by micro-emulsification creates engine performance problems [11]. Laboratory tests for motors operating on fuels generated by micro-emulsion show carbon deposits, injector needle sticking and an increment in the viscosity of the lubricating oil. Because of the limitations of this method of production and those of pyrolysis, transesterification is usually preferred [11, 70].

6.1 Transesterification

The most popular method of producing biodiesel is the transesterification of vegetable oils and renewable lipid sources. The transesterification is a process of transforming one type of ester into another. Glycerol is a by-product of the process. This process includes mixing any natural oil (vegetable oil or animal fat) with virtually any alcohol in the presence of a catalyst.

The first conventional way of producing biodiesel is using a base catalyst. In the alkali process, sodium hydroxide (NaOH) or potassium hydroxide (KOH) is used as a catalyst along with methanol or ethanol [8]. Initially, during the process, alcoxy is produced by the reaction of the catalyst with methanol or ethanol and the alcoxy is then reacted with any vegetable oil to form biodiesel and glycerol. The alcoxy reaction is as follows:

$$\text{R-CH}_2\text{OH} + \text{NaOH} \rightarrow \text{H}_2\text{O} + \text{R-CH}_2\text{ONa}$$

Glycerol being denser settles at the bottom of the container and then carefully biodiesel can be decanted. This process is the most efficient and least corrosive of all the processes, and the reaction rate is fairly high even at low temperatures of 60 °C. There may be a risk of the free acid or water contamination, and soap formation is anticipated to take place which makes the separation process complex.

The transesterification reaction with alcohol is represented by the general equation shown in Fig. 18.2. Transesterification consists of a series of three sequential reversible reactions as shown in Fig. 18.3. The initial step is the conversion of triglycerides to diglycerides, and then to monoglycerides, and ultimately to glycerol, producing one ester molecule for each glyceride at each step.

According to a comparative study of different alkaline catalysts used in the transesterification of vegetable oil with methanol under the following conditions: 238 K, 6:1 molar ratio of methanol to oil and 1% wt. Sodium methylate (CH_3ONa)

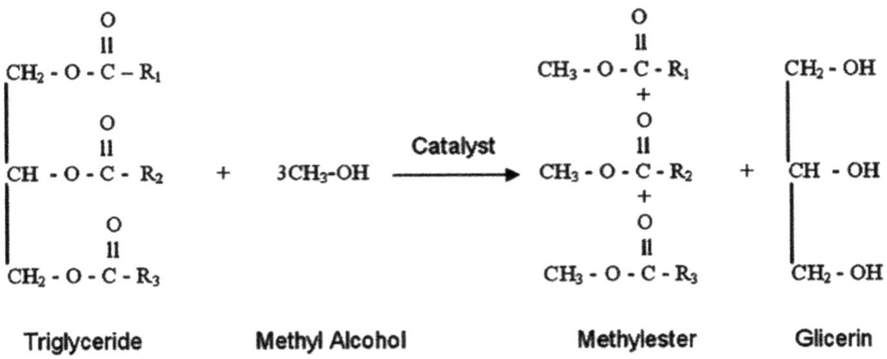

Fig. 18.2 Biodiesel production process through transesterification reaction. Where R_1, R_2 and R_3 are long chains of carbons and hydrogen atoms, are called fatty acid chains

First -step

Triglyceride (TG) + R'OH $\xleftrightarrow{\text{Catalyst}}$ Diglyceride (DG) + $R'COOR_1$

Second -step

Diglyceride (DG) + R'OH $\xleftrightarrow{\text{Catalyst}}$ Monoglyceride (MG) + $R'COOR_2$

Third -step

Monoglyceride (MG) + R'OH $\xleftrightarrow{\text{Catalyst}}$ Glycerol (GL) + $R'COOR_3$

Fig. 18.3 Transesterification reactions of oil with alcohol to esters and glycerol. Where, R is a mixture of various fatty acid chains. The alcohol used for producing biodiesel is usually methanol (where the R'-corresponds to CH_3)

catalyst gives the highest yield (98.6%) i.e. 12% higher than sodium hydroxide [72]. By this process, the flash point of biodiesel is lowered, and the cetane number is improved [73].

6.2 *Factors Affecting the Transesterification Reaction*

The transesterification reaction is influenced by factors such as molar ratio of alcohol, temperature and reaction time, agitation speed and mixing condition, free fatty acid and moisture, types and amount of catalyst, etc.

6.2.1 Molar Ratio of Alcohol to Oil

The molar ratio of alcohol (ethanol or methanol) to oil is one of the common variables affecting the ester yield. The commonly employed molar ratio for two-step transesterification is 6:1 for acid transesterification and 9:1 for alkali-catalyzed transesterification. However, the optimum molar ratio has been shown to differ a little depending on the raw oil chosen and its acid value. For single step transesterification reaction, 10:1 molar ratio has been used more often, although a molar ratio varying from 6:1 to 13:1 has been employed by other researchers [74].

6.2.2 Temperature and Reaction Time

Transesterification can occur at various temperatures depending on the type of oil employed, and it is a critical factor that affects the yield of biodiesel [11]. With sodium and potassium hydroxides, the transesterification reaction between oils and ethanol could reach a conversion efficiency of 91.6% at room temperature [75]. The commonly employed temperature ranges from as low as room temperature to up to 65 °C. Higher temperature favors saponification and hence must be avoided [76]. The conversion rate increases with reaction time [77]. The higher molar ratios result in greater ester conversion in a shorter time.

6.2.3 Agitation Speed and Mixing Condition

Agitation speed plays a significant role in the formation of the end product (mono alkyl ester) because agitation of oil and catalyst mixture enhances the reaction. At 400 rpm higher conversion of the end product was obtained. Higher stirring speed favors formation of soap [78]. The triacylglycerol (TAG) breaks down into small drops at sufficient magnitude of agitation speed, which leads to better contact with methanol phase, leading to more efficient conversion into fatty acid methyl esters (FAME) and glycerin [79]. The rate at which FAME are produced throughout the transesterification reaction is thus controlled by mass transfer conditions, which results in a lag time before start of conversion to FAME begins [80]. This condition is more prominent when the reaction is catalyzed by solid catalysts [72].

6.2.4 Free Fatty Acid and Moisture

The free fatty acids (FFAs) are the saturated or unsaturated monocarboxylic acids that usually occur in fats, oils but are not attached to glycerol backbones. During acid catalysed process, the FFA react with alcohol to produce esters but simultaneously, water is also produced which inhibits the transesterification reaction [81]. The starting materials used for base-catalyzed alcoholysis should meet some specification which is that a free fatty acid (FFA) value lower than 3% is required to carry the base catalyzed reaction to finish [72]. If the reaction conditions do not meet the particular specification, ester yields are significantly reduced [69, 77].

6.2.5 Types and Amount of Catalyst

Catalyst amount can affect the yield of the biodiesel product. Catalysts are classified as alkali and acid. Alkali-catalyzed transesterification is much faster than acid-catalyzed. Sodium methoxide was more effective than sodium hydroxide because of the assumption that a small amount of water was produced by mixing NaOH and methanol. Sodium hydroxide, however, is cheaper and is used widely in large-scale processing [11]. However, higher content of the catalyst above optimum decreases biodiesel yield a little, since excess catalyst causes more triglycerides to react with the alkali catalyst and form more soap [82].

7 Economics of Biodiesel Production

The potential of biofuels depends on its economic viability. The economics of biofuel production depends substantially on the cost of seeds/raw oil, other inputs, interstate variations in subsidies, and the economy of scale at which the processing plant is operating. Several subsidy programs and tax exemptions are also parts of the central government's effort to speed up the influenced transition from fossil fuels to biofuels [82]. Minimum Support Price (MSP) for seeds and Minimum Purchase Price (MPP) offered by the state governments and oil marketing companies, respectively, also help to act as a guarantee to the biofuel (both bioethanol and biodiesel) manufacturers against price troughs.

A few studies have pointed at the negative socioeconomic consequences of *Jatropha* cultivation on the impoverished farmers. The *Jatropha* value chain consists of several activities starting from raising of the nursery to the distribution of biodiesel to end users. The cost of production can be reduced by the utilization of by-products of the transesterification process, viz., seed hull, oil cake, and glycerin. The trading or retail price of biodiesel must exceed the feedstock cost to cover processing, transportation, marketing and profit [25]. Farm studies conducted in three major *Jatropha* producing areas of Rajasthan, Chhattisgarh, and Uttarakhand, suggested that *Jatropha* is a profitable plant species in the long run, provided the government support in the form of input subsidies, technical skills, and marketing assistance is made available throughout the initial few years.

Rajasthan, Chhattisgarh, and Uttarakhand farmers/growers incurred a cost of around Rs. 31,295 ha^{-1}, Rs. 8319 ha^{-1}, and Rs. 12,050 ha^{-1}, respectively during the first year. The farmers of the Sikar district in Rajasthan had to pay Rs. 6–10 per seedling as they did not get any grants or subsidy from the state government. The cost of seedling alone came around 35% of their total cost [84]. In contrast, Chhattisgarh producers were getting seedlings at a very subsidized rate of Rs. 0.50 per seedling and the Uttarakhand producers were provided 100% subsidies on per seedlings. Wage rate was another significant part of the cost of *Jatropha* cultivation [83].

Seed processing infrastructure is one of the essential necessities in the *Jatropha* seed-based biodiesel, production value chain and is presently the main constraint holding back the progress of the biodiesel sector in India. In the same study, the cost of biodiesel production in Rajasthan State Mines and Minerals Ltd., the facility was found to be about Rs. 40 per kg., whereas in Chhattisgarh Biodiesel Development Authority (CBDA) unit it was around Rs. 19 per kg. The reasons for the lower cost at CBDA was the lower cost of seeds at factory gate (depends on the method of procurement, availability of seeds, handling and transportation charges). The crushing capacity of the plant, higher recovery rate of oil due to the economics of scale in plant operation, etc. also contibuted for lowering the cost of biodiesel at CBDA [83]. In both the plants, the sale of by-products such as glycerol and oil cake also helped reduce the production costs. The biodiesel industry is still growing and comparatively in infancy stage in India, so as it grows to a large scale and when appropriate infrastructure is developed, the cost of production and marketing of biodiesel may decline. Biotechnology applications to increase the oil content in the crops through bio-engineering will also result in better recovery and lower down the cost of biodiesel production.

8 Life Cycle Analysis of Biodiesel

Life cycle assessment (LCA) of produce or process is a management technique to quantify the mass flow, energy, and emissions assessing the environmental aspects and potential impacts from its production chain [85]. When the afore mentioned assessment is performed in conjunction with a feasibility analysis, all economic and environmental advantages and drawbacks of a process can be quantified.

8.1 Properties and Combustion Profile of Biodiesel

Biodiesel has properties similar to petroleum-based diesel fuel and can be blended into conventional diesel fuel [11]. This interest is based on properties of biodiesel, non-toxicity, and its potential to reduce exhaust emissions [86–88]. Biodiesel is by nature an oxygenated fuel, which improves fuel combustion and reduces CO, soot, and unburnt hydrocarbon. Biodiesel is nonflammable and, in contrast to petrodiesel, is non-explosive. The flash point of biodiesel (>130 °C) is significantly higher than that of petroleum diesel (64 °C) or gasoline (−45 °C).

The biodiesel has a density of ~0.88 g/cm^3, greater than petrodiesel (~0.85 g/cm^3). It has better lubricating properties and much higher cetane ratings than today's lower sulfur diesel fuels [32]. Biodiesel blending reduces fuel system wear and in low levels in high-pressure systems extends the life of the fuel injection devices that relies on the fuel for its lubrication [89]. The calorific value of biodiesel is about 37.27 MJ/L. Variations in biodiesel energy density are further dependent on the feedstock used than the production process and properties of biodiesel from different oils [90–92] as shown in Table 18.5. Since the fundamental properties of the biodiesel are comparable to those of diesel fuel, it can be utilized in all types of diesel engines with slight modification or no change either in its pure forms or as a blend with traditional diesel [18].

8.2 Energy Balance in Biodiesel Production

The energy balance (or net energy gain) is an indicator of biodiesel production sustainability and refers to the ratio of the energy released by burning the resulting biodiesel fuel to the energy used in the process. The US Department of Energy analyzed the full biodiesel life cycle and observed that for each unit of fossil fuel used to get biodiesel, 3.2 units of energy were gained. By contrast, petroleum diesel's life cycle yields only 0.83 units of fuel product energy [92]. Net energy ratios (NER) are indicators of energy balance. Several studies have reported NER

Table 18.5 Approximate biodiesel fuel characteristics related to flammability

Biodiesel from vegetable oil	Kinematic viscosity (mm^2/s)	Cetane no.	Heating value (MJ/kg)	Flash point (°C)	Density (kg/l)
Peanut	4.9	54	33.6	176	0.883
Soybean	4.5	45	33.5	178	0.885
Palm	5.7	62	33.5	164	0.880
Sunflower	4.6	49	33.5	183	0.860
Diesel	3.06	50	43.8	76	0.855
B20 (20%blend)	3.2	51	43.2	128	0.859

Source. Chhang et al. [90] and Rao and Gopalakrishnan [91]

values in the range of 1.2–8.6 and 0.1–4.2 for *Jatropha* biodiesel (JBD) and algal biodiesel (ABD), respectively. The variation in the values depends on the methodology used for energy distribution between product and co-products as well as the cultivation process [93, 94].

8.3 Exhaust Emission Reduction by Biodiesel

The biodiesel serves in decreasing the emissions of carbon monoxide (CO), hydrocarbons (HC), oxides of sulfur (SOx), particulate matter, and smoke density [95, 96]. The benefits of biodiesel blending in its pure forms (100% or B100) and 20% (B20), in terms of percent pollutant reduction and reduction in emission in g km^{-1} for 10% and 15% blend [96] is shown in Table 18.6. According to the Environmental Protection Agency's Renewable fuel standards program regulatory impact report released in February 2010, biodiesel from soybean oil results on an average 57% decrease in greenhouse gases (GHG) compared to fossil diesel, and biodiesel obtained from waste grease results in an 86% reduction [18].

Based on engine testing, adopting the usual stringent emissions testing protocols demanded by EPA for certification of motor fuels or fuel additives in the USA, the overall ozone (smog) forming potential of the hydrocarbon emissions from biodiesel is almost 50% less than that measured for diesel fuel. The mass concentration of the particulate matter or smoke is reduced up to 33% when the engine burned 100% biodiesel as fuel, compared to the 100% petroleum diesel [97].

8.4 GHG Emissions Reduction by Biodiesel

US Department of Energy (DOE) pointed out that the extent of the life cycle CO_2 emissions highly depends on the production pathway of biodiesel. In most cases, it ranges between 80% and 95% of CO_2 saving with respect to fossil diesel [98]. Over

Table 18.6 Reduction in pollution emission with biodiesel blending

	Emissions reduction (%)[a]		Emission reduction (g/km)		
Pollutant	B100	B20	Diesel	B10	B15
Particulate matter	−30	−22	0.129	0.093	0.080
Nox	+13	+2	0.79	0.83	0.89
Carbon monoxide	−50	−20	0.77	0.65	0.62
Unburned hydrocarbons	−93	−30	0.37	0.22	0.16
Sulfur dioxide	−100	−20	–	–	–

[a](−) and (+) represents less and more % of emission respectively from biodiesel in comparison to 100% diesel

Source Planning Commission [24]; Vasudevan et al. [96]

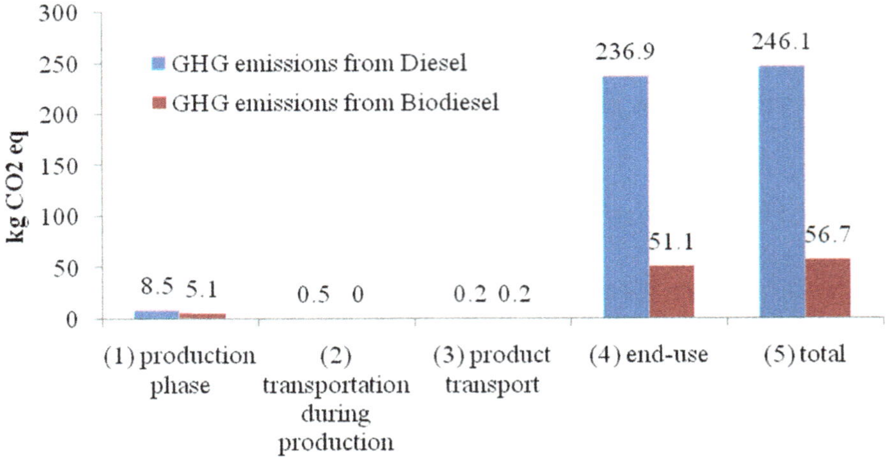

Fig. 18.4 Lifecycle GHGs emissions of biodiesel and diesel. *Source*. Kritana and Shabbir [99]

90% of GHG emissions from the life cycle of both diesel and biodiesel are from the use phase. The rate of total global warming potential from biodiesel production and use is just 23% of diesel [99]. This is because the CO_2 emissions from combustion of biodiesel in the engine during the use phase are considered GHG-neutral as they are of biomass origin and thus absorbed from the atmosphere by the *Jatropha* (biodiesel) plants during growth. Figure 18.4 shows the comparative life cycle GHG emissions of biodiesel and diesel [100]. GHG emissions reduction by *Jatropha* biodiesel and algal biodiesel, compared to fossil diesel ranges from 40% to 107% and from 35 to 140 gCO_2 eq per unit of MJ of biodiesel produced, respectively [93].

8.5 Biodiesel and Human Health

Bünger et al. [101] examined the mutagenic and cytotoxic impacts of diesel engine exhaust (DEE) from a modern traveler car with rapeseed oil methyl esters (RME) biodiesel as fuel. DEE showed a higher mutagenic potency compared to RME due to the higher content of polycyclic aromatic compounds in DEE exhaust. The application of a B20 fuel in the heavy duty diesel vehicles (HDDV) fleet is rated to reduce the per million risk of unanticipated death by exposure to air toxins in the SoCAB area of Southern California by nearly 2% and 5% respectively (Table 18.7) for the 50% and 100% HDDV fleet penetration of B20 biodiesel in the HDDV fleet emission scenarios determined with no indoor/outdoor (I/O) impacts and accounting for I/O effects on an annual mean and hourly basis [102].

Table 18.7 Average risk (out of a million) of premature death for the standard diesel base case and the 50% and 100% penetration of B20 biodiesel in the HDDV fleet emission scenarios

Scenario	Std diesel risk	50% B20 fuel		100% B20 fuel	
		Risk	(%)	Risk	(%)
No I/O effects	1950	1910	−2.1	1835	−5.9
Annual I/O effects	1284	1261	−1.8	1216	−5.3

Source. Prasad and Dhanya [32]

Scientific research proves that biodiesel exhaust has a less adverse impact on human health than petrodiesel fuel. Lower levels of PAHs and nitrited PAH compounds (potential carcinogens) in biodiesel emissions are also positive points in favor of biodiesel. In addition, about 47% decrease in particulate matter emission and 48% decrease in carbon monoxide, also reduce incidences of respiratory problems like asthma [103]. Biodiesel is the unique alternative fuel to have fully completed the health effects testing specifications of the Clean Air Act Amendments 1990, as biodiesel generates less sulfur emissions than regular petrodiesel. The public health benefits of reduced particulate matters and household air pollution (HAP) vulnerability from liquid biofuels outweigh the negligible smog consequence of relatively small NOx and permeation emissions increase from biofuels blends [104].

9 Conclusion

The approaches in this chapter allow the understanding of the developmental strategies of biodiesel production and consumption in India, based on policy, techno-economic, social, and sustainability assumptions in the senario of India's energy demand increase rate of 6.5% per annum. Due to the rapid increase in the demand for diesel and other petroleum products, India's dependency on crude oil imports is expected to rise to 92% by 2030. Energy security, climate change mitigation, and the reduction of air pollution are some of the fundamental gains of expanded biodiesel production. When especially favorable improvements in technology over the next decade are assumed, the costs of emissions from biodiesel could be approximately equal to, but unlikely less than, those of conventional fuels. Biodiesel holds the promise of yet greater environmental benefits, especially different nonedible tree borne oilseeds which are found to be more suitable for biodiesel production in India. The production of domestic biodiesel is also anticipated to improve lives of the poor by creating more rural employment opportunities and a better health. Thus biofuels production in a sustainable manner provide lot of benefits including reduction in the reliance on petroleum fuels, reduction of greenhouse gases and pollution emissions, increased energy security and rural development, and a sustainable fuel supply for the future.

References

1. US EPA (2013) International energy outlook 2013 with projections to 2040. Available at http://www.eia.gov/forecasts/ieo/pdf/0484(2013).pdf
2. Rhoads J (2007) Biodiesel and India's rural economy. Andersen PP, Cheng F, Cornell University. Available from https://www.uni-hohenheim.de/fileadmin/einrichtungen/jatropha/Jatropha_India_Case_study.pdf
3. Jain S, Sharma MP (2010) Prospects of biodiesel from Jatropha in India: a review. Renew Sustain Energy Rev 14:763–771
4. Sood PR (2012) Bioethanol—a viable answer to India's surging energy needs. World Acad Sci Eng Technol 68:1357–1361
5. Prasad S, Kumar A, Muralikrishna KS (2014) Biofuels production: a sustainable solution to combat climate change. Ind J Agric Sci 84(12):1443–1452
6. Vasudevan PT, Boyi F (2010) Environmentally sustainable biofuels: advances in biodiesel research. Waste Biomass Valori 1:47–63
7. Knothe G (2010) Biodiesel and renewable diesel: a comparison. Prog Energy Comb Sci 36:364–373
8. Demirbas A (2009) Progress and recent trends in biodiesel fuels. Energ Convers Manage 50:14–34
9. Karmakar AA, Karmakar SB, Mukherjee S (2010) Properties of various plants and animals feedstock for biodiesel production. Bioresour Technol 101:7201–7210
10. Chisti Y (2007) Biodiesel from microalgae. Biotechnol Adv 25:294–306
11. Ma FR, Hanna MA (1999) Biodiesel production: a review. Bioresour Technol 70(1):1–15
12. Azam MM, Waris A, Nahar NM (2005) Prospects and potential of fatty acid methyl esters of some non-traditional seed oils for use as biodiesel in India. Biomass Bioenergy 29:293–302
13. Barua P, Dutta K, Basumatary S, Dinesh CD, Dibakar CD (2014) Seed oils from non-conventional sources in north-east India: potential feedstock for production of biodiesel. Nat Prod Res 28(8):577–580
14. Keskin A, Guru M, Altiparmak D (2007) Biodiesel production from tall oil with synthesized Mn and Ni based additives: effects of the additives on fuel consumption and emissions. Fuel 86:1139–1143
15. Yamik H, Hamit S, Yakup I (2014) A comparative study on engine performance and emissions of biodiesel and JP-8 aviation fuel in a direct injection diesel engine. Ind J Eng Mater Sci 21(4):438–444
16. Sheehan J, Dunahay T, Benemann J, Roessler P (1998) A look back at the US department of energy's aquatic species program—biodiesel from algae, golden (CO), National Renewable Energy Laboratory, NERL/TP 580-24190
17. Subramanian KA, Singal SK, Saxena M, Singhal S (2005) Utilization of liquid Biofuels in automotive diesel engines: an Indian perspective. Biomass Bioenergy 29:65–72
18. Petracek R (2011) The advantages of using biodiesel blends. Available online at http://www.buzzle.com/articles/the-advantages-of-using-biodiesel-blends.html
19. Lin L, Zhou C, Vittayaoadung S, Shen X, Dong M (2011) Opportunities and challenges for biodiesel fuel. Appl Energy 88:1020–1031
20. Pasqualino JC, Montane D, Salvado J (2006) Synergic effects of biodiesel in the biodegradability of fossil-derived fuels. Biomass Bioenergy 30:874–879
21. GOI (2009a) India biofuels annual report 2009. Available at www.thebioenergysite.com
22. GOI (2009b) Production of oil seeds: Ministry of Agriculture, Government of India 2009
23. Biswas S, Kaushik N, Srikanth G (2006) Biodiesel: fuel for the future? Available online at http://tifac.org.in/index.php?option=com_content&view=article&id=776:biodiesel-fuel-for-the-future&catid=120:publication-bioprocess-a-bioroducts&Itemid=1380
24. Planning Commission (2003) Report of the committee on development of biofuels, GOI. Available at http://planningcommission.nic.in/reports/genrep/cmtt_bio.pdf

25. Francis G, Edinger R, Becker K (2005) A concept for simultaneous wasteland reclamation, fuel production, and socio-economic development in degraded areas in India: need, potential, and perspectives of Jatropha plantations. Nat Resour Forum 29:12–24
26. DIE (German Development Institute) (2008) Bio-diesel policies for rural development in India. Altenburg T, Dietz H, Hahl M, Nikolidakis N, Rosendahl C, Seelige K (eds) Bonn 27th May 2008
27. MoPNG (2009) Ministry of Petroleum and Natural Gas, Basic statistics on the Indian petroleum & natural gas 2008–2009. Available at http://petroleum.nic.in/petstat.pdf
28. Singh S (2009) India-Biofuels Annual. Tech. Rep, GAIN Report. USDA Foreign Agricultural Service, New Delhi, India
29. MNRE (2009) National policy on biofuels. Available at http://mnre.gov.in/file-manager/UserFiles/biofuel_policy.pdf
30. Indian Express (2008) Petrol with 20% biofuel to be mandatory by 2017. Available at http://www.indianexpress.com/news/petrol-with-20-biofuel-tobemandatory-by/333541/
31. Aradhey A (2010) India-biofuels Annual Tech Rep GAIN report. USDA foreign agricultural service, New Delhi, India, 2010
32. Prasad S, Dhanya MS. (2011) Air quality and biofuels, environmental impact of biofuels, Dr. Marco Aurelio Dos Santos Bernardes (ed), ISBN: 978-953-307-479-5, InTech, DOI:10.5772/17889. Available from http://www.intechopen.com/books/environmental-impact-of-biofuels/air-quality-and-biofuels
33. Raju SS, Shinoj P, Chand R, Kumar P, Siwa M (2012) Biofuels in India: potential, policy, and emerging paradigms. policy paper 27 NCAP. Indian Council of Agricultural Research, New Delhi
34. Miguel C (2007) U.S. Biodiesel production: recent developments and prospects. Iowa Ag Rev 13(2):8–9
35. Licht FO (2012) Earth Policy Institute—www.earth-policy.org. World Ethanol and Biofuels Report 10 (14): 281
36. TEDDY (2011) Teri Energy Data Directory and Yearbook (TEDDY) Published: 2011
37. IEA (International Energy Agency) (2009) World energy outlook. IEA Publications, Paris
38. Babu Umesh MS, Nautiyal S (2012) Socioeconomic and ecological consequences of biofuel development in India. Working paper 807, Department of International Economics, Norwegian Institute of International Affairs, Oslo, Norway
39. Shinoj P, Raju SS, Chand R, Kumar P, Siwa M (2011) Biofuel in India: future challenges. Policy Brief No 36. National Centre for Agricultural Economics and Policy Research, New Delhi
40. NRSA (2000) Wasteland Atlas of India. National Remote Sensing Agency, Department of Space, Government of India, Balanagar, p 81
41. Balusamy T, Marappan R (2008) Combustion characteristics of Thevetia Peruviana seed oil fueled in a direct injection diesel engine. Energ Sour A 7:397–403
42. Bhardwaj O, Abraham M (2008) A comparative study of performance and emission characteristics of a CRDe SUV fueled with biodiesel blends and diesel fuel. SAE publication no 2008-28-0075
43. Sharma YC, Singh B (2009) Development of biodiesel: current scenario. Renew Sustain Energy Rev 13:1646–1651
44. Ahmad M, Khan MA, Zafar M, Sultana S (2011) Biodiesel from non edible oil seeds: a renewable source of bioenergy. Available online at http://cdn.intechopen.com/pdfs-wm/17887.pdf
45. Deligiannis AG, Anastopoulos L, Karavalakis D, Mattheou FK, Zannikos S, Stournas E (2009) Castor (Ricinus communis L.) seed oil as an alternative feedstock for the production of biodiesel. Proceeding of the 11th Int Confer on Environ Science Technol. Chania, Crete, Greece
46. Nicholas Z (2007) Coproducts energy value is rising. Ethanol Producer Magazine, October 2007

47. Encinar JM, Gonzalez JF, Pardal A, Martinez G (2010) Transesterification of rapeseed oil with methanol of various co-solvents. Proceedings Venice 2010. Third Int Sympo Energy from Biomass and Waste, Venice, Italy, 8–11 November 2010
48. Kumar M, Sharma MP, Dwivedi G (2013) Algae oil as a future energy source in Indian perspective. Int J Renew Energy Res 3(4):913–921
49. Chhetri AB, Martin ST, Suzanne MB, Watts KC, Islam MR (2008) Non-edible plant oils as new sources for biodiesel production. Int J Mol Sci 9:169–180
50. Pramanik K (2003) Properties and use of Jatropha Curcas oil and diesel fuel blends in compression ignition engine. Renew Energy 28:239–248
51. Prasad S, Dhanya MS, Gupta N, Kumar A (2012) Biofuels from biomass: a sustainable alternative to energy and environment. Biochem Cell Arch 12(2):255–260
52. Kumar A, Sharma S (2011) Potential non-edible oil resources as biodiesel feedstock: an Indian perspective. Renew Sustain Energy Rev 15:1791–1800
53. Tewari DN (2007) Jatropha and bio-diesel. Oceans Books Pvt Ltd, New Delhi, pp. 1–228
54. Samantaray BB, Choudhary R, Kumar A, Akhtar W, Kumar M, Garanaik SK (2014) Formation of glycerol as byproduct and its application during biodiesel production from Pongamia pinnata. IOSR J Mecha Civil Eng 11(1):90–93
55. Kant P, Wu S, Chaliha S, Jasrotia R (2011) Going beyond Jatropha: can an expanded land and feedstock base help India meet its ambitious biodiesel target? IGREC working paper IGREC-22, Institute of Green Economy, New Delhi
56. Chen Y, Xiao B, Chang J, Fu Y, Ly P, Wang X (2009) Synthesis of biodiesel from waste cooking oil using immobilized lipase in fixed bed reactor. Energ Convers Manage 50:668–673
57. Piloto-Rodríguez R, Melo EA, Goyos-Pérez L, Verhelst S (2014) Conversion of by-products from the vegetable oil industry into biodiesel and its use in internal combustion engines: a review. Braz J Chem Eng 31(2):287–301
58. Al-Zuhair S, Dowaidar A, Kamal H (2009) Dynamic modelling of biodiesel production from simulated waste cooking oil using immobilized lipase. Biochem Eng J 44:256–262
59. Pugazhvadivu M, Jeyachandran K (2005) Investigations on the performance and exhaust emissions of a diesel engine using preheated waste frying oil. Renew Energy 30:2189–2202
60. Demirbas A (2006) Oily products from mosses and algae via pyrolysis. Energ Sour A 28 (10):933–940
61. Kumar A, Ergas S, Yuan X, Sahu A, Zhang Q, Dewulf J, Xavier Malcata F, van Langenhove H (2010) Enhanced CO_2 fixation and biofuel production via microalgae: recent developments and future directions. Trends Biotechnol 28:371–380
62. FAO (2008) The state of food and agriculture. Biofuels: prospects, risks, and opportunities. FAO, Rome
63. Dragone G, Bruno F, António AV, José AT (2010) Third generation biofuels from microalgae. Available online at http://www.formatex.info/microbiology2/1355-1366.pdf
64. Um BH, Kim YS (2009) Review: a chance for Korea to advance algal-biodiesel technology. J Ind Eng Chem 15:1–7
65. Pate R, Klise G, Wu B (2011) Resource demand implications for US algae biofuels production scale-up. Appl Energy 88:3377–3388
66. Gashaw A, Lakachew A (2014) Production of biodiesel from non-edible oil and its properties. Int J Sci Environ Technol 3(4):1544–1562
67. Tyagi OS, Atray N, Kumar B, Datta A (2010) Production, characterization and development of standards for biodiesel. J Metrol Soc 25(3):197–218
68. Ibeto CN, Ofoefule AU, Ezugwu HC (2011) Analytical methods for quality assessment of biodiesel from animal and vegetable Oils. Trends in Appl Sci Res 6:537–553
69. Anggraini AA, Wiederwertung VG (1999) Speiseolen/fetten im energetisch technischen bereichein verfahren und kassel: Witzenhausen, Germany, p 193
70. Hymavathi D, Prabhakar G, Sarath BB (2014) Biodiesel production from vegetable oils: an optimization process. Int J Chem Petrochem Technol 4(2):21–30

71. Owolabi RU, Adejumo AL, Aderibigbe AF (2012) Biodiesel: fuel for the future (A brief review). Int J Energy Eng 2(5):223–231
72. Meher LC, Dharmagadda VSS, Naik SN (2006) Optimization of alkali-catalyzed transesterification of Pongamia pinnata oil for production of biodiesel. Bioresour Technol 97:1392–1397
73. Parawira W (2009) Biotechnological production of biodiesel fuel using biocatalysed transesterification: a review. Crit Rev Biotechnol 29(2):82–93
74. Sharma YC, Singh B, Upadhyay SN (2008) Advancements in development and characterization of biodiesel: a review. Fuel 87:2355–2373
75. Encinar JM, Gonzalez JF, Rodriguez JJ, Tejedor A (2002) Biodiesel fuels from vegetable oils: transesterification of Cynara cardunculus L oils with ethanol. Energy Fuel 16 (2):443–450
76. Ramadhas AS, Jayaraj S, Muraleedharan C (2005) Biodiesel production from high FFA rubber seed oil. Fuel 84:335–340
77. Ma FR, Clements LD, Hanna MA (1998) The effects of catalyst, free fatty acids, and water on transesterification of beef tallow. Trans ASAE 41(5):1261–1264
78. Eevera T, Rajendran K, Saradha S (2009) Biodiesel production process optimization and characterization to assess the suitability of the product for varied environmental conditions. Renew Energy 34(3):762–765
79. Moser BR (2009) Biodiesel production, properties, and feedstocks. Vitro Cell Dev Biol Plant 45(3):229–266
80. Boocock DGB, Konar SK, Mao V, Lee C, Bulugan S (1998) Fast formation of high-purity methyl esters from vegetable oils. J Am Oil Chem Soc 75(9):1167–1172
81. Canakci M (2007) The potential of restaurant waste lipids as biodiesel feedstocks. Bioresour Technol 98:183–190
82. Dorado MP, Ballesteros E, De AJA, Schellert C, Lohrlein HP, Krause R (2002) An alkali-catalyzed transesterification process for high free fatty acid waste oils. Trans ASAE 45 (3):525–529
83. Shinoj P, Raju SS, Kumar P, Msangi S, Yadav P, Thorat VS, Chaudhary KS (2010) An economic assessment along the Jatropha-based biodiesel value chain in India. Agric Econ Res Rev 23:393–404
84. Nagarajan S, Chou SK, Cao S, Wub C, Zhou Z (2013) Bioresour Technol 145:150–156
85. Soares C (2008) Potenciais de Geração de Créditos de Carbono na Indústria da Palma. Dissertação
86. Iva K, Johan S (2010) Biofuels–renewable energy sources: a review. J Dispers Sci Technol 31:409–425
87. Jha SK, Fernando S, Columbus E, Willcutt H (2009) A comparative study of exhaust emissions using diesel-biodiesel-ethanol blends in new and used engines. Trans ASABE 52 (2):375–381
88. Knothe G (2006) Biodiesel and vegetable oil fuels: then and now. Paper presented at the 97th American Oil Chemists Society Annual Meeting, St. Louis, MO
89. Kushwah G (2013) Analysis of the smoke of diesel engine by using biodiesel as fuel. Int J Sci Res Publications 3(4):1–5
90. Chhang DYZ, Van Gerpen JH, Lee I, Johnson LA, Hammond EG, Marley SJ (1996) Fuel properties and emissions of soybean oil esters as diesel fuel. J Am Oil Chem Soc 73:1549
91. Rao PS, Gopalakrishnan KV (1991) Vegetable oils and their methylesters as fuels for diesel engines. Ind J Technol 29:292–297
92. US EPA (2015) Biodiesel: benefits and concerns. Available at http://www.epa.gov/region9/waste/biodiesel/benefits.html. Retrieved on 20.2.2015
93. Ajayebi A, Gnansounou E, Raman JK (2013) Comparative life cycle assessment of biodiesel from algae and jatropha: a case study of India. Bioresour Technol 150:429–437
94. Kumar S, Singh J, Nanoti SM, Garg MO (2012) A comprehensive life cycle assessment (LCA) of Jatropha biodiesel production in India. Bioresour Technol 110:723–729

95. Barnwal BK, Sharma MP (2005) Prospects of biodiesel production from vegetable oils in India. Renew Sustain Energy Rev 9:363–378
96. Vasudevan P, Sharma S, Kumar A (2005) Liquid fuel from biomass: an overview. J Sci Ind Res 64:822–831
97. Zou L, Atkinson S (2003) Characterizing vehicle emissions from the burning of biodiesel made from vegetable oil. Environ Technol 24(10):1253–1260
98. Marina E, Antonio S, Boyan K, Peder J (2002) Techno-economic analysis of Bio-diesel production in the EU: a short summary for decision-makers. Report EUR 20279 EN, European communities, 2002
99. Prueksakorn K, Gheewala SH (2006) The 2nd joint international conference on sustainable energy and environment, 21–23 November 2006, Bangkok, Thailand, pp 1–6
100. Kritana P, Shabbir HG (2006) Energy and greenhouse gas implications of biodiesel production from Jatropha curcas L. The 2nd Joint International Conference on Sustainable Energy and Environment, 21–23 November 2006, Bangkok, Thailand
101. Bünger J, Krahl J, Franke H, Munack A, Hallier E (1998) Mutagenic and cytotoxic effects of exhaust particulate matter of biodiesel compared to fossil diesel fuel. Mutat Res 415:13–23
102. Morris RE, Pollack AK, Mansell GE, Lindhjem C, Jia Y, Wilson G (2003) Impact of biodiesel fuels on air quality and human health summary report September 16, 1999–January 31, 2003. NREL/SR-540-33793
103. Sinobioenergy (2011) Approval for use in USA, Executive Summary 1. 26.02.11. Available from http://www.sinobioenergy.com/enArticleView.aspx?id=99v
104. Coleman B (2011) A brief summary of air quality and impacts of biofuels. Available from http://www.nebiofuels.org/pdfs/AQ_Summary.pdf

Chapter 19
Biofuel Policy in Indian Perspective: Socioeconomic Indicators and Sustainable Rural Development

Anuj K. Chandel, Latika Bhatia, Vijay Kumar Garlapati, Lakshmishri Roy, and Anju Arora

Abstract Biofuels are regarded as one of the most promising options to fulfill the higher energy requirement in future and simultaneously decarbonize the environment caused by excessive usage of fossil fuels. Technologies for the production of biomass based fuels particularly bioethanol and biodiesel are rapidly developing. However, successful commercialization of biofuels is still far due to high capital and operation expenditures and technical immaturities at large operations. Biofuel policies from major biofuels supporting nations might capitalize rural economic development and sustainable agricultural growth after strongly promoting biofuels. India has a biofuel policy, which foresees biofuels as potential candidates stimulating rural development by generating employment opportunities, together with environmental and economic benefits. The biofuel industrial sector has shown promising results and could serve as a potential source of substantial employment in near future. However, for the successful incorporation of biofuel policy, sustainable key indicators encompassing social and economic factors need to be evaluated properly in different scenarios. This chapter discusses the biofuel policies, key sustainable indicators, socioeconomic aspects of biofuels in Indian perspective.

A.K. Chandel (✉)
Centro de Tecnologia Canavieira (CTC), Piracicaba, Fazenda Santo Antônio, Santo Antônio, SP 13400-970, Brazil
e-mail: anuj.kumar.chandel@gmail.com; chandel_anujkumar@yahoo.com

L. Bhatia
Department of Microbiology and Bioinformatics, Bilaspur University, Old High Court Building, Bilaspur 475001, Chhattisgarh, India

V.K. Garlapati
Department of Biotechnology and Bioinformatics, Jaypee University of Information Technology (JUIT), Waknaghat, Himachal Pradesh 173234, India

L. Roy
Department of Food Technology, Techno India Group, Kolkata, West Bengal 700091, India

A. Arora
Division of Microbiology, Indian Agricultural Research Institute, New Delhi 110012, India

A.K. Chandel, R.K. Sukumaran (eds.), *Sustainable Biofuels Development in India*,
DOI 10.1007/978-3-319-50219-9_19

Keywords Biofuels • Biofuel policies • Socioeconomic indicators • Rural employment • Sustainability

1 Introduction

The rapid pace of industrial developments and burgeoning urbanization have called for environmentally sustainable energy sources like clean transportation liquid and gaseous fuels, wind and solar energy, and others [1]. Liquid transportation fuels such as ethanol made from first generation sugars (first generation ethanol) and cellulosic sugars (sugars derived from cellulosic biomass), biodiesel, bio-hydrogen, solar and wind energy provide unique environmental, economic strategic benefits and can be considered a safe and the cleanest renewable energy alternative to fossil fuels [2]. Significant advances have been made in last 2 to 3 decades towards the development of renewable energy. However, there are still several technical challenges and economical barriers, which play a pivotal role in making the renewable energy program unsuccessful at commercial scale.

The attention on sustainable biofuel production worldwide is growing rapidly. The profitability of biofuel production is significantly linked with the policies of multiple sectors such as agriculture, food and feed processing, research and development, industry and commercial trade. The uncertainties in oil prices mandating the serious evaluation of opportunities for the production and consumption of biofuels. Furthermore, because of the price instability in crude oil and overdependency on import, governments are showing keen interest in development and promotion of biofuels even leveraging the subsidies to make biofuels commercially viable. The paramount challenges of today such as energy security, concerns about trade balances, greenhouse gas (GHG) emissions, rural livelihoods, and domestic farm commodities are important for pushing biofuels in near term [3]. Biofuels are also considered as a potential contributor in the socioeconomic development of rural areas, and could play a substantial role in alleviation of poverty via creation of new employment opportunities—impacting directly or indirectly the multiple millennium development goals [4–6]. For example, sugar and ethanol industry provides direct and indirect employment to an approximately 12% of the rural population in Brazil. Moreira [7] observed that sugarcane crop in Brazil employs one million workers at various levels. In fact, this industry is very important for rendering socioeconomic services to the rural community such as improvement in infrastructure, opening of schools, colleges, hospitals, among others in the sugarcane growing countries like Brazil, India, China and others.

Cultivation of bioenergy crops on unused or degraded forest land and wastelands is advantageous for the restoration of green land [7]. Biofuels sector may provide substantial employment in India, as the country is basically rich in agriculture. Biofuels sector covering second generation ethanol, biodiesel and associated products from biorefinery may create new employment opportunities. Sugarcane cultivation and sugarcane processing industries for sugar and ethanol production are the primary agro-industrial sector in the country providing the livelihood of 50 million farmers and their dependants, (approximately 7.5% of the rural population). The sugar industry based employment further helped to save substantial foreign exchange, enabling India to strengthen infrastructural facilities with improvement in Gross Domestic Productivity (GDP). Moreover, it will reduce the dependency on the Organization of the Petroleum Exporting Countries (OPEC) countries and open a path for self-reliance in petroleum requirements.

Socioeconomic and environmental indicators (such as social wellness, energy and food security, external and internal trade, profitability, conservation of natural reserves) are the key elements to measure the sustainability of renewable energy options [8]. A balanced evaluation of sustainability indicators is imperative for commercial biofuel producers to make decisions about cultivation of biofuel crops, transportation, processing and conversion, and logistic of fuel blends concerning with the future or retrospective measures [8, 9]. For the long haul, it is very important to understand bioethanol production technologies in terms of their economic viability, environmental feasibility and empowering employment opportunities before implementing a fuel ethanol policy. The choice of the best technology for economically viable renewable fuel production directly depends on the overall economics (lowest cost), environmental benefits (lowest pollution), and least energy consumption (higher efficiencies). Correct measurement of economic viability with the right scale is essential to estimate renewable fuel production costs with precision. Critical analysis of key parameters such as (i) capital expenditures (CAPEX) constituting total capital investment, equity and leverage, interest rates, life cycle of plant machineries and their maintenances and (ii) operational expenditures (OPEX) embedding raw material, auxiliaries, residue management, (iii) fixed OPEX (operation, personnel, insurances, and servicing), and (iv) revenues from by-products are imperative for accurate economic analysis of renewable fuel production.

In this chapter, we review the biofuel implementation policies from major countries who are the strong advocates of biofuel usage and socioeconomic indicators in those countries. However, particular emphasis is given to renewable biofuel development, key sustainable and socioeconomic indicators in Indian context.

2 International Biofuel Policy Scenario

In the current scenario, when the world is developing not only industrially but also technologically, the concerns like global warming and depleting oil reserves have profound impact on setting up right biofuel policies in international arena. In fact, these problems act as driving force worldwide to explore the bio based materials which can be potentially used as feedstock for first and second generation biofuel production [10]. The interest in biofuel production got momentum due to unprecedented price hike in crude oil prices in 1970's, originally because of supply restrictions imposed by the Organization of the Petroleum Exporting Countries (OPEC) cartel. High uncertainty in oil prices and negative environmental impact on environment using oil and gas catalyzed the search for alternatives, save or replace oil with economically feasible options such as biofuels. As per the analysis by British Petroleum (BP), energy consumption is expected to increase by 34% between 2015 and 2035. Oil and gas will hold the dominant place in supply energy. However, renewable energy in the form of liquid biofuels are also growing rapidly at 6.6% per year [11]. In 2015, world ethanol production has increased by 4.1% while biodiesel decreased by 4.9% [11].

With an increased environmental concern in the 1980's and 1990's, the use of biofuel gained further popularity, as a result of which biofuel production experienced sharp hike. High oil prices, benefits to agriculture and rural areas. and diversification of sources of energy might have also contributed to this hike. Along with all these issues, biofuel production also gained momentum by the concern of various government policies such as mandates, targets, and subsidies which cover the grounds of energy security and environmental considerations. Biofuel policies not only play a major role in the development of the energy sector but also provide thrust for innovation, research, and development. The key profitability indicators of biofuel production are primarily influenced by the policies of various sectors such as agriculture, research, industry, and trade. But this optimism with respect to biofuels took a pause when the concerns were raised by the global food crisis in 2007–2008 and the policy makers were in a dilemma on food security issues generated by first generation biofuel [12].

2.1 Global Biofuel Policies

This section deals with biofuel policies in four main biofuels markets, i.e., European Union, the USA, Brazil, and India. It is based primarily on the studies from the references from [13–16]. An overview of these government biofuel policies from major biofuel producing countries is shown in Table 19.1 [17].

2.1.1 Brazil

The only mature, integrated biofuel market in practice in the world is Brazil's cane-based ethanol market. The sugarcane -based ethanol/electricity cogeneration system became a competitive energy provider at USD$35 per barrel of crude oil prices [18]. Brazil has been the world leader in the mandated blending of biofuels for over 30 years, primarily under its National Alcohol Program (Portuguese: 'Programa Nacional do Álcool') or "*Pró-Álcool*" program [19].

It was a government intervention as a response to the petroleum shortage caused by the 1973 oil crisis that initial promotion of ethanol policies took place in Brazil

Table 19.1 Government biofuel policy of various countries

Features	Brazil	USA	EU	India
Mandatory/ Indicative targets	Mandatory blend of ethanol in Gasohol: 20–25%	– Mandatory use under the Renewable Fuels Standard (RFS). Rising to 28.1 billion liters in 2012 – Specific blend requirements being implemented at State level	Indicative blending target was 5.75% by 2010. Mandatory 10% blending target by 2020. Trend towards mandatory requirements in many member states	E5 blending program in some states recently extended E10 country wide by 2017
Tax credits/ excise duty reductions	Ethanol attract a lower rate of value added tax as against gasohol	If the cellulosic ethanol qualifies for alcohol fuel tax credits, the credit amount is reduced to $0.46 per gallon	Excise duty reductions allowed until 2010, then revision. Applied in many member states ranging from small reduction to complete removal. Trend towards declining tax support	Some tax exemptions provided in the Union Budget 2007
Import tariffs	No tariff	USD 0.14/l +2.5% of the value	Very high tariffs. EUR19.3 hl for un-denatured and EUR 10.2 for denatured ethanol	198.96% ad valorem (as a percentage of the customs value of the import) on CIF value (cost, insurance, and freight) for un-denatured and 59.08% ad valorem on CIF value for denatured
Feedstock supports	No	No specific support to biofuels but possible in 2007 Farm Bill	Yes With the CAP reform set-aside land can be used for energy crops production for which farmers receive a EUR45/ton premium (maximum 2 mln hectares)	Not directly Indirectly through minimum cane pricing

Source: Jolly [18] (modified from PowerPoint presentation) Government Biofuels Policy and Sugar Crops Policy Crops Outlook, International Sugar Organization

[19]. The *Pró-Álcool* program was started in 1975, as a strong support for ethanol demand and supply and to enhance the share of domestically produced fuel. The government initiated the bioethanol development of the industry through low-interest loans and enlisted a state-owned enterprise, Petrobras [20], to incorporate the product into gasoline. Through beneficial tax treatment, ethanol was available at a price that made it competitive with gasoline, and automobile manufacturers were persuaded to produce cars that were able to use the fuel at levels above traditional gasoline-powered vehicles. The 1990s observed the termination of this government supported *Pró-Álcool* program but market regulation and tax incentives of the program continued to be supported [20, 21]. Transition phase to a full liberalization took place between 1996 and 2000. Hebebrand and Kara [22] found that the blending mandate has been met by supporting policies, including local distribution requirements, subsidized credit for ethanol storage and tax preferences for vehicles. Currently, various incentives supporting demand is maintained without direct control on ethanol production and trade.

Currently, anhydrous ethanol blending is 18–25% for gasoline. This mandate is not relevant for pure hydrous ethanol which is sold at filling stations for use in flex fuel vehicles. Ethanol–gasoline blends up to 85% of anhydrous ethanol or on up to 100% hydrous ethanol fuel can successfully run these flex fuel vehicles. The mandatory blend of ethanol (18–25%) replaces the existing gasoline dependency (50%) and showcases the success of flex-fuel vehicles allowing ethanol fuel consumption in Brazil for transport. All gas stations are required to sell both gasohol (E25) and pure ethanol (E100). Figure 19.1 shows the competitiveness in

Fig. 19.1 A gasoline filling station from Piracicaba, São Paulo state of Brazil showing ethanol and gasoline price in August, 2016

ethanol (*etanol comum*), gasoline (gasoline *comum*) from one of gasoline station in Piracicaba-São Paulo state.

Biodiesel, though not a major product in Brazil, is blended with regular diesel (B5) at 5% rate [22]. In 2013, Brazil introduced biodiesel blending targets of 5%. The Brazilian federal and state governments grant tax reductions and exemptions to reach these obligations based on the agricultural producer's size and developmental level of region. Flex fuel vehicles get the privilege of lower taxes than those on petrol-powered vehicles, and similarly, ethanol is also bestowed with favorable tax treatment at the pump relative to petrol. The Common External Tariff of Mercosur (an economic and political agreement between Argentina, Brazil, Paraguay, and Uruguay) also protects Brazilian domestic biofuel production with ethanol duties of 20% and biodiesel duties of 14% which could be eliminated under the Doha and/or the EU-Mercosur trade negotiations with non-tariff barriers on Brazilian biofuels imports. Foreign investments in ethanol distillation plants and sugarcane production, especially from Europe and the USA, provide an important explanatory factor in the growth of the ethanol sector in Brazil. International investors are motivated by lower costs for ethanol production in Brazil which is an outcome of low prices of raw materials and the high technological level of the whole production process.

The drivers for biofuels vary according to local circumstances, specific domestic policies and initiatives. These issues drastically differ between countries. Brazil is the only country where production cost of ethanol and biodiesel is competitive with petroleum sources because of some inherent advantages like land availability, quality of land for sugarcane cultivation, rain, sunlight, soya production, and availability of knowledge and skilled professionals. Ethanol attracts a lower rate of value-added tax against gasohol. Over the medium term, the ethanol–sugar interface will remain centered in Brazil. In long term, some other countries will also start producing ethanol from sugar crops, which could be triggered by the correlation between sugar and energy prices. Brazil will continue to be the key player of the oil–ethanol–sugar interface. In the years of global sugar deficit, the world market can "buy-back" ethanol from Brazil.

2.1.2 USA

US Corn-based ethanol is an integrated and fully evolved system. The main drivers for increased biofuel demand in the USA are high energy prices and incentives provided by the Energy Policy Act of 2005 (EPACT05). The EPACT05 requires a minimum of 7.5 billion gallons (approximately 28.7 billion liters) of renewable fuels (ethanol and biodiesel) to be used in the nation's motor fuel.

The main issues those motivated the USA to form biofuel policies in 1970 were the concern for country's economy, environment and domestic farm commodities, which could be a prospective source of raw material for biofuel. Dependency on imported fossil fuels and increasing greenhouse gas emission were troubling country's economy and environment respectively [23]. Energy Tax Act of 1978 introduced the tax exemptions and subsidies for the blending of ethanol in gasoline. A subsidy of 4 cents per gallon of gasohol (E10), equivalent to 40 cents per gallon of pure ethanol, was introduced in the Energy Policy Act of 1978, through a partial exemption from the federal gasoline excise tax. Tyner [24] attributes the launch of the ethanol industry to this policy. The level of ethanol subsidy has varied over the years but presently stands at 45 cents per gallon, operated through a volumetric ethanol excise tax credit (VEETC). Biodiesel blenders receive a tax credit of $1.00 per gallon. Most of the states have tax exemptions and credits for use of ethanol along with the federal tax credit. The range of programs that exist for individual states is a testament to the strength of the lobbying effort of proponents of alternative fuels and vehicles that make use of those fuels. It also illustrates the difficulty of creating an effective notification mechanism for governmental assistance [25].

It took almost 2 decades for biodiesel to have subsidies implemented on it, on the platform of Conservation Reauthorization Act in 1998. Though the US fiscal incentives and mandates differed from state to state they were complemented by those at federal level. The US biofuel policies targeted specifically on mandates in volumetric terms as a part of the Renewable Fuel Standard (RFS) program, introduced under Energy Policy Act of 2005. This act framed the policies on mandates on consumption of biofuels. This act declared purchase objective of 4 billion gallons and 7.5 billion gallons of biofuel in 2006 and 2007 respectively. With the addition of biodiesel mandates, RFS program was expanded in The Energy Independence and Security Act (EISA) of 2007. According to this act, nine billion gallons of renewable fuel will be blended into transport fuel, and this horizon will expand to 36 billion gallons in 2022. To avoid the stress on first generation grain-based ethanol, the major share of this 36 billion gallons blending fuel will be covered by advanced second or higher generation biofuels (about 60%). The 2022 mandate for biodiesel is 1 billion gallons. Output-connected measures, support for input factors and consumption subsidies are the three principal instruments for current US biofuel policies. Largest direct subsidies are the tax credits whereas indirect subsidies are mandates and it does not provide direct support for price. Tariffs on ethanol are much higher as compared to that on biodiesel (24% in former and 1% in later in ad-valorem equivalent). This high tariff on ethanol limits its import from Brazil.

Ethanol producers significantly benefit from tax credits based on biofuel blended into fuels. Fuel consumers profit from the blender's tax credit as well—in the form of a lower fuel price—when the tax credit is combined with a binding blend or consumption mandate. Fuel consumers do not benefit, however, when the tax credit is the only binding policy. In this case, the consumer fuel price does not depend on the level of the tax credit, as shown by Gorter and Just [26]. The Volumetric Ethanol Excise Tax Credit and the Volumetric Biodiesel Excise Tax Credit provide the largest subsidies to biofuels while there are some smaller additional subsidies connected to biofuel outputs both on the state and federal levels.

2.1.3 European Union (EU)

Germany is the major producer of biodiesel in EU, where tax exemptions promote it. EU ethanol production capacity in 2006 was 2.1 billion liters which quadrupled about 8.5 billion liters in 2013. In EU, wheat, corn and sugar beet derivatives are main sources for bioethanol production. Germany, France, Hungary and the Czech Republic are the major countries in EU in 2015 and 2016 for bioethanol use. In 2016, consumption of biofuels in Germany is expected to increase because of the mandates in biofuels mandates to save GHG emissions.

In Europe, biodiesel production is growing more rapidly than ethanol production. World's largest biodiesel production occurs in EU. Biodiesel represents about 80% of the total transport biofuels market. Domestic consumption and competition from import are the two major driving forces for biodiesel production in EU. In 2014, biodiesel production increased by 11%, mainly in Germany, Spain, and the Benelux because of substantially lower imports and higher domestic consumption. In 2016, biodiesel production capacity in EU is expected to be 25.2 billion liters, with a 2% reduction from 2014.

The EU Biofuels Directive (BFD) has revised the total transportation fuel consumption profile and supporting biofuels and other renewable fuels in the EU. Almost 25% greenhouse gas (GHG) emissions are caused by the gasoline consumption. The goal was set to reduce the contribution in GHG emissions (based on energy content) to 5.75% and 10% by the end of 2010 and 2020 respectively. European Parliament and Member States [30] is reviewing these goals and will decide based on the 2020 results. Most the EU countries have not met the goals of 2010 to reduce 5.75% gas emissions [27].

It was need of the hour for EU to design its biofuel policy in order to fulfill its commitment to the Kyoto targets of Green House Gas (GHG) emission as concern for rising environmental issues from the EU population. It is not a single document, but a collection of a number of records which captures the EU biofuel policy. These documents are issued by different parts of the EU governance structure [28]. There are additional national aids for the production of "ethyl alcohol of agricultural origin" and occasional tenders for surplus crops to be converted to bioethanol. EU introduced Biofuel Directive 2003/30, which targeted at the use of 2% and 5.75% of biofuel in transport sector by 2005 and 2010 respectively at the EU level. It was only Germany and Sweden that exceeded 2005 target as they respectively used 3.86% and 2.11% of biofuel in total fuel consumption.

Germany's biggest drive on renewable energy by "Energiewende" (Energy Transition), with massive energy infrastructure investments is a big push in this direction and a good example for several countries. Among the EU nations, Germany is among the top users of renewable technologies in transport, heating, and power sector. In Germany, 12.2% of energy consumption, 20% of electricity consumption, 10.4% of heating, and 5.6% of transport consumption comes from renewable sources. Germany is also strongly advocating the use of biofuels in aviation sector also [29].

To execute the strategies on biofuel consumption EU Renewable Energy Directive (2009/29) established a "20-20-20 Policy" in 2009. Under this policy, various targets are set to be accomplished by the year 2020, which includes 20% reduction of total energy consumption in the EU-27, 20% reduction in Greenhouse gas emission and the share of renewable energy is set at 20% in the total EU energy consumption. Some agreements such as Cotonou Agreement, the Euro-Med Agreements and the Generalized System of Preferences Plus (GSP+) provides duty-free access to biofuel market of EU. These free access opportunities highly benefit some countries like Guatemala, South Africa and Zimbabwe.

Common Agricultural Policy (CAP) launched in 1992 supported the bioenergy drive in the EU. This policy brought energy-crop-premium of EUR 45/ha on a maximum of 2 million ha of set-aside land. This first pillar of reformed CAP proved itself unsupportive for bioenergy production. The second pillar of CAP comprising Rural Development Policy reinforced several measures that supported bioenergy development through the modulation instrument. Promotion of renewable energy production from perennial energy crops, agricultural and forest biomass was supported by enabling investment in infrastructure.

The EU trade policies have a big impact on domestic biofuel production, reduction in production incentives, and export opportunities for foreign biofuel producers. For biodiesel, the most-favored-nation duty is 6.5%. For un-denaturated ethanol and denaturated ethanol, the ethanol tariff barriers are on the higher level, EUR 19.2/hectoliter and EUR 10.2/hectoliter, respectively. Non-tariff barriers affect the trade of biodiesel considerably. For instance, even in the situation if tariffs for biodiesel are reduced, trade would have impact due to more restrictive non-tariff barriers in the form of quality and environmental standards [23]. Excise duty reduction was effective until 2010, followed by a gradual reduction to complete removal. Under the CAP reform, set-aside land can be used for energy crops production providing a premium of EUR 45/ha to the farmer.

EU passed a legislation in 2008 to use biofuels in transportation. Adopting the "Climate Change Package" in Directive for Renewable Energy (DRE) by EU in 2009, it was mandated that in EU, 10% of the transport energy should be from renewable sources by 2020 and 20% of the overall energy requirement should come from renewable sources [30].

2.1.4 India

According to International Energy Agency [31], global oil demand is projected to be increased by 60%, i.e., 7700 billion liters in 2030 with 68% increase in China and India alone. India is the fifth largest primary energy consumer and fourth largest petroleum consumer in the world. India spends over 45% of their export earnings for importing energy [32]. Biofuels have received considerable attention in India to curb the oil dependency [33].

Food security is primary concern for India to provide food to huge population while controlling the price elevation for food commodities and improving agricultural productivity. Therefore, Indian government cannot allow the use of arable land, which is used for food and feedstock production, for the cultivation of biofuel crops. Food security is a grave concern for India as it is already one of the largest importers of vegetable oils in the world. Further, productivity of food grains like wheat, corn, rice and coarse cereals has been quite slow in recent years causing scarcity of food grains.

Water scarcity problems are also one of the major issues that India is already suffering from and this issue will only worsen as their food demand continues to grow with a rise in populations and incomes. India is exploring the possible implementation of a controversial multibillion-dollar project of inter-basin water transfers, to meet future demands [34].

The Government of India introduced Ethanol Blended Petrol Program (EBPP) in January 2003 with a vision of blending ethanol with petrol. National Policy on Biofuels formulated by the Ministry of New and Renewable Energy (MNRE) in 2009 recommended the blending at least 20% biofuels with diesel and petrol by 2017. Ethanol blending will increase the bioethanol requirement to 3.4 billion liters by 2020 [35]. As the present requirement of ethanol for the potable and chemical sector is fulfilled solely by molasses, to suffice the above need of blending there is a need to cultivate 20–23% more sugarcane (to generate molasses) than what is the present requirement of the sugar industry. However, to grow biofuel crops more land and water will be needed. To produce 736.5 million tons of sugarcane, to meet the molasses demand, an area of 10.5 million hectares will be required which seems impossible as it will deprive the production of food crops [36, 37]. Hence crop-based biofuels will exert pressure on water and land resources that already are heavily exploited or overexploited [34].

Considering pro-poor dimension as prime focus to harness the potential of biofuels sector in India, biofuel policy of India should address these concerns for the sustainable development. Subramanian et al. [38] stated that endeavor should be to produce ethanol from various feedstocks or organic matter we should not rely solely on molasses or food crops. India is blessed with abundant renewable energy resources. Provisions are being planned to encourage their use in every possible way. The Indian approach to biofuels, in particular, is somewhat different from the current international approaches which could lead to conflict with food security. It is based solely on non-food feedstocks to be raised on degraded or wastelands that are not suited to agriculture. The issue of fuel versus food security is not relevant in the Indian context. Policy support mechanisms to promote alternative feedstocks that will benefit all the stakeholders in the bioethanol supply chain in the long run while meeting the mandated requirements. In future too, it would be ensured that the next generation of technologies is based on non-food feedstocks.

In 1948, the Power Alcohol Act heralded India's recognition of blending petrol with ethanol. The main objective was to use ethanol from molasses to blend with petrol to reduce dependence on petrol imports, trim wastage of molasses and to bring down the price of sugar. *Jatropha curcas* was identified as the most suitable tree borne oilseed for biodiesel production by the National mission in April 2003 by the Government. Blending mandate was made optional in October 2004 because of scarcity of ethanol during 2004–2005. Blending mandate was resumed in October 2006 in 20 States and 7 union territories in second phase of EBPP. In December 2009, the government introduced the comprehensive national policy on Biofuels after a series of policy changes. The final policy was framed by MNRE which called for at least 10% biofuels blending with diesel and petrol by 2017 [13].

2.1.5 Salient Features of National Policy on Biofuels, 2009

- The Policy aims at mainstreaming of biofuels and, therefore, envisions a central role for it in the energy and transportation sectors of the country in coming decades. The national indicative target of 5% blending by 2012, 10% by 2017, and 20% after 2017 has been recommended in the policy. The goal of the policy is to cater biofuel demand by ensuring the availability of a minimum level of biofuels in country.
- Nonedible oilseeds crop plantations on waste/degraded/marginal lands would be encouraged while the plantation on infertile irrigated lands would not be supported. In the context of the International perspectives and National imperatives, it is the endeavor of this policy to facilitate and bring about optimal development and utilization of indigenous biomass feedstocks for production of biofuels. The policy also envisages the development of the next generation of more efficient biofuel conversion technologies based on new feedstocks.
- A Minimum Support Price (MSP) to be announced for farmers producing nonedible oilseeds used to produce biodiesel. MSP with the provision of periodic revision for biodiesel oilseeds would be announced to provide a fair price to the growers. The Minimum Purchase Price (MPP) for the purchase of bioethanol by the oil marketing companies (OMCs) would be based on the actual prices of bioethanol production and import price of bioethanol. For biodiesel, the MPP should be linked to the prevailing retail diesel price.
- Financial incentives for new and second generation biofuels, includes a National Biofuel Fund. Bio-ethanol already enjoys concessional excise duty of 16%, and biodiesel is exempted from excise duty. No other central taxes and duties are proposed to be levied on biodiesel and bio-ethanol. Custom and excise duty concessions would be provided on plant and machinery for production of biodiesel or bio-ethanol, as well as for engines and that run on biofuels for transport, stationary and other applications, except those that are not manufactured indigenously.
- "Declared goods" would include biodiesel and bioethanol for ensuring unrestricted movement of biofuels within and outside the states. According to the policy, biodiesel would be exempted from taxes and duties. Inter-Ministerial National Biofuel Coordination Committee under the Chairmanship of the Prime Minister and a Biofuel Steering Committee under the Chairmanship of the Cabinet Secretary, Government of India are proposed to set up to address these issues in India. The Government is also considering the creation of a National Biofuel Fund for providing financial incentives like subsidies and

grants for new and second generation feedstocks cultivation; advanced technologies and conversion processes and production units. Hundred percent foreign equity would be allowed for biofuel technology and projects when the fuels produced are utilized domestically only in order to attract foreign direct investment (FDI). Vision, medium term goals, strategy, and approach to biofuel development are therefore set out by the policy. It also proposes a framework of technological, financial, and institutional interventions and enabling mechanisms.

- National Biofuel Coordination Committee under the Prime Minister for a broader policy perspective and implementation.
- Under this policy a Biofuel Steering Committee was set up under the Cabinet Secretary to oversee policy implementation. Several ministries are currently involved in the promotion, development and policy making for the biofuel sector.

  Overall policies for the development of biofuels and research and technology development for its production are made by the Ministry of New and Renewable Energy (MNRE), Government of India. Marketing of biofuels and developing and implementing a pricing and procurement policy is under the purview of Ministry of Petroleum and Natural Gas. The role of Ministry of Agriculture has promoted research and development for the production of biofuel feedstock crops. The Ministry of Science and Technology supports biotechnological research in biofuel crops, specifically in the area of biotechnology. A National Biofuel Coordination Committee (NBCC) headed by the Prime Minister was set up to provide high-level coordination of multiple departments and agencies and policy guidance/review on different aspects of biofuel development, promotion, and utilization. NBDB is developing a definite road map for use of biofuels in stipulated time frame to measure the policy issues.

The other main concern of this policy is the state independence. States are given the liberty to have biofuel policy and set targets of their own. This has resulted in a varied response in different states. States such as Karnataka and Rajasthan have taken off with a good start, and at the same time, there are states which are yet to make a landmark. "National Mission on Biodiesel" was launched in 2003 to address socioeconomic and environmental concerns. Two phases were proposed: Phase 1 by 2006–2007, demonstration project and by 2011–2012, Phase 2-self-sustaining expansion. Biodiesel purchase policy issued on 1st January, 2006 looks at biofuels as a potential tool to stimulate rural development by creation of employment opportunities to reap environmental and economic benefits. Biofuels promotion also outlines research and development, capacity building, sales and distribution of biofuels. India's current ethanol production allows blending of only ~3% in gasoline [39, 40]. Hence, the target set to make India energy independent turned out to be "highly unrealistic." A solution for reducing the dependence on foreign fuel in the perspective of changing global climate can be the promotion of advanced second and third generation biofuels.

3 Rural Development, Benefits to Indian Farmers and Daily Wagers

Biofuels are advantageous in that they would promote sustainable development, supplement conventional energy sources in catering to the transportation fuel demand and meeting the rural populations energy demands. Biofuels are bound to satiate the ever increasing energy needs in an ecofriendly manner and minimize the dependence on imports of fossil fuels and consequently providing a high degree of national energy security. Biofuel growth across the globe has been promulgated by energy security and environmental concerns, a wide range of market mechanisms, incentives and subsidies that have been put in place to facilitate growth. Figure 19.2 presents socioeconomic indicators of biofuels implicated for sustainability of biofuels for designing sustainable systems. Developing countries like India, apart from these considerations, also view biofuels as a potential means to stimulate rural development and create employment opportunities.

The Indian Government's vision 2020 states that cultivation of Jatropha on 10 million ha would create 7.5 million tons of fuel a year and consequently create year round jobs for people. But despite ambitious programs, targets are likely to be missed owing to the high costs of Jatropha-based fuel and red tape [41].

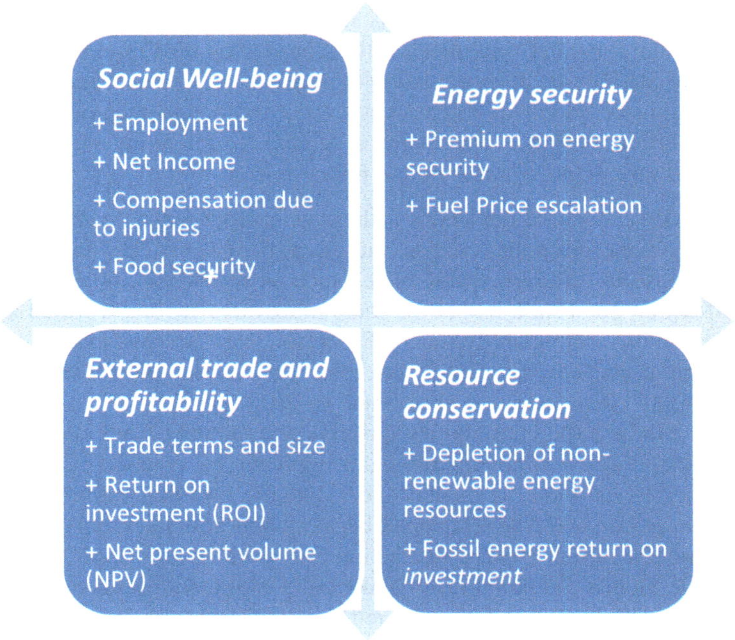

Fig. 19.2 Socioeconomic indicators of biofuels implicated for sustainability of biofuels for designing sustainable systems. These indicators have been modified for socioeconomic sustainability of bioenergy from algal biofuels systems [8]

Thus national policy of biofuels encourages biodiesel production from nonedible oilseeds, plants growing on waste, degraded, and marginal lands. This would encourage various agricultural producers to undertake plantations that provide feedstock for the biofuel industry. Over 400 species of trees are recognized in India that bear nonedible oil seeds. All of the recognized species would be evaluated for their exploitation as biofuel industry substrates by assessing their techno-economic viability. Plantations of trees bearing nonedible oilseeds will be taken up on government/community wasteland, degraded or fallow land in forest and non-forest areas. Minimum support price mechanism as proposed in the policy is also likely to promote contract farming on private wasteland. While plantations on agricultural lands would be discouraged, the corporates would be encouraged to undertake plantations by contract farming in consultation with panchayats where necessary. The employment created and provided in plantations of trees and shrubs of nonedible oilseeds would be covered under the National Rural Employment guarantee program (NREGP) [42] by Government of India. Seedlings from certified nurseries of recognized institutions identified by states would be distributed to the growers and cultivars.

The Ministry of Rural Development is specially promoting Jatropha plantations on wastelands and marginal land. Several programs nationwide have encouraged the planting of biofuel crops as well as procurement of seeds and cultivating. In this initiative, Ministry of Agriculture provides a subsidy through National Oilseeds and Vegetable Oils Development (NOVOD) Board to the farmers, ongovernmental organizations (NGOs), and individuals for the production of Tree Borne Oilseeds (TBOs) which include biofuel crops. Under this scheme, 30% credit linked subsidy is provided, with 50% term loan from the bank, and 20% beneficiary share in the form of land, labor, etc. The Ministry of Rural Development has provided financial aid of INR 490 million to 9 identified states in 2005–2006 and INR 495 million to 15 states in 2006–2007 for the plantation of *Jatropha*/*Pongamia* seedlings. Department of Rural Development of Chhattisgarh state is planting the TBOs under Mahatma Gandhi National Rural Employment Guarantee Scheme (MGNREGS). Plantation of biofuel crops is also in progress by the initiatives of several Joint Forest Management Committees (JFMCs) on forest land in the Chhattisgarh state [43]. The Chhattisgarh Biodiesel Development Authority (CBDA) provides *Jatropha* seedlings at highly subsidized rates for the promotion of planting in waste and forest lands. The authority is also encouraging the private investors to enter into contracts with local farmers for Jatropha cultivation.

Setting up of processing units by industry for bio-oil expelling/extraction and transesterification for the production of biodiesel is encouraged in the national biofuel policy. Gram/Intermediate Panchayats are also encouraged to create village level facilities for bio-oil extraction with corresponding sale to biodiesel processing units. Priority plantation of nonedible oil bearing plants, the setting up of oil expelling/extraction and processing units for production of biodiesel and creation of any new infrastructure for storage and distribution are declared as a priority sector for the purposes of lending by financial institutions and banks. National Bank of Agriculture and Rural Development (NABARD) would provide refinancing towards loans to farmers for plantations. Indian Renewable Energy Development Agency (IREDA), Small Industries Development Bank of India (SIDBI), and other financing agencies as well as commercial banks would be actively involved in financing various activities under the entire biofuel value chain, at different levels.

National Biofuel Policy will also provide support for creation of awareness about biofuels and its potential and opportunities in upgrading the transportation infrastructure to support the rural economy. Various educational institutions will be encouraged to introduce suitable curricula to cater the trained manpower in all segments of the biofuel sector. Moreover, efforts will also be directed to enhance and expand the consultancy capabilities to fulfill requirements of biofuel sector. Intensive R&D work would be undertaken in the biofuel feedstock production based on sustainable biomass through nonedible oilseed bearing plantations on wastelands.

An instrumental factor of this policy includes providing a Minimum Support Price (MSP) for oilseeds and also having a thorough check on periodic revision which will ensure a fair price to the farmers and growers. The implementation of the MSP mechanism will be worked out carefully in consultation with concerned government agencies, states, and other stakeholders. For smooth working, Biofuel Steering Committee has to be set up in accordance with the National Biofuels Co-ordination Committee under this Policy. The farmers are getting the MSP of INR 6/Kg of *Jatropha* seeds from the governmental seed procurement agencies, while private companies buy at higher price of INR 7–10/Kg. The local farmers feel that decentralized value-addition options can provide a big change in their livelihood [44].

The Government of Odisha believes in revolving fund under the coordination of Odisha Renewable Energy Development Agency (OREDA) and Odisha Forest Development Corporation (OFDC) for providing subsidies of 50% to local communities like *Pani Panchayats* and self-help groups. and 33% directly to farmers above poverty line and 50% to farmers as groups. The policy also promotes the potential interlinkage between the biofuel program and various institutions like Swarna Jayanti Gram Swarozgar Yojana, MGNREGS, Integrated Tribal Developmental Agency, Compensatory Afforestation, Backward Regions Grant Fund schemes.

The government of Andhra Pradesh introduced a biodiesel policy in the year 2005 to facilitate investors and farmers to grow oil-bearing trees, mainly *Jatropha*, *Pongamia*, and *Simarouba*. Originally, the policy has proposed a three-way partnership program between government, industry, and farmers. In this program, the provision of buy-back arrangements for seeds and credit disbursement for farmers will be moved through industries. Plantations are promoted on forest lands under contract farming arrangements between private entrepreneurs and farmers [45].

Tamil Nadu government is also promoting the planting of *Jatropha* seedlings on wastelands. Primary agricultural cooperative banks in the state are enabling subsidized loans to farmers for *Jatropha* cultivation. According to the industrial policy of Tamil Nadu, 50% subsidy is given for *Jatropha* plantation and other biofuel crops and this subsidy is extended to the agro-processing industries. The state has promoted the *Jatropha* cultivation for private companies like D1 Mohan Bio Oils Ltd., AGNI NET Biofuels Pvt. Ltd., AHIMSA, for the contract farming to farmers. Additionally, other services like training and extension support and provision of agricultural inputs are also being offered. These companies are offering buy-back arrangements with a certain market-linked price in the range of INR 5–10/kg for *Jatropha* seeds [46]. Biofuel policy of Tamil Nadu aims to bring a sustainable development and promotion for the cultivation, production and usage of biofuels crops and thus biofuels in state.

4 Socioeconomic Indicators and Governing Implementation Issues

With 4% of the world's energy consumption, India is the fifth largest consumer of energy after the USA, China, Russia, and Japan [47]. India consumes 4.4% energy resources (524.2 million tone oil equivalent (mtoe) of the world total (12,000 mtoe). In the world, primary commercial energy consumption in terms of using natural reserves (coal, oil and natural gas, nuclear and major hydro) has grown at 2.6% in the last 10 years. The growth rate of energy demand in India has grown approximately 6.8%; however, the supply has increased with a compounded annual growth rate (CAGR) of only 1% [48]. Therefore, India needs energy badly. Oil and gas constitutes about 45% share in the total energy consumption in India [48]. However, India is heavily dependent on crude oil imports, with petroleum crude accounting for about 34% of the total inward shipments (Ministry of Commerce and Industry, India). The imports have been ever-increasing, leaving the country with a growing balance of payment deficits. This necessitated India to work towards renewable and alternative fuels for energy security. Oil imports during April–February, 2014–2015, valued at US $130,848.36 million (Ministry of Commerce and Industry, India), were 12.24% lower than the oil imports in the corresponding period last year.

Though some alternatives like ocean water power, geothermal energy, wind energy, and solar energy are being explored, bioenergy is considered as a strong source of renewable energy in the coming years. The advantage of biomass is that the production of biomass for energy generation can contribute not only to climate change mitigation and energy security but also to rural development and employment generation [49].

The Government of India (GoI) approved the National Policy on Biofuels to encourage the use of renewable energy resources as alternate fuels to supplement transport fuels (gasoline and diesel for vehicles) and proposed a target of 10% biofuel blending (E10) by 2017. India produces conventional bioethanol from sugar molasses and production of advanced bio-ethanol is still in the research and development phase. However, taking the leads from R&D and process development unit, Praj Industries, Pune [50] is in the process of setting up the only one of its kind demonstration facility for cellulosic ethanol production close to Mumbai. This facility is poised to use multiple feedstock (sugarcane bagasse, corn cobs, corn stover) for cellulosic ethanol production with low CAPEX and OPEX, with low wastes generation and reduced energy consumption. Recently, India Glycols Ltd., Kashipur has also successfully demonstrated the first cellulosic ethanol production process from lignocellulosic biomass with a capacity of 10 t/day. This facility has been built with the financial support from Department of Biotechnology (DBT), Ministry of Science and Technology and its Public Sector Undertaking—Biotechnology Industry Research Assistance Council (BIRAC). DBT believes that this technology which has lowest capital and operating costs may provide cellulosic ethanol at competitive price [51].

4.1 Elements of Energy Security

The Asia Pacific Energy Research Centre (APERC) defines energy security as "the ability of an economy to guarantee the availability of energy resource supply in a sustainable and timely manner with the energy price being at a level that will not adversely affect the performance of the economy" [52].

Several factors affect the security of energy supply of a nation—the availability of new and renewable energy resources, the accessibility of these sources regarding political and economic factors, the affordability of development and utilization of these resources, and the acceptability of these resources on environmental sustainability.

4.2 Availability of Resources

Oil contributes 40% of the world's primary energy demand [52]. India is the third largest oil consumer, utilizing 3.7 million barrels a day and accounted for about 4.2% of global oil consumption in 2013 [53]. The uncertainty of existing reserves and fluctuating crude oil prices have led to political instabilities in several countries. The reliance on foreign countries for oil and competition over energy sources are being considered as the major threats to a nation's energy security. Controlled pricing of petroleum products is yet another issue faced by the Indian oil sector [54]. Moreover, the world oil production is projected to peak in the next 10–15 years [52].

The national energy policy of India is planned by the Association of Southeast Asian Nations (ASEAN) which focuses on sustainable development [55]. India's energy policy integrates measures including the enhancement of energy efficiency and saving, exploration of domestic energy reserves, strategic reduction of energy poverty, and responding to the issues of climate change and sustainable development [56]. The development and utilization of nonconventional sources of energy is getting utmost importance. Biomass, an abundant but underutilized resource, makes up 26% of India's energy resources [57]. Biomass can be used for combustion, gasification, pyrolysis, anaerobic digestion, fermentation and transesterification for the production of heat, electricity and as a substitute for petroleum for use as a transportation fuel [58]. To enhance biomass productivity and ensure biomass availability, biomass Research Centers (BRCs) have been set up in nine of the different agro climatic zones in India with an aim to develop packages of practices for fast growing, high yielding and short rotation fuel-wood tree crops suitable for degraded wastelands [59]. Development of processes for the production of the significant amount of biofuels without threatening food production needs special mention. The so-called "next generation biofuels" can tackle the "food-vs-fuel debate" and can open new avenues for promoting energy security. Several nations have implemented biofuel policies with the aim of securing future energy supplies [52]. Modernizing biomass technologies is a viable option. Also, concerted and continuous support from policy makers to have policies are a major challenge in the path towards a clean and green future.

4.3 Accessibility and Affordability of Resources

In order to attain the sustainable development goals of poverty alleviation, achieving food security, employment generation, and assumed access to affordable, reliable, sustainable, and modern energy for all, India has to significantly increase its energy availability [39, 60]. Besides availability, accessibility of resources economically is another challenge in ensuring a nation's energy supply. India, being the world's 7th largest economy [61], needs to maintain its annual growth rate at 8% to achieve its goal of energy security for sustainable development [56, 62]. However, the relatively higher energy prices result in a slow paced economic growth of the deficient energy economy of India [63]. Production and utilization of energy require both capital and labor. In order to make the inputs affordable, costs of nearly all goods and services in the economy need to be reduced, which can be affected only by lower energy prices. Indian transport sector is the largest and the fastest-growing consumer of petroleum, with 39% of petroleum products consumed [64]. The uncertainty of crude oil prices requires the transportation sector to explore new alternatives against the emerging economic challenges posed by a volatile oil market.

Renewable energy, which accounts for only ~1% of India's energy sources, is being promoted through public-private collaborations, as envisioned by the Ministry of New and Renewable Energy, GoI [57, 64]. The ministry provides budgetary support for research, development, and demonstration of renewable energy technologies, besides facilitating institutional finance and promoting private investment through fiscal incentives [64]. However, the high initial capital makes the renewable energy sources limited in access [52]. So, financial subsidies and policies are required to encourage investors. Also, transfer of technology from developed economies to developing economies is recommended to increase developing economies' accessibility to renewable energy [52]. The SAHYOG Project (Strengthening Networking on Biomass Research and Bio-waste Conversion—Biotechnology for Europe India Integration) aims to actively link research activities implemented within EU research programs and related programs by Indian national institutions [65].

4.4 Acceptability of Resources

Biofuels are envisioned for their ability to contribute to energy security via supplementing transport fuels and combating climate change. However, ineffective transportation and production technologies may result in the requirement of more energy for production than the energy that can be supplied [39]. However, non-food crop based biofuels, with a smaller carbon footprint can significantly reduce dependence on fossil fuels [66]. Interestingly, reports show that production costs of biofuels have decreased over time, thereby lowering the efficiency costs of biofuel policies in the long run [66, 67]. An integrated biofuels strategy is required for meeting the imbalances between demand and supply in the energy sector. Cost competitiveness is a major challenge in this sector. To make renewable energy affordable in the future, investments in R&D and technology transfer are recommended [52]. Risks associated with all these factors and their impacts on national political, economic, social, and cultural security need to be considered to envisage a nation's energy security.

5 Biofuel Policy Implementation: Employment Creation and Rural Development

5.1 Employment Index in Indian Sugar, Ethanol, and Other Biofuel Industries

Sugar industry serves as a good example for how ethanol and biofuel industries could drive employment scene and impact rural society. The sugar industry is the second foremost agro-based rural industry that accounts for nearly 8% of industrial investment and provides employment for about 7.4% of the industrial working force by providing direct employment to 4 lakh people and about 35 lakh people are indirectly connected with this industry. A sugar mill of 1250 tons crushing capacity per day creates an employment potentiality of around 300–350 permanent workers and an equal number of seasonal workers. Besides this, for harvesting sugarcane, 5000 male and female workers are required to be engaged during the crushing season. Likewise, around 100 tractors and 1000 bullock carts are given employment during the crushing season [68].

The Indian Sugarcane industry has more than 50 million farmers and their families' dependent on it for their livelihood. Direct and indirect employment generated out of the sugar industry caters to an estimated 12% of rural population. In addition to farmers, an estimated 0.5 million workers are directly employed as agricultural labor involved in cultivation and harvesting. Diversified ancillary activities and skills supporting local economy are also supported by the sugar industry [67].

Sugar mills have been working successfully, they have rendered considerable socioeconomic services to the rural community such as opening of schools, colleges, and hospitals. Besides, they provided numerous other facilities to the farmers in general. These activities and services are more evident in Maharashtra and Gujarat. The sugar mills have brought about a far reaching social, economic, and political transformation in the rural areas by providing various facilities like the modernization of agriculture, extension of the irrigation, employment, infrastructural facilities, education, health and recreation facilities, changing cropping pattern, and have promoted dairy and poultry activities. Thus, the sugar mills have acted as a catalyst for the socioeconomic development and these activities lead to the betterment of the economic conditions, not only of the farmers but also of landless laborers and other people in the areas.

5.2 Social Security

Employees feel insecure for many reasons such as inadequate wages, layoff and retrenchments, accidents and injuries in the course of employment, occupational hazards, sickness, old age, and total or partial disability. These factors cause anxiety and fear in their minds. Thus, the measures adopted to provide such protection to employees are known as social security measures. These measures protect workers and their families through various benefits such as compensation, maternity, sickness, and other benefits. Under Workmen's Compensation Act 1923, compensation is payable by the select sugar mills to workmen for all personal injuries caused to them by accident arising out of and in the course of his employment which disable him for more than 3 days. If the workman dies, the compensation is to be paid to his dependants. Under Employee State Insurance Act 1948, an insured person in the select sugar mills is entitled to receive benefits such as medical benefit, sickness benefit, maternity benefit, disablement benefit, dependants' benefit, funeral benefit, etc. The Employee Provident Funds and Miscellaneous Provisions Act, 1952 has made schemes for three types of benefits, viz., provident fund, family person and deposit linked insurance. Under the Payment of Gratuity Act, 1962 gratuity is payable to an employee of the sugar mill on the termination of his employment after he has rendered continuous service for not less than a 5 years. The completion of continuous service of 5 years is, however, not necessary where the termination of the employment is due to death or disablement. The Maternity Benefits Act, 1961 applies to women in factories, mines and other establishments. In the select sugar mills, married women employees having not more than 2 living children are eligible for maternity leave as per the provisions of the Maternity Benefit Act, 1961. Employees Family Pension Scheme, 1971 seeks to provide some monetary relief to the family members of employees in the select sugar mills, who die in service, that is, before superannuation. In the event of an employee's death, his family gets a pension on a graded scale depending on the employee's last salary grade [68].

5.3 Human Resources Working in Public and Private Sector

The sugar industry provides employment to a large mass of population in India. Approximately, sugar industry provides livelihood of 6 million agricultural and 0.5 million skilled and semi-skilled industrial workers. Additionally, this sector generates significant employment in ancillary and allied activities. In India, sugarcane is grown approximately of 5-million-hectare land which is roughly 3% of the gross cultivable area in the country. In reality, sucrose-alcohol manufacturing sector is one of the largest agro-based industry in the country. The annual turnover of the sugarcane and sugar related segment is in the range of INR 80–85 thousand crores. A significant portion of this turnover around INR 55–60 thousand crores accrue to the sugarcane farmers of the various states in the country [67].

5.4 Number of Sugar and Alcohol Mills in India

Sugarcane cultivation and processing is major employment in ten states and union territories. Uttar Pradesh and Maharashtra are major sugarcane growing and processing states in India. However, other states like Andhra Pradesh, Karnataka, and Tamil Nadu are also contributing considerably in sugarcane production and processing. Sugarcane processing mills are widespread in Uttar Pradesh and Maharashtra. Sugarcane mills of every size, however, mostly do sugarcane processing of 2500–5000 TCD (ton cane/day). This capacity is expanding up to 10,000 TCD. Tropical and subtropical sugarcane producing belts in India has the capacity of 12–16 million ton white sugar production per season. In Gujrat and West Bengal, two standalone sugar refineries produce refined sugars (5000 MT capacity) from imported and indigenous raw sugar. These sugar processing units aims to augment the refined sugars production particularly when sugar production is low in country in order to ensure the sugar security in the country [68].

Globally, sugar market is passing through with the transformational changes and so Indian market is also being affected severely. For example, Brazil, the largest producer of sugarcane (~673 million ton) and thus producing largest sugar (~35 million ton) and second largest producer of ethanol (~30 billion liters), in

2014–2015 faced severe crisis. In last couple of years, more than 70 sugarcane processing units got bankrupted. Since 2010, around 50% drop in revenues was observed with cutting of thousands of jobs. However, the sucro-energy sector in Brazil is looking for regain due to heating up and competitiveness in fuel market along with the perspective of bigger profits in the various links which are encouraging to industries to invest in the renewal of sugarcane cultivation and eventually in sucro-alcohol processing units. In addition to the conventional, sucro-alcohol products, sugarcane sector is also looking for cellulosic ethanol production at commercial level from sugarcane bagasse and leaves. In Brazil, Granbio, Raizen and Abengoa has set up 80 million liters/year, 40 million liters/year and 65 million liters/year respectively [69]. Centro de Tecnologia Canaviera (CTC)-Piracicaba has also set up a demonstration facility of 100-ton bagasse processing/day (~3-million-liter ethanol per year) at Usina de São Manoel in São Paulo state.

India has in total 690 registered sugar mills and amongst them, 93 sugar mills are almost on the stage of permanent closure, as these sugar mills are not in production during the last five sugarcane processing seasons. However, some new projects are coming up in Maharashtra and Karnataka with the collaboration with existing sugar mills integrated with the planning for cogeneration and distillation facilities. Figures 19.3 and 19.4 show the total 642 sugar mills in India map. Maximum sugar and ethanol mills in India are privately owned and only few are governed by state governments. Privately owned sugar mills have higher cane crushing capacity than cooperative sugar mills. For instance, out of the operational per day cane crushing capacity of 22.24 Lac TCD, the private sector owns 13.74 Lac TCD while the cooperative sector owns 7.77 Lac TCD and remaining 0.73 Lac TCD is processed by public sector undertakings (29 sugar mills). Cooperative sugars are lagging in terms of fetching upcoming projects.

Indian Sugar Mills Association [67] provided the global and domestic data on sugarcane and its byproducts and domestic agricultural market intelligence with

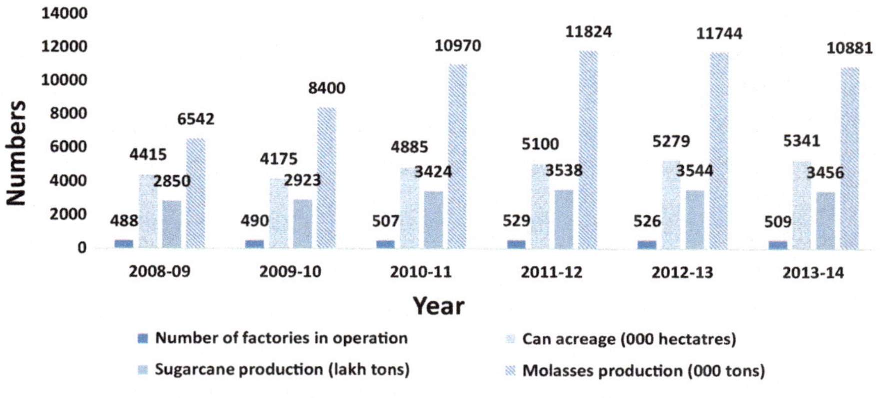

Fig. 19.3 Statistics of sugarcane production and sugarcane processing in India (Source: [67], http://www.indiansugar.com/SugarMap.aspx)

Fig. 19.4 Atlas of sugar mills in India (Source: [67], http://www.indiansugar.com/SugarMap.aspx)

Table 19.2 State wise of Indian ethanol production in 2015–2016 (adapted and modified from: [67], http://www.indiansugar.com/PDFS/List-ethanol_producers-2015-16.pdf)

Indian states	Number of Industries	Total installed capacity	
		KLPD (Kilo liter per day)	KLPA (Kilo liter per annum)
Bihar	6	335	90,450
Uttar Pradesh	35	2414	635,250
Punjab	1	60	16,200
Andhra Pradesh	11	500	144,900
Tamil Nadu	8	320	86,400
Maharashtra	68	3093	790,950
Gujarat	10	340	91,800
Karnataka	17	1225	294,300
Telangana	3	150	40,500
Other states (Uttarakhand + Sikkim + Haryana)	3	135	38,250
Grand total	162	8572	2,229,000

broad overview, key performance indicators and outlook analysis. ISMA report presents the current factual scenario of Indian sugarcane production, acreages, harvesting and crushing reports, sucrose-alcohol production, stock position and policies in states and the country [68] (Table 19.2).

6 Current Social Status and Improvement in Future

Development of villages have prominent role in holistic development of India as a large part of India's population lives in villages. Economic development in villages have a large gap than cities in India. Large rural population still do not have sustainable employment throughout the year because of poor infrastructure, non-connectivity with roads for transportation, inadequate electricity supply. Generally, for the daily routine activities and livelihood activities, rural population have to rely on polluting energy sources which have a harmful effect on the environment and their own socioeconomic development and health. Agricultural productivity has a profound impact in economic development in India. Agricultural products, byproducts and their management play a central role in renewable energy developent in India. Renewable energy development in rural areas can provide the following benefits:

- Facilitates access to drinking water;
- Allows lighting which increases security and enables the nighttime use of educational media and communication at school and home;
- Minimizes indoor pollution caused by the use of conventional fuels.

Rural population faces lack of access to affordable energy services. Biofuels produced from the raw materials grown locally in rural areas is likely to witness and aid in the alleviation of problems associated with poor standards of living. It is likely to mitigate the health issues faced by the rural population due to the traditional practices adopted to meet their regular fuel demands and also ensure energy security to the rural population.

The wasteland available in the country will be developed and utilized for cultivation of ethanol and biodiesel producing crops and will fulfill the basic requirement of fuel, fodder and food. By-products of the sugarcane ethanol industry contain valuable nutrients that have immense potential to be exploited as organic fertilizers.

The biofuels sector has the potential to serve as a source of substantial employment. Sugar industry in the major backbone of ethanol production industry in India. The sugar industry is the source of the livelihood of 50 million farmers and their dependants, comprising 7.5% of the rural population. Another half a million people are employed as skilled or semi-skilled laborers in sugarcane cultivation. This will reduce import burden to save substantial foreign exchange, enabling our country to strengthen infrastructural facilities with increased GDP. This will reduce the dependency on OPEC countries and open a path for self-reliance in petroleum requirements.

Sugar industry is likely to emerge as a significant contributor of the biofuel industry sector in catering to the energy needs of the country. This role would lead to a phenomenal transitional shift of the sugar mill into the energy complex. It is likely to meet around 20–25% of the total motor fuel requirements of the country in the near future. This would result in reduction of foreign exchange outgoing, attain energy security along with satiating the traditional demands of chemical and potable alcohol based industry. All of this is likely to open up huge employment opportunities in the near future.

Today, Government of India through National Institute for Transforming India (NITI) Aayog is also trying to deploy methanol or wood alcohol as a promising source to curb carbon emissions while savings foreign exchange via reducing imports crude oil. Currently, India annually spends INR 4.5 lakh crore on purchasing crude petroleum. Chemistry of methanol is excellently support to be blended it with petrol while dimethyl ether can be a good and cleaner alternative to diesel [70].

6.1 SWOT Analysis

In general, SWOT matrix entails the strategic planning to evaluate the Strengths, Weaknesses, Opportunities, and Threats of any business venture. Here, we do the SWOT Analysis of biofuel production and its impact on socioeconomic status in Indian context (Fig. 19.5).

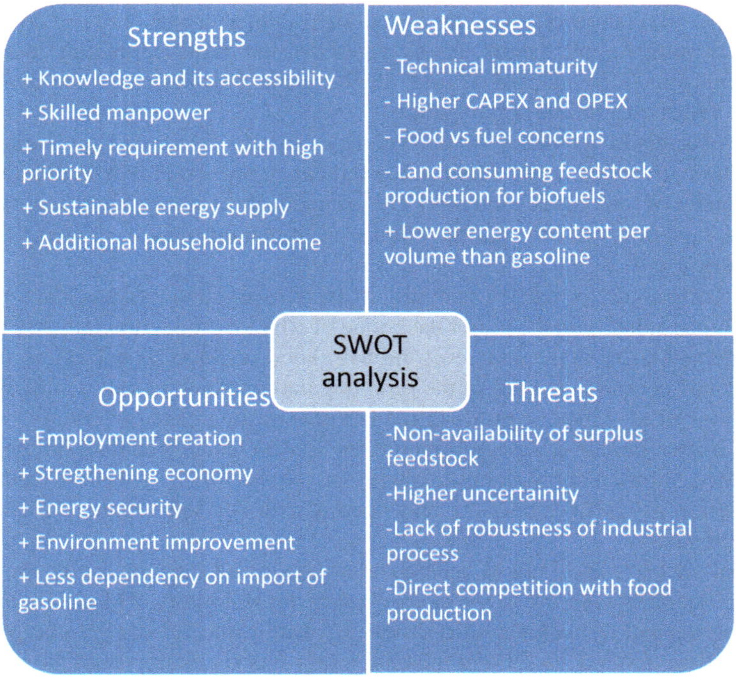

Fig. 19.5 SWOT analysis of biofuels in India

India ranks among one of the fastest growing markets for biofuels and renewable energy in the world. The Indian Government is making efforts to diversify the renewable energy portfolio, particularly in solar energy, small hydro and biomass. India has now over 8% annual economic growth rate. To maintain this growth rate, energy demand is projected to double by 2020. Therefore, India needs to harness all of the resources to get the power. Indian Government is making policies to bring foreign investment on renewables development in India. According to the HSBC estimate, India's "climate economy" will grow fivefold over the next decade, from $23 billion in 2009 to about $135 billion in 2020, roughly a 17% compound annual growth rate over that period [70]. In the last decade, private and public capital from national and international sources is flowing in through various national and international banks like International Finance Corp. and the Asian Development Bank.

7 Conclusions

Non-food crop-based biofuels, with a smaller carbon footprint, can significantly reduce dependence on fossil fuels. In the biofuel sector an integrated strategy is becoming imperative to meet the imbalances of demand and supply. Sugar industry is the emerging industry for catering to the ever increasing energy demands of the country by generation of electricity and bioethanol. These renewable energy sources hold the potential of converting the sugar mills into energy mills in due course of time. The biofuel sector has the potential to serve as a source of substantial employment. The national biofuel policy would also aid in the awareness creation of the role and importance of biofuels in the domestic market. It is also likely to promote wide dissemination of information pertaining to the potential of biofuel in upgrading the transportation infrastructure and supporting the rural economy. Substitution of petrol and diesel for transport with biofuels would witness accelerated development and promotion of cultivation of feedstock crops and production and use of biofuels. Increased utilization of biofuels in stationary and other applications and contribution to energy security, climate change mitigation, and creation of new employment opportunities is also likely to happen. By-products generated during the first and second generation biofuel production have the potential to bring down the cost of biofuels. However, constraints such as high operational expenditure (OPEX) and capital expenditure (CAPEX), and inefficient by-product and effluent management practices should be fixed in near term for sustainable development of biofuels.

This is imperative to set the biofuel policy by visualizing the situations in future using land-use pattern, alterations in crop mix, and the subsequent impact on food production. Biofuel crops could be even grown on fertile lands if oil prices keep skyrocketing. Certainly, promotion of biofuel development is highly judicious in India because of its potential of creating employment at each level, women empowerment, rural development, and augmenting entrepreneurships.

References

1. Chandel AK, Chan EC, Rudravaram R, Narasu ML, Rao LV, Ravindra P (2007) Economics and environmental impact of bioethanol production technologies: an appraisal. Biotechnol Mol Biol Rev 2:14–32
2. Silva SS, Chandel AK (eds) (2014) Biofuels in Brazil: fundamental aspects, recent developments, and future perspectives. Springer, Heidelberg. isbn:978-3-319-050195
3. Dufey A (2006) Biofuels production, trade and sustainable development: emerging issues. International Institute for Environment and Development, London
4. FAO (2005) Bioenergy and the millennium development goals. Forestry Department, FAO
5. http://www.fao.org/forestry/energy. Accessed 11 Apr 2016
6. Kammen DM (2006) Bioenergy in developing countries: experiences and prospects. In: Hazell P, Pachauri RK (eds) Bioenergy and agriculture: promises and challenges. IFPRI, Washington, DC
7. Moreira J (2005) Agreeing or disagreeing. Policy debate on global biofuels development. Renewable energy partnership for poverty eradication and sustainable development. Stockholm Environmental Institute (SEI), Sweden
8. Mandal R, Mithra P (2004) Biofuels: Indian scenario and policy issues. In: Proceedings of the International Conference on Biofuels: perspectives and prospects, organized by Winrock International India, September 16–17
9. Efroymson RA, Dale VH, Kline KL, McBride AC, Bielicki JM, Smith RL, Parish ES, Schweizer PE, Shaw DM (2013) Environmental indicators of biofuel sustainability: what about context? Environ Manag 51(2):291–306
10. Dale BE, Ong RG (2012) Energy, wealth and human development: why and how biomass pretreatment research must improve. Biotechnol Prog 28:893–898
11. Bhatia L, Sonia J, Rumana A (2012) An economic and ecological perspective of ethanol production from renewable agro-waste- a review. Appl Microbiol Biotechnol Exp 2:65
12. www.bp.com
13. Janda K, Kristoufek L, Zilberman D (2012) Biofuels: policies and impacts. Agric Econ Czech 58:372–386
14. National Policy on Biofuels (2009) Available via http://mnre.gov.in/file-manager/User Files/biofuel_ policy.pdf
15. Al-Riffai P, Dimaranan B, Laborde D (2010) European Union and United States biofuels mandates: impact on world markets. Technical Notes IDB-TN-191, Inter-American Development Bank, December
16. Ziolkowska J, Meyers WH, Meyer S, Binfield J (2010) Targets and mandates: lessons learned from EU and US biofuels policy mechanisms. AgBioForum 13:398–412
17. Serra T (2011) Volatility spillovers between food and energy markets: a semi parametric approach. Energy Econ 33:1155–1164
18. Jolly L (2015) Government biofuel policy and sugar crops. Outlook to 2015. A presentation of International Sugar Organization
19. Schmidhuber J (2006) The nutrition and the energy transition of world agricultural markets. In: Bahrs E, von Cramon-Taubadel S, Spiller A, Theuvsen L, Zeller M (eds) Unternehmenim Agrarbereichvorneuen Herausforderungen. Münster-Hiltrup, Landwirtschaftsverlag, pp 9–30
20. Goldemberg J (2006) The ethanol program in Brazil. Environ Res Lett 1:1–5
21. www.petrobras.com.br/. Accessed Sept 2016
22. Hebebrand C, Kara L (2007) An examination of U.S. and EU Government support to biofuels: early lessons. IPC Issue Brief, 26 October
23. Harmer T (2009) Biofuels subsidies and the law of the World Trade Organization. ICTSD Programme on Agricultural Trade and Sustainable Development. Issue Paper No.20, International Centre for Trade and Sustainable Development, Geneva
24. Tyner W (2008) The US ethanol and biofuels boom: its origins, current status, and future prospects. BioScience 58(7)

25. Josling T, Blandford D, Earley J (2010) Biofuel and biomass subsidies in the U.S., EU and Brazil: towards a transparent system of notification. International Food and Agricultural Trade Policy Council
26. Gorter H, Just DR (2008) Water in the U.S. ethanol tax credit and mandate: implications for rectangular deadweight costs and corn-oil price relationship. Rev Agric Econ 30:397–410
27. Banse M, Grethe H (2008) Effects of the new biofuel directive on EU land use and agricultural markets. In: Paper presented on the 107th EAAE Seminar ‘Modeling Agricultural and Rural Development Policies’, Seville, Spain
28. Jeníček V, Krepl V (2009) Energy and the European Union. Agric Econ Czech 55:1–11
29. Sengupta M, Poddar A (2013) National policy on biofuel under the scanner. Int J Emerg Technol Adv Eng 3(3):521–526
30. European Commission (2016) https://ec.europa.eu/energy/en/topics/renewable-energy/biofuels
31. International Energy Agency (IEA) (2004) World energy outlook 2004. OECD/IEA, Paris
32. Jain G (2010) Energy security issues at household level in India. Energy Policy 38 (6):2835–2845
33. Chandel AK, Kapoor RK, Narasu ML, Viswadevan V, Kumaran SSG, Ravinder R, Rao LV, Tripathi KK, Lal B, Kuhad RC (2007) Economic evaluation and environmental benefits of biofuel: an Indian perspective. Int J Global Energy Issues 28:357–381
34. Fraiture C, Giordano M, Liao Y (2008) Biofuels and implications for agricultural water use: blue impacts of green energy. Water Policy 10(1):67–81
35. Basavaraj G, Rao PP, Basu K, Reddy R, Kumar AA, Rao PS, Reddy BVS (2013) Assessing viability of bio-ethanol production from sweet sorghum in India. doi:10.1016/j.enpol.2013.01.012
36. Shinoj P, Raju SS, Chand R, Kumar P, Msangi S (2011) Biofuels in India: future challenges. Policy Brief No. 36, National Centre for Agricultural Economics and Policy Research, India
37. Ray S, Goldar A, Miglani S (2012) The ethanol blending policy in India. Econ Pol Wkly 1:23–25
38. Subramanian KA, Singal SK, Saxena M, Singal S (2005) Utilization of liquid biofuels in automotive diesel engines: an Indian perspective. Biomass Bioenerg 29:65–72
39. Biswas AK, Manocha N (2014) India’s biofuel policies: progress or boondoggle? NGI Know Exc 5(15):6–8
40. Lali A (2016) Biofuels for India: what, when and how. Curr Sci 110(4):552–555
41. Padma TV (2005) India’s biofuel plans hit roadblock: red tape and rising costs are choking India’s biofuel plans SciDev.Net, 5 September. Available via www.scidev.net/News/index
42. National Rural Development Guarantee Program (2016) http://www.nrega.nic.in/netnrega/home.aspx/. Accessed Sept 2016
43. CBDA (Chhattisgarh Biodiesel Development Authority) (2011) http://www.cbdacg.com/. Accessed Jul 2011
44. CIDL (Coffey International Development Ltd.) (2008) Vision document on biofuel for Madhya Pradesh. Department for International Development, UK
45. Raju SS, Parappurathu S, Chand R, Kumar P, Msangi S (2012) Biofuels in India: potential, policy and emerging paradigms. NCAP, New Delhi
46. Ariza-Montobbio P, Lele S (2010) Jatropha plantations for biodiesel in Tamil Nadu: viability, livelihood trade-offs and latent conflicts. Ecol Econ 70:189–195
47. USEA (2014) U.S. energy information administration report 2014. Retrieved from http://www.eia.gov/countries/analysisbriefs/India/india.pdf
48. Federation of Indian Chambers of Commerce and Industry—Price water house Coo Report (2013) Forging ties securing energy supply for a stronger economy. Retrieved from https://www.pwc.in/en_IN/in/assets/pdfs/publications/2013/forging-ties-Securing-energy-supply-for-a-stronger-economy.pdf
49. Faaij AP, Domac J (2006) Emerging international bio-energy markets and opportunities for socio-economic development. Energy Sustain Dev 10(1):7–19

50. www.praj.co.in/. Accessed Sept 2016
51. www.birac.nic.in/. Accessed Sept 2016
52. APERC (Asia Pacific Energy Research Centre Report (2007) A quest for energy security in the 21st century. Resources and constraints. APERC, Japan
53. India Outpacing China's Oil Demand (2014). Retrieved from http://blogs.wsj.com/moneybeat/2014/09/01/india-outpacing-chinas-oil-demand/
54. PHD Chamber and Little AD (2013) National energy security—coal and beyond. PHD House, New Delhi, pp. 8–18
55. Bhattacharya A, Bhattacharya T (2014) ASEAN-India gas cooperation: redefining India's "Look East" policy with Myanmar. In: Han P, Kimura F (eds) Energy market integration in East Asia: energy trade, cross border electricity, and price mechanism, ERIA Research Project Report FY2013, No. 29. ERIA, Jakarta, pp 63–109
56. Ghosh L (2012) Twenty years of political-security aspects of ASEAN–India relations: issues and challenges. In: Paper presented at International Conference to commemorate the 20th Anniversary of ASEAN-India Cooperation, Bangkok, 20–21 September, 2012
57. Debu C (2015) Why is energy security a strategic imperative for India? Retrieved from http://www.mapsofindia.com/my-india/government/why-is-energy-security-a-strategic-imperative-for-india
58. EAI Newsletter. India biomass energy. Retrieved from http://www.eai.in/ref/ae/bio/bio.html
59. Shukla PR (2010) Biomass energy in India: transition from traditional to modern. Retrieved from http://www.decisioncraft.com/energy/papers/ecc/re/biomass/bti.pdf
60. The Annual Economic Survey (2014–2015) Retrieved from http://pib.nic.in/budget2015/ecosurveyRel.aspx
61. International Monetary Fund. https://en.wikipedia.org/wiki/List_of_countries_by_GDP_(nominal)
62. Ministry of Commerce & Industry Press Release (2015) India's foreign trade (merchandise). F. No. 1(11)/2014-EPL, dated 13 Mar 2015
63. http://www.sciencespo.fr/psia/content/global-trends-renewable-energy-investments
64. TERI Energy Data Directory and Yearbook (2012/2013). Indian energy sector: an overview. 2013. TERI, New Delhi, pp 1–14
65. Global Biofuels—An Overview (2015) Retrieved from http://www.biofuelstp.eu/global_overview.html
66. Khanna M, Chen X (2013) Economic, energy security and greenhouse gas effects of biofuels: implications for policy. Am J Agric Econ 95(5):1325–1331
67. Indian Sugar Mills Association (2016) http://www.indiansugar.com/Indiansugar.com/. Accessed Sept 2016
68. Rodriguez L (2016) MBA ESALQ/University of Sao Paulo, Setor Sucro-energitico
69. http://timesofindia.indiatimes.com/city/mumbai/Methanol-is-future-of-country-Nitin-Gadkari/
70. Tan E (2011) The grid: demand, incentives makes Indian renewable attractive. Dow Jones Commentary https://www.bvp.com/sites/default/files/files/strategy-resource/india_renew_0.pdf

Chapter 20
An Assessment on Indian Government Initiatives and Policies for the Promotion of Biofuels Implementation, and Commercialization Through Private Investments

Rupam Kataki, Neonjyoti Bordoloi, Ruprekha Saikia, Debasish Sut, Rumi Narzari, Lina Gogoi, and Rahul S. Chutia

Abstract Energy has emerged as one of the most critical issues governing the economic, political, environmental, and social development of countries directly or indirectly. Availability of clean, efficient, affordable, and reliable energy is at the center of global prosperity and sustainable development. For developing counties like India, expanded access to dependable and modern energy service is a must for their fight against poverty and low living conditions of their citizens, while meeting objectives like increasing productivity, growing competitiveness, and improving economic growth at the same time.

It has already been proved that the current energy portfolio dominated by fossil fuel is insufficient to meet the requirements of the present times. Increasing concerns like shrinking supplies and rising demands of fossil fuels leading to further rise of fossil fuel prices, fossil fuel-induced climate change, as evidenced by rising temperature and environmental pollution, and energy security needs from political perspectives have promoted Governments all over the world to switch to renewable energy as it offers the promise of continued energy supply at a local level without comparable harm.

Government of India approved the National Policy on Biofuels on December 2009. The main objectives of biofuel policy are to reach the target of 20% biofuel blending (both biodiesel and bioethanol) by 2017. The government declared *Jatropha carcus* as the most suitable tree-borne oilseed for biodiesel production and encouraged new and second-generation biofuels for both biodiesel and bioethanol. The Planning Commission of India had set the target of covering

R. Kataki (✉) • N. Bordoloi • R. Saikia • D. Sut • R. Narzari • L. Gogoi
Biofuel Laboratory, Department of Energy, Tezpur University, Tezpur 784028, Assam, India
e-mail: rupamkataki@gmail.com

R.S. Chutia
Department of Energy Engineering, North Eastern Hill University, Shillong 793022, India

A.K. Chandel, R.K. Sukumaran (eds.), *Sustainable Biofuels Development in India*,
DOI 10.1007/978-3-319-50219-9_20

11.2–13.4 million hectares of land under *Jatropha* cultivation by the end of the 11th five year plan. In order to achieve the goal, the central and state governments are providing financial incentives to farmers to encourage oil seeds plantations.

This chapter discusses the role played by the Indian government in promoting biofuels particularly biodiesel and bioethanol production from nonedible oil seeds and other second-generation feedstocks. Government-funded research and challenges are also discussed in this chapter.

Keywords Biofuel • Policy • *Jatropha* • Indian government

1 Introduction

Biofuels particularly bioethanol and biodiesel have fascinated worldwide interest over the past few decades. Governments in most of the countries have been trying to endorse these fuels since it can be made from the locally available renewable sources as a substitute for expensive fossil fuels. According to a global oil company, the oil reserves will last only for 40.5 years if the rate of extraction and utilization of oil remains the same as of today [1]. In this regard, Government of India has adopted many biofuel promotion policies due to its potential contributions towards (1) energy security (2) economic development and poverty reduction, and (3) to achieve lower greenhouse gas (GHG) emissions and reduction in air pollution [2]. With increase in energy demand, India has emerged as the 4th largest energy consumer in the world after the United States, China, and Russia, according to the US Energy Information Administration (EIA), as reported on energy outlook for India 2011 [3]. According to EIA's recent report, [4], India and China lead to attribute one-half of the world's energy expansion by 2035 [5]. India's energy security depends on the development of unconventional fuels based on naturally obtained renewable feedstocks for gradual replacement of fossil fuels [6]. Further, in order to deal with future energy crisis in India, various energy efficiency improvement policies are also adopted. The National Policy on Biofuels which was planned by the Ministry of New and Renewable Energy (MNRE) was approved by the Union Cabinet in September 2008 and set a target of 20% blending of biofuels with both biodiesel and bioethanol by 2017 [7]. It is aimed to supplement the traditional energy resources with biofuels to fulfill the rapidly increasing demand of transport fuels. Use of biofuels will also help to reduce the energy requirements of India's vast rural population by using nonedible feedstocks which in turn will decrease the dependence on import of fossil fuels. The Government of India (GoI) has publicized and encouraged the production and use of (a) ethanol derived from sugar molasses/juice for blending with gasoline and (b) biodiesel derived from nonedible oils and oil waste for blending with diesel [8]. In this chapter, we have discussed about the types of biofuel, its policies, and challenges regarding biofuel production in India.

2 Development of Biofuels in India

2.1 *Bioethanol*

2.1.1 Production of Ethanol

The biofuel production in India is merely 1% of global production, which includes 380 million liters of ethanol and 45 million liters of biodiesel. In India, the main feedstock used for production of ethanol is molasses which is a by-product of sugar industry. Production of sugarcane for the period 2013–2014 is estimated at 350.02 million tonnes, which is 8.82 million tonnes more than its production during 2012–2013 (341.20 million tonnes) ([9], Ministry of Agriculture). It is estimated that from one tonne of sugarcane, 85–100 kg of sugar and 40 kg of molasses can be recovered [10]. As the production of sugarcane and sugar are of cyclical nature in India, the processing industry experiences interrupted market surpluses of sugarcane, sugar, and molasses. India ranks second in the world in terms of sugarcane production, yet its ethanol production is only around 1% of global production. In India, 70–80% of the sugarcane produced is used for sugar production and the remaining 20–30% for alternate sweetener such as jaggery and khandsari [11]. Out of the total alcohol produced, 25% is being used in industry; 30–35% is used for potable purposes and 3–4% for other uses. The remaining alcohol is being used as a fuel [12]. Currently, India ranks seventh in world in terms of ethanol production by producing 200 million liters annually. The vegetative sources (raw materials) for producing bioethanol can be classified into three classes [13]:

(a) Starch such as grain, corn, and tubers like cassava
(b) Sugar plants (sugar beet or sugar cane)
(c) Cellulose plants (trees and biomass)

2.1.2 Potential Bioethanol-Producing Substrates

Sugarcane India ranks second after Brazil in the world in terms of sugarcane production. In spite of the growth and significant levels of production in India, sugarcane is used mainly for human consumption, and bioethanol production is limited to a comparatively lower percentage for balancing the demand of imported fuels. While the area for sugarcane production has increased since 1950–1951, recently, the area and the yield have taken a downward slide. In terms of molasses production, a steady progress is observed. Ethanol-producing firms purchase molasses from sugar industries. In 2003, Government of India (GoI) made 5% blending of bioethanol with petrol as mandatory in nine states and four union territories and later extended it to other parts of the country based on the availability of ethanol; however, this was not strictly implemented due to the shortage of ethanol. This shortage in supply was caused by droughts and pest attacks and also surplus of sugarcane taxes, mill prices, and state regulations [14]. Therefore, in order to ensure

Table 20.1 Area, production, and yield of sugarcane in India

Year	Area (Lakh hectare)	Production (Million tonnes)	Yield (kg/ha)
2011–2012	50.38	361.04	71668
2012–2013	49.99	341.20	68254
2013–2014	50.12	350.02	69839

Source: Ministry of Agriculture, Government of India, Annual Report 2014–15 [9]

better availability of feedstocks, efforts are going on to diversify the feedstock base by including alternative crops, like sweet sorghum, for improving the production of bioethanol. Table 20.1 shows the area, production, and yield of sugarcane in India.

Sweet Sorghum Another potential substrate for bioethanol production is sweet sorghum. It can be grown in dry areas and needs less water than sugarcane. The amount of sugar present in it is about 15–20% which can be fermented and distilled to blend with petrol (gasoline). Though Tata Chemicals have started the production of sorghum, the commercial and large-scale production has yet to be realized to its full potential. Organizations such as ICRISAT, Hyderabad and the ICAR—Indian Institute of Millets Research (IIMR), Hyderabad have developed open pollinated varieties and photoperiod sensitive hybrids, enabling year-round production. In India, ethanol production is limited to a single source. To meet the ever rising demand for biofuel, it is necessary to promote alternative crops as feedstock to increase the bioethanol production [14].

2.1.3 Bioethanol Consumption

In India, bioethanol consumption is limited to the transportation sector only. Rural India will be benefitted from the massive investment in increasing the bioethanol value chain. In India, bioethanol production is done mostly from sugar molasses, while only a part of it is done from grains. Production of advanced bioethanol is still in its research and development stage. Ethanol production in calendar year (CY) 2016 will remain approximately 2.2 billion liters (current year's level) as the sugarcane supply is stable for the sixth consecutive year. Similarly, market penetration of ethanol fuel will be 2.9% and its 2.8% blending with gasoline looks attainable in CY 2015. Industrial sources have indicated that the oil marketing companies (OMCs) have targeted to produce nearly 800 million liters during CY 2015. Technically, this installed capacity is sufficient to meet about 8% of blending with gasoline. The price of landed-ethanol delivered at Oil Marketing Companies (OMC) depot is raised from Rs 48.50 to Rs 49.50 per liter ($0.76 to $0.78 per liter), which is an increase of 3–5% over the previous price. However, due to high production cost and low sugar prices, the sugar mills are running on negative margins, and any procedural delay in Ethanol Blending Program (EBP) could allow them to divert ethanol to chemical and potable industries. Moreover, mills could divert molasses as cattle feed or for exports if their prices are reasonable.

Ethanol consumption in CY 2016 will remain steady at this year's level of 2.3 billion liters even though one-third of total ethanol supply is estimated to blend with gasoline. Steady rise in supply along with strong demand for molasses from allied sectors will support larger ethanol consumption. During CY 2014, India imported 107 million liters of ethanol ($87 million) mostly from United States. Exports were limited to over 18 million liters ($15 million). The major exporters of ethanol to India are United States, the Netherlands, Spain, Bhutan, and Pakistan whereas Saudi Arabia, Ghana, Kenya, Nepal, Cote d Ivoire, and Cameroon are the major importers. The latest trade statistics for first quarter of CY 2015 show that India has imported 35 million liters of ethanol more than the previous one. In the same quarter, exports were approximately 2.8 million liters [15]. In India, export of biofuel is permitted only after it meets the domestic requirement and the final decision regarding it is taken by the National Biofuel Coordination Committee. There is no financial assistance for exports of biofuels by the Govt. of India. Generally, India imports ethanol only to meet deficits in demand during years of lower sugar production. The demand is mostly for consumption across the potable liquor and chemical industries than for fuel. There is no quantitative restriction on import of biofuels as well.

2.2 Biodiesel

Biodiesel can be defined as a mono-alkyl ester-based oxygenated fuel which can be produced from renewable resources like vegetable oils, algae, used cooking oil, and animal fats. It contains no petroleum products, but compatible with conventional diesel and can be blended in any proportion to form a stable biodiesel blend. Biodiesel is an alternative to conventional diesel with its properties quite similar to it. Biodiesel offers various substantial benefits like reducing GHG emission, improving fuel properties for vehicles, providing energy security, and reducing dependency on oil imports. Moreover, it can increase the employment in the agricultural sector and convert the wasteland into a productive one. Another important aspect of biodiesel is that it can be used in an unmodified diesel engine as a partial or complete petro-diesel substitute. Since it is produced from renewable sources, it is biodegradable and can burn up to 70% cleaner with 93% lower total hydrocarbon, 50% lower CO, and 45% lower particulate matter when compared with conventional diesel fuel [16]. Liquid biofuels can be classified as first and second generations based on the feedstock used for production. Use of first-generation feedstocks (such as corn and sugarcane for ethanol production and vegetable oils such as soybean oil and palm oil for biodiesel production) has been debatable, as they are directly consumed as food and using them in biofuel industry will increase their prices and, therefore, might trigger the food vs fuel crisis. Second-generation feedstocks are nonedible vegetable oils which do not have any importance as food resources and thus resolve the problems associated with first-generation feedstocks. In India, *Jatropha* and *Pongamia* are gaining significance as a feedstock for biodiesel production as they are nonedible and have better potential to substitute diesel thereby reducing the reliance on import of crude [17].

2.2.1 Oil Crops in India

The feedstock for biodiesel production includes both edible and nonedible sources. The selection of vegetable oils for biodiesel production depends on the climate as well as soil conditions. For example, in Europe, rapeseed and sunflower oils are used for biodiesel production while soybean oil in USA, coconut oil in Philippines, and palm oil in South East Asia are regarded as substitutes for petro-diesel. In India, utilization of vegetable oils such as mustard, palm, rapeseed, and soybean for biodiesel production is not possible as the country is not self-sufficient in edible oils and has to depend upon its imports to meet the large domestic demand. The use of nonedible seed oils for biodiesel production, which are extracted from trees and forest sources, does not affect the food security directly as the trees can be grown on marginal or waste land that does not compete with food production [18]. India has a rich potential for biodiesel production as more than 300 different tree species which yield oil-bearing seeds can be found in it [19]. Considering India's agricultural economy in terms of land use, production, and value, oilseeds succeed the food grains. Though the estimated potential of tree-borne oilseed (TBO) is more than 5 million tones, but as of now only 0.8–1 million tons of it can be collected. At present, the vegetable oil extracted from the potential tree-origin resources (more than 1 million tonnes) is about 15,000–20,000 tonnes. The potential oil-yielding species identified are *Jatropha curcus, Pongamia pinnata, Azadirachta indica, Diploknema botryceae (Cheura), Garcinia indica, Madhuca indica, Mesua ferrea L, Prunus armeniaca (wild apricot), Shorea robusta, Simarouba glauca, and Simmondsia chinensis (Jojoba)* among the hundreds of species available in India ([14, 20]). The nonedible oil seed plants mentioned above have the potential to produce vegetable oil and subsequently for biodiesel production besides their other commercial uses like burning, candle and soap making, illumination, etc. [21].

2.2.2 Biodiesel Production Potential

India with its abundant resources of Tree-Borne Oilseeds (TBO) has a huge possibility for biodiesel production by utilizing the existing barren resources and undertaking new plantation activities. However, the farmers are not willing to allow their suitable agricultural land for the purpose of fuel crop plantations owing to poorer yield and price. For this reason, the Forest department, under the Joint/Community Forest Management Programme, has undertaken plantation of *Pongamia* which is a nonedible oil-yielding crop with good potential in forest territories. The plantations are done along the roadways, railway lines, waterways, farm bunds, and community land in addition to the agricultural and forest land. The centrally sponsored programs like Mahatma Gandhi National Rural Employment Guarantee Scheme (MNREGS) and Watershed Development Programme are also providing subsidies to encourage *Pongamia* plantation [14]. The Ministry of Rural Development in India recognizes a land area of 552,692 km^2 out of a total area of 3.3 million km^2 as wasteland [22]. However, difficulty may arise while utilizing the whole area of wasteland for biofuel crop plantation as these lands are extensively used for livestock nurturing, fuel wood collection, etc. and people may not allow for such a switch. According to the Planning Commission of India, about 10% of the total unexploited land (24 million hectares) can be used for plantation of *Jatropha*. It is estimated that by planting *Jatropha* as a protective hedge around agriculture fields, approximately 3.0 million hectares can be brought under cultivation of the biofuel plants [13]. The Government of India has reduced its estimated land reserves area for biofuel crop cultivation to 72,000 km^2 as the wastelands get degraded due to unsustainable usage [23]. Due to their extensive adaptability and regional suitability, *Jatropha* and *Pongamia* have gained immense significance in terms of biofuel production. The National Bureau of Soil Survey and Land Use Planning (NBSS&LUP) has defined the suitability zones for *Pongamia* and *Jatropha* based on the ecological region approach considering soil, climate, and other physiographic factors. According to that approach, the arid and humid areas are classified as poor, as seed setting and yield will be poor for both *Jatropha* and *Pongamia*. On the other hand, the plantations of both *Jatropha* and *Pongamia* are considered highly suitable in the semiarid and subhumid tropical region. Among the semiarid tropical region, *Jatropha* plantation is preferred in dry semiarid while *Pongamia* in wet semiarid region [14]. Moreover, with a length of about 3 million km, India is ranked third in Asia in terms of road network. *Jatropha* and *Pongamia* can be grown along this road network and subsequently for biodiesel production. In addition, India has a railway network of around 63,140 km and the land along the track can be used for cultivation of *Jatropha curcas*. This will help in controlling the soil erosion besides improving the fertility in addition to vegetable oil for biodiesel production [21].

2.2.3 Biodiesel Consumption

With the ever rising demand for diesel and petroleum fuels, it is estimated that India's oil import will rise to 92% by 2030 [24]. It is important to encourage renewable prospects like biodiesel to substitute fossil diesel so that India's growing reliance on oil imports can be reduced and the import cost can be cut down as well. In the Table 20.2, the demand for diesel in India in near future and the amount of biodiesel required for blending in various percentages are presented. In Fig. 20.1, the land used for cultivation of *Jatropha* in different states of India is indicated. It is observed that maximum cultivation of *Jatropha* is done in Maharashtra, followed by the states Gujarat, Tamil Nadu, and Rajasthan.

Table 20.2 Diesel demand and future biodiesel requirements

Year	Diesel demand (million ton)	Biodiesel requirements for blending (million ton)		
		5%	10%	15%
2010–2011	100.5	5.0	10.0	20.1
2011–2012	106.0	5.3	10.6	21.2
2012–2013	111.8	5.6	11.1	22.3
2013–2014	118	5.9	11.8	23.6
2014–2015	124.5	6.2	12.4	24.9
2015–2016	131.3	6.6	13.3	26.2

Source: Planning Commission 2003 Report [13]

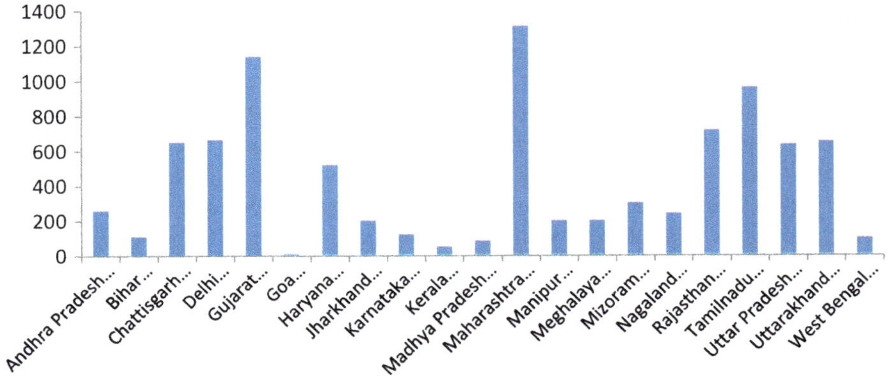

Fig. 20.1 Land Area under cultivation of *Jatropha* in various states of India. Source: Bioenergy Resource Status in India 2011 [23]

3 Sustainability of Biofuels in India

The Indian economy has been growing at a rate of approximately 7% since 2000 [25]. The demand for energy is also increasing to drive this increasing economic growth. To meet the energy requirements biofuels can be considered as one of the most promising alternative options. Bio-ethanol and biodiesel are the most commonly used biofuels. According to IPCC [26], biofuels have the capacity that can reduce GHG emissions in the transportation sector. For bio-ethanol, sugarcane, maize, sugar beet, and cassava are the most commercially used feedstocks whereas palm oil, edible oil from various oilseed crops, and *Jatropha* oil are used for the production of biodiesel [27]. Depending on the feedstock choice, second-generation biofuel production is more sustainable as compared to first-generation biofuel production [28]. Regarding GHG emissions, second-generation biofuels are thought to provide a clear benefit over first-generation options. Biofuels produced from crop residues are estimated to result in emissions of 11 g CO_{2e}/MJ_{fuel} whereas the first-generation conversion of cereals to ethanol emissions is estimated at 37–64 g CO_{2e}/MJ_{fuel} (ECOFYS) [29]. Biofuels, apart from supplementing petro-fuels, are aimed at lowering GHG emissions and reversing climate change conditions (i.e., rising average atmospheric temperature). Biofuels are more hospitable to human life in the long run. The biofuel sector has the potential to create substantial employment to the local people. Various jobs during collection and transportation of residue, biomass processing, generation of bioethanol, and by product generation can be generated with second-generation biofuel industries. Unlike fuel-free technologies (e.g., wind and solar PV), which mainly create jobs distant from their point of application, biofuel production is labor intensive at the point of feedstock growth and production [30]. For developing countries or even developed countries that seek to promote investment in rural areas, this characteristic of biofuels is of great value. As far as environmental sustainability of biofuels are concerned, the following general issues are important: GHG emission reductions; biodiversity; identification of areas of high conservation value; impacts on water; impacts on air; and impacts on soil [31–40].

In the past few decades, India has seen a rise in the demand for fuel due to increase in the numbers of motor vehicle, and it is expected to continue since it is projected that there will be 10–12% growth in number of vehicles in the near future. Thus to reduce the growing demands on fossil fuels and to bring energy security in the country, India has to embrace biofuel production on its own. Apart from this, biofuels provide various other socioeconomic and environmental benefits. Continuous depletion of oil reserves, increased demand for energy, rising concerns over global climate change, and energy security have forced to search for alternatives for fuel which are renewable in nature. In this context, biofuels have emerged as a viable and potential alternative to fossil fuels [10].

4 Indian Biofuel Initiatives and Policies

In India, biofuel program was started with an aim to explore a cleaner source of energy and to counterbalance, at least to some extent, the growing burden of import of crude oil. In a report prepared by U.S. Energy Information Administration, it was projected that the world energy consumption will increase by 53% from 2008 to 2035. Energy use rises from 505 quadrillion British thermal units (Btu) in 2008 to 619 quadrillion Btu in 2020 and 770 quadrillion Btu in 2035. According to the report, the role of coal is expected to remain important but natural gas, petroleum, and other liquid fuels will play a major role as the largest global energy source with an increase of 26.9 million barrel per day and energy consumption together in China and India will account for half of the projected increase [4]. Indian government initiated an ambitious programme addressing the multiple objectives of ensuring energy security with minimum damage to environment, enhancing income and employment opportunities for the rural communities as well as greening of wastelands and regeneration of degraded forest lands through cultivation of biofuel crops, particularly biodiesel crops like *Jatropha* and *Pongamia* [41]. National Mission on Biodiesel identified *Jatropha curcas* as the most suitable tree-borne oilseed for biodiesel production. The basic agenda of Indian government's biofuels policy is to emphasize an optimistic point of view of biofuels which indicates that these alternative fuels will bring greater energy security and independence to the nation. Intelligently designed renewable energy policies and careful planning with local input and cooperation can facilitate economic and social development for a community. Particularly in rural areas, growth of a domestic biofuel market can improve lives of the poor by creating more rural employment opportunities. When the employment rates are improved in rural areas, it will contribute to the overall improvement of the health and well-being of the population [42]. In India, several ministries are associated with the biofuel sector for policy making, regulation, promotion, and development. The Ministry of Agriculture provides subsidy through National Oilseeds and Vegetable Oils Development (NOVOD) Board to the farmers, Non-Governmental Organizations (NGOs), individuals, etc. for the production of Tree-Borne Oilseeds (TBOs), including biofuel crops, under the Integrated Development of Tree-Borne Oilseeds Scheme. Under the program of states/central sector area development for rising of *Jatropha*/*Pongamia* seedlings and their plantation, the Ministry of Rural Development provided financial assistance of Rs 490 million to 9 identified states in 2005–2006 and Rs 495 million to 15 states in 2006–2007. The ministries involved with the development of biofuels in India are shown in Table 20.3 [41]:

For the development of the energy sector, biofuel policies play a very important role. The biofuel policies significantly influence the profitability of production affecting the sectors like agriculture, research, industry, and trade [43].

The salient features of the national biofuel policy of India are as follows [44]:

1. An indicative target of 20% blending of biofuels both for biodiesel and bioethanol by 2017
2. Biodiesel production from nonedible oilseeds grown on waste, degraded, and marginal lands to be encouraged
3. A Minimum Support Price (MSP) to be announced for farmers producing nonedible oilseeds used to produce biodiesel

Table 20.3 Ministries involved with the development of biofuels in India

Ministry	Role
Ministry of New and Renewable Energy (MNRE)	Overall policy making, supporting research and technology development
Ministry of Petroleum and Natural Gas (MoPNG)	Marketing, development of pricing and procurement policy
Ministry of Agriculture (MoA)	Research and development on feedstock crops
Ministry of Rural Development (MoRD)	Identification of wastelands; promotion of biofuel plantations
Ministry of Science and Technology (MoS&T)	Biotechnological research on feedstock crops
Ministry of Environment and Forests (MoEF)	Ensuring implementation of Tree-Brone Oilseeds (TBO) crop plantations in forest wastelands and monitoring health and environmental effects of biofuels

4. Financial incentives for new and second generation biofuels, including a National Biofuel Fund
5. Biodiesel and bioethanol are likely to be brought under the ambit of "declared goods" by the Government to ensure the unrestricted movement of biofuels within and outside the states
6. Setting up a National Biofuel Coordination Committee under the Prime Minister for a broader policy perspective
7. Setting up a Biofuel Steering Committee under the Cabinet Secretary to oversee policy implementation

The Indian government launched the National Biofuel Mission (NBM) focusing to meet these objectives in 2003. In 2009, "National Policy on Biofuels" was released with an aim at mainstreaming the biofuels by setting an indicative target of blending up to 20% with petrol and diesel in the transport sector by the year 2017 [45]. The nonfood feedstocks raised on degraded or wastelands which are not suitable for agriculture and avoiding a possible conflict of food versus fuel security are the base of Indian biofuels policy. In India, molasses are the major feedstock for production of bioethanol. Nonedible oilseed crops like *Jatropha* and *Pongamia*, edible oil waste, and animal fats are the main source of biodiesel production in India [41]. The government initially intended to plant *Jatropha* on 11.2 million hectares of wasteland by 2012 and achieve a 10% blending target [46]. The Department of Land Resources under the Ministry of Rural Development has estimated that around 25 million hectares of fallow land is available in the country that can be used for growing feedstock crops, including *Jatropha*. Planning Commission aimed to cover an area of 11.2–13.4 million hectares under *Jatropha* cultivation by the end of 11th Five-Year Plan. Government of India has undertaken several policy measures to augment production and use of biofuels during the past one decade. As mentioned earlier, the National Biofuel Mission (NBM) launched in the year 2003 under the aegis of Planning Commission, Government of India, is the frontrunner of such efforts in the country. The NBM focussed mainly on phased expansion of area under biofuel feedstock crops like *Jatropha, Pongamia*, etc. Several micro-missions were also included in this mission that covers promotion of large-scale plantation of feedstock crops in forests and wastelands, procurement of seeds and oil extraction, transesterification, blending, trade, and R&D. Two integral parts of NBM, Ethanol Blended Petrol Programme (EBPP), and Biodiesel Blending Programme (BDBP) initiated blending of biofuels with transport fuels like petrol and high speed diesel on a commercial scale. In 2003, Ministry of Petroleum and Natural Gas (MoPNG) made 5% ethanol blending in petrol mandatory across in 9 states and 5 union territories. But it could not be fully implemented due to low sugarcane production in 2003–2004 and 2004–2005. In 2006, the blending mandate was extended to cover 20 states and 8 union territories. The National Biofuel Policy released in December' 2009 foresees that biofuels have the potential for rural development, employment opportunities, and environmental and economic benefits. The policy envisages the setting up of a "National Biofuels Development Board" (NBDB) to form a road map for the use of biofuels in petrol and diesel engines [41]. The government is also considering the creation of a National Biofuel Fund (NBF) for providing financial incentives in the form of grants and subsidies for advanced equipment, technologies and processes, and production units based on new and second-generation feedstocks. The policy also supports alternative feedstocks. Modifications in the National Policy on Biofuels favoring bioethanol production from alternate feedstocks like sweet sorghum will benefit all the stakeholders in the biofuels supply chain and will quicken the pace of biofuel production in the country to meet blending mandates.

5 Challenges of Biofuel Development

To reduce the dependence on imported fossil fuel and enhance energy security, biofuels are globally considered as sustainable and an eco-friendly source of energy [47]. In spite of various benefits provided by biofuels, there are certain challenges which have to be overcome to ensure a successful biofuel program implementation in India. The challenges are discussed below.

5.1 *Environmental Sustainability*

The determination of net environmental impact due to biofuels is still a challenging task to do. Biofuel's GHG reduction potential was evaluated from Life Cycle Assessment (LCA) studies, which reveals whether biofuel crops yield more energy than its need to produce (Tables 20.4 and 20.5). The LCA studies theoretically confirmed that both the first- and second-generation biofuels have potential to reduce GHG emissions (Table 4) and have higher net energy value compared to fossil fuels (Table 20.5) [2].

Table 20.4 Comparison of feedstocks in terms of GHG emission reductions

Fuel	Country	CO_2	Source
Corn ethanol	USA	2 (for E10) to 23 (for E85)	Wang [48]
Corn ethanol	USA	−30	Pimentel [49] as quoted by International Energy Agency [50]
Cassava	Thailand	63	Nguyen et al. [51]
Sugarcane	Brazil	80	International Energy Agency [50]
Oil palm	Malaysia	60	Zutphen [52]
Jatropha	India	80	Hooda and Rawat [53]
Coconut	Philippines	60	Pascual and Tan [54]

Source: Elder et al. [2]

Table 20.5 Comparison of feedstocks in terms of Net Energy Value (NEV)

Feedstocks	Country	NEV (MJ/L)	Source
Corn	US	5.89	Shapouri et al. [55]
Corn	US	−6.17	Pimentel [56]
Cassava	China	15.14	Hu et al. [57]
Cassava	Thailand	22.38	Nguyen et al. [51]
Sugarcane	Brazil	41.34	Macedo et al. [58]
Oil palm	Malaysia	37.45	Zutphen [52]
Jatropha	Thailand	3.82	Prueksakorn and Gheewala [59]
Jatropha	India	5.26	Tobin [60]
Coconut	Philippines	31.72	Tan et al. [61]

Source: Elder et al. [2]

In comparison to fossil fuel, biofuels have positive impact on environment as concluded by the Institute for Energy and Environmental Research in Germany [62]. They also concluded that bioethanol from sugarcane and biodiesel from rapeseed is more favorable than other sources. The European Union (EU) and International Energy Agency (IEA) stated that a considerable decrease in GHG emission occurs from burning of biofuels [50]. According to Paramathma et al. [63], biodiesel has a positive energy balance and emits 78% carbon dioxide (CO_2) which is a smaller amount than conventional diesel. It has been established that use of biodiesel reduce emission of particulate matter (PM), hydrocarbons (HC), and carbon monoxide (CO), adding up to reduction of carbon dioxide (CO_2) emission [64].

However, India has to consider detailed studies on the following aspects before expanding the biofuel sectors all over the country.

(a) Cropwise, location-specific impacts on energy consumption and emissions over complete production cycle
(b) Impact on biodiversity
(c) Effects on land and water resources
(d) Cost-effectiveness of biofuels [10]

Apart from positive impact on environment, some studies have also suggested that biofuels affect the environment in negative ways due to modification on using land, for example, renovation of tropical forests to farm land leads to secretion of huge amount of CO_2 which causes various environmental problems [65]. The emission of GHG from biofuels in Asia could be higher as compared to developed nations, since the use of energy during production may be less efficient. For example, India uses 4653 MJ/t of energy for corn [66], while the USA uses 4168 MJ/t [56] or 2068 MJ/t [55]. Thus, there is an urgent need to carry out lifecycle studies within an Asian context.

5.2 Food-Fuel Conflicts and Resource Availability

Steadily increasing food demand and its escalating price have created a debate among the scholars of the world over food vs. fuel. Various researchers [67–69] have reported that that there is an increase in food prices due to change of land from growing food crops to grow biomasses for producing biofuels. The world is already facing the conflict partly due to this conversion of using land [70].

According to International Food Policy Research Institute (IFPRI), prices of corn will increase by 26% and oilseeds by 18% due to the upcoming worldwide increase of biofuels [71]. A similar forecast has been made by Bhardwaj [72] regarding the palm oil prices. According to Regmi [73], due to increased primary staple food price, there will be reduction in consumption by the poor people leading to increase in the incidence of poverty and hunger.

In India, more than 70% of total inhabitants living in rural areas depend on agriculture to earn their livelihood directly or indirectly. Launching a large-scale biofuel program based on agricultural feedstocks might affect food and livelihood security of the people. Large-scale shift of area from food crops to energy crops has already created awareness among the policymakers in some developed and developing countries as this has potential not only to affect the agricultural sector but also other sectors of the economy [10]. Moreover, in India food security is a priority as around 28.3% people are below poverty line [74]. Therefore, it is necessary to ensure food security before launching any large-scale biofuel programme [10]. So, in order to minimize the conflict, it is imperative to identify newer nonfood feedstocks, manage pretreatment strategies, and investigate efficient and cheap conversion methods to produce modern biofuels.

5.3 Socioeconomic Development

Poverty reduction is one of the many benefits promised by biofuel program. Biofuel production could generate extra income and employment under conditions such as (i) adoption of labor intensive production methods; (ii) using surplus agricultural residues; (iii) using locally developed biofuel refining infrastructure; (iv) producing and consuming a significant amount of it locally; and (v) utilization of unused/marginal land for its cultivation [2]. However, According to Ankumu [75] and Friends of the Earth [76], the participation of biofuels towards scarcity reduction and sustainable rural development is not certain. They suggested that its production may be capital intensive if dominated by large producers thereby causing income inequality and unsafe or worsened working conditions among small farmers and workers. Currently, most of the private sector investor's main objective is to achieve low expenditure of production rather than reducing scarcity and using sustainable methods for production [77]. In order to prioritize sustainable development goals like scarcity reduction and employment generation through biofuel endorsement, the government needs to develop policies to encourage more labor intensive production methods which may not be cost effective.

In countries like India, promotion of biofuel development is attractive as it has the potential for creating employment opportunities, promoting entrepreneurship locally, and enhancement of women's involvement. Joint Forest Management (JFM) committees, self-help groups (SHGs), and panchayats can play a vital role to involve villager with biofuel programs. It can provide fuel for irrigation pump sets and for electricity generation. It will enhance women's involvement for raising nurseries and collection of tree-borne oilseeds. However, due to the various experience in yield and management sectors of crops, some awareness have been raised among farmers regarding the hesitation for planting of energy crops in marginal lands [10]. Therefore, it is of paramount importance to combine diverse interests of biofuel production with sustainable development objectives. While implementing the biofuel programmes, area-specific planning may be necessary keeping in mind the developmental goals of the particular area.

5.4 Technology and Feedstock Availability

Technologies used for production of ethanol from different lignocellulosic biomass follow the same method except the steps used for pretreatment of biomass. Type of enzymes and microbes employed also varies according to the type of feeedstocks. Pretreatment of biomass includes enzymatic saccharification, fermentation of the hexose and pentose sugars released by the hydrolysis, and saccharification [78]. Transesterification reaction converts an ester (vegetable oil or animal fat) into a mixture of esters of the fatty acids that makes up the oil (or fat). Biodiesel is a purified mixture of the fatty acid methyl esters (FAME). A catalyst is used to accelerate the reaction and based on nature of the catalyst, transesterification reaction can be classified as basic, acidic, or enzymatic [79].

Considering the availability of feedstock and conversion technology, the ethanol production in the country does not seem to be cost-effective. To bring down the cost of production, it is necessary to improve production technology and management practices. Some of the major technical constrains are lower plant capacity, use of batch process technology, inefficient by-product, and effluent management practices. Moreover, some biotechnological applications have to be applied in order to increase sugar content of crops, commercial use of membranes, and microbes for ethanol production. Currently in India, production of ethanol is solely based on sugarcane molasses, and hence it is necessary to search an alternate option to enhance its production. Besides molasses, sweet sorghum is a leading crop targeted for bioethanol production globally. Sorghum is best known as a grain crop; the juice, grain, and bagasse can be used to coproduce a combination of food, fodder, and ethanol in a bio-refinery approach. Its wide cultivation has proven that they are resistance to drought, saline-alkaline soils, and waterlogging condition. The comparative advantage of sweet sorghum with sugar cane is that its growing period is only 4 months against 12–16 months of sugar cane and water requirement is about four times lesser. The cost of cultivation is also about 60% lower than that of sugar cane which makes the ethanol produced from it is more cost-effective than that of sugar cane [80]. Though sweet sorghum is one alternative but much of research is still required in this direction [81].

5.5 *Cost of Biofuel Production and Prices*

At present, biofuels are costlier compared to fossil fuel worldwide, although the price gap varies widely depending on the cost of local inputs, feedstock productivity, and production technology. According to [82] and Organisation for Economic Co-operation and Development [83], biodiesel is about $0.27 per liter of diesel equivalent more expensive than usual diesel which may be attributed to cost of feedstock production [84].

There are some Asian countries where biofuel are cheaper than fossil fuel. For example, in India in 2006, the government of India set a purchase price of $0.68 per liter of diesel equivalent for the oil distribution companies compared to a retail price of $0.76 per liter of diesel oil [85]. The difference in price is due to differences in feedstock prices, farm subsidies, and fossil fuel prices.

Due to high cost of biofuels, consumers are reluctant to go for biofuels unless they are compensated for the same. Hence, to promote biofuels, a combination of subsidies, tariffs, fuel taxes (and tax exemptions), and blending mandates has been provided by the government. This extra cost borne by the government affects the effectiveness of the policy. However, if global crude oil prices increase further and biofuels become economically competitive, the rush to biofuels may accelerate without concern for environmental impacts or sustainable development [2].

6 Implementation/Commercialization

According to Sood [86], country's demand over the next couple of decades will increase at the rate of 4.8% p.a. At present, India's energy demand is dependent on fossil fuel-based products such as coal, petroleum, and natural gas. Only 25–30% of national consumption is satisfied by domestic crude oil production while the rest have to be imported. In these circumstances, biofuels play an important role in meeting India's growing energy needs [64].

Due to increasing demand for fuel and continuous depletion of nonrenewable resources, the high trend of consumption will not be limited to bioethanol but will also affect the fuels as well. At present, Indian biofuel sector is dominated by ethanol [87]. Currently, India accounts for 1% of global biofuel production. Over

3.50 billion liters of alcohol is produced every year from 320 distilleries in India. In the year 2008, out of 2.15 billion liters of ethanol produced, 280 million liters was blended with petrol. As ethanol serves various other purposes such as in potable liquor, chemical, and pharmaceutical industries, hence availability of ethanol for blending is dependent on the prevailing market prices which determine its viability for the Oil Marketing Companies (OMCs) for its use as a fuel. Due to constant shortfall in molasses availability, the government is unable to meet 5% blending target. Therefore, in April 2010, in order to increase the availability of ethanol for blending, the government decided to raise the minimum purchase price (MPP) of ethanol to Rs 27.00 per liter from Rs 21.50 per liter [88].

In India, large-scale blending of biodiesel with conventional diesel has not yet started. There are about 20 biodiesel plants which produce 140–300 million liters of biodiesel annually which is generally utilized for irrigation and generation of electricity and by the automobile and transportation companies to run their projects [89]. The Planning Commission has launched National Biodiesel Mission to expand *Jatropha* area in two phases.

The 1st phase, viz., demonstration phase, was taken up during 2003–2007 which included promotion of its large-scale plantations in forests and wastelands, procurement of seed and oil extraction, trans-esterification, blending and trade, and technological research and development.

The 2nd phase of expansion aimed at making the program self-sustainable by producing enough biodiesel to meet the 20% blending target [90]. To ensure a fair price of *Jatropha*, various state governments have fixed minimum purchase price (MPP) for *Jatropha* seeds. Apart from this, some subsidy programs and tax concessions/exemptions have also been provided by the government to enhance the production of feed stocks for biofuels. Several public institutions like National Oilseeds and Vegetable Oils Development (NOVOD) Board under the Ministry of Agriculture, Government of India, state biofuel boards, state agricultural universities and non-state sectors like nongovernmental organizations (NGOs), self-help groups (SHGs), cooperative societies, etc. are also actively supporting the biofuel program [88]. Their major work is to identify the best planting material, to develop high-yielding varieties (HYVs) with better quality and reliable seed source, intercropping trials, developing suitable package of practices, postharvest tools, and technology and detoxification. Field trials on various cultivars of *Jatropha* and *Pongamia* are being done by Central Soil Salinity Research Institute (CSSRI) under ICAR to develop site-specific genotypes that are tolerant to adverse climatic conditions. The Central Research Institute for Dry land Agriculture (CRIDA) undertakes studies on genetic diversity, variability, and other biotechnological studies on *Jatropha*. With the target to select good germplasm and develop quality planting material, the Department of Biotechnology (DBT) has initiated a "Micro mission on production and Demonstration of quality planting material of *Jatropha*."

Various projects to develop biofuels in India have been taken up by the Ministry of New and Renewable Energy. The lists of total projects (current and completed) are given below:

Projects Completed

1. Assessment of Techno-Economic Feasibility of Large Scale Seaweed Cultivation Integrated with Biofertiliser and Ethanol Production
2. Demonstration of Modular Pyrolysis Unit (20 kg/h) to produce Bio-Oil from Agro-Industrial Biomass Wastes and Methodology for Analysis, Use and Up gradation of Bio-oil
3. Integrated Technology Development for Biodiesel Production using Heterogeneous Catalyst
4. Design and Development of Dual operating Pilot scale Bio-Reactor system for comparative simulations studies on algal Cultivation
5. Development of a Hybridized Bioreactor—Open Pond Cultivation System Integrating Sinusoidal Magnetic Field technology to enhance the Qualitative and Quantitative Efficacy of Algal Biomass Production

Ongoing Projects

Sl. No.	Project name	Executing institution	Date of sanction and duration (Years)	Total cost (Rs in lakhs)	Objective
1	Design and Development of Vacuum Pyrolysis Plant to Process Various Agricultural and Agro-Industrial Biomass to Demonstrate Its Technical and Economical Feasibility	Indian Institute of Technology, Bombay, Powai, Mumbai—400 076	1.10.2010 Extended up to April, 2014	137.77	To evaluate technical and economical feasibility of the pyrolysis units for implementing the pyrolysis technology
2	Demonstration of Promising Genotype of Jatropha in Chhattisgarh State	Chhattisgarh Bio-fuel Development Authority (CBDA), MIG 33, Indrawati Colony, RAIPUR, 492007	16.03.2010 & 5 years	34.10	To study and assess the suitability and productivity of the promising genotypes of Jatropha in Chhattisgarh state
3	Demonstration of promising genotypes of Jatropha in the Karnataka state	Department of Forestry and Environmental Science, University of Agricultural Sciences, GKVK campus, Bangalore-560056, Karnataka	15.03.2010 & 5 years	34.10	To study and assess the suitability and productivity of the promising genotypes of Jatropha in Karnataka state
4	Demonstration of promising genotypes of Jatropha in the Rajasthan state	CEO, Zila Parishad-Banswara and Rajasmand	16.03.2010 & 5 years	34.10	To study and assess the suitability and productivity of the promising genotypes of Jatropha in Rajasthan state

(continued)

Sl. No.	Project name	Executing institution	Date of sanction and duration (Years)	Total cost (Rs in lakhs)	Objective
5	Demonstration of promising genotypes of Jatropha in Tamil Nadu state	Tamil Nadu Agricultural University, Coimbatore-641 003	16.03.2010 & 5 years	34.10	To study and assess the suitability and productivity of the promising genotypes of Jatropha in Tamilnadu state
6	Enhanced butanol production from lignocellulosic biomass using improved pre-treatment and integrated saccharification, fermentation and separation in a membrane bioreactor	National Environmental Engineering Research Institute (NEERI)	30.09.2011 & 3 years	39.7056	Development of improved process for Biobutanol production from lignocellulosic biomass
7	Biocrude Production: Hydrocracking of nonedible vegetable oil	Sardar Swaran Singh National Institute of Renewable Energy, Kapurthala	13.9.2011 & 3 years	68.40	Development of Process for the production of bio-crude by hydrocracking of the vegetable oils
8	Development of pretreatment strategies and bioprocess for improved production of cellulolytic enzymes and ethanol from crop byproduct for demonstration at pilot plant	Department of Microbiology, University of Delhi South Campus, New Delhi	17.1.2012 & 3 years	148.47	Development of hyper cellulase producer and a fermentation process for production of bio-alcohol from lignocellulosic biomass
9	Process development for bioethanol production from agricultural residues Phase-I: Development of process for co-fermentation of hexose and pentose sugars of agricultural residues	Sardar Swaran Singh National Institute of Renewable Energy, Kapurthala-144 601	25.1.2012 & 3 years	132.19	Development of Process for co-fermentation of hexose and pentose sugars of agricultural residues by selected isolates of co-fermenting mesophilic and thermophilic strains for ethanol production

(continued)

Sl. No.	Project name	Executing institution	Date of sanction and duration (Years)	Total cost (Rs in lakhs)	Objective
10	Hydropyrolysis of lignocellulosic biomass to value added hydrocarbons	Indian Institute of Petroleum, Dehradun	31.01.2012 & 3 years	186.40	To convert ligno-cellulosic biomass into value-added hydrocarbons/ fuels that can be used in the transportation sector and chemicals
11	Sorghum Stover based Biorefinery for Fuels and Chemicals	NIIST, Trivandrum; MNNIT, AllahabadTERI, New Delhi & IICT, Hyderabad	06.03.2012 & 3 years	184.138	To scale-up integrated technology for converting Sorghum Stover to Biofuels and value-added chemicals
12	Stabilization and up gradation of biomass derived bio-oils over tailored multifunctional catalysts in a dual stage catalytic process to produce liquid hydrocarbon fuels and its application studies	The Energy & Resources Institute (TERI), New Delhi	13.09.2013 & 3 years	164.07	To establish technical feasibility of stabilising bio-oil to increase the storability and enhance the properties to be blended with petro-fuel and upgrading bio-oils to transport fuels
13	Improved production of Biogas and Bio-CNG from Ligno-cellulosic Biomass	DBT-ICT Centre for Energy Biosciences, ICT, Matunga (E), Mumbai, Kirloskar Integrated Technologies Limited, Pune, India Glycols Limited, Noida Distt: Gautam Budh Nagar (UP)	14.11.2013 & 3 years	445.90 MNRE share 356.53	Development and up-scaling of integrated technology for production of biogas and value-added products from ligno-cellulosic biomass

Source: MNRE website (http://www.biodieseltechnologiesindia.com/recent.html)

My Eco Energy, a Pune-based biodiesel manufacturing company in 2013 announced its plan to manufacture, distribute, market, and retail biodiesel made from waste oil. They aimed to set up biodiesel pump in Maharashtra and other adjoining states in the initial stage with 20 refilling stations Maharashtra. They further planned to expand their network over 250 retail pumps in Maharashtra within a time span of 1 year. Chief Minister B S Yeddyurappa inaugurated Eco Green's Fuel biodiesel plant in Bangalore which is country's first commercial biodiesel plant, with a capacity of 5000 L per day. Apart from this, Indian Railways is ready to set up four biodiesel plants with an estimate cost of Rs 120 crore with a capacity of 30 tons per day of biodiesel [91].

With a prime objective of producing biodiesel from *Jatropha*, the West Bengal government has come with an idea of a park in Salt Lake to promote *Jatropha* plantation. Apart from growing *Jatropha* trees, the park will also house a pilot processing plant for extraction and conversion of crude *Jatropha* oil into bio-diesel. An assistance of 2.36 lakhs as first installment will be provided to Biotechnology board by West Bengal Technical University (WBTU) for this project. Similarly in Assam for promotion of *Jatropha* cultivation, a special scheme was launched by the Department of Agriculture covering 137.4 ha area. To achieve the current requirement of land for cultivation of *Jatropha* and *Karanjia*, Indian Oil Corporation (IOC) is planning to procure 50,000 ha of wasteland in Uttar Pradesh.

7 Future Prospectives

7.1 Biodiesel

Biodiesel is becoming popular—that might assist in rural employment creation and in gaining energy independence in India. National mission of biodiesel is based on production and promotion of biodiesel produced from *Jatropha*. This program also promotes the use of wastelands across India and controls soil erosion, through large-scale plantation of *Jatropha*. It is considered as an excellent source of biodiesel with a potential of producing more than 2 tons of oil per hectare per year [89]. Crude *Jatropha* oil can be used directly in small-scale diesel generators, oil lamps, and stoves. To meet the country's energy requirement, India imports a great amount of fossil-based crude oil which can be reduced by *Jatropha*-based biodiesel. The program will not only reduce the dependence on imported petroleum but it will also generate employment opportunities and will help in improving the economic status of the country. It also reduces various health hazards as the toxic pollutant emission is lower compared to that for existing fossil fuels. It is expected that blending (10–20%) of biodiesel with petrol/diesel may become mandatory in the future which will boost the production and use of biodiesel in India. Biodiesel production may also increase due to the expected incentives and subsidies from the government of India [93]. However, search for other nonedible oil seed-bearing tree/shrub species, efficiency improvements in oil extraction, catalytic conversion seed oil to biodiesel, improvements of biodiesel property, exploitation of coproducts of biodiesel production are some of the areas that need attention of all stakeholders to achieve the set goals.

7.2 Bioethanol

In order to curb country's future carbon footprint and dependence on foreign crude oil, the Ministry of New & Renewable Energy established a National Policy on Biofuels in the year 2008. The bottom line objective of this program is to lay down a roadmap for the phased implementation regarding production and commercialization of biofuel. However, shortage of ethanol for transport industry (as its demand for chemical and beverage industries has also been increasing) has limited its use for blending with petrol to less than 3%. Government's proposed target of 20% ethanol blending by 2017 still remains a big challenge which will require the development and extensive use of second generation feedstocks [94].

As earlier reported, second-generation or cellulosic ethanol can be derived from various agricultural residues such as stalks, leaves, bagasse, rice husk, wheat, wood chips, sawdust, or energy crops. Currently, molasses are the major source for commercial production of ethanol in India and is one of the 7th largest producers in the world. Hence to meet the constantly increasing demand, it has become vital to search for other renewable alternative for sourcing ethanol. Since India does not possess any surplus grains or starchy biomass to be used as fuel, therefore ligno-cellulosic biomass is the only viable option for production of biofuel. Due to uncertainty of the availability of feedstock throughout the year, it would be safer for the bioethanol plant to opt for multiple feedstocks. This uncertainty in availability may be due to prolonged drought or distributed occurrence of the resources which creates collection, transportation, and logistics problems. State- or private-owned initiatives are needed for biomass collection and supply, and this single factor will be a significant determinant of the success of any biomass-ethanol technology in India [78].

8 Conclusion

India has a long history of bio-energy development and program interventions. During the years of 1970s, national biomass policies had been originated as a component of rural and renewable energy policies. Government of India's Ministry of New and Renewable Energy (MNRE) has been started to support the promotion of bio-energy program since the mid-1990s. Further, the Government of India (GOI) on December 24, 2009 passed the approval of National Policy on Biofuels. The biofuels policy promotes an increased use of renewable energy resources as alternate fuels for transport fuels supplementation (diesel and gasoline for vehicles). It also suggests a target of 20% biofuels blending (bio-ethanol and biodiesel) by the year 2017. At present, it is increasingly difficult for the government to apply the necessary blending of 5% ethanol in petrol (gasoline) as there is no sufficient stock of sugarcane crop for bioethanol production. India practices commercial biodiesel production on a very small scale while its utilization is also primarily restricted to the unorganized sector. Due to the limited supply of *Jatropha* seeds for bio-diesel production, Indian government failed to complete the target of 20% diesel blending by the year 2011/2012. In order to make the biofuels economically feasible in India, it will take time since the research for commercialization of advanced biofuel is still under progress. Further, in order to avoid the possible conflict of fuel versus food security, Indian government needs to focus more on non-competent feedstock that can be grown in waste/degraded lands not suitable for cultivation of food crops and thereby reducing stress on agricultural lands [95]. It is expected that the new biofuels policy will provide incentives for cultivation and mass propagation of nonedible oilseeds, like *Jatropha* and *Pongamia* over about 11.2 million hectares of land, which is 30 times of present cultivation, resulting in 13.38 million tons of bio-fuel to meet its policy target of 20% blending of biofuels in transportation fuel by 2020 [96].

References

1. Beyond Petroleum (2007) BP statistical review of world energy. Beyond Petroleum, London
2. Elder M, Shivpuram P, Romero J, Matasumoto N (2008) Prospects and challenges of biofuels in Asia: policy implications. IGES White Paper, Hayama Japan, 105–132
3. http://www.thehindubusinessline.com/industry-and-economy/india-is-fourth-largest-energy-consumer-in-the-worldreport/article4525172.ece. Accessed 17 July 2015
4. International Energy Outlook (2011) Available from http://www.eia.gov/forecasts/ieo/pdf/0484 (2011).pdf
5. http://www.deccanherald.com/content/320499/india-worlds-fourth-largest-energy.html. Accessed 20 July 2015
6. http://www.teriin.org/policybrief/docs/biofuel.pdf. Accessed 3 July 2015
7. http://www.teriin.org/policybrief/docs/biofuel.pdf. Accessed 26 July 2015
8. Biofuels annual India (2013) GAIN report. http://gain.fas.usda.gov/Recent%20GAIN%20Publications/Biofuels%20Annual_New%20Delhi_India_8-13-2013.pdf
9. Annual Report (2014–2015) Department of Agriculture and Cooperation, Ministry of Agriculture, Government of India, Krishi Bhawan, New Delhi-110 001, March, 2015
10. Raju SS, Parappurathu S, Joshi PK (2009) Sustainable development of biofuels: prospects and challenges. Econ Polit Wkly 44(52):65–72
11. Patni N, Pillai SG, Dwivedi AH (2011) Analysis of current scenerio of biofuels in India specifically bio-diesel and bio-ethanol. International conference on current trends in technology, NUiCONE
12. http://www.ethanolindia.net/ethanol_demand.html. Accessed 17 July 2015
13. Planning Commission (2003) Report of the committee on development of biofuels. Government of India, New Delhi
14. http://www.pisces.or.ke/sites/default/files/Bioenergy%20Resource%20Status%20in%20India.pdf
15. Biofuels Annual Report (2015) http://www.scribd.com/doc/193826767/Biofuels-Annual-New-Delhi-India-8-13-2015. Accessed 20 July 2015
16. Dwivedi G, Sharma MP, Jain S (2013) Diesel engine performance and emission analysis using biodiesel from various oil sources—review. J Mater Environ Sci 4:434–447
17. Sharma MP, Jain S (2010) Prospect of biodiesel from *Jatropha* in India: a review. Renew Sust Energy Rev 14:763–771
18. Dwivedi G, Sharma MP, Jain S (2011) *Pongamia* as a source of biodiesel in India-A review. Smart Grid Renew Energy 2:2184–2189

19. Gogoi P (2008) *Pongamia* Oil—a promising source of biodiesel. Assam University Press, Assam
20. Agarwal AK (2007) Biofuels (alcohols and biodiesel) applications as fuels for internal combustion engines. Prog Energy Combust Sci 33:233–271
21. Dwivedi G, Sharma MP, Kumar M (2014) Status and policy of biodiesel development in India. Int J Renew Energy Res 4(2):246–254
22. http://dolr.nic.in/fwastecatg.html. Accessed 14 July 2015
23. Bioenergy Resource Status in India (2011) Working paper May 2011. Prepared for DFID by the PISCES RPC Consortium. M.S Swaminathan Research Foundation Chennai, India
24. World Energy Outlook (2012) http://www.worldenergyoutlook.org/weowebsite/2012/WEO2012_Renewables.pdf. Accessed 20 July 2015
25. EIA (2013) India country information. U.S. Energy Information Administration (EIA), Washington, DC. http://www.eia.gov/countries/cab.cfm?fips=IN. Accessed 13 Apr 2013
26. IPCC (2007) Mitigation of climate change: technical summary. Intergovernmental Panel on Climate Change (IPCC), Geneva
27. Fischer G, Hizsnyik E, Prieler S, Shah M, Van Velthuizen H (2009) Biofuels and food security. A study by OPEC Fund for International Development (OFID) and International Institute for Applied Systems Analysis (IIASA), Laxenburg
28. Eisentraut A (2010) Sustainable production of second generation biofuels: potential and perspectives in major economies and developing countries. International Energy Agency (IEA), Paris
29. ECOFYS (2012) Renewable energy progress and biofuels sustainability. Report produced for the European Commission (Tender ENER/C1/463-2011-Lot2)
30. International Renewable Energy Agency (IRENA) (2011) Renewable energy jobs: status, prospects and policies. IRENA Working Paper
31. Caldeira-Pires A, da Luz SM, Palma-Rojas S, Rodrigues TO, Silverio VC, Vilela F, Barbosa PC, Alves AM (2013) Sustainability of the biorefinery industry for fuel production. Energies 6 (1):329–350
32. Cherubini F, Bird ND, Cowie A, Jungmeier G, Schlamadinger B, Woess-Gallasch S (2009) Energy and greenhouse gas-based LCA of biofuel and bioenergy systems: key issues, ranges and recommendations. Resour Conserv Recycl 53(8):434–447
33. Fischer G, Schrattenholzer L (2001) Global bioenergy potentials through 2050. Biomass Bioenergy 20(3):151–159
34. Fischer G, Prieler S, van Velthuizen H, Berndes G, Faaij A, Londo M, de Wit M (2010) Biofuel production potentials in Europe: sustainable use of cultivated land and pastures, part II: land use scenarios. Biomass Bioenergy 34(2):173–187
35. Kendall A, Yuan J (2013) Comparing life cycle assessments of different biofuel options. Curr Opin Chem Biol 17(3):439–443
36. Leal MRLV, Horta Nogueira LA, Cortez LAB (2013) Land demand for ethanol production. Appl Energy 102:266–271
37. Mohr A, Raman S (2013) Lessons from first generation biofuels and implications for the sustainability appraisal of second generation biofuels. Energy Policy 63:114–122
38. Rowe RL, Street NR, Taylor G (2009) Identifying potential environmental impacts of large-scale deployment of dedicated bioenergy crops in the UK. Renew Sust Energy Rev 13 (1):271–290
39. UNEP (2009) Towards sustainable production and use of resources: assessing biofuels. International panel for sustainable resource management, United Nations Environment Programme (UNEP), Paris
40. Wu Y, Liu S (2012) Impacts of biofuels production alternatives on water quantity and quality in the Iowa River Basin. Biomass Bioenergy 36:182–191
41. Raju SS, Parappurathu S, Chand R, Joshi PK, Kumar P, Msangi S (2012) Biofuels in India: potential, policy and emerging paradigms. Policy Paper 27, April

42. Thomas J (2013) A fundamental study of Indian biofuel. IRJA-Ind Res J 1(1). Issue: December, 2013
43. Basavaraj G, Rao PP, Reddy CR, Kumar AA, Rao PS, Reddy BVS (2012) A review of the National biofuel policy in India: a critique of the need to promote alternative feedstocks, Working paper series no. 34 RP—Markets, Institutions and Policies
44. Mahurkar D (2007) PetroFed & UOP Symposium on biofuels: Indian biofuel policy, September
45. GoI (Government of India) (2009) National policy on biofuels. Ministry of New and Renewable Energy, New Delhi
46. GoI (Government of India) (2003) Report of the committee on development of biofuels. Planning Commission, Government of India (GoI), New Delhi
47. http://biofuelsandthepoor.com/biofuels-in-india-future-challenges/. Accessed 19 July 2015
48. Wang M (2005) The debate on energy and greenhouse gas emissions impacts of fuel ethanol. Energy systems division seminar, 3 August 2005, Argonne National Laboratory
49. Pimentel D (2001) Limits of biomass utilization. In: Encyclopedia of physical science and technology. Academic Press, San Diego, CA
50. International Energy Agency (2004) Biofuels for transport: an international perspective. Paris
51. Nguyen TLT, Gheewala SH, Garivait S (2007) Energy balance and GHG-abatement cost of cassava utilization for fuel ethanol in Thailand. Energy Policy 35(9):4585–4596
52. Zutphen HV (2007) The CO_2 and energy balance of Malaysian palm oil: current status and potential for future improvements. Paper read at 2nd symposium on sustainable resource development, 6 June, at Brussels, Belgium
53. Hooda N, Rawat VRS (2006) Role of bio-energy plantations for carbon-dioxide mitigation with special reference to India. Mitig Adapt Strateg Glob Chang 11(2):437–459
54. Pascual LM, Tan RR (2004) Comparative life cycle assessment of coconut biodiesel and conventional diesel for Philippine automotive transportation and industrial boiler application. Paper read at In LCA/LCM, 11–24 July 2004. http://www.lcacenter.org/InLCA2004/. Accessed 13 July 2015
55. Shapouri H, Duffield JA, Wang M (2002) The energy balance of corn ethanol: an update. United States Department of Agriculture, Argonne
56. Pimentel D (2003) Ethanol fuels: energy balance, economics, and environmental impacts are negative. Nat Resour Res 12(2):127–134
57. Hu Z, Pu G, Fang F, Wang C (2004) Economics, environment, and energy life cycle assessment of automobiles fueled by bio-ethanol blends in China. Renew Energy 29 (14):2183–2192
58. Macedo EDC, Leal MRLV, Silva JEARD (2004) Assessment of greenhouse gas emissions in the production and use of fuel ethanol in Brazil. Secretariat of the Environment, Government of the State of São Paulo, São Paulo
59. Prueksakorn K, Gheewala SK (2006) Energy and greenhouse gas implicationsof biofuel production from *Jatropha curcas* L. 2nd joint international conference on sustainable energy and environment, Bangkok, Thailand
60. Tobin J (2005) Life cycle assessment of the production of biodiesel from Jatropha. MSc Thesis, Energy Group, School of Construction Management and Engineering, University of Reading, Reading
61. Tan RR, Culaba AB, Purvis MRI (2004) Carbon balance implications of coconut biodiesel utilization in the Philippine automotive transport sector. Biomass Bioenergy 26(6):579–585
62. Quirin M, Gartner SO, Martin M, Reinhardt GA (2004) CO_2 mitigation through biofuels in the transportation sector: status and perspectives. Institute for Energy and Environmental Research, Heidelberg
63. Paramathma M, Venkatachalam P, Raja AS (2007) Jatropha improvement, management and production of biodiesel, Centre of excellence in biofuels. Agricultural Engineering College and Research Institute, Coimbatore
64. Babu MSU, Nautiyal S (2012) Socioeconomic and ecological consequences of biofuel development in India. Norwegian Institute of International Affairs 1–18

65. Zah R, Böni H, Gauch M, Hischier R, Lehmann M, Wäger P (2007) Executive summary of life cycle assessment of energy products: environmental assessment of biofuels. Empa, Bern
66. Ali N (2006) Energy management in agriculture: status, issues and strategy. Paper read at balancing energy, development and climate priorities in India, 30 Nov–1 Dec, New Delhi, India
67. Food and Agricultural Policy Research Institute (2005) Implications of increased ethanol production for US Agriculture, Missouri
68. Msangi S, Sulser T, Rosegrant M, Valmonte-Santos R, Ringler C (2006) Global scenarios for biofuels: impacts and implications. International Food Policy Research Institute, Rome
69. Rajagopal D, Zilberman D (2007) Review of environmental, economic and policy aspects of biofuels. Policy Research working paper. The World Bank, Washington, DC
70. Schmitz A, Seale JL Jr, Schmitz TG (2003) Sweetener-ethanol complex in Brazil, the United States and Mexico: do prices matter? Int Sugar J 105(1259):505–513
71. Braun JV (2007) The world food situation: new driving forces and required actions. CGIAR Annual General Meeting, Beijing
72. Bhardwaj M (2007) Global palm oil prices to rise sharply. Reuters 23 September
73. Regmi A (2001) Cross-country analysis of food consumption patterns. In: Regmi A (ed) Agriculture and trade report. Washington, DC: USDA – ERS, Market and Trade Economics
74. Bamji SM (2007) Nutrition Secure India—how do we get there? Nutrition conclave discusses the way forward. Curr Sci India 93(11):1473–1475
75. Ankumu G (2007) The dark side of biofuels. Climate Network Africa. http://www.cnaf.or.ke/Publications/THE%20DARK%20SIDE%20OF%20%20BIOFUELS. Accessed 14 July 2015
76. Friends of the Earth (2008). Losing ground: the human rights impacts of oil palm plantation expansion in Indonesia. In: Dilworth A, Baird N, Kirby J (eds). Friends of the Earth
77. Hazell P, Braun JV (2006) Biofuels: a win-win approach that can serve the poor. IFPRI Forum. http://www.ifpri.org/pubs/newsletters/ifpriforum/if15.pdf. Accessed 13 July 2015
78. Sukumaran RK, Surender VJ, Sindhu R, Binod P, Janu KU, Sajna KV, Rajasree KP, Pandey A (2010) Lignocellulosic ethanol in India: prospects, challenges and feedstock availability. Bioresour Technol 101:4826–4833
79. Romano SD, Sorichetti PA (2011) Introduction to biodiesel production. In: Dielectric spectroscopy in biodiesel production and characterization. Springer, London
80. Rao PP, Bantilan MCS (2007) Emerging biofuel industry: a case for pro-poor agenda with special reference to India. ICRISAT strategic assessment and development pathways for agriculture in the semi-arid tropics. Policy Brief No. 12, Pattancheru, Hyderabad
81. TERI (2005) Liquid biofuels for transportation: India country study on potential and implications for sustainable agriculture and energy, Report submitted to German Technical Cooperation
82. Duncan J (2003) Costs of biodiesel production. Energy Efficiency and Conservation Authority, Wellington
83. Organisation for Economic Co-operation and Development (2007) Biofuels for transport: policies and possibilities. OECD Policy Brief, Paris
84. Kojima M, Mitchell D, Ward W (2007) Considering trade policies for liquid biofuels. World Bank, Energy Sector Management Assistance Program (ESMAP), Washington, DC
85. Ministry of Petroleum and Natural Gas (2005) Biodiesel purchase policy. Government of India, New Delhi
86. Sood PR (2012) Bioethanol—a viable answer to India's surging energy needs. World Acad Sci Eng Technol 68:1357–1359
87. http://www.rncos.com/Report/IM124.html. Accessed 06 July 2015
88. Shinoj P, Raju SS, Chand R, Kumar P, Msangi S (2011) Biofuels in India: future challenges. Policy Brief. NCAR, New Delhi
89. USDA (United States Department of Agriculture) (2010) Biofuels annual, GAIN report IN 1058. Foreign Agricultural Service, New Delhi

90. NCAER (National Council of Applied Economic Research) (2007) Biodiesel production: institutional constraints. Macrotrack, New Delhi
91. Biodiesel Technologies (2013) Biodiesel scenario in India, October 2013. http://www.biodieseltechnologiesindia.com/recent.html
92. Comprehensive Jatropha Report (2015) http://www.biozio.com/ref/report/jat/Comprehensive_Jatropha_Report_Preview.pdf. Accessed 21 July 2015
93. Datta A, Mandal BK (2014) Use of *Jatropha* biodiesel as a future sustainable fuel. Energy Technol Policy 1:8–14
94. http://www.business-standard.com/content/b2b-chemicals/can-second-generation-ethanol-fuel-india-s-growth-113111500167_1.html. Accessed 11 July 2015
95. MNRE (2009) National policy on biofuels. Ministry of New and Renewable Energy (MNRE), Government of India, New Delhi
96. http://ebtc.eu/pdf/111031_SNA_Snapshot_Biofuels-and-Bio-energy-in-India.pdf. Accessed 1 July 2015

Chapter 21
Perspectives on Climate Change, Water and Land Use, Carbon Accumulation, and Effects on Ecosystem after Biofuel Implementation in India

K. R. Sheetal, S. Prasad, and Anuj K. Chandel

Abstract Concerns over the increasing dependence on energy imports and growing emissions owing to demographic and economic factors have led India to adopt alternative forms of renewable and clean energy, especially biofuels. Though biofuel development provides opportunities for energy security and climate change mitigation, the cultivation of energy crops for biofuel has raised concerns due to the demand for fertilizers, pesticides, water, land, its possible impacts on ecosystems, and also rising food or fuel debate. This chapter tries to assess the implications that production and use of bioethanol and biodiesel have caused in India in the past decade through relevant studies. The greater focus that the current government and the IPCC have put on biofuels also demands that we understand both sides of these fuel resources before they are further expanded. Enhanced scientific understanding, supporting policies, and better technology can go a long way in ensuring that the biofuel implementation achieves its intended purpose.

Keywords Biofuels • Water usage • Land use • Climate change • Mitigation • India

---

K.R. Sheetal (✉)
Regional Research Station, ICAR-Central Arid Zone Research Institute, Bikaner, Rajasthan 334004, India
e-mail: sheetalr69@gmail.com

S. Prasad
Centre for Environment Science and Climate Resilient Agriculture, ICAR-Indian Agricultural Research Institute, New Delhi 110012, India

A.K. Chandel
Centro de Tecnologia Canavieira (CTC), Fazenda Santo Antonio s/n-Caixa Postal 162 CEP, 13400-970 Piracicabaz-Sao Paulo, Brazil

A.K. Chandel, R.K. Sukumaran (eds.), *Sustainable Biofuels Development in India*,
DOI 10.1007/978-3-319-50219-9_21

1 Introduction

Energy is indispensable for industrial and economic growth and activities that provide essential services to improve the quality of life. Today, most of the energy we use comes from fossil fuels, which raise serious environmental concerns [1–3]. Over the years, anthropogenic activities have released large amounts of greenhouse gases into the atmosphere, the majority coming from fossil fuels (Fig. 21.1). Other activities contributing to GHGs include deforestation, industrial activities, and agricultural practices [4, 5]. Hence, sustenance of the social, economic, and environmental growth in the country requires a continuous supply of eco-friendly energy. Substitution of traditional fossil fuels by biomass-based energy has been proposed as a means of reducing greenhouse gas emissions [6, 7]. In this regard, biofuels are a potentially significant energy source for the society in this period of increasing demand for energy, progressive depletion of fossil fuels, and increasing societal risks from climate change.

Bioenergy systems are thought to be carbon neutral since only the CO_2 captured by the feedstock during its growth is returned to the atmosphere on combustion. The hidden environmental costs include that embedded in inputs such as fertilizers and fuel required for biomass production, but there are also savings in the sequestration of carbon in the soil [8, 9]. The reduced dependence on imported oil, improved energy efficiency, lower emissions of health-damaging pollutants, higher water use efficiency, increased opportunities for carbon storage, and carbon credits are all benefits that may be obtained by the proper development and utilization of this bioenergy [1, 10].

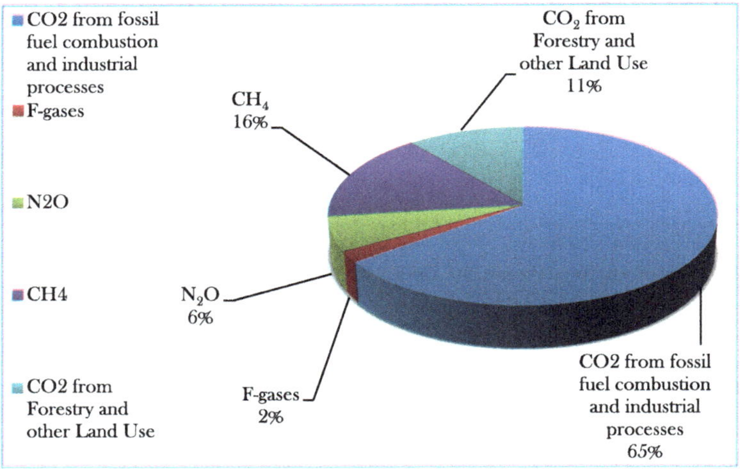

Fig. 21.1 Total annual anthropogenic GHG emissions (Gt CO_2eq/year) by groups of gases 2010. *Source*: IPCC [1], Fifth Assessment Report

For a developing country like India, the introduction of a domestic, eco-friendly, and competitive source of energy is a must so as to meet the multiple challenges of climate change and pollution, at the same time considering the burden on the economy. With India meeting 77.6% of its demand through imports in 2013–2014 (as per MoPNG) and import expenditure increasing manyfold in the past few years due to escalation in demand and prices, biofuels are expected to be pressed into service [2, 3]. Keeping in the view particularly welfare of the country's economic and environmental security, the Indian Government has implemented several policies, viz. Ethanol Blended Petrol Programme (EBPP), National Mission on Biodiesel, and National Policy on Biofuels [11]. The adoption of these policies was a milestone in the march towards sustainable energy, providing incentives to the idea of biofuels and further research in this field. Increasing efforts to mitigate climate change imply an impetus to biofuels, at the same time increasing the complexity of interactions among land use, water use, energy, and biodiversity; but understanding and management of these interactions remain inadequate.

2 Trends in Transport Sector in Developing Nations

The transport sector accounts for the fastest growth of greenhouse gas emissions [1, 12]. The World Business Council for Sustainable Development [13] projects that there would be more vehicles in the developing than in developed nations within the next 15–20 years. If the motorization levels of developing countries increase beyond limits, the absolute number of vehicles may overthrow any minor advances made by cleaner fuels. Regardless of the growth in number of vehicles and thereby emissions, the transport sector has produced rather few mitigation strategies within the mechanisms of the Kyoto Protocol [14].

2.1 Trends in Transport Sector in Indian Cities

A large share of petroleum products in India goes to the transportation sector. A survey conducted by the Petroleum Ministry has revealed that 70% of diesel and 99.6% petrol are consumed by this sector alone [3], with demand expected to grow at 6–8% over the years with the expanding number of vehicles [15]. Indian cities have experienced tremendous growth in registered motor vehicles in the last decade. The total numbers of motor vehicles increased from 52.37 million in 2000 to 159.5 million as in 2012, an average growth rate of 10.5% between 2002 and 2012 [16, 17]. In India, the transport sector emits an estimated 261 Tg of CO_2, of which 94.5% was contributed by road transport [17, 18]. The transport sector in India consumes about 17% of total energy and is responsible for 60% of the greenhouse gas from various activities [19]. The massive size of the nation's motor vehicle numbers has raised concern over the addition of CO_2 to the atmosphere and increase in demand for fuels and its potential for global climate change, thus further showcasing the importance of low carbon-releasing fuels.

3 Global Climate Change Scenarios

The greenhouse effect, the natural phenomenon by which the Earth traps and holds warmth from the sun, is vital to our survival. Primary GHGs, i.e., carbon dioxide, methane, nitrogen oxides, ozone, and others, in the atmosphere absorb and retain heat before it escapes into space [20]. Without it, the Earth's surface temperature would be about 33 °C cooler and unable to support life. Being the dominant GHG, CO_2 is more responsible for the greenhouse effect.

According to the fifth climate change assessment report [1], anthropogenic activity has caused an imbalance in the natural phenomenon of the greenhouse effect and related processes. Retention of additional heat by the increasing concentration of GHGs over the years (Table 21.1) traps more heat, resulting in changes in climatic processes as evidenced by an increase in temperatures, changes in rainfall patterns, rising sea levels, increased ocean acidity, and melting of glaciers. The report [1] reveals that total anthropogenic GHG emissions have continued to grow, with larger increases towards the end of 2010. Despite the large number of mitigation policies, annual GHG emissions rose by 1.0 Gt carbon dioxide equivalent (Gt CO_2eq) (2.2%) per year from 2000 to 2010 compared to 0.4 Gt CO_2eq (1.3%) per year from 1970 to 2000, with highest recorded emission of 49 (±4.5) Gt CO_2eq/year in 2010 [1].

With the increase in the global population and increase in standards of living, emissions are expected to increase, if additional efforts to counter this are not taken. Increases in global mean surface temperature from 3.7 to 4.8 °C compared to preindustrial levels have been predicted in 2100 if additional mitigation measures are not adopted [1], which may lead to adverse impacts on human systems like agriculture and health and leave irreversible effects on the Earth and the ecosystem as a whole. Climate change, in particular, appears to be altering the functions and stability of the Earth's ecosystems [21].

Table 21.1 Recent tropospheric greenhouse gas (GHG) concentrations

GHGs	Pre-1750 level	Recent level	GWP (100-year time horizon)	Increased radiative forcing (W/m^2)
CO_2	280	395.4 ppm	1	1.88
CH_4	722	1893/ 1762 ppb	28	0.49
N_2O	270	326/324 ppb	265	0.17
O_3	237	337 ppb	n.a.	0.40
CFC-11	zero	236/234 ppt	4660	0.061

Source: Blasing [90]

3.1 India's Climate Change Scenarios

India is the sixth largest greenhouse gas emitter, contributing almost 3% of the world's total emissions. India's per capita CO_2 emission is projected to increase to 1.6 tons by 2030. India's huge population, however, aggravates the net emissions into the atmosphere [22]. Global warming is expected to hit India severely and it is also one of the most disaster-prone nations in the world, with much of its population living in regions vulnerable to hazards such as floods, cyclones, and droughts. Hence it assumes the most importance that we limit our emissions to the prescribed levels by using environment-friendly technologies and other options to control this problem.

4 Biofuels as a Leading Alternative to Fossil Fuel

In light of factors including increased occurrence of climate change impacts, increasing prices of crude oil and its declining availability, several alternate fuels are evolving rapidly. Fossil fuels such as coal, petroleum-based products, and natural gas meet most of our current energy demands [2, 3]. With the rapid depletion of fossil fuels, biofuel as a renewable source of energy affords immense potential [23]. Globally, biofuels have attracted much attention since they have become a leading alternative to fossil fuel, are produced domestically by many countries, and require only minimal changes to retail distribution and end-use technologies. By 2022, global ethanol production is expected to increase by 70% compared to the average of 2010–2012 and reach some 168 billion by 2022 [24, 25]. Figure 21.2 shows the projected regional distribution of world ethanol production and use in 2022. Moreover, they have the potential to spur rural development [26].

An advantage of biofuels is their compatibility with the petrol-fuels to be used as blends in existing internal combustion engines (ICE) [27, 28]. Liquid biofuels are similar to petrol/diesel in terms of engine performance and refueling times. The other side includes limits on the percentage of blends and the difficulty in its distribution using existing fuel pipelines. The sustainability of available biomass resources is a serious issue for some biofuels [1].

In its fourth climate change assessment report, biofuels were identified as a "key mitigation strategy" [29]. However, increasing concerns about diversion of food to make fuels, land and water use, replacement of forests, etc. have acted as incentives for the development and implementation of sustainability criteria and frameworks [28]. Furthermore, the support for advanced biorefinery and next-generation biofuel options are driving biofuel to be more sustainable [1].

Recent years have witnessed a rapid expansion of biofuel production worldwide. In India, the Biofuel Policy, dealing with bioethanol and biodiesel, aims to channelize biofuels into the energy and transport sector to address topics like energy

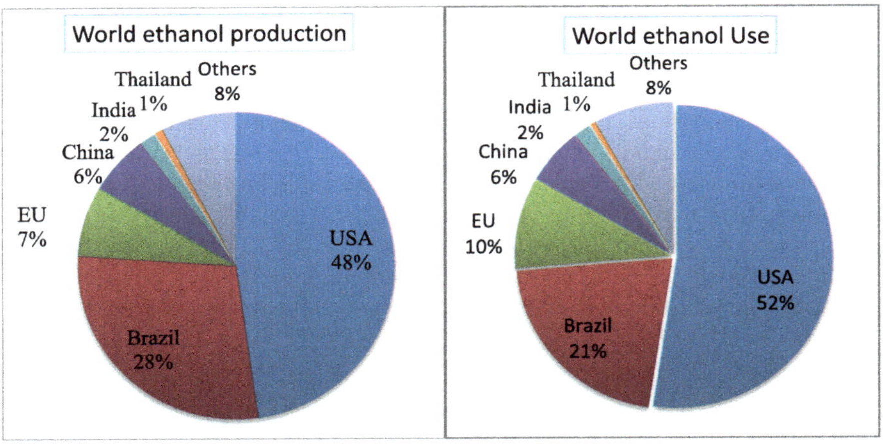

Fig. 21.2 Projected regional distribution of world ethanol production and use in 2022. *Source*: FAO [24, 25]

security, climate change, and rural development [30]. In India, biofuel is one of the best alternatives as the next best substitute for petrol and diesel. India being endowed with significant potential for generating renewable energy, the GOI is promoting the production of both liquid biofuels, i.e., ethanol derived from sugarcane molasses and biodiesel derived from inedible and waste oils for blending with petrol and diesel, respectively [31].

Though the biofuel industry in the country is still in developmental stages, ethanol is the most widely produced biofuel being produced from the abundantly available molasses. Currently, there are 140 distilleries in India, with the capacity to distil around 2 billion liters of conventional ethanol per annum [31, 32]. The production of biodiesel is still commercially negligible. Even though India has an installed biodiesel capacity of 1.2 Mt per annum, presently only 7% is utilized because of restrictions [33].

There is a clear need to identify suitable feedstock, optimum energy conversion methods and policies, and management systems which suit the varying social, economic, and environmental situations existing in different agroecological zones of India [34, 35]. Ensuring energy and food to the population, while keeping in mind international commitments towards climate change, provides an opportunity for the strengthening of the existing linkages between science and policy. Besides direct benefits as fuels, indirect benefits, such as employment generation, reduction of dependence on oil import, carbon sequestration, and rural development, can be additionally gained [34].

4.1 Global Road Transport Share of Biofuels

Ethanol and biodiesel constitute the liquid biofuels for global road transport. Biofuel use increased at an average annual growth rate of 1.5%, 12.1%, and 15.4% for solid biomass, liquid, and gaseous forms, respectively, between 1990 and 2008 [28]. In 2009, biofuels contributed nearly 3% of global road transport, with the production of ethanol and biodiesel increasing by 10% and 9%, respectively [28]. Biofuels, especially the second-generation fuels, are expected to play a crucial role in meeting the needs for transport fuel in the coming years.

4.2 India's Road Transport Share of Biofuels

Biofuels continue to play a significant and increasing role in meeting India's energy requirements. There is an enormous potential of marketing for biodiesel, considering the ever-increasing oil prices. Besides reducing GHG emissions and ensuring energy security, biodiesels from feedstock like *Jatropha* and algae can earn carbon credits under Kyoto Protocol [23]. India being the fourth largest global contributor to carbon emissions, GOI transport policy has targeted EURO-III and EURO-IV vehicle emission norms for vehicles, requiring the adoption of clean and green fuel [28]. The lack of proper baseline and regulation mechanisms has prevented our biofuel projects from contributing to the CDM projects in India [34]. The GOI promotes production and blending of ethanol derived from sugarcane molasses, and biodiesel derived from TBOs and non-edible and waste oils for blending with diesel. The success of GOI's current target of 5% blending of ethanol in petrol is highly correlated with the sugarcane production. The mandate has been partially effective in years of surplus sugar production and unfilled when sugarcane production declines [36]. Even in years with good sugarcane production, failure in setting ethanol pricing formula and procedural delays by state governments have been reported to cause delays in procurement under Ethanol Blending Program.

4.3 Biofuel Blending Mandates

Imposing the quantitative targets in the form of blending mandates, supported by policy measures, research, and development, is the key driver in the development and growth of the biofuel industry. Currently, these targets mandating 5% ethanol blending in petrol are being implemented in the country. In 2008, the target was tried to be raised to 10% blends. In fact, even the 5% blending target has yet to be accomplished [15]. An indicative target of minimum 20% ethanol-blended petrol across the country has been set for the year 2017 through the National Policy on Biofuels formulated by the Ministry of New and Renewable Energy (MNRE).

Mandatory targets for biofuels have led to their rapid adoption worldwide, but problems have emerged connected to its large-scale production. Better technologies that produce biofuels sustainably and economically are currently under scrutiny.

Annual sugarcane production in 2012–2013 was estimated to be around 341 million tons [37]. In a recent report, the US Department of Agriculture (USDA) pegged India's ethanol production at 2170 million liters, of which 400 million liters have been estimated to be blended with gasoline in 2012 [38]. According to Cabinet Committee on Economic Affairs (CCEA), the ethanol blending program is presently being implemented in 20 states and 4 UTs [2, 3] with blending level of about 2%. Currently, the ethanol supply for years 2013 and 2014 is projected to be sufficient to satisfy 2.9% and 2.1% of blending target, respectively [31].

On the other hand, production of biodiesel mainly from Jatropha seeds is commercially negligible. Lower availability of Jatropha seeds (feedstock) due to slow progress in the planting of Jatropha for biodiesel production has led to most of the biodiesel units operating in India to shift to alternative feedstocks. Use of edible oil waste (unusable oil portions), animal fat, and inedible oils enables the biofuel production units to continue operations throughout the year at 40% of their capacity [15, 23, 31].

5 Mitigation Potential of Biofuels in Indian Context

Conversion of biomass feedstock to biofuels and its use as a supplement to petrol fuels are more environmentally friendly than petrol-fuels alone. When we use ethanol instead of petroleum fuel, we help reduce atmospheric CO_2 by avoiding the emissions directly associated with the use of petrol, not releasing CO_2 stored in the fossil fuels, and contributing a mechanism for CO_2 absorption by growing new biomass. Because of their compatibility with the natural carbon cycle, biofuels offer the most beneficial alternative for reducing greenhouse gases from the transportation sector [39].

Currently most biofuel systems, including liquid biofuels, result in GHG emission reductions, while advanced biofuels or next-generation biofuels could provide even higher GHG mitigation [28]. Land use changes and corresponding emissions and removals affect the GHG balance. Biofuel can lead to avoided GHG emissions from residues and wastes in landfill disposals and coproducts; the combination of biofuel with carbon capture and storage (CCS) may provide further reductions. The GHG implications related to land management and land use changes in carbon stocks, however have considerable uncertainties [9].

Advanced conversion technologies including the use of biomass residues and long-lived plantations can lead to 80–90% reduction in GHG emissions compared to the fossil fuel energy usage levels [28]. Figure 21.3 shows a range in reductions of GHG emissions per vehicle-km (v-km) obtained from various studies. Larson et al. [40] assessed emissions from petroleum fuels, first-generation biofuels (sugar- and starch-based ethanol, oilseed-based biodiesel), and some second-generation

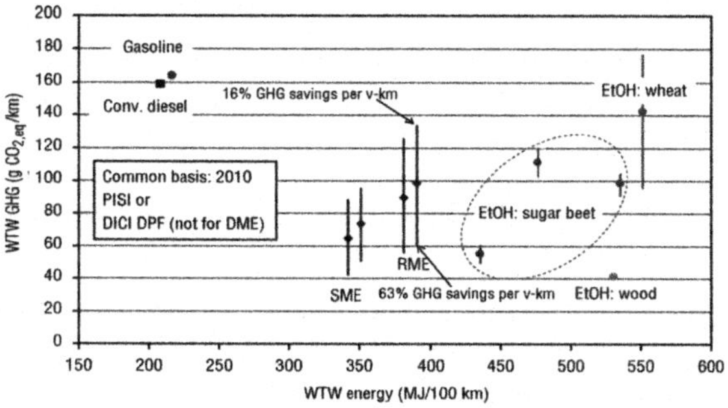

Fig. 21.3 Well-to-wheels energy requirements and greenhouse gas emissions for conventional biofuel pathways compared with gasoline and diesel pathways, assuming 2010 vehicle technology. Note: *EtOH* ethanol, *SME* soy methyl ester, *RME* rape methyl ester, *PISI* port injection spark ignition, *DICI DPF* direct injection compression ignition with diesel particulate filter. *Source*: [40]

biofuels derived from lignocellulosic biomass (ethanol and Fischer-Tropsch diesel) on a well-to-wheel basis.

Substituting biofuels for petroleum-based fuels was found to have the potential to reduce emissions directly linked with the fuel supply mechanism. Second-generation biofuels (with life cycle GHG emissions between −10 and 38 g CO_2eq/MJ) were reported to provide greater mitigation potential over first-generation biofuels (with life cycle GHG emissions between −19 and 77 g CO_2eq/MJ) compared to 85–109 g CO_2eq/MJ for petroleum fuels [40]. Estimates of GHG emissions vary for both biofuels and petroleum fuels, largely due to assumptions about technological issues and where and how the feedstocks are produced. Land and forest conversions for biofuel plantation lead to massive losses of carbon stock effects that can lessen the net positive GHG mitigation impacts [41].

5.1 Carbon Cycle and Biofuels

Consistent with our knowledge of the lifetime of a molecule of CO_2, the most efficient form of climate change mitigation is to suppress carbon emissions from all sources. The assumption that emissions can be offset by the uptake of CO_2 in land systems in the long term is to be changed, and utmost care should be taken by each individual to reduce fossil fuel use and thus ultimately emissions [42].

CO_2 is part of the Earth's natural carbon cycle, which circulates carbon through the atmosphere, biosphere, hydrosphere, and lithosphere, maintaining a life-sustaining and delicate natural balance between storing and releasing carbon. By using biofuels such as bioethanol and biodiesel for transportation, we can help restore the natural balance of CO_2 in the atmosphere. Besides displacing fossil fuels, the biofuel feedstock uses CO_2 to grow. Thus, much of the CO_2 released when biomass is transformed into a biofuel and burned in vehicle engines is recaptured when new biomass is grown to produce more biofuels [39].

5.2 *Biofuel Net Energy and Carbon Balances*

The actual benefits which may be realized from biofuels depend on the energy and carbon balances, indicating the magnitude of fossil fuel inputs (and relative GHG emission) relative to fossil fuel savings (and avoided GHG emissions) due to its use as alternative fuels [28]. A study compared life cycle impacts of a fossil fuel reference system with Jatropha biodiesel system (based on small-scale, low-input Jatropha system grown on degraded land, which was unsuitable for cultivation of food crops). This study showed that the production and use of Jatropha biodiesel trigger an 82% decrease in non-renewable energy requirement and a 55% reduction in global warming potential (GWP) compared to the fossil-fuel-based system [43]. The another study by Confederation of Indian Industry [44] calculated the net energy balance and carbon balance for selected categories of biofuels which are summarized in Table 21.2. The study shows that the net energy and carbon balance per year is highest for Jatropha-based biodiesel. The significant energy contribution from the co-products (seed husk, seed cake, and glycerol) obtained during biodiesel production contributes almost half of the total energy generated during the end-use stage. Ethanol from sweet sorghum was observed to be most efficient in input-to-output energy conversion [44].

Some studies show that change in land use from native vegetation to biofuel crops leads to significant greenhouse gas emissions and negative carbon balance or carbon debt [39, 45, 46]. Only the low-carbon ecosystems, when cleared for food-based biofuel crops, have a payback time of less than 10 years, because crops like sugarcane are highest yielding crops [46]. Reducing GHG emissions, thus

Table 21.2 Net energy balance and carbon balance for selected categories of biofuels

Biofuel type	Feedstock	Net energy ratio	Net energy balance (GJ/kl)	Net carbon balance (tons CO_2eq/kl)	% C emission reduction
Bioethanol	Molasses	4.57	19.11	−1.1	75
	Sweet sorghum	7.06	21.57	−1.4	86
	Cellulosic biomass (bagasse)	4.39	25.41	−1.7	70
	Cellulosic biomass (rice straw)	3.32	22.79	−1.6	68
Biodiesel	Jatropha-transesterification	5 3.41	63.76	−4.0	30
	Jatropha—SVO	4.38	66.73	−4.5	50

Source: CII [44]

contributing to the common goal of climate change mitigation, is now widely recognized as an important driver of biofuel development. Many target-based biofuel policies specify that biofuels counted towards targets must lead to net GHG emission reductions.

5.3 Greenhouse Gas Mitigation Potential of Ethanol

The influence of biofuels on the slowdown of climate change also provides motivation to its production. Theoretically, the net greenhouse gas emissions from biofuels may approach zero because the carbon emitted while burning was sequestered throughout the photosynthesis process. Most studies have found that the use of first-generation biofuels results in emission reductions of 20–60% of CO_2eq relative to fossil fuels. Expected reductions for future commercialized second-generation biofuels are in the range of 70–90% of CO_2eq relative to fossil fuels [47]. In a study by Kadam [48], the potential greenhouse values for burning as a disposal option versus conversion to ethanol of excess of sugarcane bagasse were compared. Lower net values for CO, hydrocarbons (except methane), SOx and NOx, particulates, CO_2, and methane were observed. LCA performed in this study demonstrated the potentially significant benefits of using ethanol derived from excess bagasse.

While the first-generation biofuels had minor contributions to carbon emission mitigation and increased the food vs. fuel debate, the second generation of biofuels is expected to provide more benefits. The use of cellulosic biomass for energy production is projected to result in significantly higher carbon sequestration compared to starch- and sugar-based biofuel [49, 50]. However, the environmental impact of biofuels can have a positive or adverse effect relative to gasoline, depending on how the fuel is grown, processed, and then used [49]. To distinguish these cases, and the myriad of other feedstock to fuel pathways, clear standards, guidelines, and models are needed.

5.4 Greenhouse Gas Mitigation Potential of Biodiesel

Several studies have underlined that using Jatropha biodiesel reduces greenhouse gas (GHG) emissions by about 8–88% compared to the use of fossil diesel [43, 51, 52]. However, if Jatropha is cultivated on former primary forest land, the impacts are most likely negative and lead to carbon debts [53, 54]. Sudhakar et al. [23] have given a general formula for the amount of the CO_2 emission by biodiesel and petrodiesel as diesel generates 2.67 kg of CO_2 per liter; biodiesel produces 0.73 kg of CO_2 per liter and hence CO_2 emission reduction by biofuel substitution = 1.94 kg of CO_2 per liter.

USEPA [55] calculated the emissions from soy-biodiesel as the percentage of petrodiesel emission and found a significant reduction in carbon monoxide, particulate matter, PAH, and unburnt hydrocarbons; no emission of SOx; and a slight increase in NOx. In a study by Gmunder et al. [56] in Andhra Pradesh, the environmental impact of Jatropha biodiesel value chain was benchmarked with fossil diesel, following the ISO 14040/44 life cycle assessment procedure. Primarily, the study showed that the use of Jatropha biodiesel reduces the global warming potential and the non-renewable energy demand as compared to fossil diesel [56]. A comprehensive evaluation of the use of 7.7% ethanol-blended fuel on the bus fleet under the Karnataka State Road Transport Corporation (KSRTC) was conducted, estimating the reduction in air pollution emissions and its effect on the air quality of the environment. The results showed considerable reduction in particulate matter, CO, CO_2, and NOx [57, 58].

5.5 Greenhouse Gas Mitigation Potential of Algal Biofuels

Ensuring sustainability of use of biofuels poses a challenge to our nation. The boon of different climatic conditions providing alternative oilseed sources, such as Jatropha, Pongamia, and crops such as tropical sugar beet, is confronted by the problem of pressure on land for food and fuel production. This coupled with the ever-growing demand for renewable cleaner energy gives the perfect opportunity to discover new biomass [59].

In this regard, algae have been reported as a very promising source of biomass for biodiesel production. Apart from sequestering carbon from the atmosphere and industrial gases during growth, it also utilizes nutrients from industrial effluents and wastewater efficiently [60]. Another added advantage of growing algae is its lack of fertile land requirement, being able to grow under unfavorable conditions using brackish and saline waters unfit for irrigation, thus avoiding competition for agriculturally important resources [60, 61]. Consequently, algal biomass cultivation provides twofold benefits of providing biomass for the biofuel production and also mitigating air and water pollution by absorbing carbon for growth and reducing the GHG that might have occurred if the wastewater was conventionally treated [62]. This is further proved by life cycle assessment (LCA) of algal biofuels, which substantiates its eco-friendliness over fossil fuels; but further improvements are still needed for the economic utilization of this resource in India [63].

6 Impact of Biofuels on Land, Water Use, Ecosystems, and Carbon Accumulation

The ongoing climate change concerns have spurred substantial interest in biofuels, and supporters of biofuels recognize that they are considered naturally carbon neutral. Critiques, on the another hand, cry that the large-scale generation of biofuels can not only strain agriculture but also threaten future food security [64]. People who live in the drylands of India are often faced with challenges and limitations of poverty. Inconsistent rainfall and droughts, unreliable irrigation water supply, and poor soil conditions are major challenges facing people living under marginal environmental conditions.

Research has shown that the potential CO_2 emission from land conversion to first-generation biofuel crops is likely to be greater than the gains expected from the initial years of growing biofuel crops [41]. However, minimal conversion of habitat for biofuel production including waste utilization, cover crops, and using wastelands can lead to biofuels having a more positive effect on mitigating climate change and enhancing environmental quality [41]. Further research, creating awareness among people, enforcement of viable production, and management practices and policies can go a long way in making this renewable source an everlasting energy source [65].

It is also important to promote exploration of technologies and practices that could lead to minimizing the GHG emissions and other adverse effects on land conversion and use for biofuel production over the long term.

6.1 Land Use and Biofuels

Biofuels can play a pivotal role in bringing increased energy security in India. However, the growing population and its needs for food and pastures cause friction with the notion of diversion of productive lands for fuel crops [66]. On the other side, large-scale biofuel cultivation could also provide benefits in the form of employment and rural development [67]. However, these opportunities depend on the security of land tenure. Studies have emerged which show that this policy has been used to divert areas already occupied by Prosopis trees (though listed as marginal lands) into Jatropha plantations in the name of energy and environment security [68]. In these contexts, where competing resource requirements exist between rural community, authorities, and biofuel producers, under unsuitable conditions, the ill-managed allotment of land for biofuel cultivation may result in loss of their livelihood sources by poor [69]. Moreover, it has significant undesirable effects on food security and economic and social life of people in such areas.

Meeting the 20% blending mandate of the biofuel policy implemented by the Government of India (GOI) requires setting aside 140,000 km^2 of land, where at present, fuel-yielding plants cover less than 5000 km^2. The controversy between land area requirement and land category for growing biofuels is at the heart of the debate on the environmental and economic impacts of biofuel production. Land used for biofuel production was estimated to be approximately 13.8 Mha in 2004, accounting for about 1% of current cropped area [70], which increased to 26.6 Mha in 2007 in an estimate by Ravindranath et al. [41]. One of the salient features of the biofuel policy is its aim to produce biofuel from non-feedstock grown on degraded or wastelands not suited to agriculture, thus avoiding any possible conflict with food

Table 21.3 Comparison of oil yield of biodiesel crops and land required to meet targets of biofuel program

Crop	Oil yield (l/ha)	Land area needed (mha)
Corn	172	1594
Soybean	446	594
Canola	1190	223
Jatropha	1892	140
Coconut	2689	99
Oil palm	5950	45
Microalgae	58,700–136,900	2–4.5

Source: Sudhakar et al. [23]

security. Thus, meeting the blending mandates without a further significant increase in requirement for land requires careful selection of suitable oil-yielding crops since each crop varies in its oil yield, time needed, etc. [23]. Table 21.3 provides a summarized comparison of oil yield of biodiesel crops and land required to meet targets of biodiesel program.

While the ethanol production has shown some progress, the biodiesel industry in India has remained in its infancy, despite the fact that demand for diesel is five times higher than that for petrol. In India, biodiesel production is focused on Jatropha. Of the 55.3 Mha of land in India classified as wastelands, 32.3 Mha have been calculated to be suitable for Jatropha cultivation. The insufficient production of Jatropha seeds led to the failure of the government mandate of 20% diesel blending by 2012 [71]. Currently, 65–70% of the 0.5 million hectares of wastelands occupied by Jatropha are new plantations of less than 3 years [71]. Even if Jatropha is promoted on marginal land, existing activities such as livestock grazing and gathering of forest products by local communities will be displaced [56]. The loss of related ecosystem services affects particularly subsistence farmers and the rural poor. On the other hand, Jatropha can help prevent desertification and improve the ecosystem functions of marginal land. Proper analysis of the capability of the land and the effect should be understood before resorting to large-scale modifications of existing land uses.

Wastelands in some parts of India may be more suitable for the cultivation of other oilseed crops such as Pongamia or Neem [34]. However, the allocation of a major fraction of wastelands for the establishment of Jatropha not only creates a significant dependence on wastelands for biofuel production but also constrains the use of other potentially better suited species. There are at present about 5–6 plants with the capacity to produce 10,000–250,000 metric tons of biodiesel per year in India able to use alternative feedstock such as edible oil waste, animal fats, and inedible vegetable oils, at times of deficiency of Jatropha feedstock [71]. Comprehensive studies on land use changes and its effects, due to biofuel production and related impacts, are not yet available.

Sugarcane in India is principally a food crop used for sugar production, and thus may have less potential for meeting the biofuel requirement. Currently, the ethanol production in India is mostly, low sugar molasses based. Growing sweet sorghum is a feasible option for marginal lands, though the yields are likely to be low [6]. Land-use change resulting from biofuel production may also contribute considerably to emissions, becoming a major driver of climate change.

Such interests have led to the exploration of other biofuel sources. Microalgae have reported highest potential producing 15–300 times more biodiesel than oilseed crops on the area basis [61]. They have added advantage that they can be grown on domestic or industrial wastewaters, without the requirement of the extra land area [62]. Bajhaiya et al. [72] further reported that microalgae cultivation on less than 2–3% of the total land area can fulfill the nation's biofuel mandate. Algal biodiesel is also reported to be more stable compared to edible and non-edible oil-based fuels for a longer period of time [73]. It may be possible to produce a large proportion of fuels using an optimum combination of advanced biofuel technologies, also keeping in view the land area availability and potential of different biofuel sources without affecting food supply [24, 25].

6.2 *Water Use and Biofuels*

As per the Ministry of Agriculture, the primary land usage is for agricultural crop production followed by forests and wastelands [6]. Irrigated land is just 40% of the total area. Changes in the hydrological cycle and rainfall patterns due to climate change can lead to many consequences and risks, and they are conditioned by and interact with non-climatic drivers of change and water management responses. Water is the critical factor that delivers many of the impacts of climate change on society, for instance, to the energy, agriculture, and transport sectors [1]. Large-scale biofuel production affects water quality in many ways: primarily by the direct use in growing crops and its processing into fuels, and at the same time polluting surface and groundwater by the release of chemical fertilizers, pesticides, and other process by-products. Hence, ambitious plans to increase biofuel production should be carried forward after considering water demands.

Sugarcane, which is a major feedstock for generating ethanol in India, also puts stress on the already depleting water resources of our country, having a water requirement of 20,000–30,000 $m^3ha^{-1}crop^{-1}$ [15]. On the other hand, Jatropha can grow on marginal lands with only a little rainfall. It is generally cultivated in monoculture plantation, by means of intercropping or around another crop field as a hedge. Although Jatropha has the potential to grow rain-fed in water-scarce areas, increased yields can be obtained through irrigation, and consequently use relatively large amounts of water, thus competing with resources for agriculture [56, 74]. A study in Tamil Nadu reported that water demands of Jatropha can potentially exacerbate the conflicts and competition over water access in its villages [75].

The lift-irrigation model followed in the dryland area of India to grow food and other crops can be utilized for the expansion of biofuel crops that has the potentiality to eliminate poverty among farming societies if appropriate sustainable development measures are carefully implemented [64].

6.3 Ecosystem Impact of Biofuels

The importance of ecosystem-based management for biodiversity conservation, instead of establishing tree plantations, was highlighted in the Eleventh Plan (2007–2012). The introduction of large-scale monospecific plantations of alien species like Jatropha, in the biodiversity-rich states of Chhattisgarh, Uttaranchal, Mizoram, etc., is of great concern. The establishment of such plantations has the potential to replace forests and common property resources such as village forests, marginal grazing lands, and watershed drainage which provide essential substances and services to landless people in the area [76].

Due to the high population pressure, the extent of true "wastelands" remaining unused in India is also unclear. There is a possibility that these lands are actually being used, not for the cultivation of agricultural crops but for other purposes like subsistence crops or livestock grazing [15, 68]. Even if Jatropha is promoted on marginal land, the displacement of existing land-use patterns by locals, such as grazing and gathering of firewood, will change, affecting poor and subsistence farmers.

Even though there may be benefits to climate change mitigation efforts and energy security, Jatropha cultivation is reported to have the higher impact on acidification and eutrophication than by the use of fossil fuels [43, 52]. Terrestrial acidification is caused by ammonia and nitrogen oxide emissions due to fertilizers and combustion of the biofuel, respectively [56]. Thus, there are chances for the environmental burden to shift from global warming to other impacts if biofuels replace fossil fuels. There are also some studies suggesting that allelo-chemicals present in leaf and ovary wall of Jatropha are slightly inhibitory, even affecting native flora [77, 78]. Hence, there is the need for further analysis of damaging effects due to the accumulation of these undesirable components to be studied on large-scale plantations.

Jatropha plantations, at the same time, can also create environmental benefits such as the protection of crops or pasture lands, and hedge for erosion control and as a source of organic manure [34]. On the other hand, the development of more efficient biomass conversion technologies may make it economically feasible to replace natural forest lands with biofuel crops. If this happens, biofuel production will pose a serious threat to biodiversity. This highlights the importance of continual research and development on policies concerning biofuel production, use, and trade. In particular, policy instruments to enhance the traceability of biofuel feedstock need to be developed to ensure that they are produced in environmentally and socially responsible ways [79].

The environmental sustainability of Jatropha production and processing is still not clear as evidenced by the huge variation of different LCA studies. Studies by several researchers have shown that the answer highly depends on the former land use, intensity of the cultivation, efficiency of processing, and the usage of Jatropha products and by-products [9, 56, 80].

6.4 Carbon Accumulation Potential Through Biofuels

Combining biomass conversion technologies to biofuels with carbon capture and storage intensifies the possibility of achieving GHG removal from the atmosphere, necessary for substantial GHG emission reductions in the long term. Growth and carbon sequestration are an important distinction between biomass and fossil fuels, as sequestration offsets the emissions released during conversion of biomass to energy. However, the net emission associated with biomass use for energy services is a complex issue due, in particular, to potential changes in land use and land management [81–83].

A study by Gmunder et al. [56] found that Jatropha under intensive cultivation system accumulated a carbon stock of about 10-ton carbon per hectare over 20 years, similar to another study [84] of carbon stock of a Jatropha plantation in southern India. Another study by Francis et al. [22] has projected a CO_2 sequestration potential in Jatropha biomass of 4.6 and 22.9 Mt year^{-1} if 2 and 10 mha of wastelands in India are cultivated with Jatropha. Such plantations, when managed with regular pruning, could store 8–10 tons C ha^{-1} in above-ground biomass and litter in India [84]. The researchers also estimated an average annual carbon sequestration rate of 2.25 CO_2 tons ha^{-1} year^{-1} for Indian wastelands planted with Jatropha [22].

In March 2015, Bangalore-based enterprise VayuGrid entered into an MoU with New Renewable Energy Development Corporation of AP (NREDCAP) for Andhra Pradesh state to develop Panchayat-level bioenergy zones (BEZs) for green power generation on wastelands, to generate 10 MW and 14 million liters of biodiesel. Each zone is forecasted to become a huge carbon sink, sequestering nearly 30,000 tons of carbon annually, thus making each BEZ eligible for carbon credits and become a significant player in the cap and trade scheme [85]. While Jatropha being planted on previously, forested land is alleged to lead to carbon debts [53, 54], and biofuels made from waste or plants grown on degraded lands incur little or no carbon debt and can offer immediate GHG advantages [45]. There are also several studies to assess the potential of microalgae for carbon sequestration, utilizing nutrients from effluents and wastewater at industrial sites. Sahoo et al. [86] reported that use of macroalgae could incorporate an average of 0.26×10^6 tons C into the harvested biomass annually. However, different types of algae vary in their carbon sequestration potential. Thus these algal resources can be used for biofuel production and climate change mitigation.

7 Intergovernmental Panel on Climate Change Recommendations

The Intergovernmental Panel on Climate Change reports over the years have played a role in reflecting the research and development trends of biofuels in the world. The Fourth Assessment Report claimed biofuels to be a "key mitigation strategy." The use of biomass to replace fossil fuels was viewed to have a huge role in reducing GHG emissions from the transport sector. This report also highlighted that land-use conversion and cultivation of biofuel crops could have significant (positive and negative) implications for food security, biodiversity, and water [29]. However, in its latest report, the IPCC shifted stands on biofuels, conceding that the value of reducing greenhouse gas emissions may be overlooked due to the negative impacts [87].

The IPCC Fifth Assessment Report confirms that biofuel has a massive positive role to play in society but requires more research to realize its full potential. Commercially available liquid and gaseous biofuels already provide co-benefits together with mitigation options that can be increased by technological advances. The report also confirms the beneficial role of biofuel in stimulating economic development by raising and diversifying farm incomes and generating rural employment [1]. It is clear that biofuels were never going to be a panacea. At best they were going to contribute 10% of the world's energy needs, because it would be unlikely to produce sufficient crops for biofuel, given the need to use the land for other purposes.

The report concludes that existing uncertainties about biofuel should not preclude society from pursuing beneficial biofuel options that are available. There is potential for biofuels to provide a lower carbon solution to energy needs and be part of the arsenal of strategies to overcome the risks of climate change. However, past research confirms the challenges ahead if biofuels are to meet this promise without worsening other global problems.

8 Future of Biofuels

As technology evolves, research improves, and policy responds, the competitiveness of biofuel with other alternative technologies will transform as a function of economics and environmental concerns beyond those considered now. Ethanol, which initially replaced MTBE as a groundwater-friendly oxygenate, is now competing with gasoline as an energy carrier. Cellulosic ethanol may eventually compete for vehicle miles with bio-electricity. Price signals to small-scale farmers could significantly increase both yields and incomes, securing real, long-term poverty elimination in countries that have a high dependence on agricultural commodities [67].

The Government of India has also shown considerable interest and support to remove impediments to the growth of the biofuel industry, with the Union Cabinet chaired by the Prime Minister, Mr. Narendra Modi, approving its direct sale by private manufacturers and suppliers as a measure to encourage production and marketing of biofuels [88]. The Prime Minister also sought collective R&D effort and collaboration towards the clean energy path at the G20 summit held in Brisbane, Australia, in 2014. He also met the Australian researchers to understand the technological advancements in agriculture and production of bio-based fuels in Australia [89]. Also with the consideration regarding the allowance of 5% biodiesel blending by bulk users such as railways and defense establishments, the blending situation is expected to improve and in turn reduce India's dependence on crude oil to some extent [33].

The ambitious target of blending 20% of biofuel by 2017 could be met by cultivating the wide varieties of oil-yielding crops suitable to the particular climatic conditions in different parts of India. The Clean Development Mechanism benefits (Carbon Credits) form a critical part of the biofuel projects. A truly integrated strategy will have to be followed to bring out a sustainable balance between the targets of biofuel mission and the consideration for India's social and environmental concerns.

9 Conclusion

The long-term predictions for biofuels lie in uncertainty. The current rush to meet policy targets has sometimes distorted agricultural food production and may have significant adverse environmental and social consequences. On the other hand, new technologies hold the promise of much more efficient biofuels in the future. However, these technologies remain as yet unrealized, and may not be technologically or economically viable in the near future. Moreover, we cannot simply base the present stress on biofuels on the potential for future scientific progress. Apparently, biofuels will play a significant role in the suite of alternative energy technologies used to combat climate change and obviate overreliance on crude oil, as new research makes it economically and technologically feasible.

References

1. IPCC (2014) Summary for policymakers. In Edenhofer O, Pichs-Madruga R, Sokona Y, Farahani E, Kadner S, Seyboth K, Adler A, Baum I, Brunner S, Eickemeier P, Kriemann B, Savolainen J, Schlomer S, von Stechow C, Zwickel T, Minx JC (eds) Climate change 2014, mitigation of climate change. Contribution of working group III to the fifth assessment report of the intergovernmental panel on climate change. Cambridge University Press, Cambridge
2. MoPNG (2014a) Energizing the nation. Annual report 2013–14. GOI. Available at http://petroleum.nic.in/docs/Annual_Report/AR13-14.pdf

3. MoPNG (2014b) Press Information Bureau. The Government of India. Ministry of Petroleum & Natural Gas. Available at http://pib.nic.in/newsite/PrintRelease.aspx?relid=102799
4. Lal R (2014) Biofuels and carbon offsets. Biofuels 5(1):21–27
5. Prasad S, Kumar A, Muralikrishna KS (2014) Biofuels production: a sustainable solution to combat climate change. Indian J Agric Sci 84(12):1443–1452
6. Dubey A, Neetu M, Vishakha B (2014) Biofuels: impact on food productivity, land use, environment, and agriculture. Int J Environ Res Dev 4:9–16
7. Hall DO, Rosilio-Calle F, Williams RH, Woods J (1993) Biomass for Energy: Supply prospects. In: Johansson TB, Kelly H, Reddy AKN, Williams RH (eds) Renewable energy: sources for fuels and electricity. Island Press, Washington, DC
8. Brown RA, Rosenberg NJ, Hays CJ, Easterling WE, Mearns LO (2000) Potential production and environmental effects of switchgrass and traditional crops under current and greenhouse-altered climate in the central United States: a simulation study. Agric Ecosyst Environ 78:31–47
9. Kammen DM, Farrell AE, Plevin, RJ, Jones AD, Nemet GF, Delucchi MA (2008) Energy and greenhouse impacts of biofuels: a framework for analysis. Report from UC Berkeley Transportation Sustainability Research Center. Institute of Transportation Studies, UC Berkeley. Available at http://escholarship.org/uc/item/7zg2x23t
10. Attal MM, Dave MC, Mahalle A, Sunkavally S, Shaik M, Babu A et al (2014) Experimental investigation on performance of a diesel engine using *Garcinia indica* (Kokam) and rice bran oil based biodiesels as fuels. IOSR J Mech Civ Eng 11(5):1–7
11. Basavaraj G, Rao PP, Reddy CH, Kumar AA, Rao PS, Reddy BVS (2012) A review of the national biofuel policy in India: a critique of the need to promote alternative feedstocks. Working Paper Series no. 34 RP—Markets, Institutions, and Policies. ICRISAT, Patancheru, Andhra Pradesh. Available at http://oar.icrisat.org/6520/1/WPS_34.pdf
12. Wright (2004) The limits of technology: achieving transport efficiency in developing nations. University College London. Available at: http://discovery.ucl.ac.uk/108/1/Lloyd_Wright,_Bonn,_Germany,_Transport_and_climate_change.pdf
13. WBCSD (2004) Mobility 2030: meeting the challenges to sustainability. World Business Council for Sustainable Development, Geneva.
14. Wright L, Fulton L (2005) Climate change mitigation and transport in developing nations. Transp Rev 25(6):691–717
15. Bandyopadhyay KR (2015) Biofuel promotion in India for transport: exploring the grey areas. TERI Policy Brief. Issue 16. pp 1–11. Available at http://www.teriin.org/policybrief/docs/biofuel.pdf
16. MoSPI (2015) Available at http://mospi.nic.in/Mospi_New/upload/SYB2015/CH-20-MOTOR%20VEHICLES/MOTOR%20VEHICLE-WRITEUP.pdf. Retrieved on 9 Apr 2015
17. Sharma RD, Jain S, Singh K (2011) Growth rate of motor vehicles in India-impact of demographic and economic development. J Econ Soc Stud 1(2):137–150
18. Vinish K (2005) Vehicular pollution control—concept note. Centre of Excellence in Environmental Economics, Madras School of Economics, Chennai. Available at http://coe.mse.ac.in/dp/vinish_vehi.pdf
19. Tedoy R (2008) National Urban Policy and its implication, vol 3. Transport Research, New Delhi, ND, pp. 13–17
20. Steinfeld H, Gerber P, Wassenaar T, Castel V, Rosales M, de Haan C (2006) Livestock's long shadow: environmental issues and options. Food and Agriculture Organization of the United Nations, Rome
21. Lovelock J (2009) The vanishing face of Gaia. Basic Books, New York
22. Francis G, Edinger R, Becker K (2005) A concept for simultaneous wasteland reclamation, fuel production, and socio-economic development in degraded areas in India: Need, potential, and perspectives of Jatropha plantations. Nat Res Forum 29:12–24
23. Sudhakar K, Anand T, Srivastava T, Premalatha M (2013) Assessment of carbon mitigation potential of various biofuels in Indian context. Int J ChemTech Res 5(5):2456–2461

24. FAO (2013a) Biofuels and sustainability challenge: a global assessment of sustainability issues, trends, and policies for biofuels and related feedstocks. Available at http://www.fao.org/docrep/017/i3126e/i3126e.pdf
25. FAO (2013b) Biofuels. OECD-FAO Agricultural Outlook 2013. Available at http://www.fao.org/fileadmin/templates/est/COMM_MARKETS_MONITORING/Oilcrops/Documents/OECD_Reports/OECD_2013_22_biofuels_proj.pdf
26. Rajagopal D, Sexton SE, Roland-Holst D, Zilberman D (2007) Challenge of biofuel: filling the tank without emptying the stomach? Environ Res Lett 2(4):1–9
27. Planning Commission (2003) Report of the committee on the development of biofuel. GOI, New Delhi. Available at http://planningcommission.nic.in/reports/genrep/cmtt_bio.pdf
28. SRREN (2011) IPCC special report on renewable energy sources and climate change mitigation. Summary for Policymakers. 11th Session of Working Group III of the IPCC, Abu Dhabi, United Arab Emirates. Available at https://www.ipcc.ch/pdf/special-reports/srren/SRREN_FD_SPM_final.pdf
29. IPCC (2007) Mitigation of climate change: technical summary. IPCC, Geneva
30. Dayal A (2013) The curious case of India's bioethanol (biofuel) policy: doomed for failure? Oxford India Policy Blog. Available at http://policyblog.oxfordindiasociety.org.uk/2013/04/02/the-curious-case-of-indias-bioethanol-biofuel-policy-doomed-for-failure/
31. USDA GAIN Report (2013) India biofuels annual. Available at: http://gain.fas.usda.gov/Recent%20GAIN%20Publications/Biofuels%20Annual_New%20Delhi_India_8-13-2013.pdf
32. India Biofuels Annual (2013) Available at http://www.agrochart.com/en/news/news/071113/india-biofuels-annual-aug-2013/
33. Economic Times (2014) Oil ministry circulates cabinet note on allowing 5 per cent blending of environment-friendly biofuels. Available at http://articles.economictimes.indiatimes.com/2014-11-06/news/55835735_1_bio-diesel-project-bio-diesel-use-bio-diesel-association
34. Das S, Priess JA, Schweitzer C (2010) Biofuel options for India—perspectives on land availability, land management, and land-use change. J Biobased Mater Bioenergy 4:1–13
35. Kammen DM, Farrell AE, Plevin RJ, Jones AD, Nemet GF, Delucchi MA (2008) Energy and greenhouse impacts of biofuels: a framework for analysis. OECD research roundtable—energy and greenhouse gas impacts of biofuels. Available via http://its.berkeley.edu/sites/default/files/publications/UCB/2008/TSRCRR/UCB-ITS-TSRCRR-2008-1.pdf
36. USDA GAIN Report (2014) India biofuels annual. Available at: http://gain.fas.usda.gov/Recent%20GAIN%20Publications/Biofuels%20Annual_New%20Delhi_India_7-1-2014.pdf
37. CACP (2014) Price policy for sugarcane: 2015–16 Sugar season. Department of Agriculture and Cooperation , Ministry of Agriculture, Government of India, New Delhi Available at http://cacp.dacnet.nic.in/ViewReports.aspx?Input=2&PageId=41&KeyId=536
38. Economic Times (2012) India's 2012 ethanol output likely to rise 29%. Available at http://articles.economictimes.indiatimes.com/2012-06-27/news/32441337_1_ethanol-output-ethanol-production-sugar-production
39. US DoE 1999 Biofuels a solution for climate change. U.S. Department of Energy Office of Energy Efficiency and Renewable Energy Office of Transportation Technologies. DOE/GO-10098-580 Revised Sept 1999
40. Larson E (2006) A review of life-cycle analysis studies on liquid biofuel systems for the transport sector. Energy Sustain Dev 10(2):109–126
41. Ravindranath NH, Manuvie R, Fargione J, Canadell JG, Berndes G, Woods J, Watson H, Sathaye J (2009) Greenhouse gas implications of land use and land conversion to biofuel crops. In: Howarth RW, Bringezu S (eds) Biofuels: environmental consequences and interactions with changing land use. Scientific Committee on Problems of the Environment (SCOPE), Island Press, New York. Online access: http://cip.cornell.edu/biofuels
42. Mackey B, Prentice IC, Steffen W, House JI, Lindenmayer D, Keith H, Berry S (2013) Untangling the confusion around land carbon science and climate change mitigation policy. Nat Clim Chang 3:552–557

43. Achten WMJ, Almeida J, Fobelets V, Bolle E, Mathijs E, Tewari DN, Singh VP, Verchot LV, Muys B (2010) Life cycle assessment of *Jatropha* biodiesel as a transportation fuel in rural India. Appl Energy 87(12):3652–3660
44. CII (2010) Estimation of energy and carbon balance of biofuels in India. Confederation of Indian Industry. Available at http://www.cii.in/webcms/Upload/Energy_Balance.pdf
45. Fargione J, Hill J, Tilman D, Polasky S, Hawthorne P (2008) Land clearing and the biofuel carbon debt. Science 319:1235–1238
46. Gibbs HK, Johnston M, Foley J, Holloway T, Monfreda C, Ramankutty N, Zaks D (2008) Carbon payback times for crop-based biofuel expansion in the tropics: The effects of changing yield and technology. Environ Res Lett 3(10pp)
47. Bessou C (2010) Greenhouse gas emissions of biofuels: improving Life Cycle Assessments by taking into account local production factors. Earth Sciences. AgroParisTech, 2009. Available at http://unfccc.int/secretariat/momentum_for_change/items/7184.php
48. Kadam KL (2002) Environmental benefits on a life cycle basis of using bagasse-derived ethanol as a gasoline oxygenate in India. Energy Policy 30:371–384
49. Farrell AE, Plevin RJ, Turner BT, Jones AD, O'Hare M, Kammen DM (2006) Ethanol can contribute to energy and environmental goals. Science 311:506–508
50. Tilman D, Hill J, Lehman C (2006) Carbon-negative biofuels from low input high-diversity grassland biomass. Science 314:1598–1600
51. Prueksakorn K, Gheewala SH (2006) Energy and greenhouse gas implications of biodiesel production from *Jatropha curcas* L. Proceedings of the 2nd joint international conference on sustainable energy and environment, Bangkok
52. Reinhardt G, Gärtner SO, Rettenmaier N, Munch J, Falkenstein E (2007) Screening life cycle assessment of Jatropha biodiesel. Technical report. IFEU, Heidelberg
53. Ndong R, Montrejaud-Vignoles M, Girons OS, Gabrielle B, Pirot R, Domergue M, Sablayrolles C (2009a) Life cycle assessment of biofuels from Jatropha curcas in West Africa: a field study. GCB Bioenergy 1(3):197–210
54. Ndong R, Montrejaud-Vignoles M, Girons OS, Gabrielle B, Pirot R, Domergue M, Sablayrolles C (2009b) Life cycle assessment of biofuels from *Jatropha curcas* in West Africa: a field study. GCB Bioenergy 1(3):197–210
55. USEPA (2002) A comprehensive analysis of biodiesel impacts on exhaust emissions printed on recycled paper. Draft technical report. Available at https://www3.epa.gov/otaq/models/analysis/biodsl/p02001.pdf
56. Gmünder S, Singh R, Pfister S, Adheloya A, Zah R (2012) Environmental impacts of Jatropha curcas biodiesel in India. J Biomed Biotechnol 2012:623070 Available at http://www.hindawi.com/journals/bmri/2012/623070/
57. UMI (2010) Urban mobility India expo and conference. Presentation on best clean development mechanism project in KSRTC. Available at http://www.urbanmobilityindia.in/Upload/Conference/1b57a546-910d-4420-adaf-601a1fe5968e.pdf
58. UNFCCC (2015) Use of ethanol blended diesels in Karnataka State Road Transport (KSRTC). Available at:http://unfccc.int/secretariat/momentum_for_change/items/7184.php. Retrieve on 12 Apr 2015
59. Singh A, Pant D, Olsen SI, Nigam PS (2012) Key issues to consider in microalgae-based biodiesel production. Energy Educ Sci Technol Part A Energy Sci Res 29(1):687–700
60. Swain KC (2014) Biofuel production in India: potential, prospectus, and technology. J Fundam Renew Energy Appl. 4:129. doi:10.4172/2090-4541.1000129. Available at http://omicsonline.com/open-access/Biofuel-Production-in-India-Potentia-Prospectus-and-Technology-2090-4541-4-129.php?aid=28675#10
61. Chisti Y (2007) Biodiesel from microalgae. Biotechnol Adv 25:294–306
62. Chanakya HN, Mahapatra DM, Sarada R, Abitha R (2012) Algal biofuel production and mitigation potential in India. Mitig Adapt Strateg Glob Change. doi:10.1007/s11027-012-9389-z

63. Singh A, Olsen SI (2011) A critical review of biochemical conversion, sustainability and life cycle assessment of algal biofuels. Appl Energy 88:3548–3555
64. Govindasamy A, Hsu MJ, Chaudhary S, Shieh P (2009) Can biofuel crops alleviate tribal poverty in India's drylands? Appl Energy 86:S118–S124
65. Robertson GP, Dale VH, Doering OC, Hamburg SP, Melillo JM, Wander MM, Parton WJ, Adler PR, Barney JN, Cruse RM, Duke CS, Fearnside PM, Follett RF, Gibbs HK, Goldemberg J, Mladenoff DJ, Ojima D, Palmer MW, Sharpley A, Wallace L, Weathers KC, Wiens JA, Wilhelm WW (2008) Sustainable biofuels redux. Science 322:49–50
66. Agoramoorthya G, Hsu MJ, Chaudhary S, Shieh PC (2009) Can biofuel crops alleviate tribal poverty in India's drylands? Appl Energy 86:S118–S124
67. Cotula L, Dyer N, Vermeulen S (2008) The implications of biofuels for land tenure and land policy. Land Tenure Working Paper 1. Bioenergy and Land Tenure. Available at http://www.fao.org/nr/lten/abst/lten_080901_en.htm
68. Baka J (2014) What wastelands? A critique of biofuel policy discourses in South India. Geoforum 54:315–323
69. Hermida J (2008) Biofuels boom and poor people's access to land. Available at http://asiadhrra.org/wordpress/2008/06/20/biofuels-boom-and-poor-peoples-access-to-land/
70. IEA (2006) World Energy Outlook. OECD/IEA.
71. Zafar S (2013) Biodiesel program in India: an analysis. Bioenergy consult. Available at http://www.bioenergyconsult.com/tag/biodiesel-production-in-india/
72. Bajhaiya AK, Mandotra SK, Suseela MR, Toppo K, Ranade S (2010) Algal biodiesel: the next generation biofuel for India. Asian J Exp Biol Sci 1:728–739
73. Kumar M, Sharma MP (2014) Status of biofuel production from microalgae in India. J Integr Sci Technol 2(2):72–75
74. Gerbens-Leenesa W, Hoekstraa AY, van der Meerb TH (2009) The water footprint of bioenergy. Proceedings of the National Academy of Sciences of the USA. Available at http://www.pnas.org/content/106/25/10219.full.pdf+html
75. Ariza-Montobbio P, Lele S (2010) Jatropha plantations for biodiesel in Tamilnadu, India: viability, livelihood trade-offs, and latent conflict. Ecol Econ 70(2):189–195
76. Awasthi K, Mishra S (2006) Jatropha plantation policy has gaping holes. Down to Earth news. Available at http://www.downtoearth.org.in/news/jatropha-plantation-policy-has-gaping-holes-8376
77. Oudhia P. Bare facts about poisonous *Jatropha curcas*. Available at http://ecoport.org/ep?SearchType=earticleView&earticleId=877&page=-2. Retrieved on 14 Apr 2015
78. Tomar NS, Agarwal RM (2013) Influence of Treatment of *Jatropha curcas* L. Leachates and Potassium on Growth and Phytochemical Constituents of Wheat (*Triticum aestivum* L.). Am J Plant Sci 4(5):1134–1150
79. Koh LP, Ghazoul J (2008) Biofuels, biodiversity, and people: understanding the conflicts and finding opportunities. Biol Conserv 141:2450–2460
80. Kumar S, Singh J, Nanoti SM, Garg MO (2012) A comprehensive life cycle assessment (LCA) of *Jatropha* biodiesel production in India. Bioresour Technol 110:723–729
81. Popp A, Lotze-Campen H, Leimbach M, Knopf B, Beringer T, Bauer N, Bodirsky B (2011) On the sustainability of bioenergy production: integrating co-emissions from agricultural intensification. Biomass Bioenergy 35:4770–4780
82. Searchinger T, Heimlich R, Houghton RA, Dong F, Elobeid A, Fabiosa J, Tokgoz S, Hayes D, Yu TH (2008) Use of U.S. croplands for biofuels increases greenhouse gases through emissions from land-use change. Science 319:1238–1240
83. Wise M, Calvin K, Thomson A, Clarke L, Bond-Lamberty B, Sands R, Smith SJ, Janetos A, Edmonds J (2009) Implications of limiting CO2 concentrations for land use and energy. Science 324:1183–1186
84. Bailis ROB, McCarthy H (2011) Carbon impacts of direct land use change in semiarid woodlands converted to biofuel plantations in India and Brazil. GCB Bioenergy 3(6):449–460

85. Biofuels Digest (2015) VayuGrid signs MOU for community-based biofuel projects. Available at http://www.biofuelsdigest.com/bdigest/2015/03/05/vayugrid-signs-mou-for-community-based-biofuel-projects/
86. Sahoo D, Elangbam G, Devi SS (2012) Using algae for carbon dioxide capture and biofuel production to combat climate change. Phykos 42(1):32–38
87. Energy Blog (2014) Milestone IPCC climate report shifts on biofuels. Available at http://energyblog.nationalgeographic.com/2014/04/01/milestone-ipcc-climate-report-shifts-on-biofuels/
88. PIB (2015) Direct sale of bio-diesel by manufacturers/suppliers other than OMCs. Press Information Bureau. The Government of India. Available at http://pib.nic.in/newsite/PrintRelease.aspx?relid=114701
89. BioSpectrum (2014) Modi's Australia visit: big focus on R&D in agribiotech, biofuels. Available at: http://www.biospectrumindia.com/biospecindia/news/219610/modis-australia-visit-big-focus-r-d-agribiotech-biofuels#sthash.dIiRPq0Q.dpuf
90. Blasing TJ (2014) Recent greenhouse gas concentrations, Feb 2014. Carbon Dioxide Information Analysis Center (CDIAC) DOI: 10.3334/CDIAC/atg.032 Available electronically at http://cdiac.ornl.gov/pns/current_ghg.html

Chapter 22
Indian Biofuel Progress, GHG Emission and GHG Savings by Biofuels: Comparative Assessment with World

Alok Satlewal, Jitendra K. Saini, Ruchi Agrawal, Anshu Mathur, Deepak K. Tuli, and Mukund Adsul

Abstract Indian biofuel research and development is increasing day by day. Indian government made a plan to make 20% ethanol/biofuel blending by 2020. Few companies are already working on the biodiesel/lignocellulosic ethanol research and production, but they are not yet commercially working. The ethanol from molasses is available only for making 5% blending with gasoline, and appropriate policies are necessary to increase the ethanol production not only from one source but from other various renewable resources. Research institutions are working to make lignocellulosic biofuels available at lower price. The production and use of biofuel not only helps India at economic level but also may help to reduce some greenhouse gas emissions. At present, India is one of the most CO_2-emitting countries but behind the USA and China. The recent report indicates the slowdown of global CO_2 emission and that may be because of several reasons, but it is a good beginning.

Keywords Biofuels • Greenhouse gas emissions • Lignocelluosic ethanol • Indian biofuel Research

1 General

India is one of the developing countries and its economy is vastly growing with time. Due to increase in the industrialization, urbanization and population, its energy demand is growing continuously. For transportation fuel, India is almost

---

Authors Alok Satlewal and Jitendra K. Saini are equally contributed authors.

A. Satlewal • R. Agrawal • A. Mathur • D.K. Tuli • M. Adsul (✉)
DBT-IOC Centre for Advanced Bioenergy Research, Indian Oil R and D Centre, Sector-13, 121007 Faridabad, Haryana, India
e-mail: madsul@gmail.com

J.K. Saini
Department of Microbiology, Central University of Haryana, Mahendergarh 123029, Haryana, India

A.K. Chandel, R.K. Sukumaran (eds.), *Sustainable Biofuels Development in India*,
DOI 10.1007/978-3-319-50219-9_22

dependent on gulf countries and in the near future has to look for the alternative sources to become partially independent. Biofuel is one of the option to look as alternative for transportation fuel. At present, India produces only ethanol form molasses which can be used for blending with gasoline. The work on biodiesel and second-generation cellulosic or algal biofuels is going on. The biofuel not only helps the Indian economy but also decreases the greenhouse gas emission. In this chapter, the biofuel progress in India with respect to ethanol, biodiesel, number of companies involved, etc, was discussed along with the overall greenhouse gas emissions and their comparisons with the world leading countries.

2 Ethanol Production in India from Molasses

India as a developing country has high energy demands which is increasing day by day. Most of its energy demand is compensated through different fossil fuels such as coal, gas and oil. These nonrenewable sources of energy generate a lot of environmental pollution which is a prime concern. Maximum energy demand in India comes from industry, followed by the transportation sector (about 16.9% of the total energy in 2005–2006). In case of transportation sector, the consumption of motor gasoline increased from 7.01 million tonnes in 2001–2002 to 11.26 million tonnes in 2008–2009 and that of high speed diesel from 36.55 million tonnes to 51.67 million tonnes. As India's population is expected to grow by 10–12% per year, the energy demand will escalate over time. Hence securing a long-term supply of energy sources and prioritizing development are critical to ensuring the country's future energy requirements are met. Thus, India is looking forward to liquid biofuels in order to meet the transportation sector's energy demands [1].

In the year of 2003, the Government of India (GOI) has mandated for the 5% blending of ethanol in gasoline. The Ethanol Blending Programme (EBP) is heavily dependent on ethanol produced from sugarcane molasses. Since sugarcane production in India is cyclical, ethanol production also varies accordingly and therefore does not assure optimum supply levels needed to meet the demand at any given time [1]. Thus, ethanol and alcohol production depends largely on the availability of sugar molasses. At times, lower availability of sugar molasses and resultant higher molasses prices affect the cost of production of ethanol, thereby disrupting supply of ethanol for the blending programme at pre-negotiated fixed ethanol prices. Table 22.1 shows the calculated requirement of ethanol in India for 5 and 10% blending (year 2006–2008) as well as the projected demand for 2017 based on the demand for gasoline in the transport sector.

India has great potential to supplement its ethanol production supply from renewable energy sources. GOI has taken a number of major initiatives for promoting the development and production of (a) ethanol derived from sugar molasses/juice/lignocellulosic biomass for blending with gasoline and (b) biodiesel derived from inedible oils and oil waste for blending with diesel.

Table 22.1 Annual demand for gasoline in India and the projected ethanol demand for blending

S.No.	Year	Motor gasoline demand (in billion litres)	Ethanol demand (@ 10% blending)
1	2006	12.80	1.28
2	2007	14.17	1.42
3	2008	15.30	1.53
4	2017	22.21	2.22

IEA's Energy Statistics of Non-OECD Countries

Table 22.2 Ethanol production, capacity and feedstock utilization (million litres)

Year	2010	2011	2012	2013	2014
Actual Production	1522	1681	2154	2064	1906
Production capacity					
No of Refineries	115	115	115	115	115
Capacity	1500	1500	2000	2000	2000
Capacity Use (%)	101	112	108	103	95
Feedstock use (1000 MT)					
Molasses	6342	7004	8975	8602	7940

Source: GAIN report

The nameplate capacity of distilleries in country to distil conventional ethanol per year is sufficient to meet the demand for 5% blending with petrol (Table 22.2). Production of advanced bioethanol is in its research and development stage.

In 2009, the supply of ethanol was sufficient to meet the total demand for EBP at the rate of 5% blending. Despite this, the blending target could not be met since Oil Marketing Companies (OMCs) were unable to procure the required fuel ethanol at prevailing market prices that are lower than alcohol prices for different uses [2]. Other reasons for this were most of the domestic ethanol producers or suppliers were disqualified to supply the ethanol and non-finalization of ethanol pricing formula and procedural delays by various state governments delayed the procurement for EBP.

3 Ethanol Production in India from Biomass

In India, lignocellulosic biomasses such as agricultural residues, forest wood and crops wastes, etc., are available in plenty. But the conversion of these materials into biofuels requires complex steps such as pretreatment, enzymatic deconstruction and then microbial fermentation. At present stage, the few feasible technologies are under development or some are developed to produce liquid biofuels.

Lignocellulosic biomass predominantly consists of cellulose (C6 or glucose polymer-35–55%), hemicelluloses (C5 or xylose polymer-20–40%) and lignin (phenolics polymer-10–25%). At present, the second-generation commercial production of ethanol, i.e. from lignocellulosic biomass, is not in practice in India. There are multiple challenges in the production of bioethanol from lignocellulosic biomass which requires large amount of R&D efforts for improved process economics as well as better conversion efficiency [3]. The cost of the cellulolytic enzymes is one of the bottlenecks in the conversion of biomass into ethanol or other fuels. The world leading enzyme producers such as Dupont-Genencor and Novozyme are developing low-cost cellulolytic enzymes to get minimum of $1/gallon cellulosic ethanol.

In the pretreatment step, biomass is treated in the presence of different catalysts like alkali or ammonia or acid at high temperature. Such pretreatment helps to remove partially or completely lignin or the hemicelluloses sugars and makes the cellulose amenable to enzymatic attack. Enzymatic hydrolysis (saccharification) is performed with different lignocellulolytic enzymes to liberate fermentable sugars from cellulose and hemicellulose. Finally, the fermentation of sugars is carried out to produce alcohol. GOI supports the use of renewable biofuels as an alternative to fossil fuel. As per India's 12th Five Year Plan (2017), the gasoline has to be supplied with 20% renewable fuel blending by 2020. The following points are the salient features of India's Biofuel Policy [4]:

- To avoid possible conflict of fuel versus food, the biofuel has to be derived from non-food feedstock grown on degraded soils or wastelands which are not otherwise suited to agriculture.
- Encouraging the use of renewable energy resources to strengthen India's energy security through the supplementation of biofuels in motor transport fuels. An indicative 20 % target for blending of biofuel for both biodiesel and bioethanol is proposed by the end of 12th Five Year Plan (IFY 2012/2013 through fiscal 2016/2017).
- Minimum support price (MSP) mechanisms for inedible oilseeds to provide fair prices to oilseed growers (subject to periodic revision).
- Oil manufacturing companies should purchase ethanol at a minimum purchase price (MPP) based on the actual cost of production and import price of ethanol. In the case of biodiesel, the MPP should be linked to the prevailing retail diesel price.
- If required, GOI proposes to consider creating a National Biofuel Fund for providing financial incentives, including subsidies and grants, for new and second-generation feedstocks, advanced technologies and conversion processes and production units based on new and second-generation feedstock.
- Need and thrust for multi-institutional, indigenous and time-bound innovation, research and development on second-generation biofuel as well as biofuel feedstock production.
- Meet the energy needs of India's vast rural population by stimulating rural development and creating employment opportunities and addressing global concerns about containment of carbon emissions through the use of environment-friendly biofuels.

- GOI may bring biofuels under the ambit of "Declared Goods" so as to ensure their unrestricted interstate and intrastate movement. Except for a concessional excise duty of 16% on bioethanol, no other central taxes and duties are proposed to be levied on biodiesel and bioethanol.
- Biofuel technologies and projects would be allowed 100 % foreign ownership through automatic approval to attract foreign direct investment (FDI), provided the biofuel is for domestic use only and not for export. Plantations of inedible oil bearing plants would not be open for FDI participation.
- Setting up of National Biofuel Steering Committee (NBSC) under Prime Minister to provide policy guidelines.

Several ministries have been allocated specific roles and responsibilities as to deal with different aspects of biofuel development and promotion [4], which are listed in Table 22.3.

In order to foster research and development in the area of biofuels, three dedicated research centres (DBT-IOC Center for Advanced Bioenergy, Faridabad, Haryana; DBT-ICGEB Centre, New Delhi; and DBT-ICT Centre, Mumbai) were developed with the support of the Department of Biotechnology, Ministry of Science and Technology, Govt. of India. These three centres work in close synergy with a mandate of bioethanol commercialization.

ICT-DBT Centre has developed a pilot-scale technology and has recently established a 10 tonnes/day pilot at India Glycols limited, Kashipur. IOC (R&D) has also set up a multipurpose pretreatment pilot plant of 0.25 tonnes/day with assistance of National Renewable Energy Laboratory, DOE of USA. ICGEB centre mandate is to concentrate its attention to development of cellulase producing strains for enzymatic treatment. All these coordinated efforts are aimed at developing a

Table 22.3 Specific roles and responsibilities of ministries (India) with respect to biofuels

Ministry	Responsibilities
New and Renewable Energy	Policymaking and overall coordination concerning biofuels
Petroleum and Natural Gas	Responsible for marketing biofuels as well as develops and implements pricing and procurement policy
Agriculture	R&D of biofuel feedstock through Indian Council for Agricultural Research and Indian Agricultural Research Institute
Rural Development	Plantation of *Jatropha* on wastelands. Integrates biodiesel programme with rural development schemes
Science and Technology	Supports research on biofuel crops through biotechnology
Road Transport and Highway	Plantation along highway rights-of-way and use biofuel blended fuel
Railways	Undertakes plantation of *Jatropha* over wastelands along rail rights-of-way and trials of biodiesel blended fuel on railroad locomotives
Environment and Forest	Ensures plantation of *Jatropha* and tree-borne oilseeds in forest wastelands

technology package for lingocellulosic ethanol production which is best suited to Indian conditions.

4 Biodiesel Production in India

Biofuels are going to play an extremely important role in meeting India's energy needs. As bioethanol blending with petrol is well known and is in current practice in India, similarly biodiesel which can be produced by the trans-esterification of algal lipids or vegetable oils can be blended with diesel. Such types of blending may decrease the requirement of diesel from petroleum sources. As compared to petroleum diesel, the biodiesel has good fuel properties. The cetane number of *Jatropha* oil-based biodiesel is 52 which is higher than the cetane number of petroleum-based diesel (i.e. 42–48). Biodiesel contains 10% built-in oxygen which helps to burn it completely. Also the biodiesel contains no sulphur, very less carbon monoxide, particulate matter and unburned hydrocarbons, which make it better than petroleum-based diesel. They are also good lubricants because they contain esters of long chain fatty acids. Apart from these advantages, there are some technical difficulties which need to be solved such as biodiesel has high viscosity and contamination problem if it is to be stored for a long time.

To meet the 20% of India's diesel requirement, GOI has developed an ambitious National Biodiesel Mission. *Jatropha Curcas* has been projected as a potential biodiesel feedstock; it has a variety of advantages as compared to other plants such as high seed yield, pest resistant, easy to propagate, requirement of less water as well as fertilizer, etc. The India's Biodiesel Mission has been implemented in two steps: First stage was at demonstration project (2003–2007) which has cultivated around 400,000 ha of land to get around 3.75 tonnes of oilseeds/hectares/year. Second stage was decided by government to develop transesterification plant and commercialize the biodiesel during the year 2007–2012, but it has not been successful [5, 6].

Afterwards, strong leap has been taken by India, which includes improvement of *Jatropha* varieties for high yield, development of jatro nurseries, helping biodiesel manufacturers by setting up pilot-plant and testing their biodiesel in buses and public transport vehicles. It was estimated that the yield of *Jatropha* oil will be around 15 Mt/year by cultivating 13.4 million hectares of land. To implement the full-fledged 5% biodiesel blending, a large cultivation of *Jatropha* was needed apart from the new infrastructure for stogare, blending places, marketing, extraction of oil from seed, transesterification, etc.

The following is an overview of the work being carried out on biodiesel development in India [5, 6]:

- Development of high oil-yielding varieties of *Jatropha* by the Department of Biotechnology, the Aditya Biotech Research Centre (Raipur), the Indira Gandhi Agriculture University (Raipur) and the Bhabha Atomic Research Centre (Trombay).

- Plantation of *Jatropha* and *Pongamia Pinnata* (Karanja) by (1) the National Afforestation and Eco-development Board (NAEB) under the guidance of the Ministry of Environment and Forests; (2) the National Oilseed and Vegetable Oil Development (NOVOD) Board under the guidance of the Ministry of Agriculture; (3) the Central Salt and Marine Chemicals Research Institute (Bhavnagar); and (4) a number of NGOs such as Uthan (Allahabad) and Sutra (Karnataka); the Institute of Agriculture and Environment (Jind, Haryana); the Bharatiya Agro Industries Foundation (BAIF) Development (Pune, Maharashtra); Pan Horti Consultants (Coimbatore); Classic *Jatropha* Oil (Coimbatore); and Renulakshmi Agro Industries (Coimbatore); etc.
- Pilot plants on transesterification set up by Indian Oil Corporation (R&D), Faridabad; the Indian Institute of Technology (IIT), Delhi; the Punjab Agricultural University (PAU), Ludhiana; the Indian Institute of Chemicals Technology (IICT), Hyderabad; the Indian Institute of Petroleum (IIP), Dehradun; the Indian Institute of Science (IIS), Bangalore; and Southern Railways, Chennai.
- Different trial runs on a variety of transport modes using 5% biodiesel blends, including railways (a locomotive used biodiesel on a regularly scheduled train ride—the Shatabdi Express—from Amritsar to Delhi on 31 December 2002); tractors tested by Mahindra & Mahindra Co.; Mercedes cars tested on Daimler Chrysler; public transport buses tested by Haryana Roadways and Bombay Electric Supply and Transport (BEST); and trial marketing of 5% diesel blends through some retail outlets is being conducted by the oil company Bharat Petroleum Corporation Limited (BPCL).

5 Companies Involved in Biodiesel and Ethanol Production, Their Present Status

The fluctuating and uncertainties in the supplies of biofuel production and less support from the GOI are among the few factors for demotivating the set-up of plants by private sector in India. A list of companies contributing to the development of bioethanol and biodiesel in India is discussed in Table 22.4.

It was decided to install a 90,000 t/year biodiesel plant by Naturol Bioenergy (a partnership between the investment firm Fe Clean Energy and Austrian biodiesel firm Energea GmbH). The Andhra Pradesh State Government allocated land for cultivation of *Jatropha* to the firm. Southern Online Biotechnologies also has large project [5]. At present, commercial biodiesel plants are not yet fully fledged in India.

Table 22.4 List of companies involved in biofuel research or production in India

S. No.	Name of organization	Vision/contribution
1	Abellon Clean Energies	Abellon CleanEnergy is an integrated sustainable energy solutions provider with a vision to contribute to clean energy generation through focus on bioenergy, including bio-pellets, biofuels, biopower and other forms of clean energy generation
2	Adi Biotech Pvt Ltd	Biodiesel, biogas, bioethanol—*Jatropha* plantation, training, plants
3	Advanced enzymes	Enzymes for cellulosic ethanol
4	Advanta Ltd	Sweet sorghum for ethanol (seedstock development)
5	Bharat Biodiesel	Manufacturer and supplier of biodiesel, cashew nut shell oil, crude glycerine, palm fatty acid distillate and palm stearin
6	Bharat Petroleum	Biodiesel value chain, *Jatropha* biodiesel and ethanol production
7	Bharat renewable energy Ltd	Consortium of Nandan Biomatrix + BPCL + Shapoorji Pallonji & Co, Jatropha, Pongamia Biodiesel
8	Biodiesel technocrats, Kolkata, India	Designing, fabricating, assembling and commissioning of biodiesel processing plants of various capacities
9	Biodiesel Technologies	Biodiesel producing units from multiple feedstocks
10	Chemcel Biotech Ltd	Biodiesel from Jatropha and Pongamia
11	Chemical construction International	Turnkey biodiesel engineering plants, collaboration with Lurgi, Germany
12	D1 oil UK India	Development of new energy crops and *Jatropha* plantation
13	Gomti Biotech	Biodiesel, natural glycerine, essential fatty acids
14	Green earth biotech	Oil Palm plantation
15	Indian Oil	Technology for biodiesel production
16	Indian oil Creda Biofuels Ltd	Biodiesel from non-food energy crops (joint venture with Chhattisgarh govt)
17	Kaveish Bioenergy Pvt Ltd	Developing portable domestic biogas plants for household (biogas from waste available in households)
18	Labland Biodiesel	*Jatropha* tissue culture, biomass development, oil extraction for biodiesel
19	Novozyme South Asia	Enzymes for cellulosic ethanol, starch-based ethanol
20	Petroleum Bazaar.com (India) Pvt. Ltd.	Auction, physical buy and sell of Bio diesel products and logistics. Marketing and help in marketing, consultation on plantation, import and export of seed and oil, etc., through its initiative The Petroleum Next
21	Praj	Ethanol and cellulosic ethanol research
22	Qteros	Cellulosic ethanol
23	Reliance industries	*Jatropha* plantation, Algal biodiesel, ethanol
24	SAIP (Solena Absi India Pvt Ltd)	Technology services in gasification from waste biomass (collaborator of Solena, USA)
25	Sea 6 Energy	Biofuel from Macroalgae
26	Shirke Biofuels	*Jatropha* high variety development, farm management services, technical consultancies, seed, oil, biodiesel producers
27	Tata Chemicals	Biodiesel, bioethanol
28	Vrindavan Biofuels Ltd	*Jatropha* elite cultivars plantation (better growth, high oil), mycorrhizal association for desired traits

6 Comparison of Total Biofuel Production in India with World Leaders

Brazil uses pure ethanol in about 20% of their vehicles and a 22–26% ethanol–petrol blend in the rest of their vehicles. The USA and Australia use a 10% ethanol blend. With a normal production rate of 1900 million litres a year, India is the world's fourth largest producer of ethanol after Brazil, the USA and China. The continental status of biomass for transport has been discussed. The Americas remain on top with 70% followed by Europe (24%), and Asia is on third place with 5% [7]. In the year 2012, approx. 106 billion litres of liquid biofuels were produced globally [7]. An estimate of the total liquid biofuel production for different regions has been tabulated in Table 22.5. The total contribution of bioethanol produced has been estimated in Table 22.6. The USA remained in top with 50.4 billion litres of bioethanol.

In India, ethanol till now has been produced from molasses which is the by-product of sugar production, and hence, the availability of ethanol for gasoline blending is directly linked to the sugarcane production. As is well known, the sugarcane production in India is cyclical in nature and is linked to the purchase price of the cane. This has led to serious disruptions in fulfilling the Government mandate of blending 5% ethanol into gasoline. Though there are efforts to allow up to 10% ethanol in gasoline, but there are serious concerns on the uninterrupted supply of ethanol. Therefore, Indian scientific community has a challenge to develop alternate bio-based sources like lingocellulosic materials for ethanol production. Biofuels can become an answer to petroleum crisis only if the India is able to overcome technological challenge.

Table 22.5 Liquid biofuel production for different countries (billion litres)

Year	USA	Brazil	Europe	China	India
2000	5.52	10.7	0.97	–	0.13
2005	15.6	12.7	5.46	–	0.19
2009	40.9	23.7	20.5	2.20	0.27
2010	45.1	24.9	23.4	2.10	0.30
2011	47.3	22.7	22.0	2.15	0.34

Source: IEA statistics [25]

Table 22.6 Production of bioethanol by different countries (billion litres)

Year	USA	Brazil	Europe	China	India
2005	15.0	15.0	0.9	1.0	0.3
2009	41.0	26.0	3.6	2.1	0.2
2010	49.0	28.0	4.5	2.1	–
2011	54.2	21.0	4.3	2.1	–
2012	50.4	21.6	4.2	2.1	0.5

REN21 global status report [26]

7 Greenhouse Gases

Greenhouse gases (GHGs) are the natural and anthropogenic gases of the atmosphere. These GHGs absorb and release infrared radiations emitted by atmosphere of Earth and Earth's surface. GHGs absorb the sun's radiated energy and help in warming the lower part of the atmosphere and cause greenhouse gas effect. As a result of growing anthropogenic activities including industrialization and urbanization, GHG emission has been amplified which consequently has resulted in increased GHG effect during the past few decades. Major greenhouse gases are methane (CH_4), nitrous oxide (N_2O), carbon dioxide (CO_2), hydrofluorocarbons (HFCs), perfluorocarbons (PFCs), chlorofluorocarbons (CFCs) and sulphur hexafluoride (SF_6), out of which CO_2 has the most dominant global warming effect (77% of total CO_2 equivalents). Some of the synthetic gases mentioned above had no existence in atmosphere approximately 100 years ago. CF_4 is present inside ice cores and, thus, has limited natural presence. Inter-Governmental Panel on Climate Change (IPCC), established in the 1988, and United Nations Framework Convention on Climate Change (UNFCCC) were instrumental in making momentum to GHG estimation by quantitative methods. The atmospheric concentrations of CO_2, CH_4 and N_2O were 280 ppm, 700 ppb and 270 ppb between the period 1000 and 1750 AD [8]. The abundance of these three major GHGs have been increased to 379 ppm, 1774 ppb and 319 ppb, respectively, in 2005 [9]. According to the IPCC, CO_2 concentration has achieved unprecedented levels in the atmosphere greater than any time in the last 650,000 years. Thus, it has become a major and the fastest growing factor in climate change [10]. According to the Energy Information Administration of the U.S. Department of Energy, global CO_2 emissions (in billion tonnes) have been projected to rise from 26.9 in 2004 to 33.9 in 2015, and 42.9 in 2030, at an average growth rate of 1.8% per year [11]. CO_2 being a major GHG accounted for about 76% of total anthropogenic GHG emission in 2010. The maximum CO_2 emission is from fossil-fuel combustion and industrial processes. Among other GHGs, 16% comes from CH_4, 6.2% from N_2O and 2.0% from fluorinated gases [12].

7.1 Comparative Account of GHG Emission in India

CO_2 represents only 0.03% of the atmospheric gases, but due to continuous combustion of fossil fuels it has been increased by about 25% as compared to the pre-industrial times. India has the second largest population in the world and is the third largest emitter of CO_2 worldwide after China and the USA with 2011 annual emission of 1.6 Gt CO_2 [13]. India contributes only about 5.3% of the total global emissions. Ironically, India, China and other developing countries of Asia have been blamed by the Western world for their "large" contributions to GHG emissions and thus global warming. Higher levels of energy consumption have contributed to the degradation of the environment. Among all Indian cities, Chennai

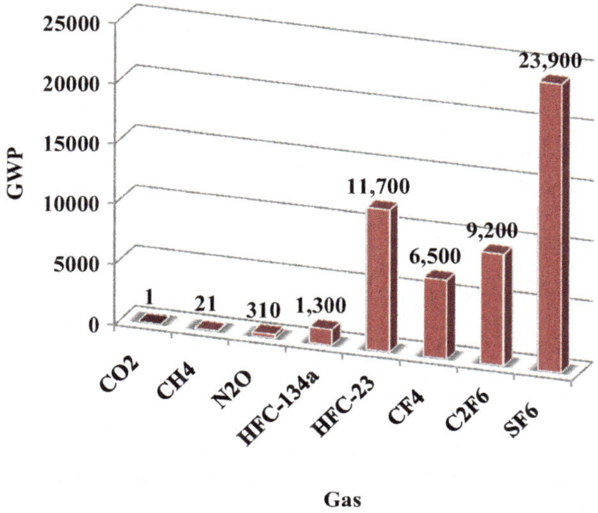

Fig. 22.1 Global warming potential (GWP) of GHGs used for CO_2 equivalent on hundred year timescale [14]

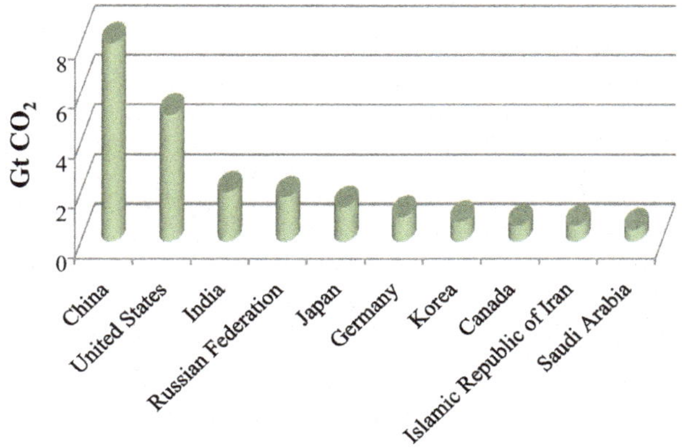

Fig. 22.2 Top ten CO_2-emitting countries of the world in the year 2012 [16]

emits highest CO_2 equivalent emissions per capita (4.79 tonnes) followed by Kolkata (3.29 tonnes). Chennai is also at top in terms of CO_2 equivalent emissions per GDP (2.55 tonnes CO_2 eq./Lac Rs.) followed by Greater Bangalore (2.18 tonnes CO_2 eq./Lac Rs). GHG Footprint of all the major cities in India is very crucial in improving inventories of national level emission [9] (Fig. 22.1).

According to IEA, nearly two-thirds of global emissions for 2012 originated from just ten countries, with China (26%) and the USA (16%) being the top emitters having combined production of 13.3 Gt CO_2 [15]. The top-10 GHG-emitting countries in decreasing order of their emissions are China, USA, India, Russian Federation, Japan, Germany, Korea, Canada, Islamic Republic of Iran and Saudi Arabia (Fig. 22.2). Globally, out of 31.7 Gt CO_2 of total emissions these top ten countries alone produce 21 Gt CO_2. Furthermore, as different regions/countries

Table 22.7 Comparison of India's CO_2 emissions (year 2012) with rest of the world based on total emissions from fuel combustion, emissions/GDP and emissions/capita [15]

Countries	CO_2 emissions: sectoral approach (million tonnes of CO_2)	CO_2 emissions/GDP using exchange rates (kilograms CO_2/US dollar using 2005 prices)	CO_2 emissions/ population (tonnes CO_2/ capita)
United States	5074.1	0.36	16.15
China	8250.8	1.73	6.08
European Union—28	3504.9	0.24	6.91
India	1954.0	1.41	1.58
World	31,734.3	0.58	4.51

Table 22.8 Comparison of the contribution of different sectors in total CO_2 emissions (% of *total CO_2 emission*) from fuel combustion in India and rest of the world[a] [15]

Countries	Electricity and heat production	Other energy ind. own use	Manufacturing industries and construction	Transport		Other sectors	
				Total	Road transport	Total	Residential
United States	41.1	5.6	9.8	13.2	8.1	10.7	5.9
China	50.1	3.6	31.0	8.6	6.9	6.7	3.8
European Union—28	37.5	4.8	15.0	24.6	23.2	18.1	11.5
India	53.4	3.4	24.2	11.1	10.3	7.8	4.1
World	42.1	4.9	20.3	22.6	16.9	10.0	5.7

[a]Based on total CO_2 emission values from fuel combustion of Table 22.7

have different economic and social structures, the representation changes significantly emissions are expressed differently. Thus, emissions per capita or per GDP for one region or country will be different from the absolute emissions by that particular region or country. Trends in CO_2 emission intensities show that per-capita emissions exhibited a global 13% increase from 1990 to 2012, but showed contrasting results for the top five emitting countries [15]. Moreover, all top five emitters cut down their emissions per unit of GDP during this period, while emissions per capita increased. As an example, China tripled its per-capita emissions, while India more than doubled theirs, as did some other rapidly expanding economies. Conversely, per-capita emissions decreased significantly in both the Russian Federation (21%) and the USA (17%), although following very different patterns. Therefore, it is essential to completely understand the factors driving CO_2 emission trends in order to plan efficient policies related to emission reductions.

As per the India's second report to the United Nations Framework Convention on Climate Change (UNFCCC), the total GHG in 2007 (including emissions from the land use, land-use change and forestry sector) were 1772 million tonnes of CO_2 equivalent which is 36% higher than year 2000 estimation [17]. Recent estimation of CO_2 emission by sectoral approach was 1954 million tonnes in 2012 which is lower than the USA, China and European Union (Table 22.7). Other detailed CO_2 emissions by each sector, i.e. heat and electricity production, manufacturing industries and construction, transport, etc., are given in Table 22.8. The increase in the CO_2 emission by India in 2012 was mainly caused by a 10% increase in the coal consumption. The power production by using coal accounts for almost 70% of the total of India's coal-related CO_2 emission which grew by about 13% in 2012 [18].

8 Recent Decline in Worldwide CO_2 Emission: A Ray of Hope for Sustainable Future

In 2012, global CO_2 emissions (majorly due to burning of fossil fuel and from small-scale industries) saw only 1.1% annual increase (downward 0.3% correction for leap year) as compared to average annual increase rate of 2.9% (for the last decade). Among the top CO_2 emitters (Japan, Russian Federation, India, EU, USA and China with 4%, 5%, 6%, 11%, 16% and 29% of global share, respectively), the emission in 2012 decreased by 4%, 1.6%, 1% and 6% by the USA, EU, Russian Federation and Japan, respectively. But in the same year, the CO_2 emission by China and India increased by 3% and 7%, respectively.

Energy-related human activities are the major cause of CO_2 emission which in turn depended on economic growth during the past decade, especially in developing economies. The uncoupling of the CO_2 emission increase and global economic growth (in GDP) in 2012 has been supposed to lead towards less fossil-fuel intensive activities, more use of renewable energy and increased energy saving. This is the first sign of a slowdown in the global CO_2 emission increase. No doubt the further decrease in the global CO_2 emission is possible only if the major CO_2 contributing countries such as the USA continue to use more gas and renewable energy, China shifts towards gas and Members of European Union agree on restoring the effectiveness of the EU Emissions Trading System [19].

9 Mitigation of GHG Emission as Key Factor for Global Bioenergy Initiatives

Advances in formulation of energy policies and progress in energy sector (especially, bioenergy and biofuels) have in recent years become top priorities globally. Climate change has been a major global concern in recent decades. This has led to several initiatives, such as UNFCCC and Kyoto Protocol (KP). Energy sector being major emitter of GHGs has also motivated the political developments towards improved energy supplies with enhanced efficiency as far as their eco-friendly nature is concerned. Global energy policies are being driven by some main interlinked factors: alteration of the climate, security of supply and energy for developmental progress and poverty alleviation. In the next 15 years or so, EU and the USA are preparing to satisfy their 25–30% fuel needs (transport sector) with sustainable and low CO_2-emitting renewable biofuels. Thus, these nations have already planned to secure their increased energy demands, create new opportunities for biomass providers and biofuel processing industry and mitigate CO_2 emissions. The National Biofuels Policy (NBP) of India has an ambitious target of achieving biofuel blending at the rate of 20%, by the year 2020 [20]. This seems to be very hard to achieve target within this time frame. However, Indian Government has significantly increased its efforts in achieving this target, and it seems that even if the target is not met by the deadline, the biofuel blending capacity in Indian scenario will be much enhanced by that time in comparison to earlier times.

9.1 Potential Benefits of Biomass-Based Bioenergy

Biofuels can help in mitigating carbon emissions compared to gasoline and diesel as transportation fuels. This is an additional advantage of using biofuels, beyond utilization of a renewable substrate. It is important for India to replace some percentage of fossil-fuel dependence by using biomass. India has around 328 million hectares land and produces about 450–500 million tonnes of biomass per year of which 200 million tonnes is surplus. At present, biomass is partly used as fodder, some part is used as household energy source and most of the part is burned in the field to make a land ready for next crop. Instead of burning biomass in field, it can be diverted to various applications such as production of heat and electricity and may be used for conversion of transportation fuel/biofuel to replace fossil fuel. The production of biofuel from renewable biomass is extremely low in India.

9.2 Biofuels for Improvement in Air Quality

Biofuels have been seen as important alternative fuel sources that can help in controlling GHG emissions and improving air quality in addition to their renewable nature. Biofuels cause no or very less pollution, are locally available, are easy to get to and are dependable. Biomass being a carbon pool is another option for petroleum-based fuels and does not contribute to atmospheric CO_2 when its production and consumption for biofuel are carried out in a sustainable manner.

9.2.1 Exhaust Emission Reduction by Bioethanol

Bioethanol contains 34.7% wt. O_2 which enhances fuel combustion and consequently reduces exhaust emission and air pollution. A blend of 10% ethanol in gasoline can decrease emission of hydrocarbon (HC) by 18%, carbon monoxide (CO) by 18% and nitrogen oxides (NOx) by 10%, which result due to incomplete combustion of gasoline. The use of bioethanol reduces 21% of CO_2 equivalent production as compared to 2.44 kg CO_2 eq. per litre production from gasoline. Studies on LCA of ethanol production and use have shown that blending 5% anhydrous ethanol in gasoline can result in net savings of GHG emissions between 2.0 (agricultural feedstocks) and 2.5 (waste lignocellulosic biomass) kg of CO_2 eq. per litre of ethanol incorporated to gasoline (when compared to unblended gasoline, on similar performance scale) [21]. However, low-level blends of ethanol with gasoline can increase the emissions of volatile compounds (VOCs) and NOx, which favour ozone formation. Emissions of aldehydes (mostly acetaldehydes) and peroxyacetyl nitrate (PAN) also increase to an extent that depends on weather conditions. The use of catalytic converters reduces the emissions of aldehydes. Reduction in the vapour pressure of gasoline, which is blended with ethanol, can prevent VOC emissions. Studies on different percentages of ethanol diesel blends show significant advantages concerning PM, NOx and CO. However, no evidence is given for improvement of HC emissions. Ethanol is more corrosive than gasoline and diesel and, at high concentration, can damage fuel system components. For low-level blends, these concerns are limited and E5 or E10 can operate on existing vehicles without violating most of manufacturers' warranties. For high concentrations of ethanol, compatible materials are used in dedicated designed vehicles. Second-generation ethanol consumption in India could reduce road transport GHG emissions from fossil gasoline by 47–69%. Furthermore, CO_2 and methane emissions would also decrease as biomass residues would no longer be burned or decompose in the field [22].

As far as the growth of cellulosic ethanol industry is concerned, it has the potential to create jobs, catalyse rural development and reduce CO_2 emissions. Based on the availability, Indian biomass has the potential to create around a million aggregated jobs predominantly in rural areas, which can enhance India's agricultural sector and providing impetus to inclusive growth. Moreover, up to $15 billion–$20 billion of annual revenues in India could be generated by 2020, specially helping the rural economy [22, 23].

9.2.2 Exhaust Emission Reduction by Biodiesel

In comparison to the conventional diesel, biodiesel has higher octane number, oxygen content, lower sulphur content and lower aromatics due to which it is used as biofuel in transportation sector. It shows reduction in PM, HC, CO and PAH and a marginal increase in NOx (1–6%). E20 biodiesel blend does not require any engine modification and does not reduce torque output. Algae-based biodiesel is a promising renewable energy source in this regard. Its production has low impact on land use and the food chain, utilizes waste streams (such as xylose stream after dilute acid pretreatment of biomass) as nutrients and has rapid growth rate and high oil content. Its production and use on large scale has according to an estimate the potential of replacing diesel and reducing around 13–14% of global GHG emissions [23, 24].

9.2.3 Exhaust Emission Reduction by Biogas

Biogas production from liquid manure reduces the emission of methane and reduces the release of CO_2 from fossil fuels and presents better characteristics than the natural gas. Though NOx emissions are increased by burning of biogas, these are still within the EU norms

9.2.4 Exhaust Emission Reduction by Hydrogen

Another cleaner environment-friendly fuel source is biologically produced hydrogen. It is the cleanest and most efficient biofuel. Hydrogen produces no harmful by-products upon combustion. If it is used in fuel cell, it produces energy and clean water when reacted with water. Biohydrogen is not extensively used currently, but research and development is going on for its production in a safe, economical and environmental friendly manner [23].

Acknowledgements All authors acknowledge Indian Oil Corporation and Department of Biotechnology (DBT), Govt. of India, for providing financial support.

References

1. Gopinathan MC, Sudhakaran R (2009) Biofuels: opportunities and challenges in India. In: Lakshmanan P, Songstad D (eds) Vitro Cell Dev Biol Plant 45: 350–371
2. Basavaraj G, Rao PP, Reddy CR, Kumar AA, Rao PS, Reddy BVS (2012) A review of the national biofuel policy in India: a critique of the need to promote alternative feedstocks. Working Paper Series no. 34 RP—Markets, Institutions and Policies. ICRISAT. Patancheru, Andhra Pradesh
3. Sukumaran RK, Surender VJ, Sindhu R, Binod P, Janu KU, Sajna KV, Rajasree KP, Pandey A (2010) Lignocellulosic ethanol in India: prospects, challenges and feedstock availability. Bioresour Technol 101:4826–4833
4. Aradhey A (2014) India biofuels annual. USDA foreign agricultural service. Global Agricultural Information Network. GAIN Report Number: IN4055. Dated 7/1/2014
5. Gonsalves JB (2006) An assessment of the biofuels industry in India. United Nations Conference on Trade and Development
6. Mandal R (2004) Indian scenario & policy issue for successful implementation of biodiesel program. Paper presented at the international Conference on Biofuels, New Delhi, 16–17 Sept

7. Bharadwaj KV (2014) WBA global bioenergy statistics 2014. World Bioenergy Association, Sweden. http://www.worldbioenergy.org. Accessed 14 Aug 2014
8. Inter-Governmental Panel on Climate Change (IPCC) (2001) Climate change 2001: impacts, adaptation and vulnerability. http://www.grida.no/publications/other/ipcc_tar/?src=/climate/ipcc_tar/wg2/. Accessed 12 Mar 2015
9. Ramachandra TV, Bharath H, Aithal K et al (2015) GHG footprint of major cities in India. Renew Sust Energ Rev 44:473–495
10. IPCC (2007) Revised 1996 IPCC guidelines for national greenhouse gas inventories. http://www.ipcc-nggip.iges.or.jp/public/gl/invs1.html. Accessed 15 Mar 2015
11. Energy Information Administration (EIA) (2007) Annual energy outlook 2007 with projections to 2030. Energy Information Administration, Office of Integrated Analysis and Forecasting U.S. Department of Energy Washington, DC. www.eia.doe.gov/oiaf/aeo/. Accessed 26 Mar 2015
12. IPCC (2014) In: Field CB, Barros VR, Dokken DJ et al (eds) Climate change 2014: impacts, adaptation, and vulnerability. Part A: global and sectoral aspects. Contribution of working group II to the fifth assessment report of the intergovernmental panel on climate change. Cambridge University Press, Cambridge, 1132 pp. https://www.ipcc.ch/pdf/assessment-report/ar5/wg2/WGIIAR5FrontMatterA_FINAL.pdf. Accessed 12 Mar 2015
13. Soam S, Kumar R, Gupta RP et al (2015) Life cycle assessment of fuel ethanol from sugarcane molasses in northern and western India and its impact on Indian biofuel programme. Energy 83:307–315. doi:10.1016/j.energy.2015.02.025
14. IPCC (1995) IPCC second assessment: climate change 1995: a report. https://www.ipcc.ch/pdf/climate-changes-1995/ipcc-2nd-assessment/2nd-assessment-en.pdf. Accessed 12 Mar 2015
15. IEA (2014) CO_2 emissions from fuel combustion: highlights. https://www.iea.org/publications/freepublications/publication/CO2EmissionsFromFuelCombustionHighlights2014.pdf. Accessed 12 Mar 2015
16. International Energy Agency (IEA) (2011) CO_2 emissions from fuel combustion: highlights. https://www.iea.org/publications/freepublications/publication/CO2emissionfromfuelcombustion HIGHLIGHTS.pdf. Accessed 21 Mar 2015
17. Ministry of Environment and Forests (MoEF) of Government of India (2012) Jawaharlal Nehru National Solar Mission—an update. http://mnre.gov.in/file-manager/UserFiles/presentations-pwc-workshop-06092012/MNRE.pdf. Accessed 12 Mar 2015
18. Saikia K, Sarkar BC (2013) Coal exploration modelling using geostatistics in Jharia Coalfield, India. Int J Coal Geol 112:36–52
19. Olivier Jos GJ, Janssens-Maenhout G, Muntean M et al (2013) Trends in global CO_2 emissions: 2013 report. PBL Netherlands Environmental Assessment Agency. http://edgar.jrc.ec.europa.eu/news_docs/pbl-2013-trends-in-global-co2-emissions-2013-report-1148.pdf. Accessed 26 Mar 2015
20. Ministry of New & Renewable Energy (MNRE) of Government of India (2009) National policy on biofuels. http://mnre.gov.in/file-manager/UserFiles/biofuel_policy.pdf. Accessed 12 Mar 2015
21. Gnansounou E, Dauriat A (2005) Ethanol fuel from biomass. J Sci Ind Res 64:809–821
22. Krishnan GS (2013) Can second generation ethanol fuel India's growth? Available via http://www.business-standard.com/content/b2b-chemicals/can-second-generation-ethanol-fuel-india-s-growth-113111500167_1.html. Accessed 18 Mar 2015
23. Prasad S, Dhanya MS, Gupta N, Kumar A (2012) Biofuels from biomass: a sustainable alternative to energy and environment. Biochem Cell Arch 12:255–260
24. Sudhakar K, Bansoda D, Premalatha M (2013) Potential to replace petro diesel and reduce greenhouse gas emissions in India by integration of power plants with algae cultivation. Int J Environ Sci Develop Monit 4:4–7
25. IEA Statistics. http://www.iea.org/statistics/. Accessed 20 Mar 2014
26. REN21 global status reports http://www.ren21.net/REN21Activities/GlobalStatusReport.aspx. Accessed 6 May 2014

Lightning Source UK Ltd.
Milton Keynes UK
UKOW06n1132121217
314335UK00003B/85/P

9 783319 502175